THE BIOLOGY OF TAURINE
Methods and Mechanisms

Recent Volumes in this Series

Volume 212
ACUTE RENAL FAILURE: Clinical and Experimental
Edited by Alberto Amerio, Pasquale Coratelli, Vito M. Campese, and Shaul G. Massry

Volume 213
MECHANISMS OF LYMPHOCYTE ACTIVATION AND IMMUNE REGULATION
Edited by Sudhir Gupta, William E. Paul, and Anthony S. Fauci

Volume 214
THE NEW DIMENSIONS OF WARFARIN PROPHYLAXIS
Edited by Stanford Wessler, Carl G. Becker, and Yale Nemerson

Volume 215
OXYGEN TRANSPORT TO TISSUE IX
Edited by I. A. Silver and A. Silver

Volume 216 A
RECENT ADVANCES IN MUCOSAL IMMUNOLOGY, Part A: Cellular Interactions
Edited by Jiri Mestecky, Jerry R. McGhee, John Bienenstock, and Pearay L. Ogra

Volume 216 B
RECENT ADVANCES IN MUCOSAL IMMUNOLOGY, Part B: Effector Functions
Edited by Jerry R. McGhee, Jiri Mestecky, Pearay L. Ogra, and John Bienenstock

Volume 217
THE BIOLOGY OF TAURINE: Methods and Mechanisms
Edited by Ryan J. Huxtable, Flavia Franconi, and Alberto Giotti

Volume 218
CORONAVIRUSES
Edited by Michael M. C. Lai and Stephen A. Stohlman

Volume 219
REGULATION OF OVARIAN AND TESTICULAR FUNCTION
Edited by Virendra B. Mahesh, Dharam S. Dhindsa, Everett Anderson, and Satya P. Kalra

---

THE BIOLOGY OF TAURINE

Methods and Mechanisms

Edited by

Ryan J. Huxtable
University of Arizona
Tucson, Arizona

Flavia Franconi
University of Sassari
Sassari, Italy

and

Alberto Giotti
University of Florence
Florence, Italy

PLENUM PRESS • NEW YORK AND LONDON

Library of Congress Cataloging in Publication Data

The Biology of taurine.

Proceedings of a symposium held Oct. 6–9, 1986 in San Miniato, Italy.
Includes bibliographies and index.
1. Taurine—Physiological effect—Congresses. I. Huxtable, Ryan J. II. Franconi, Flavia. III. Giotti, Alberto. [DNLM: 1. Taurine—congresses. QU 60 B6157 1986]
QP801.T3B56 1987 615′.78 87-20335
ISBN 0-306-42665-X

Proceedings of a symposium on Sulfur Amino Acids, Peptides, and Related Compounds, held October 6–9, 1986, in Firenze-San Miniato, Italy

A Division of Plenum Publishing Corporation
233 Spring Street, New York, N.Y. 10013

Printed in the United States of America

THE BIOLOGY OF TAURINE

Methods and Mechanisms

Edited by

Ryan J. Huxtable
University of Arizona
Tucson, Arizona

Flavia Franconi
University of Sassari
Sassari, Italy

and

Alberto Giotti
University of Florence
Florence, Italy

PLENUM PRESS • NEW YORK AND LONDON

Library of Congress Cataloging in Publication Data

The Biology of taurine.

Proceedings of a symposium held Oct. 6–9, 1986 in San Miniato, Italy.
Includes bibliographies and index.
1. Taurine—Physiological effect—Congresses. I. Huxtable, Ryan J. II. Franconi, Flavia. III. Giotti, Alberto. [DNLM: 1. Taurine—congresses. QU 60 B6157 1986]
QP801.T3B56 1987 615′.78 87-20335
ISBN 0-306-42665-X

Proceedings of a symposium on Sulfur Amino Acids, Peptides, and Related Compounds, held October 6–9, 1986, in Firenze-San Miniato, Italy

A Division of Plenum Publishing Corporation
233 Spring Street, New York, N.Y. 10013

Printed in the United States of America

The starfish, rich in taurine, with its five arms, may indeed be a good symbol for this symposium.

A. Giotti (This volume).

(Photo courtesy of Lorenzo Giotti)

PREFACE

I was pleased and at the same time filled with some misgivings when Professors Alberto Giotti and Ryan Huxtable asked me to introduce this book. The book is the outcome of the Symposium held in Firenze-San Miniato (PI), October 6-9, 1986. The symposium was entitled "Sulfur Amino Acids, Peptides and Related Compounds" and was the 7th international symposium on taurine and associated substances.

It is always difficult to introduce, with the right brevity and emphasis, a topic which has been studied in depth by numerous experts. Nevertheless, I shall do my best to give a historical perspective of the subjects of the meeting which I consider to be very important for the frontiers of research on taurine. The following topics have also become coherent areas of study during the development of research on taurine: metabolism, nutrition, neurochemistry, cardiovascular regulation. Although taurine was isolated in 1827 by Tiedman and Gmelin, its only biochemical role known at the time was the synthesis of bile salts in mammalian tissue.

There has been an increasing interest in the biological action of taurine from metabolic aspects to other biological aspects (nutrition, development, etc.). In 1975 it was first demonstrated that taurine deprivation produced retinal degeneration in cats; more recent studies showed that a taurine-free diet or the administration of taurine transport inhibitors caused retinal degeneration in other mammlas. More recent studies have pointed out the role of taurine in development, and the first part of this book is dedicated to these topics.

From the pioneer work of Read and Welty, which showed the antiarrhythmic action of taurine, particular attention has been focussed on the effect of taurine on cardiovascular regulation. One important issue is inotropism and the cardioprotective effect of taurine. Although pharmacological studies are in progress on structur-action relationships, there are few electrophysiological studies, and thus the action of taurine on ion currents has yet to be clarified.

Neurochemistry, with neuropharmacology, is widely represented in this book. The data are interesting although the classical question "Is taurine a neurotransmitter?" is still without reply, at least for mammalian tissue.

I should like to point out the multidisciplinary approach of the symposium and consequently of the book, a fact also demonstrated by the different methodological approaches represented. In the symposium a taurine antagonist was introduced and in the interests of research, I hope that it will be very selective. Although much progress has been made in discovering the biological role of taurine no insight on its mechanisms is available, although the unifying hypothesis of Ryan Huxtable is a good point at which to start new research.

I hope that this book will be a good guide to the state of the art of research on taurine and related compounds.

Flavia Franconi

ACKNOWLEDGMENTS

The editors are grateful to all those who in ways big and small contributed to the success of the meeting and the production of this book. These include the members of the local Organizing Committee, Fabrizio Ledda, Silvana Romanelli, Paola Failli, Isabella Stendardi, Rosanna Matucci and Federico Bennardini. A suitably international flavor was supplied by the other members of the Scientific Committee, Simo Oja and Kenji Yamaguchi.

The symposium was sponsored by the Universita degli Studi di Firenze, with additional support provided by the Consiglio Nazionale delle Richerche, Fidia Farmaceutici, Sandoz Prodotti Farmaceutici, Sigma-Tau, A. Menarini, Istituto Gentili, Roussel Maestretti, Cassa di Risparmio di San Miniato, Banca Popolare dell'Etruria and Banca Toscana. We thank all these for their help and generosity.

Last, but not least, we thank Harriet Larkin who, with the help of Doni Garcia, retyped the majority of manuscripts in this volume. We are especially appreciative of the help they gave us in this painstaking task of editing.

Ryan J. Huxtable

Flavia Franconi

Alberto Giotti

CONTENTS

SECTION I: TAURINE AND CARDIOVASCULAR PHARMACOLOGY

SECTION II. BIOCHEMISTRY, NUTRITION AND DEVELOPMENT

SECTION III. PHARMACOLOGY OF EXCITABLE TISSUE

SECTION IV: NEUROCHEMISTRY

SECTION I: TAURINE AND CARDIOVASCULAR PHARMACOLOGY

CARDIOVASCULAR PHARMACOLOGY AND EXPERIMENTAL THERAPEUTICS OF TAURINE AND RELATED COMPOUNDS

Alberto Giotti

Department of Preclinical and Clinical Pharmacology "Mario Aiazzi Mancini", University of Florence, viale G.B. Morgagni 65, 50134 Florence, Italy

Ladies and Gentlemen,

Let me thank the members of the Scientific Committee, Drs. Flavia Franconi, Ryan J. Huxtable, Simo S. Oja and Kenji Yamaguchi for having selected me to lecture in this opening ceremony. The main subject of this lecture is taurine, 2-aminoethane sulfonic acid, an amino acid with zwitterion characteristics, probably a very old molecule widely represented in the animal kingdom, important for osmoregulation in primitive sea animals and also for feeding behavior, in so far as it is a agonist for antennular receptors in the lobster (27), as we shall have the opportunity to discuss in the Symposium (33).

The substance, however, was discovered more then one hundred and fifty years ago in the bile of <u>Bos taurus</u> where, as well as in the bile of other animal species, it exerts an important, although apparently not obligatory, function in the secretory process of biliary acids.

Why our interest in taurine? Because in the late Fifties, a period in which we had to build our own apparatus - this was the case with a cathodic oscillograph - we were interested in the electrophysiological aspects of pharmacological actions. We could, through intracellular recordings, give the first (at least so we considered it) demonstration of histamine activation of pacemaker currents in cardiac sinoatrial cells (46). Our interest was not only in physiological compounds like histamine but also in cardiac glycosides and quinidine (7,30).

We were aware of the controversial interpretation of the possible role of isethionic acid - a derivative of taurine after deamination (see 16 for an updated version of taurine metabolism) - in the physiology of squid axon (45). We had, therefore, great interest in reading the work of Read and Welty (51) on the possible role of the compound in regulating cardiac cell excitability. Welty and Read had shown also that taurine antagonized the arrhythmogenic action of cardiac glycosides as well as that of epinephrine (57). For a pharmacologist coming from a medical background the main problem was: does taurine directly or through its eventual metabolite(s) also affect the cardiac positive inotropic effect - the main pharmacological action of cardiac glycosides which is the basis, together with the negative chronotropic effect, of their wide - and difficult - therapeutic use?

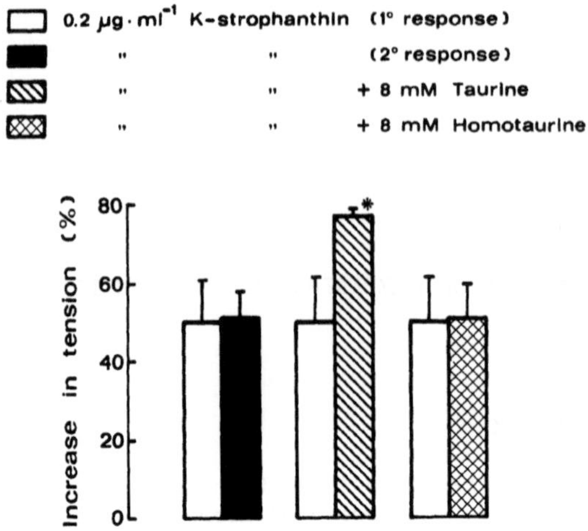

Fig. 1. Effect of taurine and homotaurine on the positive inotropic effect of K-strophanthin in spontaneously beating isolated guinea pig auricles. Values are means ± SEM of 10 experiments. *P<0.001. Modified from (37).

A second but equally important problem, especially from the physiological point of view, was: does taurine also affect the cardiac inotropic effect of epinephrine - the hormone of emergencies - or only its arrhythmogenic properties? The answers to these two questions were not present in the papers of Read and Welty nor elsewhere in the literature. A simple experiment was therefore devised using the auricles of guinea pig, a species which responds to cardiac glycosides with a positive inotropic effect (38). It was found that taurine, in a mM range in vitro, actually potentiates the positive inotropic effect of K-strophanthin - an effect which is not shared by homotaurine, the homologue, or by isethionate (Fig. 1). The auricles, which were responding positively to K-strophanthin and showed the

Table 1. Levels of taurine ($mg.l^{-1}$ of intracellular water)[a] in isolated guinea pig auricles and in the superfusion medium.

	Taurine Tissue Content $mg.l^{-1}$	Released $mg.l^{-1}$
A) Fresh preparation	(9) 3.09 ± 0.11	--
B) 120 min. incubation in Tyrode (washed every 15 min.)	(10) 2.12 ± 0.09*	(10) 0.95 ± 0.03
C) 90 min. in Tyrode and 30 min. with 8 mM Taurine	(7) 3.09 ± 0.19	--

[a]The water was measured by the antipyrine method (6). The extracellular space was assumed to 26% (34). Data are means ± SEM for numbers of experiments given in parenthesis. *Significantly different from both A and B. Modified from (38).

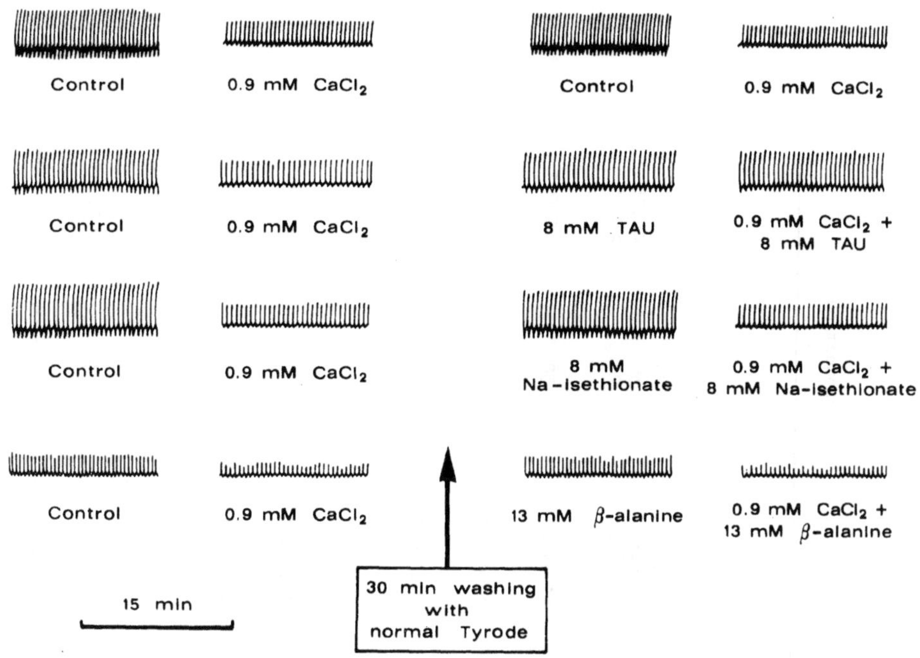

Fig. 2. Typical mechanogram of the effect of low calcium medium and low calcium medium plus taurine or Na-isethionate, or Beta-alanine on tension measured in spontaneously beating isolated guinea pig auricles. The experimental procedure is described in Guidotti and Giotti, 1969 (38).

potentiation of this action by taurine, had lost a labile pool of taurine which was entirely recovered in the medium and which was restored if taurine was present in the bath fluid (Table 1) (38,37).

Indeed, this lecture is dedicated to H.B. Van Dyke and A. Gilman who taught me (when a student at Columbia University) the rule of always following chemically in the tissue the compound which I was studying functionally.

The now classic review of Jacobsen and Smith (44) helped me avoid losing some of the best years of my life in trying to collect the scanty literature on taurine.

Thus, taurinology was born also in Florence, and it represented a good link between basic medical sciences like cytochemistry, biochemistry, physiology and pharmacology and the clinical sciences. The starfish, rich in taurine, with its five arms may indeed be a good symbol for this Symposium.

After explaining my personal interest in taurine I would like to present some information on the rapid (probably too rapid) passage of taurine from basic sciences and pharmacology to clinical use, as was realized at the end of the Sixties in this country. In our experiments on the effects of taurine in vitro it became evident that a direct positive inotropic effect of taurine could be seen if the concentration of calcium in the bath fluid was reduced below about 2 mM, a concentration which is routinely used in vitro experiments. The positive inotropic effect of taurine, even in reduced calcium conditions, was not shared by beta-alanine (Fig. 2). Taurine

Table 2. Effect of taurine and related compounds on oxygen consumption inhibition induced by administration (7.2 $mg.kg^{-1}$ sc) of KCN in rat

Treatment	N.	Oxygen consumption ($mM.Kg^{-1}.h^{-1}$) before treatment	after treatment (drug+KCN)	Inhibition %	Protection %	P
Physiol. Soln.	46	78.6±1.3	56.4±1.6	31.7±2.2	0	-
Taurine						
100 $mg.100g^{-1}$	16	73.9±2.0	67.2±1.5	6.9±4.4	78	<0.001
50 $mg.100g^{-1}$	10	79.9±2.0	64.8±1.0	20.4±4.5	36	<0.05
25 $mg.100g^{-1}$	8	76.2±5.4	60.0±10.7	25.0±4.7	20	>0.05
Monomethyltaurine 111 $mg.100g^{-1}$	10	73.8±2.2	63.0±2.0	13.6±5.0	58	<0.001
Isethionic acid 100 $mg.100g^{-1}$	9	82.2±3.4	63.8±3.4	22.0±3.2	31	<0.02
Dimethyltaurine 122 $mg.100g^{-1}$	9	78.4±5.4	53.9±3.1	28.5±6.0	11	>0.05
Homotaurine 112 $mg.100g^{-1}$	4	74.7±4.2	54.2±4.2	27.0±4.9	15	>0.05
-Alanine 72 $mg.100g^{-1}$	12	79.4±1.8	59.0±1.7	24.0±4.1	24	>0.05
Glycine 60 $mg.100g^{-1}$	11	80.8±7.5	52.7±4.6	35.2±6.5	0	>0.05
Cysteine 25 $mg.100g^{-1}$	7	78.4±3.4	72.5±4.0	5.8±2.8	82	<0.001
Thiosulphate 100 $mg.100g^{-1}$	7	69.5±3.6	66.3±3.4	7.3±5.1	74	<0.001

Data are means ± SEM. From Guidotti et al., 1967 (35)

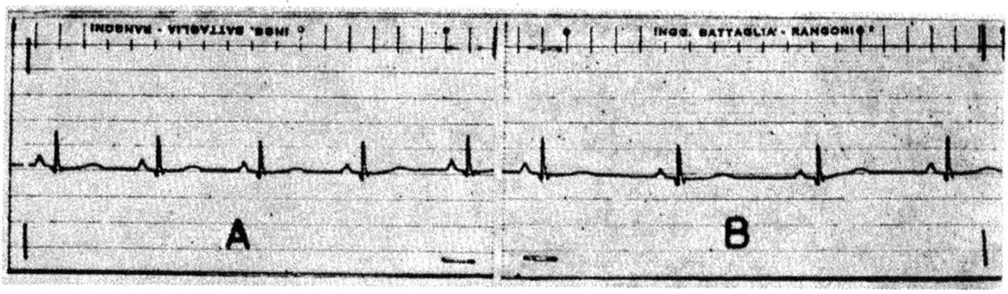

Fig. 3. Effect of taurine on the electrocardiogram of a dog anesthetized with nembutal. A) Control; heart rate 160 beats. min^{-1}. B) After administration of taurine (1.34 $g.kg^{-1}$ iv); heart rate 120 beats.min^{-1}. Taken from Guidotti and Giotti, 1970 (39).

injected in bolus or intravenously perfused in normal anesthetized rats, cats and dogs appeared to be well tolerated: the most prominent cardiovascular effect was bradycardia (Fig. 3). It must be noted that taurine had been injected in humans at the beginning of the century without unwanted reactions (53) except perhaps the exacerbation of skin lesions in psoriatic patients (52).

Many pharmacological-clinical findings on taurine were collected at the end of the Sixties and the early Seventies using a pharmaceutical preparation O-DUE FALORNI (now still on the market as O-DUE NATIVELLE). We shall limit ourselves to considering some of the observations related to the cardiovascular system:

a) In 10 subjects an acute intravenous pretreatment with 4 g taurine prevented the decrease of kininogen to 50-60% of the normal values in venous blood collected from an antecubital vein during a local ischemia of 5 min (3).

b) A chronic treatment with taurine 50 mg/kg per os activated fibrinolysis: the effect, which was absent after placebo, was maintained throughout the 5 weeks of treatment (15).

c) The arterial-venous difference in the percentage saturation in O_2 of hemoglobin, measured before and after treatment with taurine 6 g per os and 5 g iv twice a day for 15 days was increased by the treatment in 10 patients with peripheral ischemic diseases (58).

Many papers by Italian authors also appeared in this period on the therapeutic action of taurine in acute or chronic coronary as well as peripheral artery diseases; these authors seemed convinced of a therapeutic action of taurine but the studies generally did not include a control group (4,12,28,29,50,54,59).

At the end of this early period we wrote a review in which the potential therapeutic actions of taurine in the cardiovascular system were discussed (39). In writing about the protective effect of taurine we were also motivated by two sets of findings not directly related to the cardiovascular system.

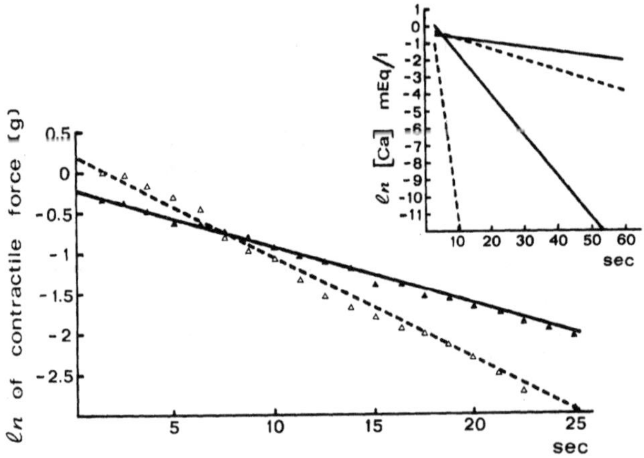

Fig. 4. Effect of taurine on tension decrease induced by perfusion with a calcium-free solution. The isolated guinea pig heart was paced at 240 beats.min^{-1}. Each point is the mean of 5 experiments. (△) Control; (▲) taurine-treated heart. The inset shows rates of calcium efflux for a slow-exchanging pool (upper lines) and a fast-exchanging pool (lower lines). The abscissa is the time from the beginning of calcium-free perfusion. (---) Control: (———) 8 mM taurine treated heart. From Dolara et al., 1973 (14).

In the rat, pretreatment with taurine 1 g/kg ip (but not with homotaurine or dimethyl taurine) prevented toxic symptoms following KCN injection at the dose of 7.2 mg/kg sc and also prevented the inhibiting effect on oxygen consumption. The protective effect was still present, although reduced, after monomethyl taurine and also after isethionate; it was absent after beta-alanine or glycine. Thiosulfate, which was an effective antagonist of KCN (as well as cysteine) increased further the urinary excretion of thiocyanate, differing from taurine which reduced it significantly. Taurine added in vitro, 10^{-2} - 10^{-4} M, did not antagonize the inhibitory effect of KCN on oxygen consumption in liver homogenates. However liver homogenates taken from animals pretreated 20 min before with taurine 1 g/kg ip showed a 50% reduction of the inhibiting effect of KCN on O_2 consumption (Table 2). The protective effect of taurine was not due to an increase of methemoglobin (35,36). Although not interpreted, the effect of taurine - possible related to its specific metabolic transformations - suggested a profound influence at the level of basal mechanisms of cell function (anion transport?).

A second experiment, left however at an anecdodal level, involved the healing power of taurine treatment of frostbite (32). The relationships between antihypoxic and antihypothermic effects are still of great interest (40).

Retrospectively we may say that the studies done in Florence, the one done in France by Dietrich and Diacono* (13) as well as the already mentioned studies of Read and Welty (51) on the effect of taurine on toxic arrhythmias, stimulated further studies all over the world. Our own studies took the

---

* These studies also have the merit to have emphasized that from the point of view of species-sensitivity, taurine and cardiac glycosides show similar behavior: both determine a negative inotropic effect in the rat.

Table 3. Actions of taurine on the heart and circulation

- Positive inotropy at low calcium concentrations
- Negative inotropy at high calcium concentrations
- Potentiation of digitalis inotropy
- Antiarrhythmia
- Antagonism of calcium paradox
- Retardation of lesion development in calcium overload myopathy
- Osmotic agent
- Hypotensive (central and peripheral)

From Huxtable and Sebring, 1983 (42).

direction of Ca^{2+}-taurine interactions. In isolated guinea-pig hearts it was shown that the decrease of contractility which follows an abrupt calcium deprivation is less sustained in taurine pretreated preparations. The effect of taurine on contractility corresponds to a slowing of the loss of calcium from two of the three compartments with different kinetics which were identified in the calcium pool (14) (Fig. 4).

A general conclusion from the pioneer studies in the Sixties and the early Seventies was that taurine and possibly related compounds would be worth studying not only physiologically but also pharmacologically as a basis for their eventual use in therapeutics; moreover controlled clinical studies were desirable. This approach was followed in the Seventies in a large number of laboratories all over the world: most of them are represented by the researchers in this Symposium.

Huxtable and Sebring (42) summarized at the beginning of the Eighties the cardiovascular action of taurine as is shown in Table 3. Therefore I will limit my presentation to discussing some contributions of our group to the complex picture of the actions of taurine and related compounds on the cardiovascular system. The experiments were done, whenever possible, in preparations from guinea pigs, a species which becomes easily sensitive to the inotropic action of cardiac glycosides as well as taurine. One of the preparations used routinely has been a couple of electrically paced strips

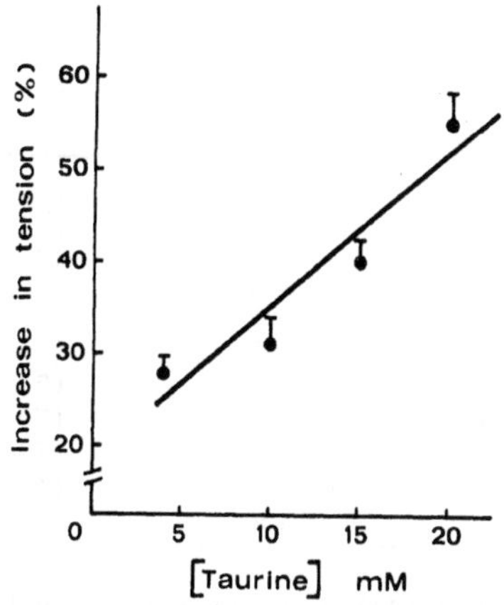

Fig. 5. Effect of different taurine concentrations on tension in guinea pig ventricular strips in low calcium medium (0.9 mM). The strips are paced at 120 beats.min^{-1}. The values are means $\pm$ SEM of at least 10 experiments.

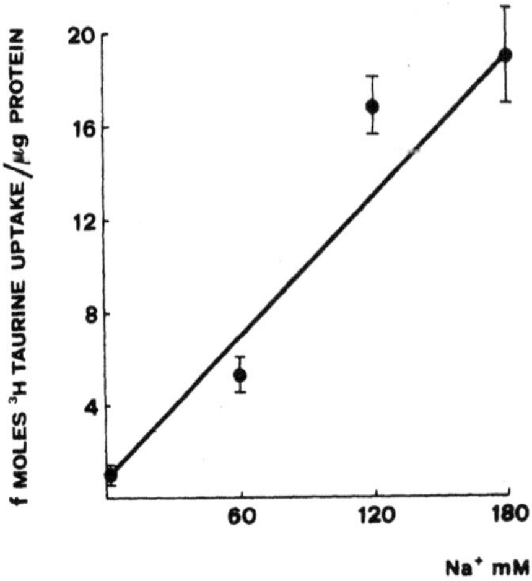

Fig. 6. Effect of Na^+ on taurine uptake measured in guinea pig cardiac sarcolemmal membranes. The experimental procedure is described in Franconi et al., 1981 (20). Results are means ± SEM of 4 experiments performed in triplicate.

from the right ventricle. In such preparations taurine, if the calcium concentration in the bath fluid is low, exerts a positive time- and dose-dependent action in the mM range (Fig. 5, ref. 21). Ventricular sarcolemmal vesicles take up taurine; the uptake is sodium-dependent (Fig. 6) but analogs of taurine may interfere with its uptake (Table 4, ref. 20). Beta-alanine, which interferes with taurine uptake, also interferes with its inotropic effect (Fig. 7, ref. 25).

In the ventricular strips, tension as well as the calcium level in the tissue are a linear function of external calcium (Fig. 8). The preparation

Table 4. Effect of taurine analogs on ^{3}H-taurine uptake in cardiac sarcolemmal membranes obtained from guinea pig heart

Analogs			Percent uptake
None			100
β-Alanine	10 mM	(6)	8
"	1 mM	(5)	6
Hypotaurine	10 mM	(5)	29
"	1 mM		--
Homotaurine	10 mM	(6)	17
"	1 mM	(5)	29
Guanidoethane sulfonate	10 mM	(5)	24
"	1 mM	(6)	28
Isethionic acid	10 mM	(6)	121*
"	1 mM	(6)	141*

Membranes were incubated with taurine analogs and ^{3}H-taurine (0.5 µM) for 30 min at 20 °C. In brackets are the number of experiments; each experiment was performed in triplicate. Only isethionic acid stimulated taurine uptake (* P<0.01). From Franconi et al., 1981 (20).

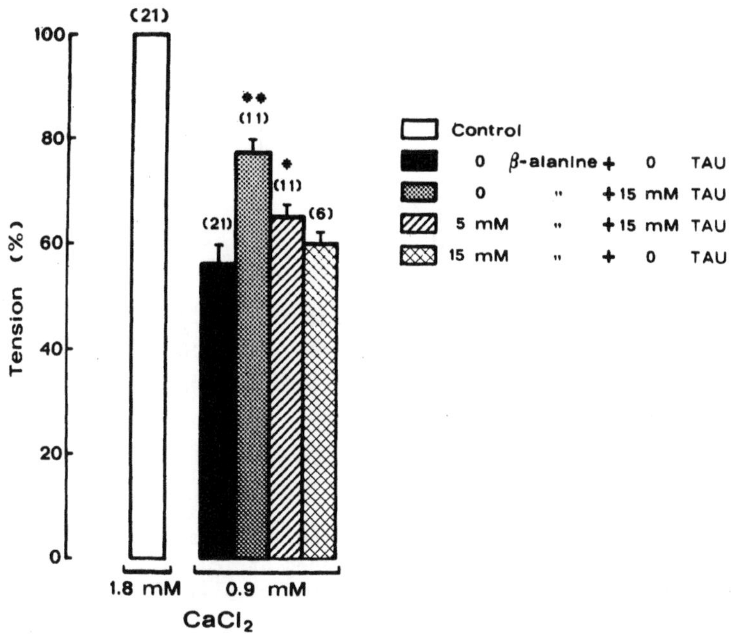

Fig. 7. Effect of beta-alanine on positive inotropic effect induced by taurine in guinea pig ventricular strips paced at 120 beats.min^{-1}. Tension is measured at 0.9 mM $CaCl_2$. Values are means ± SEM; in brackets the number of experiments. * 0.05>P>0.01; ** 0.01>P>0.001. Modified from Franconi et al., 1984 (25).

is indeed a good titrator of Ca^{2+} in the medium. This conclusion, however, no longer holds in the presence of taurine: the relationship between tension and calcium becomes a complex one, characterized by a dose-dependent positive inotropic effect with a maximum of about +40% when taurine is 20 mM and calcium is 0.9 mM ("low calcium"), a point of "no-effect" when calcium is 2 - 2.2 mM ("physiological calcium"); a negative dose-dependent inotropic effect with a maximum of about -40% when taurine is 20 mM and calcium is 2.7 mM or more ("high calcium") (Fig. 9).

The above described linear relationship between tissue calcium and bath calcium is also transformed into a curvilinear one in the presence of taurine (Fig. 9). As a matter of fact, Ca^{2+} transfer into the cell is reduced when its concentration is above 2.2 mM and taurine is present in concentrations between 4 and 20 mM (higher concentrations have not been explored). This phenomenon is not due to fortuitous calcium precipitations although some forms of physicochemical interaction between calcium ions and taurine itself in the sarcolemma may be suspected. In this ventricular model, as was the case in the primitive experiments with the isolated auricles, taurine added to the perfusing fluid restores the taurine lost in the incubation period; the increse of taurine in the tissue is also calcium-dependent but the relationship between taurine in the tissue and calcium in the bath is complex with a point of inversion again at 2.2 mM calcium (Fig. 9). We could say that as taurine prevents an excess of calcium in the tissue, calcium prevents an "excess" of taurine.

Obviously any interpretation of this phenomenon should take into consideration that taurine in the presence of "low" concentrations of

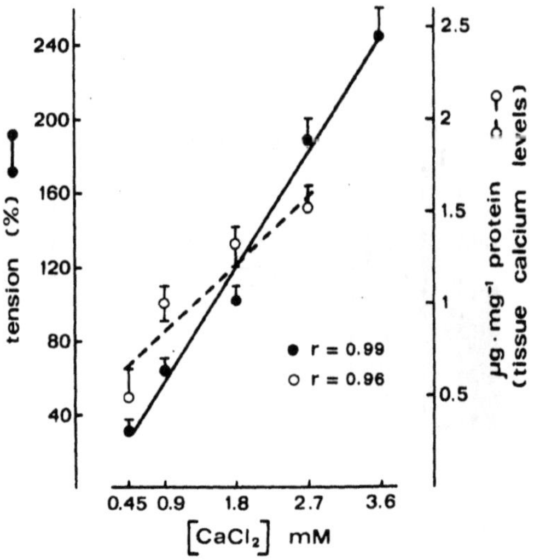

Fig. 8. Effect of external calcium on tension and tissue calcium levels measured in guinea pig ventricular strips paced at 120 beats.min^{-1}. The experimental procedure is described in Franconi et al., 1982 (21). Values are means ± SEM of at least 9 experiments.

calcium, increases dose-dependently the calcium in the tissue while calcium in low concentration favors dose-dependently the increase of taurine in the tissue.

The inactivity of taurine on contractility when calcium is in the physiological range is independent of the rate of the stimulation which, conversely, strongly influences contractility in the presence of an alpha-adrenergic stimulant (to be discussed later). In considering these results I would like to modify the table from the review of Huxtable and

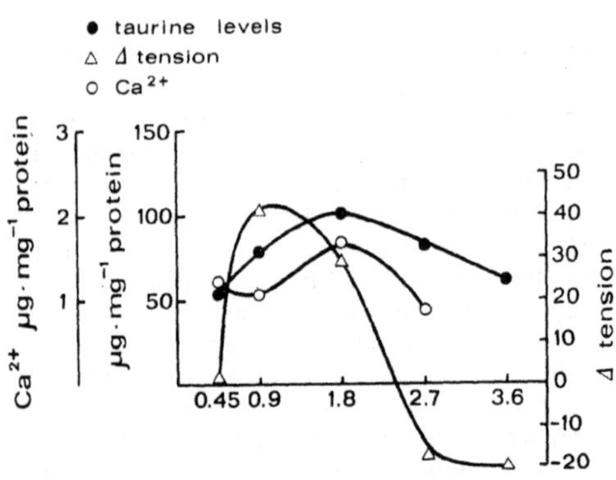

Fig. 9. Effect of 10 mM taurine on cardiac taurine and calcium levels and tension as a function of calcium concentration (abscissa). The experiments are performed in guinea pig ventricular strips paced at 120 beats.min^{-1}. Values are means of at least 5 experiments. The experimental procedure is described in Franconi et al., 1982 (21).

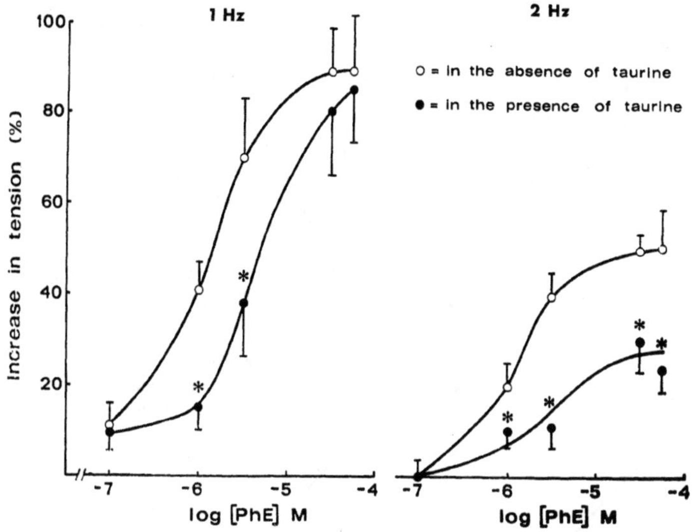

Fig. 10. Dose-response curve of phenylephrine in the presence and absence of 20 mM taurine at 2.2 mM $CaCl_2$. Values are means ± SEM of at least 5 experiments. * 0.05>P>0.01. From Franconi et al., 1986 (17).

Sebring (42), substituting paragraphs 1, 2 and 3 with the following ones:

1. Positive inotropy at low calcium concentrations in taurine-depleted preparations is linked to a restoration of the taurine content of the tissue and an increase of calcium in the tissue.

2. Negative inotropy at high calcium concentrations in taurine-depleted preparations is linked to a codepression of taurine and calcium transport.

3. There is no effect at physiological calcium concentrations in taurine-depleted preparations although at such calcium concentrations taurine enrichment in the tissue is maximal.

In conclusion, taurine behaves as a buffer towards variations of cardiac contractility above or below the "physiological" value when the variations are evoked by external calcium. It would be stimulating to speculate that taurine is "physiologically" fundamental in stabilizing heart function towards variations of ionized calcium: obviously, the hypothesis, to be checked in vivo, requires adequate means to modulate the taurine content of the heart; research is in progress in this field (43). We have mentioned that taurine in the mM range does not influence the contractility of ventricular strips when the calcium concentration is in the physiological range. However, phenylephrine, in the presence of a beta-blocker, exerts a positive inotropic effect which appears with a long latency (different from beta-adrenergic effects), depends on the rate of stimulation (at 1 Hz greater than at 2 Hz) and is antagonized by alpha-blockers.

Taurine depresses the alpha-adrenergic inotropic effect at concentrations which do not influence the beta-adrenergic one (Fig. 10). Taurine interferes with alpha-receptors as shown by its influence on prazosin binding; this effect is not shared by analogs like beta-alanine and requires preloading to become evident (Fig. 11, ref. 17). We shall see later on that

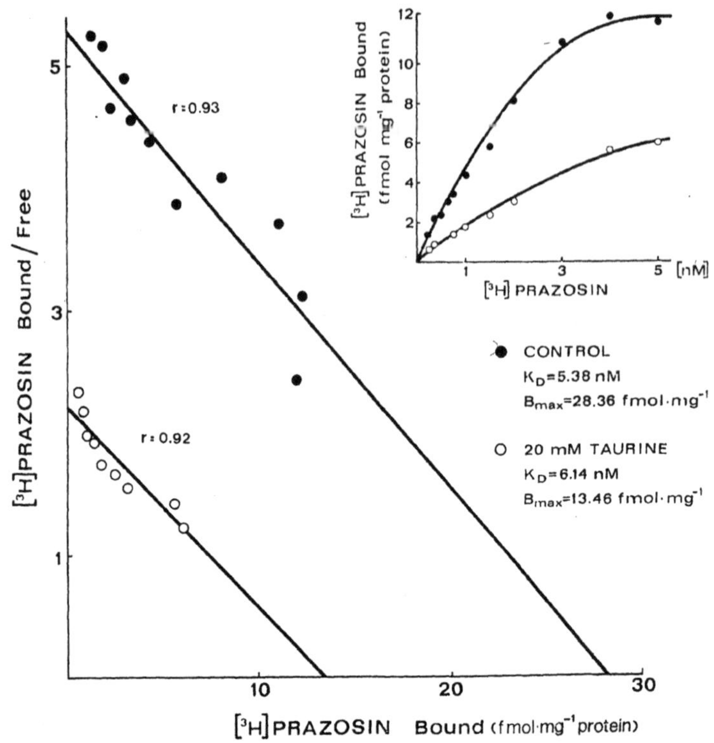

Fig. 11. Scatchard analysis of ^{3}H-prazosin binding to guinea pig ventricular membranes preincubated with or without 20 mM taurine for 30 min. In the inset are the saturation curves. Values are means of at least 6 experiments performed in triplicate. From Franconi et al., 1986 (17).

the structural requirements for the α-antagonist effect of taurine towards phenylephrine inotropy are more stringent that those for the direct positive inotropic effect of taurine in low calcium medium. At physiological calcium concentrations, taurine, in a concentration which is not per se effective on contractility, exerts a dose-dependent positive inotropic effect if contractility is depressed by methoxyverapamil: the action is characterized by a time course analogous to that of the action of taurine in low calcium concentrations (Fig. 12, ref 24).

Interestingly, taurine, which per se has a negative chronotropic action on isolated auricles of guinea pig (as well as other animal species and also in vivo) antagonizes the negative chronotropic effect of methoxyverapamil (Table 5). That the action of taurine is specifically linked to the site(s) of action of methoxyverapamil and of prazosin is further substantiated by the ineffectiveness of taurine on nitrendipine binding sites (Table 6). One is inclined to envisage two adrenergically regulated calcium mechanisms: one operated by beta-receptors on the outer face of the sarcolemma membrane with a prompt response and a low frequency dependence not influenced by taurine, the other operated by the alpha-1-receptors on the inner side of the sarcolemma membrane with a slow response and a high frequency dependence, strongly regulated by taurine. The loss of taurine (and, perhaps, potassium) during hypoxia (11,23) would represent a molecular mechanism for alpha-adrenergic activation, a condition which at the cardiac level is

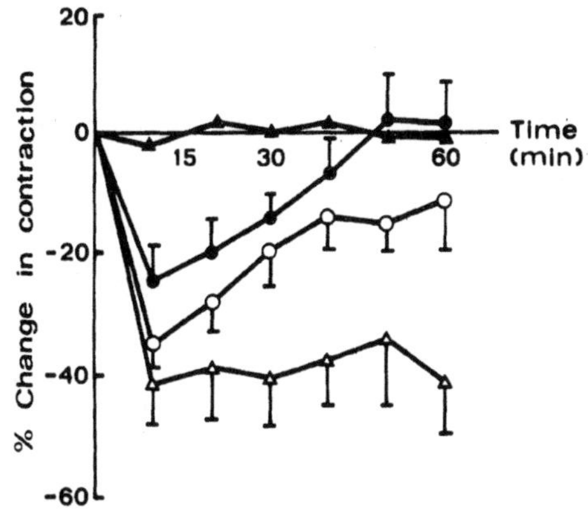

Fig. 12. Effect of D600 and taurine on tension in guinea pig ventricular strips. Results are means ± SEM of at least 6 experiments. ▲ Control; ● 10^{-10} M D600 + 20 mM taurine; ○ 10^{-7} M D600 + 10 mM taurine; △ 10^{-7} M D600. From Franconi et al., 1982 (24).

characterized by positive inotropy and negative chronotropy (and possible an activation of anerobic glycolysis). Evidence in favor of a relationship between hypoxia and alpha-adrenergic expression has been presented (10, 31). If this is the case, the role of taurine in cardiac pathophysiology would be indeed fundamental. However, its administration as a therapeutic agent in non-severe ischemic conditions is perhaps questionable.

On the other hand, in severe hypoxia the suggested mechanism (i.e. activation of alpha-adrenergic mediated biochemical and functional parameters caused by the decrease of taurine concentration in the tissue determined by taurine release) could reach an excessive level of activation: as a matter of fact, in severe hypoxia alpha-blockers as well as taurine exert a protective effect (23).

We must also consider that nitrendipine binding sites, which are not affected by taurine in physiological conditions, are strongly affected by hypoxia followed by reoxygenation (Kd = 0.14 nM; Bmax = 603 $fmol.mg^{-1}$ protein in physiological conditions; Kd = 0.13 mM; Bmax = 186 $fmol.mg^{-1}$ protein after hypoxia followed by reoxygenation) (5, 48). Taurine, however, protects nitrendipine binding sites from oxygen toxicity after hypoxia (Fig.

Table 5. Effect of D600 and taurine on the frequency of isolated spontaneously-beating atria in guinea pig

Conditions	$Beats.min^{-1}$	
Control	111 ± 4	(23)
20 mM taurine	100 ± 4	(6)
10^{-7} M D600	81 ± 4 *	(6)
10^{-7} M D600 + 20 mM taurine	97 ± 3	(6)

Results are means ± SEM; the number of experiments are given in parenthesis. *P<0.001. From Franconi et al., 1982 (24).

Table 6. Effect of 20 mM taurine on specific ^{3}H-nitrendipine binding to membranes isolated from guinea pig heart

	^{3}H-Nitrendipine bound (fmol.mg^{-1} protein)
Control	236.6 ± 54.1
20 mM Taurine	240.6 ± 48.9

The membranes were incubated for 60 min at 25°C, in the presence or absence of 20 mM taurine, with 1 nM ^{3}H-nitrendipine. Results are means ± SEM of 7 experiments.

13). The protective effect of taurine extends more generally to myocardial cells: the protection is shown on lactate dehydrogenase leakage as well as on functional parameters and calcium overload (Fig. 14, ref. 23). A protective effect of taurine is also evident in a model of ischemia-reperfusion in the anesthetized guinea pig (Table 7, ref. 47).

In addition to its possible regulatory role in very general functions of cellular life taurine also possesses a profound relaxing effect on potassium depolarized arteries. The relaxing effect is dose and calcium dependent (18,19).

The therapeutic effects of taurine observed in many models like the cardiomyopathic hamsters of Welty and McBroom (49,56) or rabbits in congestive heart failure (2) or our laboratory's guinea pigs with cardiac damage by ischemia-reperfusion may be related to the sum of elementary mechanisms on calcium kinetics, on cytoprotection, on arterial relaxation as well as on sinoatrial cell pacemaker currents: the bradycardiac effect in critical conditions of oxygen disposal may become relevant to survival (Fig. 15). The trial of Azuma et al. on the effect of taurine treatment on congestive heart failure (1) - a trial which can be considered the first controlled study on the effect of taurine in humans, at least at

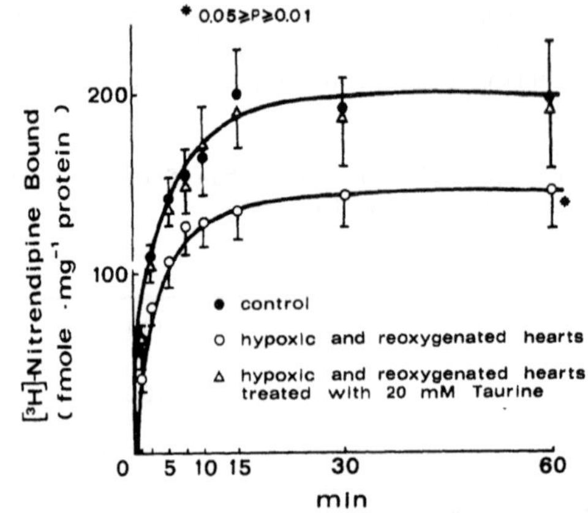

Fig. 13. Time-course of specific ^{3}H-nitrendipine binding (1nM) to membranes obtained from control, hypoxic and reoxygenated heart and hypoxic and reoxygenated heart treated with 20 mM taurine. Values are means ± SEM of at least 6 experiments. (Unpublished results).

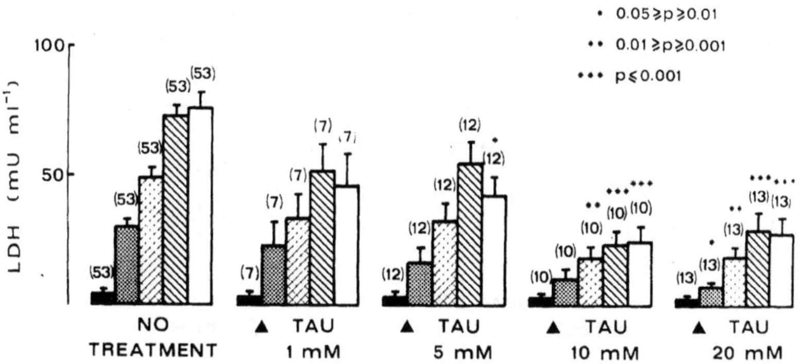

Fig. 14. Effect of taurine on lactate dehydrogenase (LDH) release in guinea pig heart submitted to 30 min of hypoxia and 30 min of reoxygenation: (■) last 5 min of stabilization; (▩) 10-15 min hypoxic phase; (▨) 25-30 min hypoxic phase; (▧) 5-10 min reoxygenation phase; (□) 15-20 min reoxygenation phase. Data are means ± SEM for the number of experiments in parenthesis. From Franconi et al., 1985 (23).

the cardiovascular level - shows that the results of experiments with taurine on laboratory animals had a predictive power for clinical results. However, some recent findings in our laboratory advise care in the interpretation and use of analogic reasoning.

Structural requirements for the inotropic effects of taurine related compounds will be extensively discussed in San Miniato (22): let me anticipate these to some extent. Among a relatively large series of taurine analogs, two (l-cysteic acid and orthanilic acid) have been found to possess taurine-like properties in the simple test of activation of contractility in vitro. The inotropic effects of l-cysteic acid and orthanilic acid are, as was shown for taurine, calcium dependent and are not related to variations of taurine content in the tissue (see Table 2 in ref. 26).

As was shown for taurine, the effects of l-cysteic and orthanilic acids are null when the calcium concentration is around 2.2 mM. The possible generalization: l-cysteate = orthanilate = taurine as far as inotropic activity is concerned does not survive if the analogs are checked by two different tests although they are related to inotropism. The antagonistic effect of taurine towards the alpha-adrenergic positive inotropic effect is indeed not shared by orthanilic acid and l-cysteic acid. In the test which measures the reversion of the negative inotropic effect of methoxyverapamil, orthanilic acid is as active as taurine whereas the action of l-cysteic acid is at the limits of statistical significance (see Table in ref. 22). Thus although the inotropic activity on ventricular strips is equal for taurine, l-cysteic acid and orthanilic acid (and the equi-effectiveness with taurine is indeed very interesting for orthanilic acid where the sulfonic group and the amino group are on a benzene ring), the modulation of alpha-adrenergic activated inotropy (an action which we consider at an internal side of sarcolemma) is absent in the higher MW compounds. If the modulation of alpha-adrenergic activated inotropy is important in some therapeutic effects obviously the taurine "analogs" would result in being inactive.

Colucci, Wright and Braunwald (8,9) have included taurine in a group of "several non-glycoside, non-sympathomimetic positive inotropic agents" (amrinone, milrinone, sulmazole, coenzyme Q, etc.) which have been administered in humans and have been shown to cause salutary short-term

Table 7. Effect of ischemia and reperfusion on the recovery of normal electrical activity, on lactate dehydrogenase (LDH) and pH variation in control and taurine-treated anesthetized guinea pigs

	Control	Taurine
% of normal electrical activity recovery		
ISCHEMIA	0	71*
REPERFUSION	0	57*
Plasma LDH ($mU.ml^{-1}$)		
BASAL	154.8 ± 14.3	183.2 ± 9.9
ISCHEMIA	303.0 ± 35.2	253.1 ± 19.3*
REPERFUSION	590.3 ± 74.5	355.8 ± 53.4
Blood pH		
BASAL	7.37 ± 0.02	7.35 ± 0.01
ISCHEMIA	7.28 ± 0.01	7.38 ± 0.01*
REPERFUSION	7.29 ± 0.02	7.39 ± 0.01*

The animals were anesthetized with pentobarbital sodium (70 $mg.Kg^{-1}$) and the ischemia was performed with the left anterior descending coronary artery ligated for 30 min according to Clark et al., J.Pharmacol.Meth. 3, 357, 1980. Taurine (2 $g.kg^{-1}$ ip) was administered 30 min before the beginning of ischemia. The Chi-square test was used to detect significant differences between control and treated preparation for the recovery of normal electrical activity: in the other cases the statistical analysis was performed with Student's t test. Values are means ± SEM of 7 experiments. * $P<0.01$. From Masini et al., unpublished results (47).

hemodynamic changes in resting patients with congestive heart failure. However, several important questions about these new drugs remain to be answered: " . . . their biochemical mechanisms of action are incompletely understood, and although such knowledge is not crucial to their clinical effectiveness, it is likely to provide important insights with respect to their optimal use".

The long history of scientific research on cardiac glycosides reminds us that the progress of cardiovascular pharmacology is strictly linked to that of cardiovascular physiology.

We have stressed the difficulties of transferring in vivo the results of experiments performed on simplified models. As a physiologist I would wish that taurine might be a link between models and animals in the field of calcium cellular regulation as ouabain has been a fundamental molecule in unraveling the mystery of sodium-potassium cellular regulation. The importance of the taurine-calcium ratio and the remarkable differences in taurine content in different species suggest that many experiments aimed at defining the role of calcium in cellular functions in vitro (where cellular taurine can be lost in the perfusing fluid) or in vivo (where taurine content can be modified by dietary and neurohumoral influences) should be repeated, measuring taurine and not only calcium content and kinetics. As a pharmacologist I do not hesitate to say that drugs with taurine-like properties would be very promising to protect the heart from calcium overload

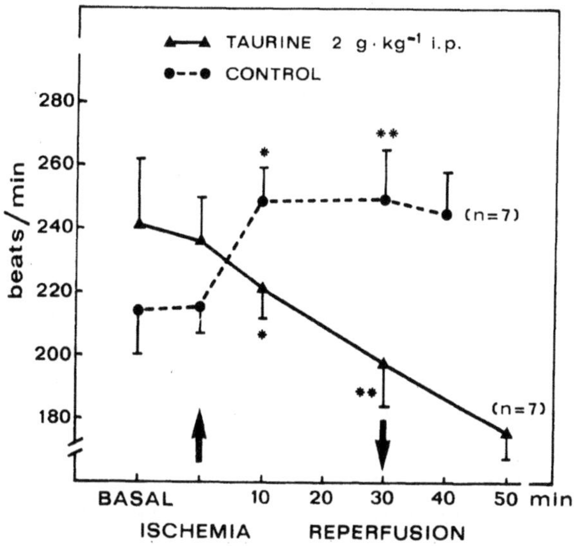

Fig. 15. Effect of taurine (administered 30 min before the beginning of ischemic phase) on heart rate variation in anesthetized guinea pig (70 $mg.kg^{-1}$ pentobarbital sodium) during eschemia and reperfusion. Values are means $\pm$ SEM for the number of experiments given in parenthesis. $*P<0.05$; $**P<0.01$. From Masini et al., unpublished results (47).

caused by calcium-paradox, ischemia-reperfusion, excessive digitalization, or hyperadrenergic activation. This differs from known synthetic organic compounds with calcium ion entry blocking properties in that taurine-like compounds would not exert a cardiodepressant activity.

Ryan Huxtable, in a recent symposium on taurine, entitled his talk "Does Taurine Have a Function?" (41). I would suggest that this question is today, after the work of so many researchers, too provocative although I agree that the function of a Symposium in its original meaning of a "drinking party" is at least as understandable as taurine's probable primordial function as an osmotic regulator (54). Certainly, to focus on the actions of taurine and related compounds on the cardiovascular system, as has happened to myself in this lecture, is like considering only a part of a picture instead of the whole. Fundamental data on taurine's actions and physiology come from scientists working on sensory organs, the central nervous system, endocrine glands, nutrition and so on. The starfish with the five arms I proposed at the beginning should evoke its potential to increase its number of arms to represent taurinology.

Finally, I would like to thank, also on behalf of my coworkers and students, the Rector Magnificus of Florence University, Professor Franco Scaramuzzi; the National Research Council of Italy and particularly its President, Professor Luigi Rossi Bernardi; the officers of the Department of Pharmacology and particularly the Dean, Professor Fabrizio Ledda and the Administrative Assistant, Mrs. Silvana Romanelli Soldani, for the support they offered to our group. Thank you for your attention.

REFERENCES

1. Azuma, J., Sawamura, A., Awata, N., Hasegawa, H., Ogura, K., Harada, H., Ohta, H., Yamaguchi, K., and Kishimoto, S., 1983, Double-blind randomized crossover trial of taurine in congestive heart failure, Current Ther. Res. 34:543-557.
2. Azuma, J., Takihara, K., Awata, N., Ohta, H., Sawamura, A., Harada, H., and Kishimoto, S., 1984, Beneficial effect of taurine on congestive heart failure induced by chronic aortic regurgitation in rabbits, Res. Comm. Chem. Path. Pharm. 45:261-270.
3. Bavazzano, A., Sidell, N., Michelacci, S., Sicuteri, F., 1969, Local ischemia as a kininogen depressant: effect of cortisone, taurine and kallikrein inhibitor, Pharmacol. Res. Comm. 1:137-138.
4. Becattini, V., 1976, Prospettive di un nuovo farmaco nella terapia delle vasculopatie: la taurina, La Settimana Medica 55:1355-1365.
5. Bennardini, F., Franconi, F., Giotti, A., Matucci, R., and Stendardi, I., 1985, Binding of ^{3}H-nitrendipine during hypoxia and subsequent reoxygenation, Br. J. Pharmacol. 86:623P.
6. Brodie, B.B., Axelrod, J., Soberman, R., and Levy, B.B., 1949, The estimation of antipyrine in biological materials, J. Biol. Chem. 179:25-29.
7. Buffoni Nardini, F., and Giotti, A., 1955, Titolazione dei glucosidi digitalici nella cavia, Il Farmaco, Ed. Pr. 11:615-642.
8. Colucci, W.S., Wright, R.F., and Braunwald, E., 1986, New positive inotropic agents in the treatment of congestive heart failure, First of two parts, N. Engl. J. Med. 314:290-299.
9. Colucci, W.S., Wright, R.F., and Braunwald, E., 1986, New positive inotropic agents in the treatment of congestive heart failure, Second of two parts, N. Engl. J. Med. 314:349-358.
10. Corr, P.B., and Grafforol, A.W., 1981, Enhanced alpha-adrenergic responsiveness in ischemic myocardium: role of alpha adrenergic blockade, Am. Heart J. 102:605-611.
11. Crass, M.F., III and Lombardini, J.B., 1977, Loss of cardiac muscle taurine after acute left ventricular ischemia, Life Sci 21:951-958.
12. De Rango, F., and Del Corso, P., 1968, Modificazioni elettrocardiografiche indotte dalla taurina nelle cardiopatie coronariche, Recenti Progressi in Medicina 14:19-51.
13. Dietrich, J., and Diacono, J., 1971, Comparison between ouabain and taurine effects on isolated rat and guinea-pig hearts in low calcium medium, Lif Sci 10:499-507
14. Dolara, P., Agresti, A., Giotti, A., and Pasquini, G., 1973, Effect of taurine on calcium kinetics of guinea-pig heart. Eur. J. Pharmacol. 24:352-358.
15. Fanciullacci, M., Sicuteri, F., Giotti, A., and Brunetti, S., 1970, La taurina come fibrinolotico nell'uomo: prime osservazioni terapeutiche, La Clinica Terapeutica 53:3-8.
16. Fellman, J.H., Roth, E.S., Avedovech, N.A., and McCarthy, K.D., 1980, The metabolism of taurine to isethionate, Arch. Biochem. Biophys. 204:560-567.
17. Franconi, F., Bennardini, F., Matucci, R., Stendardi, I., Failli, P., Manzini, S., and Giotti, A., 1986, Functional and binding evidence of taurine inhibition of alpha adrenoceptor effects on guinea-pig ventricle, J. Mol. Cell Cardiol 18:461-468.
18. Franconi, F., Giotti, A., Manzini, S., Martini, F., Stendardi, I., and Zilletti, L., 1982, The effect of taurine on high potassium and noradrenaline-induced contraction in rabbit ear artery, Br. J. Pharmac. 75:605-612.
19. Franconi, F., Manzini, S., Stendardi, I., Bennardini, F., Antonini, G., Failli, P., Matucci, R., and Giotti, A., 1982, Differential inhibitory

effect of taurine on contractile responses to potassium and noradrenaline in rabbit ear artery, J. Pharm. Pharmacol. 35:46-48.
20. Franconi, F., Martini, F., Manghi, N., Galli, A., Bennardini, F., and Giotti, A., 1981, Uptake of ^{3}H-Taurine into myocardial membranes, Biochem. Pharmacol. 30:77-80.
21. Franconi, F., Martini, F., Stendardi, I., Matucci, R., Zilletti, L., and Giotti, A., 1982, Effect of taurine on calcium levels and contractility in guinea-pig ventricular strips, Biochem. Pharmacol. 31:3181-3185.
22. Franconi, F., Stendardi, I., Failli, P., Fazzini, A., and Giotti, A., Inotropic activity of orthanilic and L-cysteic acid on isolated guinea-pig ventricular strips, In this volume.
23. Franconi, F., Stendardi, I., Failli, P., Matucci, R., Baccaro, C., Montorsi, L., Bandinelli, R., and Giotti, A., 1985, The prospective effects of taurine on hypoxia (performed in the absence of glucose)and on reoxygenation (in the presence of glucose) in guinea-pig heart, Biochem. Pharmacol. 34:2611-2615.
24. Franconi, F., Stendardi, I., Martini, F., Zilletti, L., and Giotti, A., 1982, Interaction between organic calcium-channel blockers and taurine in vitro and in vivo, J. Pharm. Pharmacol. 34:329-330.
25. Franconi, F., Stendardi, I., Matucci, R., Failli, P., Bennardini, F., Antonini, G., and Giotti, A., 1984, Inotropic effect of taurine in guinea pig ventricular strips, Eur. J. Pharmacol. 102:511-514.
26. Franconi, F., Failli, P., Stendardi, I., Matucci, R., Bennardini, F., Baccaro, C., and Giotti, A., 1986, Positive inotropic effect of some taurine-related compounds on guinea-pig ventricular strips perfused with low calcium medium, Eur. J. Pharmacol. 124:129-133.
27. Fuzussery, Z.M., Carr, W.E.S., and Ache, B.W., 1978, Antennular chemosensitivity in the spiny lobster, Panulinus Argus: studies of taurine sensitive receptors, Biol. Bull. 154:226-240.
28. Gagliarducci, U., 1971, Possibilita' terapeutiche di un aminoacido solforato nelle eritrocitosi ipo-ossiche, Clin. Ter. 58:145-150.
29. Gagliarducci, U., 1974, Sul trattamento delle cardiopatie ischemiche acute con taurina, Osservazioni cliniche su trenta casi, Clin. Ter. 68:261-270.
30. Giotti, A., 1959, Rapporti fra tempo diastolico, concentrazione degli elettroliti ed effetti della chinidina nel miocardio atriale di mammifero, Lo Sperimentale 109:129-188.
31. Giotti, A., Ledda, F., Franconi, F., Mantelli, L., and Mugelli, A., 1984, Cardiac alpha adrenergic receptors: do they have a role? In: Frontiers in Cardiology for the eighties, ed. by L. Donato and A. L'Abbate, pp. 127-129. Academic Press Inc. Ltd. (London).
32. Giotti, A., Personal observations, Unpublished.
33. Gleeson, R.A., A taurine receptor model: taurine-sensitive olfactory cells in the lobster, This volume.
34. Goodford, P.J. and Lullmann, H., 1962, The uptake of ethanesulphonate-S ions by muscular tissue, J. Physiol. 161:54-61.
35. Guidotti, A., Buffoni, F., and Giotti, A., 1967, Azione della taurina e di alcuni suoi derivati sulla riduzione del consumo di ossigeno prodotta dal cianuro nel ratto, Boll. Soc. Ital. Biol. 43:1390-1392.
36. Guidotti, A., Buffoni, F., and Giotti, A., 1967, Ricerche sull'interazione taurina-cianuro nel ratto, Boll. Soc. Ital. Bio. Sper. 43:1392-1395.
37. Guidotti, A., Badiani, G., and Giotti, A., 1971, Potentiation by taurine of inotropic effects of straphanthin-K on guinea-pig isolated auricles, Pharmacol. Res. Comm. 3:29-38.
38. Guidotti, A., and Giotti, A., 1969, Digitalis inotropic effect on the auricular myocardium of taurine, In: Medicaments et metabolism du myocarde, ed. by D. Lamarche et Royer, Nancy, pp. 487-491.
39. Guidotti, A., and Giotti, A., 1970, Taurina e sistema cardiovascolare, Recenti Prog. Med. 49:61-97.

40. Hochachka, P.W., 1986, Defense strategies against hypoxia and hypothermia, Science 231:234-240.
41. Huxtable, R.J., 1979, Does taurine have a function? TIPS 1:V-VII.
42. Huxtable, R.J. and Sebring, L.A., 1983, Cardiovascular actions of taurine. In: Sulfur Amino Acids: Biochemical and Clinical Aspects, Liss, A.R., Inc., New York, pp. 5-37.
43. Huxtable, R.J., Laird II, H.E., and Lippincott, S.E., 1979, The transport of taurine in the heart and the rapid depletion of tissue taurine content by guanidinoethyl sulfonate, J. Pharmacol. Exp. Ther. 211:465-470.
44. Jacobsen, J.G., and Smith, L.H., Jr., 1968, Biochemistry and physiology of taurine and taurine derivatives, Physiol. Rev. 48:424-491.
45. Koechlin, B.A., 1954, Isethionic acid, the major anion in axoplasm of squid giant nerve, Federation Proc. 13:80.
46. Levi, R., and Giotti, A., 1967, Effect of Histamine on sinoatrial node cells of rabbit heart, Experientia 23:1-4.
47. Masini, E., Failli, P., Baccaro, C., Franconi, F., and Giotti,A., Unpublished results.
48. Matucci, R., Bennardini, F., Sciammarella, M.L., Baccaro, C., Stendardi, I., Franconi, F., and Giotti, A., 1983, ^{3}H-nitrendipine binding in membranes obtained from hypoxic and reoxygenated heart, 1987, Biochem. Pharmacol., 36:1059-1062.
49. McBroom, M.J., and Welty, J.D., 1977, Effects of taurine on heart calcium in the cardiomyopathic hamster, J. Mol. Cell. Cardiol 9:853-858.
50. Montanini, R., Zibetti, A., and Gasco, P., 1971, Osservazioni preliminari sull'impiego della taurina nel trattamento delle vasculopatie cerebrali, Clin. Ter. 59:321-329.
51. Read, W.O., and Welty, J.D., 1965, Taurine as a regulator of cell potassium in the heart. In: Electrolytes and Cardiovascular Diseases, Ed. by E. Bajusz, S. Karger, Press Basel, New York, pp. 70-85.
52. Roe, D.A., 1966, Taurine intolerance in psoriasis, J. Invest. Derm. 46:420-430.
53. Schmidt, C.L.A., Von Adelung, E., and Watson, T., 1981, On the elimination of taurine administered to man, J. Biol. Chem. 33:501-503.
54. Thurston, J.H., Hauhart, R.E., and Naccarato, E.F., 1981, Taurine: possible role in osmotic regulation of mammalian heart, Science 214:1373-1374.
55. Sicuteri, F., Fanciullacci, N., Franchi, G., Giotti, A., and Guidotti, A., 1970, Taurine as a therapeutic agent in vascular pain, Clin. Med. 77:21-32.
56. Welty, J.D., and McBroom, M.J., 1985, Effects of verapamil and taurine administration on heart taurine and calcium in bio 14.6 cardiomyopathic hamsters, Res. Comm. Chem. Path. Pharmac. 49:141-144.
57. Welty, J.D., and Read, W.O., 1964, Studies on some cardiac effects of taurine, J. Pharmacol. Exp. Ther. 144:110-115.
58. Zanelli, G., Ferrero, S., and Patrone, P., 1971, La taurina nella terapia delle arteriopatie obliteranti degli arti inferiori, Nota I. Risultati clinici nei diversi stadi evolutivi, Minerva Cardioangiol 19:605-615.
59. Zanelli, G., Quaglia, A., and Zino, P., 1971, La taurina nella terapia delle arteriopatie obliteranti degli arti inferiori, Nota II. Variazioni emossimetriche nei diversi stadi evolutivi, Minerva Cardioangiol. 19:616-619.

SECTION II: BIOCHEMISTRY, NUTRITION AND DEVELOPMENT

INTRODUCTION: BIOCHEMISTRY, NUTRITION AND DEVELOPMENT

Ryan J. Huxtable[1] and Flavia Franconi[2]

[1] Department of Pharmacology, College of Medicine, University of Arizona, Tucson, Arizona 85724
[2] Cattedra di Saggi e Dosaggi Farmacologici, Universita di Sassari, Sassari, Italy

Although the basic outlines of sulfur amino acid metabolism in mammals have been known for some years, there is a surprising ignorance of many details of enzymology. This ignorance extends into the biochemistry of taurine, where, despite numerous investigations, the regulatory mechanisms which control the proportion of sulfur flowing through taurine versus the proportion flowing through sulfate are not understood. This relative ignorance is a consequence of the numerous oxidation states which can be assumed by sulfur, and the ready oxidizability of sulfur at states between -2 and +4 or +6. Sulfur in such states tends to be chemically and biochemically reactive, and to have available both monomeric and dimeric pathways available for further reaction. These factors and others account for the complex biochemistry of sulfur (1).

Taurine biochemistry starts with the enzyme cysteine dioxygenase. This enzyme oxidizes cysteine to cysteine sulfinic acid, a change in oxidation from -2 to +2. This enzyme is the rate limiting point in both sulfur catabolism and taurine biosynthesis, cysteine sulfinic acid either being oxidized to sulfate via aspartate aminotransferase, or to taurine via cysteine sulfinic acid decarboxylase. As befits its metabolic importance, hepatic cysteine dioxygenase is regulated by a number of factors, including the dietry content of sulfur amino acids. In the first article in this section, Yamaguchi and Hosikawa describe the isolation and properties of a proteinaceous stabilizer of cystine dioxygenase. It is of interest to note that this stabilizer, protein A, has a widespread extra hepatic distribution, being found in tissues not containing cysteine dioxygenase activity. This suggests that it may have other functions.

Cysteine sulfinic acid decarboxylase converts cysteine sulfinic acid to hypotaurine. The enzymology of the oxidation of hypotaurine to taurine was for many years one of the more profound mysteries of taurine metabolism. Although in vivo hypotaurine was readily converted to taurine, on attempts to purify the responsible enzyme, activity would "softly and suddenly vanish away" (2). Now the elusive enzyme has been hunted down, pursued not "with forks and hope" like the ethereal snark, but by careful attention to co-factor requirements and problems of oxidizability (3). The enzyme is an NADPH-dependant oxidase, and is not rate limiting in the biosynthesis of taurine. Further work by the same group reported herein indicates that this oxidation involves the novel intermediacy of a disulfone. This is an unexpected interposition of a dimeric intermediate in a monomeric pathway.

The monomeric oxidation of sulfur-containing metabolites involves intermediate oxidation states of 0, 2, and 4. However, the formation of a disulfone allows one electron oxidation of the sulfur. That is to say, hypotaurine (oxidation state +2) is oxidized via an intermediate disulfone (oxidation state +3) to taurine (oxidation state +4). The demonstration of this intermediate carries not only metabolic significance, but suggests that the free radical mechanism of disulfone formation may be related to the high concentration of hypotaurine found in male assesory sex tissues. The mM concentrations of hypotaurine in ejaculate and oviductal fluid may serve to protect the highly unsaturated membranes of sperm from oxidation in the female genital tract (4).

The chapter by Chesney et al. is concerned with the regulation of taurine reabsorbtion by the kidney. The kidney regulates total body turnover of taurine. Thus, the half life of exchange of taurine between various tissues of the rat is faster than excretion from the body (5). It has been shown that alteration of dietary taurine content in the rat has no effect on the overall rate of taurine biosynthesis, but does modify kidney excretion in such a way that total body taurine pools remain constant (6). Similarly, Chesney et al. show that a decreased level of methionine in the diet leads to an increase in renal taurine reabsorption. This is due to a change in the capacity rather than the affinity of the transporter. The reabsorption of the sulfur amino acids methionine and cysteine are unaffected by these various dietary manipulations. Although the functionality of the regulation of taurine reabsorption is clear, the mechanism of regulation is not. Chesney et al. find no changes in the phosopholipid composition or biophysical properties of renal membranes corresponding to the dietary changes.

The liver is another organ responsible for removal of taurine from the body, in this case by metabolism to bile salts rather than by direct excretion. A number of novel functions have been elucidated for taurine recently. Gaull and Wright now suggest that taurine has a further function in the detoxification of certain secondary bile acids produced by gut microflora. Thus, chenodeoxycholate and deoxycholate are cytotoxic to lymphoblastoid cells in culture. However, the corresponding taurine conjugates of these acids are not toxic. Although taurine conjugation has a primary function in lipid digestion, detoxification may be a significant secondary function.

The second group of articles in this section is concerned with brain development. It is well known that there is a profound alteration in taurine metabolism in mammalian brain during development, in that the initially high concentrations of taurine in the brain decrease rapidly towards the adult levels. Both the mechanism mediating this fall and the functionality of the high levels in developing brain are unclear, although the literature contains a number of suggestions. Kuriyama and his co-workers have reproduced the same phenomena in cultured cerebral neurons. In this simplified system, taurine content falls with the growth of neurons in culture over the first several days, suggesting that the alterations seem in the brain in development may be intrinsic to the developing neuron.

One of the minor curiosities of taurine metabolism is that although the levels of taurine in the brain fall with development, the activity of the enzyme cysteine sulfinic acid decarboxylase shows the opposite developmental pattern. The same pattern is seen in cultured neurons, enzyme activity increasing by about six fold over fourteen days of culture. The bulk of the evidence suggests that taurine in rat brain derives primarily from the diet and biosynthesis in the liver, and that the endogenous capacity of the brain for synthesis is low, (although the taurine synthesized in the brain may be functionally high significant). It is to be expected, therefore, that the fall in brain taurine content with development is primarily a phenomenon of transport rather than biosynthesis, and may be related to the known alterations in phospholipid content of neuronal membranes with development. However, the factors which regulate this change in brain content are still unresolved. Kuriyama et al. find that the transport capacity of cultured neurons increases with time under culture. Kontro and Oja have similarly addressed the problem of taurine efflux with development, and find efflux rates are greater in immature brain. The resolution of the problem of brain regulation of taurine content must, therefore, await further studies.

Taurine is intimately involved in the structural and functional integrity of the retina. The structural necessity for taurine appears to be independent of its electrophysiological functions. Elucidation of the latter in the retina presents the same problems as in the brain. Pasantes-Morales et al. examine the action of taurine as an antioxidant in the retina. The study of the antioxidant properties of taurine and its precursors is a novel and growing area, which may be related to the membrane stabilizing properties of taurine. Pasantes-Morales et al. explore how the neuromodulatory, osmotic, and antioxidant actions of taurine may be separate but integrated phenomena which maintain functionality of the retina.

The physiological functioning of the retina is so dependent on its taurine content that it is reasonable to examine whether or not alterations in taurine biochemistry are implicated in neurogenetic retinal disorders in humans such as retinitis pigmentosa or Refsum disease. Wright et al. have used cultured lymphoblastoid cells and fibroblasts from humans with such disorders to examine for genetic alterations in taurine metabolism and biochemistry. As the essentiality of taurine becomes incorporated into our thinking on cell functioning and human health, one can anticipate that this type of investigation will become more common. Indeed, the final group of articles in this section are directly concerned with the involvement of taurine in mammalian nutrition and disease, with particular emphasis on the relevance of taurine in development.

Sturman et al. continue their exciting studies of taurine deprivation in felines. The cat is a convenient model for taurine depletion because it is unable to synthesize taurine but is an obligate conjugator of taurine with cholic acid in vivo (although in vitro liver extracts are able to conjugate glycine, this has no biological significance). In this particularly vulnerable species, females made taurine deficient by dietary depletion have serious reproductive problems. The offspring which do survive show severe developmental abnormalities, including skeletomuscular difficulties, defective cerebellar maturation, and abnormalities in the visual cortex.

The significance of taurine in development has been futher studied in primates. Neuringer et al. raised rhesus monkeys from birth on human infant formulas low in taurine content. By examination of the visual acuiety of developing monkeys, a more sensitive index of retinal dysfunctioning than electroretinogram measurement, they established the essentiality of taurine for normal visual development of primates.

The final two papers of this section are concerned with studies in humans. Bohles et al. present further data on taurine levels in human hearts at autopsy. In view of the differing degrees of dehydration, fibrosis, and lipid content of autopsy samples, studies such as this present problems of how to report the data. Bohles et al. have chosen to report their values in terms of per gram wet weight tissue, and, by this measure, find that the highest taurine concentrations are associated with the presence of congestive heart failure.

In the last article of this section, Michalk et al. take issue on the need for taurine supplimentation of human infant formula. This is part of a transatlantic dispute between the laissez-faire policies of Europe and the activism of America, where taurine levels in infant food formulas have recently been raised to the levels found in human breast milk. Michalk et al. argue that cysteine sulfinic acid decarboxylase activity in infant liver, although low, is sufficient to synthesize the required amounts of taurine. However, the enzyme rates used are based on in vitro conditions in which 0.5 - 2 mM cysteine sulfinic acid, an artificially high range, was used (7). In vivo, the low substrate concentrations would make for much lower rates of synthesis. However, the authors raise an important issue; one that merits further debate. Our assumptions should always be open to re-evaluation.

REFERENCES

1. Carroll, L., 1976, "The Hunting of the Snark: An Agony in Eight Fits" Folio Society, London
2. Fellman, J.H., and Roth, E.S., 1985, The biological oxidaiton of hypotaurine to taurine: hypotaurine as an antioxidant, "Taurine: Biological Actions and Clinical Perspectives" Oja, S.S., Ahtee, L., Kontro, P., and Paasonen, M.K. eds., Alan R. Liss New York pp. 71-82.
3. Huxtable, R.J., 1986, Biochemistry of Sulfur, Plenum Press, New York, pp. 445.
4. Huxtable, R.J., and Lippincott, S.E., 1982, Relative contribution of diet and biosynthesis to the taurine content of the adult rat, Drug-Nutrient Inter. 1:153-168.
5. Jacobsen, J.G., Thomas, L.L., Smith, L.H., Jr., 1964, Properties and distribution of mammalian L-cysteine sulfinate carboxylases, Biochim. Biophys. Acta 85:103-116.
6. Meizel, S., Lui, C.W., Working, P.K., and Mrsny, R.J., 1980, Taurine and hypotaurine: their effects on motility, capacitation and the acrosome reaction of hamster sprm in vitro and their presence in sperm and reproductive tract fluids of several mammals, Devel. Growth Differ. 22:483-494.

THE STABILIZING PROTEIN, PROTEIN-A, OF CYSTEINE DIOXYGENASE

Kenji Yamaguchi and Yu Hosokawa

Division of Maternal and Child Nutrition, National Institute of Nutrition, 1-23-1, Toyama, Shinjuku-ku, Tokyo, Japan

Cysteine dioxygenase [EC.1.13.11.20] catalyzes the oxygenation of cysteine to cysteine sulfinic acid, a key intermediate of cysteine metabolism to hypotaurine, taurine, isethionic acid and sulfate in mammalian tissues (1,3,8,10,11,18,19,22). The rat liver cysteine dioxygenase is obtained as an inactive form, which is activated by preincubation with L-cysteine under anaerobic condition, but the activated enzyme is rapidly and irreversibly inactivated during aerobic assay conditions. This inactivation during assay is prevented by a distinct cytoplasmic protein, called protein-A (13,14,18).

Cysteine dioxygenase of rat liver is induced by the injection of hydrocortisone or sulfur amino acids such as cysteine or methionine and is highly responsive to the dietary intake of sulfur amino acids (2,9,16,21) and protein (9), and also to the nutritional value of protein (21). In addition, the urinary excretion of taurine (9,21) and hepatic glutathione levels (21) are proportionally changed with hepatic cysteine dioxygenase activity. It appears likely that hepatic cysteine dioxygenase activity regulates the cellular levels of sulfur amino acids, including glutathione and taurine. On the other hand, fetal and neonatal rat liver and retina lack enzyme activity, suggesting that taurine and sulfate are essential nutrients in these stages (4,5,7,12).

This paper deals with the purification and some properties of protein-A, and how significance does protein-A involve in the regulation of cysteine dioxygenase.

THE OXYGEN ELECTRODE METHOD FOR ASSAY OF PURIFIED ENZYME ACTIVITY

Purified enzyme activity can be estimated with an oxygen electrode (model K-ICT-C, Gilson Co.) (7). Using this method, the activation procedure is separately performed in Thunberg tube.

The activation mixture consists of 1 µmol L-cysteine, 10 µmol glycine-NaOH at pH 9.0, 50 µg bovine serum albumin and 10 to 30 units of enzyme in a total volume of 0.1 ml. Activation is performed as mentioned above. The reaction mixture includes 1.50 µmol of substrate, 150 µmol of the glycine buffer, the appropriate amount of protein-A and 0.5 mg of bovine serum albumin in total volume of 1.5 ml. An aliquot, 10µl, of activated enzyme solution is added with a microsyringe into the electrode reaction cell mixture to initiate the reaction. With this method, cysteine content derived from the preactivation system is negligible.

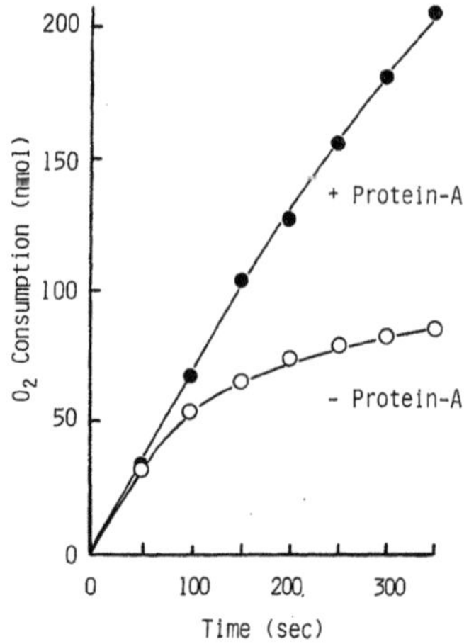

Fig. 1. Effect of protein-A on the inactivition of cysteine dioxygenase during assay. Enzyme assays were performed under standard assay conditions with or without 100 ng of protein-A.

Oxygen consumption is determined in a closed system. Enzyme activity is calculated from the rate of oxygen consumption.

ASSAY OF PROTEIN-A ACTIVITY

The oxygen electrode method, noted above, is most suitable for assay of the stabilizing activity of protein-A, since the reaction velocity is readily calculated from the slope of oxygen consumption versus the reaction time. If protein-A is omitted from the assay mixture, the slope of cumulative oxygen consumption (P) is nonlinear from the beginning of the reaction as shown in Fig. 1. Thus, the instantaneous reaction velocity ($P/sec=V_t$) at a given time, t, decreases during the course of assay. For a first order decay of activity the instantaneous reaction velocity should obey an equation of the form $V_t=V_o \cdot e^{-kt}$, as shown in Fig. 2. The true initial velocity

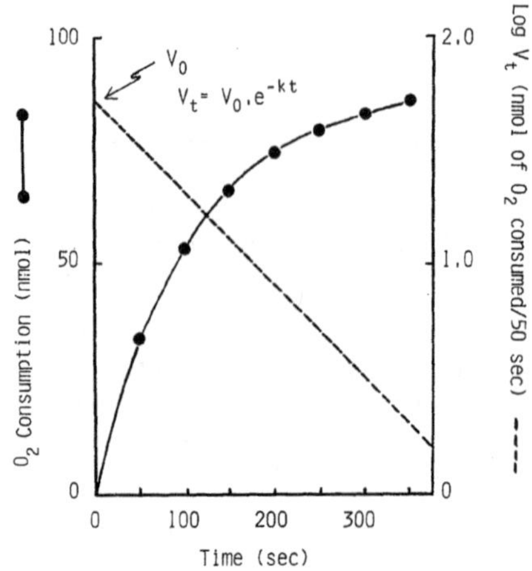

Fig. 2. Change in activity during assay of cysteine dioxygenase. Enzyme assays were performed under standard assay conditions without protein-A.

(V_o) and the time to reach half of V_o ($t_{1/2}$) can be obtained from the equation. The $t_{1/2}$ value(in sec) was proportionally increased as a function of protein-A concentration as shown in Fig. 3 and the V_o value was increased with enzyme concentration, as shown in Fig. 4. However, the $t_{1/2}$ value was not changed by enzyme concentration and the V_o value was not changed by protein-A concentration. Accordingly, the specific stabilizing activity of protein-A can be expressed as the $t_{1/2}$ value prolonged per mg of protein-A. The $t_{1/2}$ value in the absence of protein-A is about 75 s, as shown in Fig. 3.

PURIFICATION OF PROTEIN-A

Step 1. Extract

The preparation obtained in step 1 of the enzyme isolation as previously reported (18) is used as starting material for protein-A.

Step 2. Acetone treatment

The extract is treated with acetone at -20°C to give a concentration of 50%(v/v). After standing for 15 min, the precipitate is collected by centrifugation and the pellet dissolved in a minimum volume of 10 mM sodium phosphate buffer, pH 6.8. After standing for 10 min, denatured protein is removed by centrifugation.

Step 3. Diethylaminoethyl cellulose

The clear supernatant is applied to a DEAE-cellulose column (2.5 x 20 cm) previously equilibrated with 10 mM sodium phosphate buffer, pH 6.8. The column is eluted with the same buffer, and 10 ml fractions are collected. The fractions showing protein-A activity (18-36) are collected. After the addition of 1% volume of 1 M sodium acetate buffer, pH 5.2, the pH of the solution is adjusted to 5.2 with 1 N acetic acid. After standing for 10 min, denatured protein is removed by centrifugation.

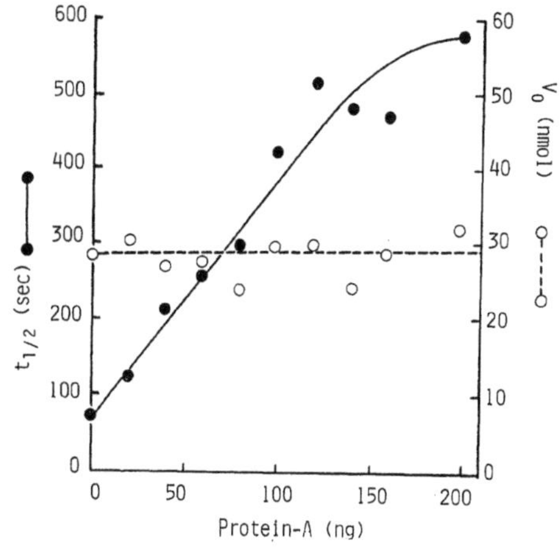

Fig. 3. Effect of protein-A concentration on the nonlinearity of the assay. Enzyme assays were performed under standard assay conditions at varying protein-A concentration with 3 units of cysteine dioxygenase.

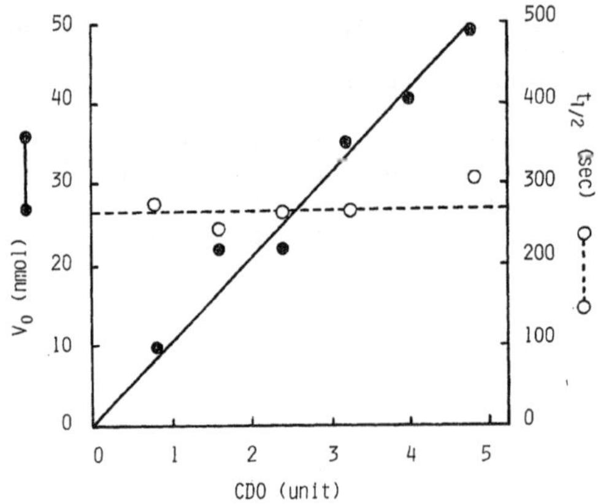

Fig. 4. Effect of enzyme concentration on the nonlinearity of the assay. Enzyme assays were performaed under standard assay conditions at varying enzyme concentration in the presence of 50 ng of protein-A.

Step 4. Sulfonate sepharose

The clear supernant is applied to a sulfonate sepharose column (Pharmacia S-Sepharose; 3.5 x 12 cm) equilibrated with 10 mM sodium acetate buffer, pH 5.2. After washing with 2 l of the same buffer, elution with a linear salt gradient of 0-1 M NaCl (400 ml) in the equilibration buffer is carried out at a flow rate of 30 ml/h, 5 ml fractions being collected. Fractions with stabilizing activity (38-43) are combined. After the addition of 2% of the volume of 0.5 M sodium phosphate at pH 6.8, the pH of solution is adjusted to 6.8 with 0.1 N NaOH.

Step 5. Toyopearl HW-55

The solution is concentrated to about 3 ml by ultrafiltration with an Amicon PM-30 membrane, and the denatured protein is removed by centrifugation. The enzyme solution is applied to a Toyopearl HW-55 column (2.5 x 100 cm) equilibrated with 2.5 mM sodium phosphate at pH 6.8, and the column is

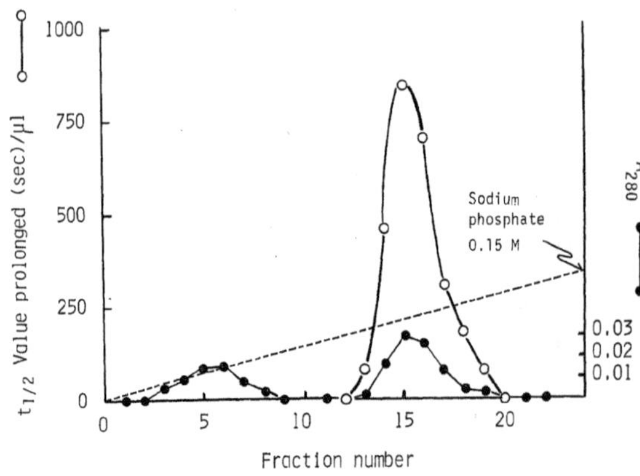

Fig. 5. Hydroxyapatite fractionation of protein-A. Enzyme assays were performed under standard conditions with 0.3 µl of each fraction.

Table 1. Summary of purification of protein-A

Step	Volume	Total activity	Total protein	Specific activity	Enrichment	Yield
	ml	$x10^5$s	mg	s/µg		%
1. Homogenate supernatant	800	1349	13760	9.8	1.0	100
2. Acetone precipitation	200	1228	6196	19.8	2.0	91
3. DEAE-cellulose	180	982	1320	74.4	7.6	80
4. S-Sepharose	30	341	20.3	1678	202	25
5. Toyopearl HW-55	30	214	5.92	3615	369	16
6. Hydroxyapatite	12	64	0.452	14181	1447	4.8

Enzyme assays were performed under standard assay conditions. Stabilizing activities of protein A are expressed as $t_{1/2}$ value prolonged in s.

eluted with the same buffer at a flow rate of 25 ml/h. Fractions of 5 ml were collected. Active fractions (38-43) are pooled.

Step 6. Hydroxyapatite

The solution is applied to a hydroxyapatite column (1 x 5 cm) equilibrated with 2.5 mM sodium phosphate at pH 6.8. After washing with 500 ml of the same buffer, elution with a linear gradient of 100 ml of 2.5-250 mM sodium phosphate buffer at pH 6.8, is carried out at a flow rate of 12 ml/h, collecting fractions of 3 ml. The active fractions appeared in tubes 14-17, as shown in Fig. 5. The preparation can be stored for at least 2 months in

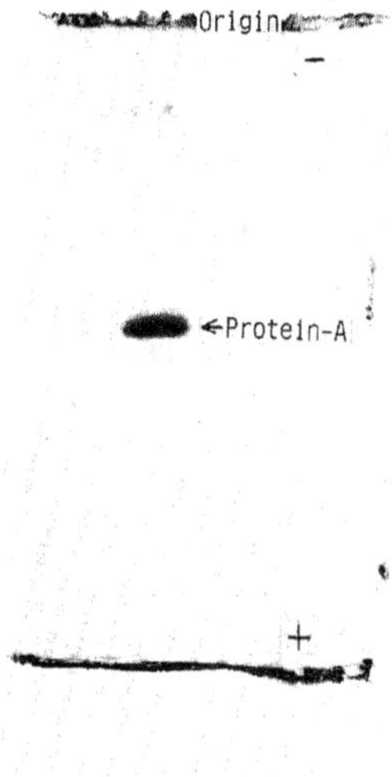

Polyacrylamide: 15%

Fig. 6. SDS-polyacrylamide gel electrophoresis of protein-A.

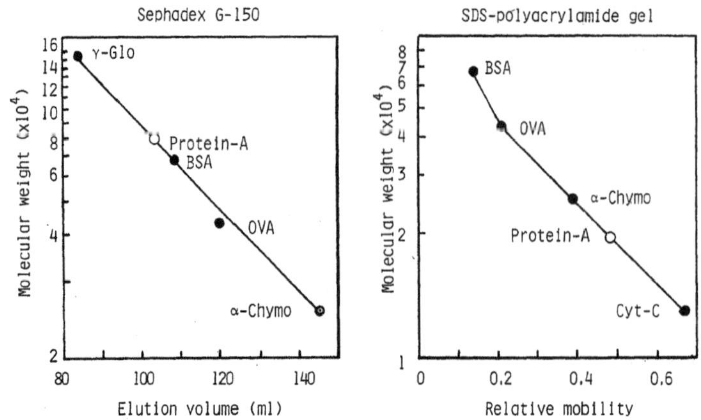

Fig. 7. Determination of the molecular weight of protein-A. γ-Glo: human γ-globulin; BSA: bovine serum albumin; OVA: ovalbumin; α-Chymo: α-chymotrypsinogen A; Cyt C: Cytochrome C.

in the presence of 1 mg/ml bovine serum albumin at -45°C without significant loss of stabilizing activity.

The purification of protein-A is summarized in Table 1.

MOLECULAR PROPERTIES, SPECIFICITY AND BIOLOGICAL ACTIVITY OF PROTEIN-A

The preparation obtained in step 6 is homogeneous by SDS-polyacrylamide gel electrophoresis as shown in Fig. 6, and a single peptide band having an MW of 19,500 is observed as shown in Fig. 7. However, the MW estimated by molecular sieving on a Sephadex G-200 column was 78,000 as shown in Fig. 7. This fact suggests that protein-A is a tetramer consisting of the same subunit, having a MW around 19,500. The isoelectric point, determined by the electrofocusing method, is at pH 7.30.

Protein-A was not substituted for by other proteins examined such as bovine serum albumin, bovine serum globulin, hemoglobin, lactoalbumin, histone, protamine, and cytochrome C. Protein-A exhibits no biological activity other than that of stabilizing cysteine dioxygenase. It appears unlikely that protein-A is a glycoprotein, since no periodic acid-fucsin reaction was detected in purified protein-A preparations.

DISTRIBUTION OF CYSTEINE DIOXYGENASE AND PROTEIN-A

Table 2. Distribution of Protein-A in rat tissues

	Prolongation of $t_{1/2}$ (s/mg protein)					
Tissue	Liver	Heart	Pancreas	Kidney	Lung	Brain
Protein-A activity	19250	9150	4738	3450	2895	1717

The supernatant of homogenate (180 mg of protein) at 105,000 x g centrifugation of each tissue was applied to a S-Sepharose column (1 x 15 cm), and eluted with a 100 ml linear gradient of NaCl (0-1.0 M) in 10 mM sodium acetate buffer, pH 5.2.

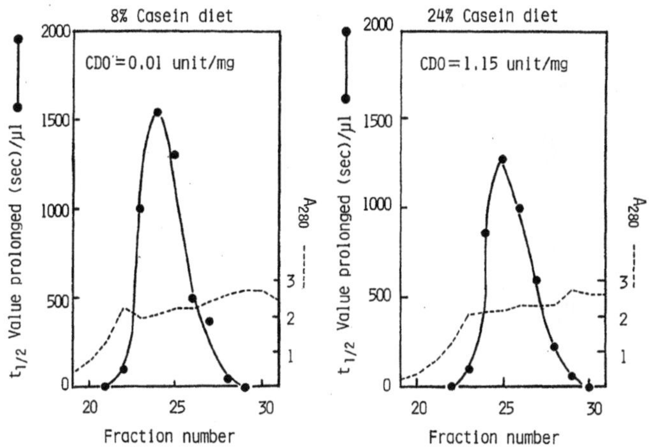

Fig. 8. The effect of dietary protein contents on protein-A. The rats were fed on the basal diet (18% casein diet) for 6 days, and then experimental diets for 10 days. The supernatant of homogenate (1.5g of protein) at 105,000xg centrifugation was applied on a S-Sepharose column (3.5 x 10 cm), and eluted with a 300 ml linear gradient of NaCl from 0 to 1.0 M in sodium acetate, pH 5.2. The assays were performed under standard assay conditions with 0.1 µl of each fraction.

Cysteine dioxygenase activity is found in the liver of various species of mammals, with decreasing activity: mouse, rat, pig, cow, dog, rabbit (7,18). In contrast, protein-A is widely distributed in tissues such as liver, brain, heart, kidney and spleen of rat (7,18), but the content of protein-A of each tissue is quite different. As shown in Table 2, the highest content of protein-A activity was found in liver, and the lowest activity was observed in brain.

UNRESPONSIVENESS OF PROTEIN-A TO INDUCERS OF CYSTEINE DIOXYGENATE

As previously reported, the hepatic cysteine dioxygenase activity is remarkably induced by high intake of protein, or sulfur amino acids such as L-cysteine, L-cystine or L-methionine (9). As shown in Fig. 8, the hepatic enzyme activities of rats fed a high protein diet was induced by over one hundred times of those of rats fed a low protein diet, whereas the protein-A contents in both group was almost equal.

EFFECTS OF PROTEIN-A ON THE KINETICS OF CYSTEINE DIOXYGENASE

As previously reported, the V_0 value in the absence of protein-A was similar to the value in the presence of protein-A. In addition, protein-A did not affect the apparent Km value of the enzyme for cysteine (18).

INDEPENDENCE OF PROTEIN-A FOR ACTIVATION OF CYSTEINE DIOXYGENASE

Preincubation with L-cysteine in the presence of protein-A, or in the absence of protein-A, yielded the same enzyme activity, suggesting that protein-A did not participate in the activation of enzyme by L-cysteine (18,23).

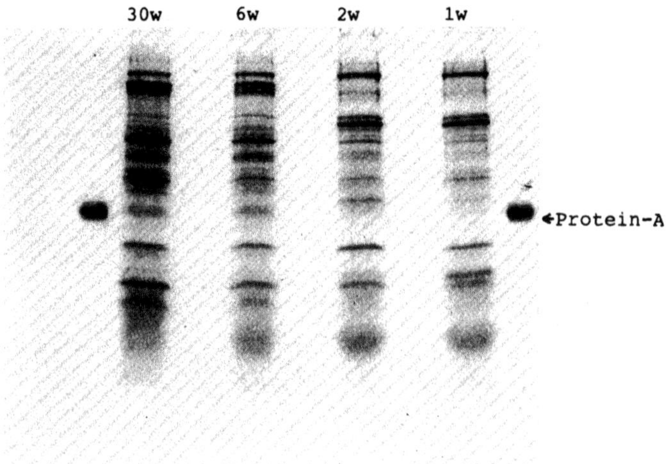

Fig. 9. The change of protein-A contents in growing rats. After S-Sepharose treatment, protein-A fraction was analyzed by SDS-polyacryamide gel electrophoresis.

THE HEPATIC CONTENTS OF PROTEIN-A OF GROWING RATS

Since hepatic cysteine dioxygenase activity is not observed in neonatal and fetal rat but a remarkable increase is seen after one week of life (5), changes of protein-A contents in livers of growing rats were examined. As shown in Table 3, the protein-A content also increased with age. Fig. 9 shows the protein profile separated by SDS-polyacrylamide electrophoresis. A protein band corresponding to the protein-A subunit also clearly increased with age.

DISCUSSION

It has been demonstrated that the hepatic cysteine dioxygenase is an enzyme whose induction is mediated by sulfur amino acids (9,19), or high protein intake (9) in the presence of permissive quantities of glucocorticoids (19), and that this enzyme seems to be a regulating enzyme in the conversion of cysteine to taurine or sulfate, since the urinary taurine excretion is well correlated with the hepatic enzyme activity in rat. On the

Table 3. Cysteine dioxygenase and protein-A activity in the growing rat

Age (week)	1	2	6	30
Cysteine dioxygenase (nmol/h/mg protein)	37	78	594	402
Protein-A (prolonged $t_{1/2}$ s/mg protein)	3410	5565	19350	16385

The supernatant of homogenate (100 mg of protein) from 105,000 x g centrifugation was applied on a S-Sepharose column (1 x 10 cm), and eluted with a 100 ml linear gradient of NaCl (0-1.0 M) sodium acetate buffer, pH 5.2.

other hand, this enzyme has a specific stabilizing protein by which the irreversible inactivation of enzyme during assay is delayed. The question may be raised whether or not protein-A plays a significant role in regulation mechanisms mediated by cysteine dioxygenase in liver.

The contents of protein-A of rat liver maintained with high (24% protein) protein diet and low (8% protein) protein diet were examined after fractionation of protein-A using with S-sepharose column. The contents of protein-A in livers of rats fed on high protein diet is almost equal to those fed on low protein diet, while the hepatic cysteine dioxygenase activity of high protein diet group was ramarkably higher than that of low protein diet group. These findings strongly suggest that protein-A does not participate in the regulation of cysteine metabolism mediated by cysteine dioxygenase. In addition, neither the initial reaction velocity nor the apparent Km value are altered by protein-A, but the inactivation rate during assay is prolonged. These facts suggest that protein-A does not participate in a catalytic process but acts to protect the enzyme from denaturation. The stabilizing mechanism still remains to be elucidated.

Protein-A content in the neonatal period is low and increases with age after birth in the same manner as cysteine dioxygenase. These facts suggest that cysteine metabolism may be regulated by either a catalytic protein or a stabilizing protein, although the most important regulating step should be in the catalytic process.

REFERENCES

1. Chapeville, F., and Fromageot, P., 1955, La formation de l'acide cysteinesulfinique a partir la cysteine chez le rat, Biochim. Biophys. Acta, 17:275.
2. Daniels, K.H., and Stipanuk, M.H., 1982, The effect of dietary cysteine level on cysteine metabolism in rats, J. Nutr., 112:2130-2141.
3. De Marco, C., Mosti, R., and Cavallini, D., 1966, Sulla ossidazion della cisteina e della cisteamina a dirivati solfinice, catalizata dal fegato di ratto, Boll. Soc. Ital. Biol. Sper., 42:94-96.
4. Di Giorgio, R.M., Tucci G., and Macaione, S., 1975, Cysteine oxidase activity in rat retina during development, Life Sci., 16;429-436.
5. Hosokawa, Y., Yamaguchi, K., Kohashi, N., Kori, Y., Fujii, O., and Ueda, I., 1980, Immaturity of the enzyme activity and response to inducers of rat liver cysteine dioxygenase during development, J. Biochem., 88:389-394.
6. Hosokawa, Y., Kohashi, N., and Yamaguchi, K., 1980, Study of cysteine dioxygenase, Sulfur-containing Amino Acids, 1:251-263.
7. Hosokawa, Y., Fujii, O., Nasu, S., and Yamaguchi, K., 1980, Study of the stabilizing protein of cysteine dioxygenase, Sulfur-containing Amino Acids, 3:237-243.
8. Jacobsen, J.G., and Smith Jr., L.H., 1968, Biochemistry and physiology of taurine and taurine derivatives, Physiol. Rev., 48:424-511.
9. Kohashi, N., Yamaguchi, K., Hosokawa, Y., Kori, Y., Fujii, O., and Ueda, I., 1978, Dietary control of cysteine dioxygenase in rat liver, J. Biochem., 84:159-168.
10. Lombardini, J.B., Turini, P., Biggs, D.R., and Singer, T.P., 1969, Cysteine oxygenase: 1. General properties, Physiol. Chem. & Physics, 1:1-23.
11. Lombardini, J.B., Singer, T.P., and Boyer, P.D., 1969, Cysteine oxygenase: 2. Study on the metabolism of the reaction with ^{18}oxygen, J. Biol. Chem., 244:1172-1175.
12. Loriette, C., and Chatagner, F., 1978, Cysteine oxidase and cysteine sulfinic acid decarboxylase in developing rat liver, Experientia, 34:981-982.

13. Sakakibara, S., Yamaguchi, K., Ueda, I., and Sakamoto, Y., 1973, Two components of cysteine oxidase in rat liver, Biochim. Biophys. Res. Commun., 52:1093-1099.
14. Sakaibara, S., Yamaguchi, K., Hosokawa, Y., Kohashi, N., Ueda, I., and Sakamoto, Y., 1976, Purification and some proerties of rat liver cysteine oxidase(cysteine dioxygenase), Biochim. Biophys. Acta, 422:273-279.
15. Sorbo, B., and Ewetz, L., 1965, Enzymatic oxidation of cysteine to cysteine-sulfinate in rat liver, Biochem. Biophys. Res. Commun., 19:359-363.
16. Stipauk, M.H., 1979, Effect of excess dietary methionione on the catabolism of cysteine in rats, J. Nutr., 109:2126-2139.
17. Tabachnik, M., and Turver, H., 1955, The conversion of methionine-^{35}S to cystathionine-^{35}S and taurine-^{35}S in the rat, Arch. Biochem. Biophys., 56:115-121.
18. Yamaguchi, K., Hosokawa, Y., Kohashi, N., Kori, Y., Sakakbara S., and Ueda, I., 1978, Rat liver cysteine dioxygenase, J. Biochem., 83:479-491.
19. Yamaguchi, Y., Sakakibara, S., Koga K., and Ueda, I., 1971, Induction and activation of cysteine oxidase of rat liver, Biochim. biophys. Acta, 273: 502-512.
20. Yamaguchi, K., Sakakibara, S., Asamizu, J., and Ueda, I., 1973, Induction and activation of in vivoactivity of cysteine oxidase of rat liver:II. The measurement of cysteine metabolism in vivo and the activaton of in vivo activity of cysteine oxidase, Biochim. Biophys. Acta, 297:48-59.
21. Yamaguchi, K., Hosokawa, Y, Niizeki, S., Tojo J., and Sato, I., 1985, Nutritional significance of cysteine dioxygenase on the biological evaluation of dietary protein in growing rats in: "Taurine: Biological Actions and Clinical Perspectives", S.S. Oja, L. Ahtee, P. Kontro and M.K. Passonen eds., Alan R. Liss, Inc., New York, pp. 23-32.
22. Yamaguchi, K., 1981, Cysteine dioxygenase and its possible role on taurine formation in rats, in: "The Effects of Taurine on Excitable Tissues", S.W. Schaffer, S.L. Baskin, and J.J. Kocsis Eds., Spectrum Pub. Inc., New York, pp. 5-20.
23. Yamaguchi, K., 1980, Cysteine dioxygenase of rat liver: some properties and its significance on cysteine metabolism of rat in: "Natural Sulfur Compounds", D. Cavallini, G.E. Gaull and V. Zappia eds., Plenum Publ. Co., New York, pp. 175-186.

THE OXIDATION OF HYPOTAURINE TO TAURINE: BIS-AMINOETHYL-α-DISULFONE, A METABOLIC INTERMEDIATE IN MAMMALIAN TISSUE.

J.H. Fellman, T.R. Green and A.L. Eicher

Department of Biochemistry, Oregon Health Sciences University
Portland, Oregon 97201, USA

The primary mammalian metabolic pathway leading to taurine is generally recognized as beginning with the oxidation of cysteine to cysteine sulfinate via cysteine dioxygenase (19). This rate limiting enzyme is carefully regulated thus setting the rate for the catabolism of sulfur amino acids (4). A major portion of cysteine sulfinate is subject to unidirectional transamination leading to pyruvate and sulfite; a smaller fraction of cysteine sulfinate is decarboxylated by cysteine sulfinate decarboxylase to hypotaurine (8). The ratio of the two pathways appears to be relatively constant; what controls the pathway bifurcation remains unknown. Although a small fraction (10%) of the hypotaurine is subject to unidirectional transamination to acetaldehyde and sulfate (4,6), the bulk of hypotaurine is oxidized to taurine. A possible alternate metabolic route involves initial oxidation of cysteine sulfinate to cysteic acid by an unspecified mechanism followed by decarboxylation of cysteic acid to taurine. The latter is an established capacity of the enzyme cysteine sulfinate decarboxylase (10).

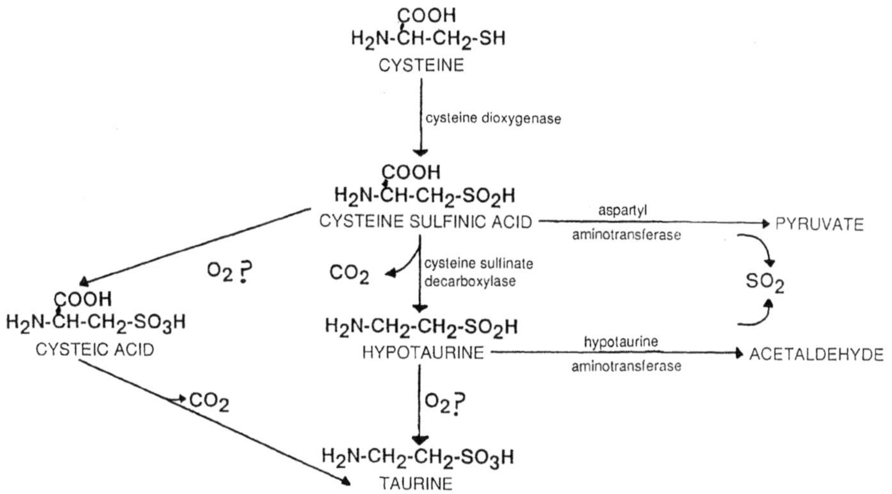

Fig. 1. Metabolic pathway: cysteine to taurine

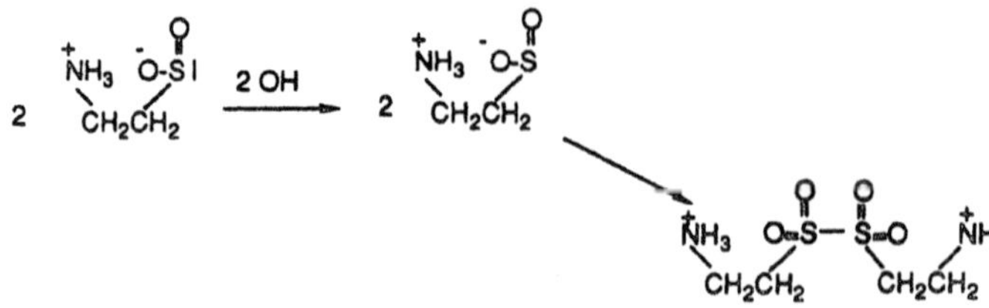

Fig. 2. Hypothetical radical pathway for α-disulfone formation

The nature of the oxidation of hypotaurine to taurine is uncertain. Several dedicated enzymatically directed oxidases have been claimed, but when scrutinized carefully could not be validated (17), or exhibited such trivial and inconsistent activities in the face of such high "blanks" so as to raise serious doubts as to the validity of such claims (15).

We recently reported oxidation of hypotaurine to taurine by a liver microsomal NADPH oxidase system which appeared to operate by generation of hydroxyl radicals. Hypotaurine behaved as a hydroxyl radical quenching agent leading to taurine. We had proposed that the hydroxyl radical generated by a number of biologically important systems abstracted an electron from the sulfinate moiety of hypotaurine; two of which combine to form the corresponding bis-aminoethyl-α-disulfone (5).

We report here evidence that the α-disulfone intermediate can be formed in vitro, and can be found in biological tissue. We have characterized it by FAB mass spectrometry, GC/MS, and proton NMR spectrometry. We propose that the α-disulfone is subject to chemical and enzymatic hydrolysis leading to one mole of taurine and one mole of hypotaurine. Furthermore, using electron paramagetic resonance spectroscopy (EPR), we report that chemically or enzymatically generated hydroxyl radicals are quenched by hypotaurine which rapidly oxidizes to the α-disulfone. Finally, we show that male sexual tissue which is recognized to contain high hypotaurine and taurine concentrations (11,12), also contains the α-disulfone. We report that sperm cells can quench hydroxyl radicals in a manner similar to hypotaurine.

MATERIALS AND METHODS

Hypotaurine, hypoxanthine, xanthine oxidase, horseradish peroxidase, hydrogen peroxide and 5,5-dimethyl-1-pyrroline-n-oxide (DMPO) were all purchased from Sigma Chemical Company; AG-50W X 8 ion-exchange resin was obtained from Bio Rad Corporation.

Sperm cells were isolated from bull testicles obtained fresh from the local slaughter house. The tunicate was carefully dissected away from the testis and the caudate epididymis, and approximately 3 inches of the vas deferens were cautiously separated from the remainder of the testes. A blunt needle was placed in the vas deferens and clamped securely with a hemostat. The needle was attached to a 60 ml plastic syringe containing 3 drops of india ink in a 75% sucrose solution and mounted vertically in a specially constructed wooden clamp. Pressure was applied by hand to a lever which depressed the syringe and back flushed the vas deferens and epididymis with the ink/sucrose solution. One of the epididymal tubules was pierced and while pressure was applied to the syringe, the semen ran out and was collected. The semen was centrifuged at 4°C for 10 min at 5000 x g. The supernatant was discarded, and the pellet (1.5 ml) was washed two times with 3 ml sperm suspension buffer (50 mM HEPES, 45 mM KCl, 105 mM NaCl, 5 mg/ml

BSA pH 7.5). The final pellet was resuspended in an equal volume 0.05 M KH_2PO_4, pH 7.8, and assayed promptly.

Gas Chromatography/Mass Spectrometry

Dried samples to be derivatized were placed in a sealed tube to which 0.1 ml of trifluoroacetic anhydride and 0.1 ml of trifluoroacetic acid was added. The samples were heated at 75°C for 30 min, cooled, and the derivatizing agents were removed by a gentle stream of nitrogen. The residue was dissolved in 0.1 ml of ethyl acetate and 2 µl injected into a 30 m X 0.25 mm id. fused silica capillary GC column directly connected to the ion source of the HP-5970 mass spectrometer. After an initial hold at 50°C for 8 min the oven temperature was programmed at 8°C/min to 280°C. The MS was screened repetitively over the mass range 50-500 at 800 Da/sec and the data stored. The ion source was operated in EI mode at 200°C with an electron energy of 70 ev.

FAB Mass Spectrometry

Mass spectra were obtained using a V.G. 7070E double focusing mass spectrometer instrument equiped with a fast atomic bombardment ion source and sample introduction system. Samples were suspended in either glycerol or a 1:5 mixture of dithiothreitol:dithioerythritol ("magic bullet") and loaded onto a probe with the instrument modified for fast atomic bombardment of the sample. The mass spectra were recorded using an 11-250B data acquisition system.

Proton Nuclear Magnetic Reasonance

Proton NMR were recorded on a JEOL FX-900 instrument. Samples were analyzed using deuterated water as the solvent and chemical shifts reported in ppm.

Electron Paramagnetic Resonance Spectra

EPR spectra were measured on a Varian E 109 spectrometer with a 9 inch magnet at room temperature. Instrument conditions for detecting hydroxyl radical and thiyl radical adducts of DMPO were as follows: microwave power, 15 mwatts; modulation amplitude, 2.0; time constant, 0.25 s; field set, 3400g; scan rate, 12.5 G/min. Specimens were analyzed on a standard 60 X 10 X 0.25 mm flat cell quartz EPR cuvette.

Synthesis of bis-aminoethyl-α-disulfone

a. Chemically-generated hydroxyl radical: To 10 µmoles of hypotaurine containing 0.36 µCi of ^{3}H-hypotaurine (10) was added a mixture of 2.5 µmoles of ferrous sulfate and 30 µmoles of hydrogen peroxide. The final volume was 1.0 ml. The mixture was then treated with 0.1 ml of 20% sulfosalicylic acid, to pH 2-3, and placed on a 0.7 X 5.0 cm column of Bio Rad AG-50W X8 (200-400 mesh, acid form) ion-exchange resin. The column was eluted first with distilled water, which was discarded. It was then eluted with 12.0 ml of 0.01 N HC1 which also was discarded. Finally, the column was eluted with 2.0 N HC1. The first 2.0 ml of eluate was discarded; the second 2.0 ml contained the α-disulfone. This fraction was collected and lyophilized for storage and characterization.

b. Enzyme generated hydroxyl radical: One unit of xanthine oxidase was added to a mixture containing 100 µmoles of an equimolar mixture of ferrous sulfate and ethylenediaminetetracetic acid (EDTA), 10 µmoles hypoxanthine, 10 µmoles of hypotaurine, and 0.36 µCi of ^{3}H-hypotaurine in a final volume of 1.0 ml. The enzyme reaction was conducted at pH 7.8 at room temperature for one minute. The reaction was stopped by the addition of 0.1 ml of 20% sulfosalicylic acid and microcentrifuged for 1 min. The supernate was placed on a 0.7 X 5.0 cm column of Bio Rad AG-50W X 8 (200 - 400 mesh, acid form) ion-exchange resin. The column was eluted with 12.0 ml 0.01 N HCl which was discarded. This was followed by 2.0 N HCl; the first 2.0 ml was discarded, the second 2.0 ml which contained the α-disulfone was collected and lyophilized for storage and characterization.

RESULTS

Hypotaurine was observed to undergo oxidation to the α-disulfone by both ezymatically and chemically generated hydroxyl radicals. The α-disulfone was characterized by mass spectrometry and Fourier transformed NMR.

Mass spectrometry

A. FAB - Mass Spectrometry: Using fast atomic bombardment, gentle bombardment of the α-disulfone in both glycerol or dithiothreitol:dithioerythritol ("magic bullet") yielded no molecular ions, but a characteristic major peak at M/z 173 Da for the monoaminoethyl-α-disulfone fragment was observed. Its glycerol adduct was observed at M/z 265 Da. However, when "magic bullet" was the suspending media on the probe, a major peak at M/z 215 Da, M-1 was observed for the parent ion.

B. Gas Chromatography/Mass Spectrometry: An alternate approach was undertaken using trifluoracetyl derivatization of the α-disulfone employing a Hewlett Packard GC/MS. A sharp peak emerging at 20.4 min was observed.

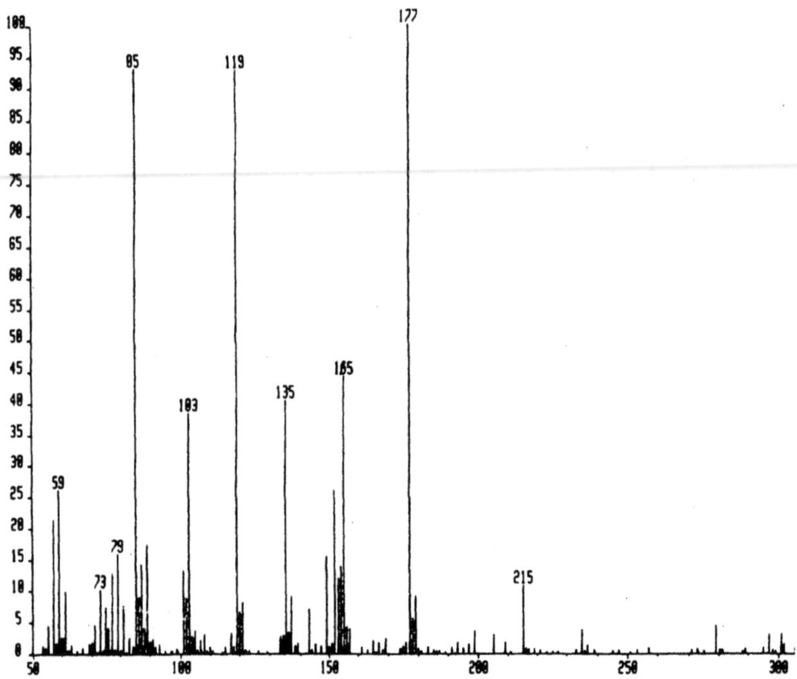

Fig. 3. FAB mass spectrum of α-disulfone suspended in "magic bullet"

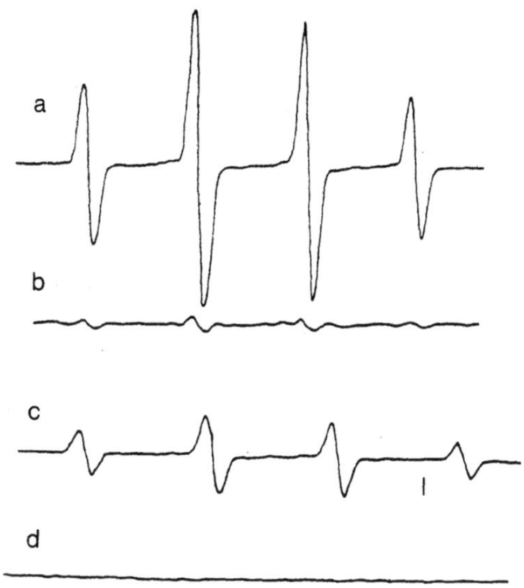

Fig. 4. EPR spectra of DMPO adducts

The base peak of the mass spectrum at M/z 217 Da, M + 1, was assigned to the parent ion less the trifluoroacetyl groups. This was additional confirmation of the structure of the bis-aminoethyl-α-disulfone. Secondary fragmentation was observed at M/z 204 Da for the symmetrical cleavage of the parent bis-N-trifluoroacetylaminoethyl-α-disulfone.

Proton NMR

The spectra from both hypotaurine and bis aminoethyl-α-disulfone were obtained using deuterated water as the solvent. Two sets of triplets were observed for each substance. The α-disulfone pattern centered at = 3.38 assigned to the CH_2 group adjacent to the sulfone, and = 2.75 assigned to the CH_2 group adjacent to the amino group. Hypotaurine exhibited triplets centered at = 3.17 assigned to the CH_2 group adjacent to the sulfinate, and = 2.30 assigned to the CH_2 adjacent to the amino group.

EPR Spectra

Hypotaurine, and sulfinates in general, appear by EPR analysis to be powerful scavengers of hydroxyl radicals. In Fig. 4a the characteristic EPR spectrum for the hydroxyl radical adduct of DMPO, DMPO-OH (7), was observed upon mixing DMPO with ferrous sulfate (80 uM) and hydrogen peroxide (69 μM) in 50 mM sodium phosphate buffer, pH 7.8, also supplemented with 1 mM EDTA. With inclusion of 41 mM hypotaurine (Fig. 4b), or a comparable concentration of p-toluene sulfinate (Fig. 4c) in the reaction medium, the formation of the DMPO-OH adduct was sharply suppressed. Taurine was ineffective in quenching DMPO-OH formation (data not shown). In the absence of hydroxyl radical generation, no DMPO-OH adduct was detected. This is shown by the flat baseline scan seen with omission of either iron or hydrogen peroxide from the reaction mixtures (Fig. 4d).

Since the hydroxyl radical is a powerful oxidant, and since it is limited in its capacity to abstract no more than one electron from

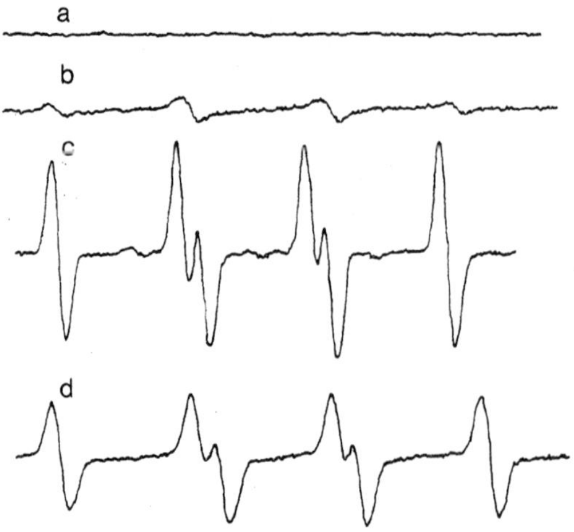

Fig. 5. EPR spectra of further DMPO adducts

appropriate electron donors, we postulated that sulfur-centered radicals were likely being formed with exposure of hypotaurine to hydroxyl radicals, and that the thiyl radicals might then condense as dimers to yield the α-disulfone product. However, attempts to trap a thiyl radical intermediate of hypotaurine were not easily accomplished. To test whether thiyl radicals might be successfully trapped by DMPO, EPR spectral scans were conducted on hypotaurine and p-toluene sulfinate oxidized by horseradish peroxidase. This method of generating a thiyl radical adduct of DMPO has been used by Harman, Mottley, and Mason in trapping the free thiyl radical adduct of L-cysteine (9). A thiyl DMPO adduct for the p-toluene sulfinate was easily observed (Fig. 5c) which matched very closely the EPR spectrum for the cystiene thiyl adduct of DMPO (Fig 5d). A very weak spectrum of ill-defined origin was seen with oxidation of hypotaurine by horseradish peroxidase (Fig. 5b). This latter spectrum appears distinctly different from that seen with p-toluene sulfinate or cysteine; it is most likely not DMPO-OH. Control experiments with exclusion of hypotaurine from the reaction mixtures yielded flat baselines (Fig 5a), and thus formation of this adduct appears to be dependent upon the presence of hypotaurine in the reaction mixture. Although it could be the spectrum of the DMPO thiyl adduct for hypotaurine, there was not sufficient material formed to allow us to confirm its structure by chemical analysis. On the other hand, we infer from the similar EPR spectra between the p-toluene sulfinate and cysteine adducts of DMPO (Fig. 5c and 5d) that the spectrum of the p-toluene sulfinate DMPO adduct is that of a thiyl intermediate of the type previously characterized by Harman, Mottley and Mason for cysteine. Hence, it appears that free radical thiyl intermediates of sulfinates can be formed, certainly in the case of p-toluene sulfinate, through peroxidation of the sulfur center.

One reason for the low levels of DMPO adduct formed by exposure of hypotaurine to horseradish peroxidase is the fact that hypotaurine is an exceedingly poor substrate of the peroxidase. Table 1 compares the relative rates of peroxidation of hypotaurine with cysteine sulfinate, p-toluene sulfinate, and cysteine. It is evident that hypotaurine is an exceedingly poor substrate with a relative rate of peroxidation of less than 1/2000th of that seen with cysteine. Even p-toluene sulfinate was peroxidized at only 1/200th of the rate of cysteine. Cysteine sulfinate, like hypotaurine, was also a very poor substrate of horseradish peroxidase.

Table 1. Relative peroxidation rates of sulfur compounds by horseradish peroxidase

Substrate	H_2O_2 (μmoles/min/mg horseradish peroxidase)
Cysteine (10 mM)	62.8
p-Toluene Sulfinate (2.5 mM)	0.58
Cysteine Sulfinate (10 mM)	0.028
Hypotaurine (10 mM)	0.028

*Stock horseradish peroxidase exhibited approximately 7100 μmoles/min/mg protein enzyme activity with pyrogallol as substrate. Assays were conducted in 50 mM phosphate, 1 mM EDTA, pH 7.8, containing 6.9 mM H_2O_2. Enzyme activity is expressed in terms of H_2O_2 consumed as followed at 240 nm with a molar extinction coefficient for H_2O_2 of 43.6.

Chromatographic properties of the α-disulfone

A. Thin layer chromatography: The α-disulfone, like taurine hypotaurine, can be detected using ninhydrin. However, the α-disulfone yielded a pink rather than purple chromaphore product. Furthermore, palladium chloride did not react with the α-disulfone while it yielded an intense yellow chromophore with hypotaurine.

Biological Systems

A. Oxidation of hypotaurine to taurine by rat liver microsomes: The complete system contained 3 μmoles NADPH, 1 μmole ^{3}H-hypotaurine (0.5 μCi), 0.1 ml rat liver microsomes (200 mg wet wt.), 83 μmole KH_2PO_4 and 10 μmole K pyrophosphate at a final pH of 7.8, in a final volume of 1.0 ml. The reaction was initiated by the addition of the microsomes, incubated at 37°C for 30 min, and terminated with 0.4 ml 1 M HPO_3. The mixture was microfuged for 3 min, and the supernate applied to a 0.7 x 5.0 cm AG-50W x 8 ion-exchange resin column. The taurine was then eluted with one column volume (3-5.0 ml) H_2O, and a fraction of the eluate counted in 5.0 ml Buhler's scintillation fluid.

B. The effect of sperm, hypotaurine and other radical quenching agents on hydroxyl radical oxidation of benzoic acid: the complete system consisted of 1 umole hypoxanthine, 1.0 unit xanthine oxidase, 1.0 μmole equimolar mixture of ferrous sulfate and EDTA, 10 μmole ^{14}C-benzoic acid (0.125-0.195 μCi) and 0.05 M KH_2PO_4, pH 7.8, in a final volume of 1.0 ml. The reaction was carried out in 20 ml glass vials with foil lined caps; the caps were equiped with a 20 gauge x 1.5 inch needle upon which was held a 1.27 cm filter paper disc saturated with 20 μl hyamine hydroxide. The reaction was

Table 2. Thin layer chromatography of α-disulfone on cellulose matrix

	Solvent System		
	A	B	C
		Rf value	
Hypotaurine	0.24	0.12	0.61
Taurine	0.12	0.09	0.34
-Disulfone	0.15	0.15	0.57

Solvent A = n-Butanol:acetic acid:water 4:1:1
B = t-Butanol:formic acid:water 75:15:10
C = Phenol:water 4:1

Table 3. Oxidation of hypotaurine by liver microsomes: stimulation by azide; quenching by DMSO and mannitol

	umoles taurine/h/mg wet wt ± SD
Complete System (CS)	.285±.03
CS + Azide 1mM	.685±.001
CS + Azide + Mannitol 30mM	.495±.04
CS + Azide + DMSO 30mM	.264±.02

initiated by the addition of the xanthine oxidase, incubated at 37°C for one min; and terminated with the addition of 0.4 ml 1 M HPO_3 with a 1.0 ml disposable syringe through the needle in the cap. The post-reaction incubation (which allowed for CO_2 equilibration) was 30 min at 37°C. Finally, the filter paper was transferred to 7 ml vials containing 5.0 ml Buhler's scintillation fluid, and $^{14}CO_2$ counted.

That sperm are vulnerable to toxic oxygen specie has been well established; presumable then, vulnerable membrane lipids are the target of such oxidants (see below). In Table 4, we demonstrated that benzoate decarboxylation by hydroxyl radicals generated by a xanthine oxidase system is diminished by the addition of sperm cells. The addition of hypotaurine and other radical quenching agents markedly diminished the rate of decarboxylation.

C. The identification of the α-disulfone in rat male sexual tissue: The testes and accessory sexual tissue of a rat were homogenized in an equal volume of 0.05 M phosphate buffer pH 7.8; 20% sulfosalicylic acid was then added to a final concentration of 3%. The material was centrifuged and the supernate was lyophilized. In a sealed tube, 0.1 ml trifluoroacetic acid and 0.1 ml trifluoroacetic anhydride residue. This tube was heated at 80°C for 20 min. The reaction mixture was freed of the derivatizing reagents by a gentle stream of nitrogen. The residue was taken up in dry ethyl acetate. The extract was centrifuged to separate the insoluble salts, and 2 ul of the extract was injected into the port of a HP-5970 gas GC/MS as described in the methods section. Like the synthetic authentic α-disulfone, it exhibited a gas chromatographic peak and identical base peak at 217 M/z at 20.4 min retention time for the bis-aminoethyl-α-disulfone and a minor fragment at 204 M/z assigned to the monotrifluoroaminoethyl-α-disulfone.

DISCUSSION

The oxidation of hypotaurine to taurine has resisted understanding since no specific oxidase has been observed. We have presented evidence in

Table 4. Radical oxidation of benzoic acid: quenching by sperm, hypotarine, DMSO and mannitol

	μmoles CO_2/min/unit enzyme ± SD
Complete system (CS)	32.2±0.1
CS With Sperm*	28.6±0.1
CS With Sperm + HT 30 mM	1.5±0.4
CS With Sperm + Mannitol, 30 mM	10.5±0.5
CS With Sperm + DMSO 30 mM	1.1±0.9

*Sperm = 5.2 mg protein/vessel

the present communication and in our previous work (5), that hypotaurine is oxidized by a hydroxyl radical via the sulfinate radical and α-disulfone to taurine.

The data in Table 1 comparing relative peroxidation rates of hypotaurine with cysteine sulfinate, p-toluene sulfinate, and cysteine suggests that hypotaurine is not metabolized by a peroxidative pathway in tissues, but rather by alternative oxidative mechanisms. On the contrary, it now seems most probable that hypotaurine is principally oxidized by hydroxyl radicals in situ. This is suggested by the efficiency with which hypotaurine quenches DMPO-OH formation (Fig. 4b). We envision that as hypotaurine is oxidized, it is converted into a thiyl radical intermediate which then condenses to form the α-disulfone. The α-disulfone in turn is hydrolyzed to yield one mole of hypotaurine and one mole of taurine from an initial two moles of hypotaurine.

Hydroxyl radicals may be the by products of a number of oxidases including liver microsomal NADPH oxidases (18), activated neutrophils (16), and xanthine oxidase (2). Hydroxyl radicals can theoretically be generated by other means such as the Haber-Weiss reaction (superoxide/hydrogen peroxide dismutation), the Fenton reaction (reduction of hydrogen peroxide by ferrous salts), and the homolytic cleavage of hydrogen peroxide. The hydroxyl radical is extremely reactive and capable of abstracting an electron from a number of Lewis bases. Hypotaurine is particularly suitable as a radical quencher since the sulfinate radical generated rapidly forms the stable α-disulfone. It is interesting to note in this connection that hypotaurine is not subject to oxidation by horseradish peroxidase in contrast to p-toluene sulfinate or cysteine (Table 1). Thus hypotaurine appears to be ideally suited as a scavenger of hydroxyl radicals as it does not appear to be susceptible to significant oxidation by peroxidases.

We predict the ultimate fate of the -disulfone is hydrolysis to taurine and hypotaurine. We have not yet explored this aspect of metabolism in any detail. We have observed that the bis-aminoethyl-α-disulfone is stable to acid and base, but is rapidly converted to taurine when added to dialyzed rat liver homgenates.

Identification of the α-disulfone in biological tissue

Our interest in sexual tissue results largely from the known (11, 12) high concentration of hypotaurine in such tissue. We hypothesize that hypotaurine plays an important role in such tissue protecting the vulnerable highly liquid membrane of sperm from toxic oxygen specie such as the hydroxyl radical. The mandatory exposure of sperm cells to ambient oxygen during requires protection against any toxic oxygen specie. A number of studies have revealed that sperm are vulnerable and lose motility when unprotected against oxygen. Oxidation of unsaturated membrane lipids are held, at least in part, to be the site of this vulnerability (1,14). We studied this by examination of the effect of hydroxyl radicals generated by xanthine oxidase on sperm cells and their protection by hypotaurine and other hydroxyl radical quenching agents (Table 4).

The observation the rat sexual tissue (testes + seminal vesicles) contains the α-disulfone clearly extablishes the biological importance of this intermediate. Thus hydroxyl radicals generated in such tissues lead to dimerization of hypotaurine as we have hypothesized. These observations suggest that the biological role of hypotaurine is to act as a hydroxyl radical scavenger in those tissues where one may anticipate a high flux of this toxic oxygen metabolite. It may also act as an antioxidant of the hydroxyl radical in tissues such as lung, neutrophils and sexual tissue.

REFERENCES

1. Alvarez, J.G., and Storey, B.T., 1983, Taurine, hypotaurine, epinephrine and albumin inhibit lipid peroxidation in rabbit spermatozoa and protect against loss of motility, Biol. Reprod. 29:548-555.
2. Brawn, K., and Fridovich, I., 1981, DNA strand scission by enzymatically generated oxygen radicals, Arch. Biochem. Biophys. 206:414-419.
3. Fellman, J.H., 1981, The synthesis of [2-$^{3}H_{2}$] taurine and [2-$^{3}H_{2}$] hypotaurine, J. Label Comp. Radiopharm. 18:765-768.
4. Fellman, J.H., and Roth, E.S., 1982, Hypotaurine aminotransferase, Adv. Exp. Med. Biol. 139:99-133.
5. Fellman, J.H., and Roth, E.S., 1985, The biological oxidation of hypotaurine to taurine: hypotaurine as an antioxidant, Prog. Clin. & Biol. Res. 179:71-82.
6. Fellman, J.H., Roth, E.S., Avedovech, N.A., and McCarthy, K.D., 1980, Mammalian hypotaurine aminotransferase: isethionate is not a product, Life Sci. 17:1999-2004.
7. Green, M.J., and Hill, H.O., 1984, The chemistry of dioxygen, Meth. Enzymol. 105:3-22.
8. Griffith, O., 1983, Cysteinesulfinate metabolism, J. Biol. Chem. 258:1591-1597.
9. Harman, L.S., Mottley, C., and Mason, R.P., 1984, Free radical metabolites of L-cysteine oxidation, J. Biol. Chem. 259:5606-5611.
10. Jacobsen J.H., Thomas, L.L. and Smith, L.H., Jr., 1964, Properties and distribution of mammalian cysteine sulfinate decarboxylases, Biochem. Biophys. Acta 85:103-116.
11. Kochakian, C.D., 1973, Hypotaurine: Regulation of production in seminal vesicles and prostate of guinea pig be testos-terone, Nature, (Lond) 241:202-203.
12. Kochakian, C.D., 1981, Free amino acids of sex organs of the mouse: regulation by androgen, Am. J. Physiol. 228:1231-1235.
13. Kohashi, N., Yamaguchi, K., Hosokawa, Y., Kori, Y., Fujii, O., and Ueda, I., 1978, Dietary control of cysteine dioxygenase in rat liver, J. Biochem. 84:159-168.
14. Leibfried, M.L., and Bavister, B.D., 1982, The effect of epinephrine and hypotaurine on the in vitro fertilization in the golden hamster, J. Reprod. Fertil. 66:87-94.
15. Oja, S.S., and Kontro, P., 1981, Oxidation of hypotaurine in vitro by mouse liver and brain tissues, Biochem. Biophys. Acta 677:350-357.
16. Sagone, A.L., Jr., Decker, M.A., Wells, R.M., and Democko, C., 1980, A new method for the detection of hydroxyl radical production by phagocytic cells, Biochem. Biophys. Acta 628:90-97.
17. Sumizu, K., 1962, Oxidation of hypotaurine in rat liver, Biochem. Biophys. Acta 63:210-214.
18. Winston, G.W., and Cederbaum, A.I., 1982, Oxidative decarboxylation of benzoate to carbon dioxide by rat liver microsomes: a probe for oxygen radical production during microsomal electron transfer, Biochemistry 21:4265-4270.
19. Yamaguchi, K., Sakakibara, S., Koga, K., and Ueda I., 1971, Induction and activation of cysteine oxidase of rat liver. I. The effects of cysteine, hydrocortisone and nicotinamide injection on hepatic cysteine oxidase and tyrosine trasaminase activities of intact and adrenalectomized rats, Biochem. Biophys. Acta, 237:502-512.

RENAL TAURINE TRANSPORT - RECENT DEVELOPMENTS

Russell W. Chesney, Israel Zelikovic, Aaron L. Friedman, Shermine Dabbagh, Shirley Lippincott, Naomi Gusowski, and Elizabeth Stjeskal-Lorenz

Pediatric Renal Disease Laboratory
Department of Pediatrics
The University of Wisconsin-Madison
Madison, Wisconsin

and

The University of California-Davis
School of Medicine
Davis, California, USA

The renal proximal tubule, with its brush border surface is the site of the active accumulation of all amino acids, including taurine (19). Following filtration at the glomeru-lus, taurine resides in the ultrafiltrate of plasma which passes by the apical or brush border surface of the tubule. As in many other biological membranes, the transfer of taurine across the brush border membrane is Na^+-dependent (5,20) and requires external Cl^- (6). This transport of taurine can best be studied by employing isolated renal brush border membrane vesicles (BBMV) which permit an evaluation of factors which influence the uptake of taurine both in vitro by changing the chemical composition of the incubation medium and in vivo by exposure of animals to various manipulations (4-5). Our group has been interested in establishing how renal amino acid transport mechanisms adapt to changes in the level of taurine or its precursors in the diet. To carry out these studies, a rat model has been used and the accumulation of taurine by BBMV has been measured.

BASIC CHARACTERISTICS OF RENAL TAURINE TRANSPORT

Brush border membrane vesicles are prepared from renal cortex homogenates by a $CaCl_2$ precipitation step which aggregates all membranes other than the brush border surface. Using γ-glutamyl-transferase, a brush border surface marker enzyme, an 11-fold enrichment of enzyme activity relative to starting homogenate and a decline in all other membrane marker enzymes establishes the purity of this preparation (7). Essentially all studies of Na+ coupled taurine co-transport are carried out at 10 μM external taurine over a time course of 45 min, but 15 sec, uptake is used to examine initial rate characteristics and kinetics (4-7,24). Equilibrium values are achieved at 45 min, and uptake at 45 min, in the presence of the non-permeant sugar sucrose indicates that uptake is inversely proportional to external osmolarity; hence uptake into the vesicle is occurring (5). A Na^+-taurine co-transport or symport is indicated by the finding of an "overshoot" of taurine uptake into BBMV in the presence of an external NaCl gradient (5,20). This symport does not occur with external K^+, Li^+, choline or mannitol. Because of the relatively low Na^+ conductance across the apical surface of the proximal convoluted tubule, this taurine symport and other Na^+-organic solute symporters represent an important mechanism by which Na^+ may enter the tubule cell (24). Various ionophores may influence the movement of cation-dependent solute transport processes. Accordingly we examined several ionophores and found that valinomycin enhances Na^+-taurine symport in the presence of an outwardly directed K^+ gradient (KCl preloaded vesicles), that the electrogenic Na+-channel ionophore gramicidin reduces Na+-taurine uptake and the nonelectrogenic K^+-H^+ exchange ionophore, nigericin, augments Na^+-taurine symport, but only in the presence of external KCl (6). Agents which inhibit Na^+/H^+ antiport and electrogenic Na^+-glucose symport, i.e., amiloride, or an internal acidic pH (pH 5.5 inside the vesicle, pH 7.5 outside) do not alter Na^+-taurine uptake. Harmaline, which blocks antiport also blocks Na^+-taurine symport, but this may involve inhibition of NaCl-dependent taurine accumulation or the type of inhibition which occurs with Na^+-glucose or Na^+-alanine uptake.

The anion employed is critical in Na^+-taurine symport. Since the data above employing ionophores imply that Na^+-taurine symport is electrogenic, uptake should be higher with more permeant or reduced with less permeant anions. Of anions tested, including more permeant (SCN^- or NO^{3-}) and less permeant (SO_4^{2-}), only chloride supports the full over-shoot (6). The diuretics furosemide and bumetanide and the disulfonic stilbenes, SITS and DITS, inhibit coupled NaCl support and electroneutral anion exchange (17). Taurine uptake is inhibited to a minor degree (10%) by the diure-tics but DITS inhibits uptake by 50 percent. It is of interest that these diuretics and stilbenes inhibit cell volume changes which depend upon NaCl or Na-K-2Cl entry mechanisms at different regions of the nephron (12).

A role for Cl^- activation of Na^+-amino acid symport has been reported for Na^+-glycine co-transport in teleost and rat vesicles (1,11) and for Na^+-taurine transport in basal lateral membranes from the winter flounder (16). It is noteworthy that the flounder secretes taurine into its urine in conjunction with cell volume regulation. We have been interested in whether chloride activation of Na^+-taurine symport required external Cl- and whether other halide anions could mimic Cl^-. Internal LiCl or choline Cl, after vesicle preloading, blocked full expression of Na^+-taurine symport whereas internal $Li_2 SO_4$, $LiNO_3$ and LiSCN did not block taurine entry (6). The halide ion, bromide, fully supported Na^+-taurine symport, but iodide and fluoride did not.

On the results reported above, a model for taurine entry across the apical surface of the proximal tubule can be constructed (Fig. 1). With this background one can approach an understanding of the role of change in sulfur amino acid intake in regulating Na^+-taurine symport.

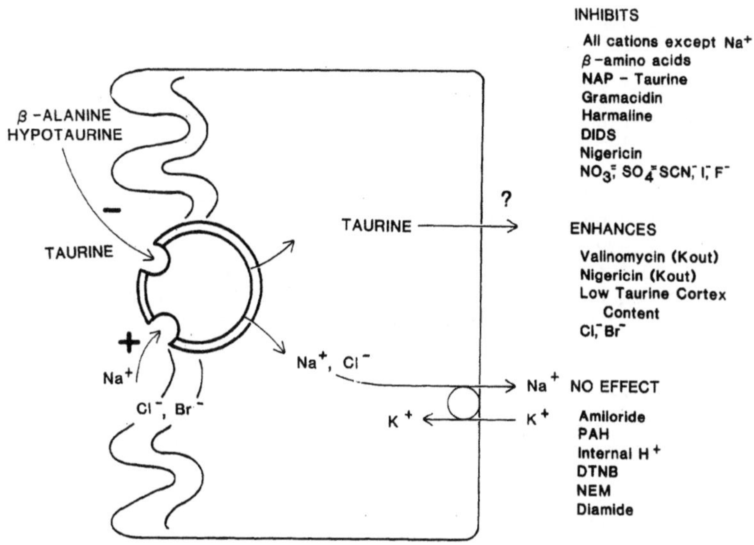

Fig. 1. Factors which inhibit, enhance or have no influence on the taurine transporter in the brush border membrane of the rat (6,7,9).

RENAL ADAPTIVE RESPONSE TO ALTERED DIET

Renal tubular epithelium conserves ions and amino acids during periods of decreased dietary intake and excretes them following dietary excess (13). In rats and mice fed reduced amounts of sulfur amino acids (0.2% methionine), the renal tubular reabsorption and transport of taurine into collagenase-isolated tubules is increased (14,21). The reabsorption of taurine increases by 90-95% (5,4,7,13,14) and tubule uptake increases by 50 to 100% (13,14). This response has been termed "the renal adaptive response to altered dietary intake of sulfur amino acids". A diet (supplemented with 3%) taurine results in massive taurinuria and decreased tubule uptake (13,14). The fractional excretion of taurine is 1-2% in rats fed the low sulfur amino acid diet (LTD), 9-10% in rats fed a normal sulfur amino acid diet (NTD) and 62-68% in rats fed a high (3%) taurine diet (HTD). Tubule uptake falls by to 50-70% after exposure to the HTD.

Taurine uptake by BBMV prepared from rats fed each diet is different. BBMV from animals fed the LTD has a higher overshoot than in animals fed the NTD and uptake is reduced in vesicles from animals fed the HTD.

Kinetic analysis of uptake over a variety of concentrations reveals the following points: (1) The LTD results in an increase in the initial rate Vmax of uptake from 49.8 $\pm$ 2.6 pmole/mg protein/15 sec, in NTD to 59.5 $\pm$ 1.6 pmole/mg protein/15 sec, in LTD ($p<.01$); (2) The Vmax falls in HTD-fed from 49.8 $\pm$ 2.6 to 32.8 $\pm$ 2.2 pmole/mg protein/15 sec, ($p<.005$); (3) No change in Km or affinity of the taurine transport site is found regardless of diet; (4) The adaptive change is expressed only for the initial uptake of the Na^+ energized portion of transport or for the Na^+-symport; (5) At equilibrium (45 min.) the amount of taurine within BBMV is comparable on each of the three diets; (6) At high taurine concentrations (>500μM) and at the high Km, Na^+-independent uptake site, uptake is the same regardless of diet.

The signal for and the membrane related events underlying the adaptive response are unclear and their elucidation forms the major focus of this research. The problem of the signal underlying renal taurine conservation is germane since whole body taurine content is regulated by the kidney (8). The plasma concentration (21) or the concentration of taurine present in the renal cortex may govern this adaptive response (5). Placing animals on the LTD reduces plasma taurine values to 185 ± 15 μM from a normal value of 230 ± 20 μM in NTD fed rats. The plasma taurine concentration in HTD animals is increased to 875 ± 150 μ moles/L. Thus changes are found in plasma taurine that could act as the signal for the adaptive response.

The influence of diet upon tissue taurine content is shown in Fig. 2. The greatest changes are in urinary taurine values, but the LTD also causes significant changes in the levels of taurine in leg muscle and kidney cortex while the HTD causes a rise in cardiac, leg, and neck muscle, and in kidney. Brain taurine values are stable, regardless of diet. An inverse correlation exists both between plasma taurine concentration and BBMV uptake (r = 0.76, p<.001) and between renal cortex taurine content and BBMV uptake (r = 0.56, p<.01) (5).

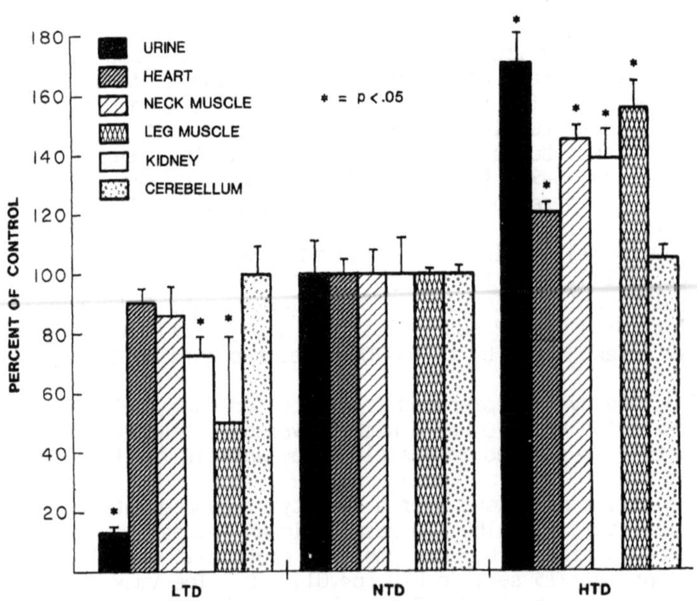

Fig. 2. The influence of dietary manipulation on urine and several tissue concentrations of taurine expressed as a percent of control. Each value represents the mean ± SD of 12 determinations.

The β-amino acid, β-alanine, can deplete tissue taurine levels without significantly changing plasma taurine values (22). We placed a group of rats receiving each diet on 3% β-alanine in their drinking water as well. Animals received one of three basic diets for 14 days (LTD, NTD, or HTD); half of each group of rats on these diets received 3% β-alanine in the drinking water from days 8-14. Another group of rats on the three basic diets received either 3% glycine or 3% methionine in their drinking water from days 8-14 (7).

β-alanine feeding lowered renal cortex taurine value as compared to diet appropriate control. Increased at 15s and 60s uptake of taurine at 15s and 60s occurred after each manipulation which reduced renal cortex taurine. Feeding 3% glycine did not alter plasma, renal cortex, urinary taurine concentrations or BBMV uptake of taurine. Feeding 3% methionine did raise taurine in plasma and urine, presumably since it is a precursor of taurine synthesis. Renal cortex taurine was unchanged as was initial rate BBMV uptake.

The increase in BBMV uptake correlates highly with the decline in renal cortex and plasma taurine concentration. However, since the 3% methionine feeding actually raised plasma taurine without altering BBMV uptake, it is more likely that the change in BBMV uptake and the adaptive response expressed at the brush border surface relates to changes in renal cortex taurine content.

Experiments performed where rats on each diet were fasted for 72 h gave the results of a reduction in renal cortex taurine content and augmented BBMV uptake. Thus these studies are also consistent with the idea that tissue taurine content governs Na^+-taurine symport.

It should be noted that brain taurine values are unchanged by β-alanine feeding and by fasting. Thus despite changes in urine plasma and renal cortex taurine content, brain taurine values remain stable, which suggests that this renal adaptive response maintains constancy of taurine levels where taurine serves as a neuromodulator.

It is possible that the changes in sulfur amino acid intake brought about by the three diets could change the transport of the sulfur amino acids methionine and cystine and inorganic sulfate (SO_4^{2-}). The dietary manipulations did not influence the time course or kinetic constants (Km, Vmax) of methionine uptake or cystine binding to BBMV since uptake or binding was similar in all dietary groups (9). The uptake and kinetics of Na_2SO_4 by BBMV was also unchanged by the dietary manipulations. However, the LTD did enhance β-alanine accumulation by vesicles.

Membrane-related events could be responsible for the increase or decrease in Na^+-taurine symport in response to dietary change. We first examined the rapid uptake of $^{22}NaCl$ by BBMV at 3,5,7,15 and 60s. Using the conditions of 40 mM KCl inside the vesicle as well as pH 5.5, both of which lead to enhanced K^+ and H^+ efflux and thereby maximize Na^+ uptake, no diet-induced change in ^{22}Na accumulation was evident. Diet could potentially change the physical characteristics of the membrane, thereby changing access of substrate (taurine) to the transport system (Na^+-taurine symport) which inserts into the brush border membrane (15). We used the lipid-binding fluorescent probe of membrane fluidity, 1,6-diphenyl-1, 3,5-hexatriene and measured fluidity in vesicles isolated from animals on each diet between 4° and 56°C. Curve-fitting showed that the data fitted

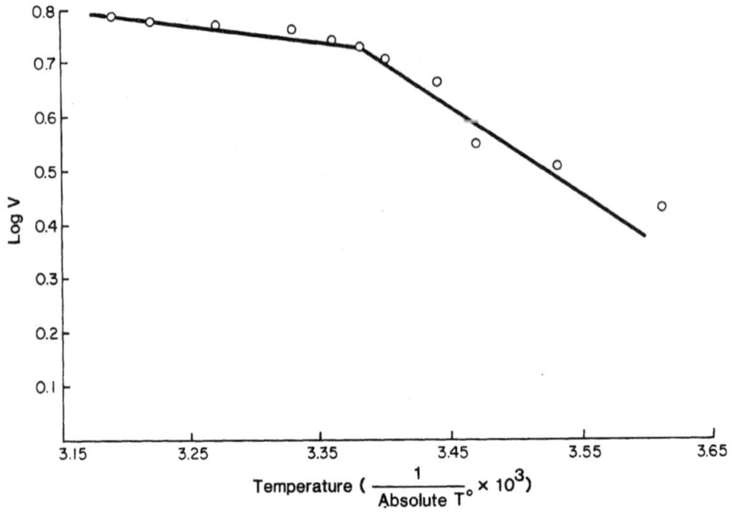

Fig. 3. The uptake of taurine (V) versus reciprocal absolute temperature (Arrhenius plot) note the single break point at 23°C.

better to a single break, two linear slope model than to a 1) single linear slope; 2) two break, three linear slope; or (20) parabolic model. The break point was at 22-25°C. No change in the upper slope, lower slope or break point was found in relation to diet. We determined the breakpoint of a standard Arrhenius plot and showed that a two phase linear model also gave the best fit. The breakpoint determined in membranes from rats fed each of the three diets was similar at approximately 23°C; thus, a dietary effect is not evident. Changes in physical state of the membrane often correspond to liquid-gel phase transitions in a variety of membranes including the intestine (2,3). The observation that a clear change in slope in the Arrhenius plot for Na^+-dependent taurine uptake occurs around 23°C provides evidence that a liquid → gel phase transition may be occurring at this temperature. Further, it is similar to the nonlinear Arrhenius plot for alkaline phosphatase in rabbit kidney brush border membranes which occurs at 26°C (18). Whatever are the molecular events responsible for the Na^+-taurine symport being responsive to changes in the physical state of the membrane lipids, it is clear that the change in the slope of the Arrhenius plot occurs at "room temperature". Hence, it is appropriate that all experiments were performed within a narrow range around this temperature. Finally, even though this liquid → gel transition may occur at a critical temperature for taurine uptake, the dietary manipulations that we have employed do not alter this change occurring at "room temperature" (Fig. 3). Finally the phospholipid composition of membranes is not altered by diet. Thus, there is little evidence that changes in the lipid composition of the membrane result in adaptive response expressed at the brush border surface.

Our group has been interested in the aminoaciduria of the neonatal period for the past decade (8). Taurinuria is increased in young rats at ages 7,14,21 and 28 days as compared to adult rats. When urine taurine excretion was examined in relation to body weight (μ mol/mg creatinine/kg body weight), an increased urinary excretion was found for all immature groups. In other words, the amount of taurine excreted per unit body weight falls only after age 28 days (Fig. 4) (10). In addition, the renal cortex taurine content is elevated in young rats suggesting that this taurinuria of immmaturity relates to either a block in efflux of taurine across the basal-lateral surface of the tubule or to a block in uptake across the vesicle (8). We, therefore, wanted to determine if membranes isolated from immature animals could show the adaptive response.

We first examined the levels of taurine in the milk fed to these rat pups. As can be seen in Fig. 5, the initial rate of taurine uptake is influenced by diet in 14- and 21-day-old rat pups, albeit uptake is lower than in adults. Not shown, uptake in 28-day-old rat membranes is entirely comparable to that found in adult rats fed each diet. By contrast, uptake in vesicles prepared from 7-day-old pups does not show the adaptive response and is markedly reduced in comparison to values in adults. These results suggest two separate issues for further study. First, since the adaptive response is found in 14-, 21- and 28-day-old rat membranes, why is taurinuria present. It may well be that there is a maturation of the basal-lateral membrane in terms of efflux. In order to establish the ontogeny of the basal-lateral membrane, we must first isolate this surface and then study the characteristics of taurine transport there.

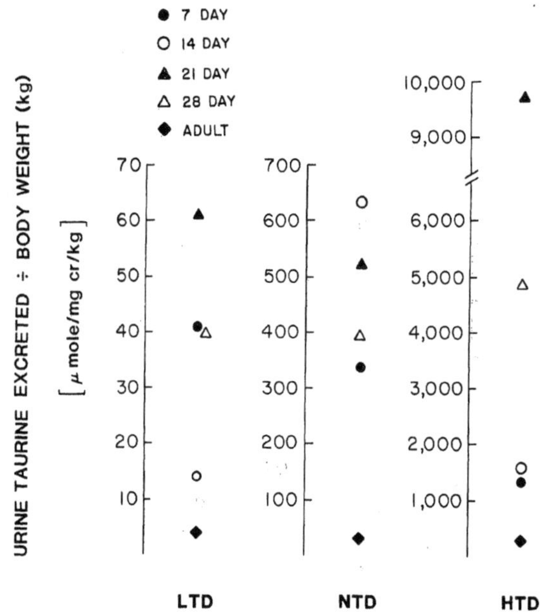

Fig. 4. Urine taurine excretion per kg body weight. Each point is the mean of 12 determinations.

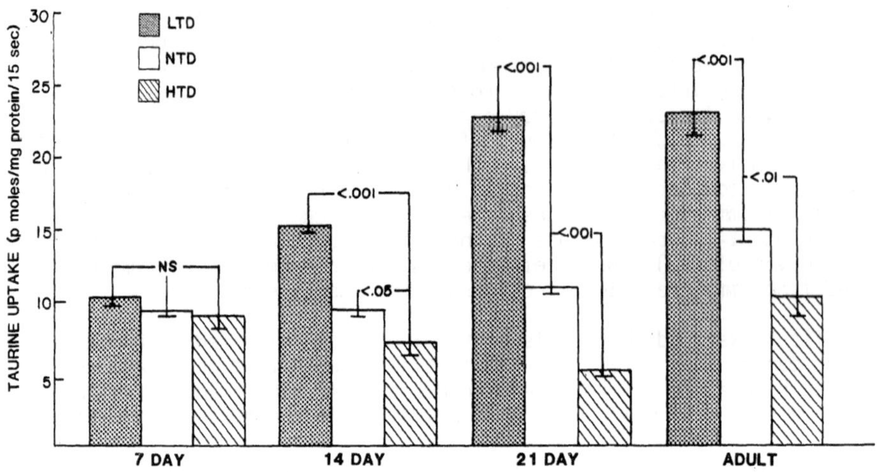

Fig. 5. The uptake of taurine (10 μM) into BBMV after 15 sec, of incubation. Each point represents the mean ± SE of 4 determinations performed in triplicate (Reproduced with permission from Pediatric Research).

Second, the membranes prepared from 7-day-old rats are potentially leaky (10). Thus it will be necessary to study $^{22}NaCl$ uptake and Na^+-glucose or Na^+-proline uptake at that site to compare to Na^+-taurine symport.

These studies allow us to put forward certain statements concerning the nature of the adaptive response. These are shown in Tables 1 and 2.

Table 1. The Renal Adaptive Response to Reduced Sulfur Amino Acid Intake

Taurine urinary conservation on a low taurine diet
Enhanced tubule and isolated brush border membrane vesicle uptake of taurine
Represents enhanced initial rate of Na^+-taurine symport ("up regulation")
Vmax of uptake is increased
No change in Km of uptake
No change in Arrhenius plot, membrane fluidity, phospholipid composition, or in the uptake of methionine, cystine or SO_4^{2-}

The overall question to be asked is why does the kidney conserve or hyperexcrete taurine in response to altered dietary load? The finding of stable brain taurine levels despite dietary and other manipulations suggests a role for the renal adaptive response. Our hypothesis states that the renal adaptive response serves to maintain the constancy of brain taurine content. The brain is a site of neural activity requiring a variety of neurotransmitters, neurostimulators and neuroinhibitors. Taurine is clearly a neuromodulating substance and has antiseizure activity (8). As well, and just as important, the brain requires a stable volume since it is encased within the skull which has little or no capacity to enlarge if brain cell volume increases as in cerebral edema. Taurine

Table 2. The Renal Adaptive Response to a High Taurine Diet

Taurine urinary hyperexcretion on a high taurine diet
Reduced tubule and brush border membrane vesicle uptake of taurine
Represents reduced initial rate of Na^+-taurine symport ("down regulation")
Vmax of uptake is diminished
No change in Km of uptake
No change in Arrhenius plot, membrane fluidity, phospholipid composition, or in the uptake of methionine, cystine or SO_4^2

thus serves to regulate brain cell volume under periods of undue stress, such as hypernatremia and, indeed, can be viewed as the "idiogenic osmolar substance" which regulates cell volume (8,23). These two important roles for taurine in the brain require a homeostatic mechanism to maintain brain taurine content. It is our contention that the kidney, and more specifically the renal tubular epithelium and its brush border surface, functions to maintain taurine homeostasis under periods of altered dietary intake of taurine and its precursors. What is unclear is what is the signal that eminates from a reduction in renal cortex taurine concentration that alters the Na^+-taurine symport system and at the same time serves to stabilize brain taurine content.

ACKNOWLEDGEMENTS

This research was supported by USPHS Grants number K04 AM00421, AM19489, AM31682 and AM37232.

REFERENCES

1. Boge, G., and Regal, A., 1981, A chloride requirement for Na^+-dependent amino acid transport by brush border membrane vesicles isolated from the intestine of a mediterranean telost (Boops salpa), J. Biol. Chem. 16:455-461.
2. Brasitus, T.A., Tall, A.R., and Schachter, D., 1980, Membranes studied by differential scanning calorimetry and fluorescence polarization, Biochemistry 19:1256-1261.
3. Chapman, D., Gomez-Fernandez, J.C., and Goni, F.M., 1979, Intrinsic protein-lipid interactions. Physical and biochemical evidence, FEBS Lett. 98:211-223.
4. Chesney, R.W., Friedman, A.L., Albright, P.W., and Gusowski, N., 1982, Fasting reverses the renal adaptation to altered dietary amino acid intake, Proc. Soc. Exp. Biol. Med. 170:493-501.
5. Chesney, R.W., Gusowski, N., and Friedman, A.L., 1983, Renal adaptation to altered dietary sulfur amino acid intake occurs at luminal brush border membrane, Kidney Int. 24:588-594.
6. Chesney, R.W., Gusowski, N., Dabbagh, S., Theissen, M., Padilla, M., and Diehl, A., 1985, Factors affecting the transport of β-amino acids in rat renal brush border membrane vesicles: The role of external chloride, Biochim. Biophys. Acta. 812:702-712.
7. Chesney, R.W., Gusowski, N., and Dabbagh, S., 1985, Renal cortex taurine content regulates renal adaptive response to altered dietary intake of sulfur amino acids, J. Clin. Invest. 76:2213-2221.
8. Chesney, R.W., 1985, Taurine: biological role and clinical perspectives, Adv. Pediatr, 32:1-42.
9. Chesney, R.W., Gusowski, N., Padilla, M., and Lippincott, S., 1986, Altered intake of dietary sulfur amino acids: Effect on renal brush border membrane transport of several amino acids and sulfate. Am. J. Physiol. 251:F125-F132.
10. Chesney, R.W., Gusowski, N., Zelikovic, I., and Lippencott, S., 1986, Developmental aspects of renal β-amino acid transport V: Brush border membrane transport in nursing animals - effect of age and diet, Pediatr. Res. 2:890-894.
11. Corcelli, A., and Storcelli, C., 1983, The role of potassium and chloride ions on the Na^+-acidic amino acid co-transport system in rat intestinal brush border membranes, Biochim. Biophys. Acta. 732:24-31.
12. Eveloff, J., Fong, J., and Calamin, J., 1986, Effects of hyperosmolarity and phorbolesters in cation fluxes in medullary thick ascending limb cells, Kidney Int. 29:346.
13. Friedman, A.L., Albright, P.W., and Chesney, R.W., 1981, Dietary adaptation of taurine transport in rat renal epithelium, Life Sci. 29:2415-2419.
14. Friedman, A.L., Albright, P.W., Gusowski, N., Padilla, M., and Chesney, R.W., 1983, Renal adaptation to alteration in dietary amino acid intake, Am. J. Physiol. 245:F159-F166.
15. Jain, M.K., 1980, Phase properties of bilayers, in "Introduction to Biological Membranes," Jain, M.K., and Wagner, R.C., (eds.) Wiley, New York, pp. 53-175.
16. King, P.A., Bayenbach, K.W., and Goldstein, L., 1982, Taurine transport by isolated flounder renal tubules, J. Expl. Zoo. 223:103-114.

17. Larsen, M., and Spring, K.R., 1983, Bumetanide inhibition of NaCl transport of Necturus gall bladder, J. Memb. Biol. 74:123-132.
18. LeGrimellec, C., Giocondi, M.C., Carriere, B., Carriere, S., and Cardinal, J., 1982, Membrane fluidity and enzyme activities in brush border and basolateral membranes of the dog kidney, Am. J. Physiol. 242:F246-F253.
19. Mitch, W.E., and Chesney, R.W., 1983, Amino acid metabolism by the kidney, Mineral Electrolyte Metab. 9:190-206.
20. Rozen, R., Tenenhouse, H.S., and Scriver, C.R., 1979, Taurine transport in renal brush border membrane vesicles, Biochem. J. 180:245-248.
21. Rozen, R., and Scriver, C.R., 1982, Renal transport of taurine adapts to perturbed taurine homeostasis, Proc. Natl. Acad. Sci., U.S.A., 79:2101-2105.
22. Schaffer, J.E., and Kocsis, J.J., 1981, Taurine mobilizing effects of β- alanine and other inhibitors of taurine transport, Life Sci. 28:2727-2736.
23. Trachtman, H., Barbos, R., Sturman, J., and Finberg, L., 1986, Taurine is an osmoprotective molecule, no longer idiogenic in chronic hypernatremic dehydration, Pediatr. Res. 20:335A.
24. Wright, S.H., Krasne, S., Klippen, I., Wright, E.M., 1981, Na^+ dependent transport of dicarboxylic acid cycle intermediates by brush border membranes, effects of fluorescence of a potential sensitive cyanide dye, Biochim. Biophys. Acta. 64:767-778.

TAURINE CONJUGATION OF BILE ACIDS PROTECTS HUMAN CELLS IN CULTURE

G. E. Gaull[1] and C. E. Wright

New York State Institute for Basic Research in
Developmental Disabilities Staten Island, NY 10314 USA

Primary bile acids such as cholate and chenodeoxycholate are the end products of hepatic cholesterol catabolism. In most healthy animals not more than trace amounts of free bile acids occur in bile and plasma, even though unconjugated forms are present in gallstones and feces. Most placental animals including man, can conjugate bile acids with either glycine or taurine (5). Since the affinity of taurine is higher for the conjugating enzyme, bile acid-CoA: amino acid H-acyltransferase (4), taurine is generally considered the preferred substrate for bile acid conjugation (4,6). After conjugation bile acids are excreted into the bile and subsequently released into the intestinal tract where they are subject to deconjugation and 7 α-dehydroxylation. These reactions are catalyzed by intestinal strains of <u>Clostridium</u>, <u>Enterococcus</u>, <u>Bacteroides</u>, and <u>Lactobacillus</u> (5). Removal of the 7 α-hydroxyl group results in the formation of the secondary bile acids: deoxycholate from cholate and lithocholate from chenodeoxycholate. Except for lithocholate which is normally excreted in the feces, all other free and conjugated bile acids are absorbed in the ileum and recycled to the liver.

Various diseases of the liver and the biliary system, or certain nutritional and physiological insults can induce deleterious effects on bile acid metabolism, shift the composition of the bile acids in the bile, or increase concentrations of bile acids in the blood and other body fluids. Thus, abnormal accumulation of bile acids may result in localized and/or systemic cytotoxic effects.

A hypothesis recently has been advanced suggesting that taurine exerts its biological effects by acting as a detoxifying agent (3). This function of taurine as a detoxifier has been clearly demonstrated in human cell culture systems using excess retinoids (7), the oxidant-generating combination of ferric ion and ascorbate (8), the oxidant hypochlorous acid (10), and certain xenobiotics (2) . Further it has been proposed that the chemical reactivity of the amino group with the potentially toxic compound, as well as the chemical properties of the sulfonic acid end of the molecule, allows taurine to exert its detoxifying action (3,10).

---

[1]Present Address: Department of Nutritional Sciences, The NutraSweet Company, Skokie, Illinois 60076 U.S.A.

The human lymphoblastoid cell culture system has been successfully used to examine the cytotoxicity of several diverse compounds (7,8,10). The present studies employ this in vitro model to systematically study the potential cytotoxic effects of primary and secondary bile acids. These studies provide evidence that conjugation of bile acids with taurine or glycine on the carboxylic acid group, and sulfate on the hydroxyl group, protects cells against cell damage by unconjugated bile acids, especially secondary bile acids.

MATERIALS AND METHODS

Cholate, chenodeoxycholate, deoxycholate, lithocholate, dehydrocholate, ursodeoxycholate, glycocholate, glycolithocholate, taurocholate, taurolithocholate, lithocholate-3-sulfate, glycolithocholate-3-sulfate, taurolithocholate-3-sulfate, lithocholate methyl ester, glycine, taurine and fatty acid-free bovine serum albumin (BSA) were obtained from Sigma Chemical Co. (St. Louis, MO). All other chemicals were of reagent grade. The origin and growth conditions of the human lymphoblastoid cell line NB76 have been described previously (9).

All cell incubations were performed at 37° C in Kreb's Ringer phosphate buffer (pH 7.2) containing glucose. The effect of the test compounds on the viability of the human lymphoblastoid cell line were performed as follows: approximately 1 x 10^6 viable cells were washed in buffer, centrifuged at 600 x g for 10 minutes, then resuspended in buffer containing various concentrations of the test compound. This mixture was then incubated at 37°C for the times indicated. In experiments in which the cholates were bound to albumin, the conditions were as follows: 2.5 mg of cholate was dissolved in 1.5 ml absolute ethanol, to this mixture 1.0 ml of bovine serum albumin (10 mg/ml) was added dropwise with continuous stirring, 1 drop of 1N NaOH was added to clarify the solution, ethanol was evaporated under a nitrogen stream and the solution was lyophilized. Albumin-bound cholates were then reconstituted in buffer. Cell viability was determined by trypan blue exclusion. All data points represent the mean of at least three separate experiments.

RESULTS AND DISCUSSION

The potential cytotoxic effect of fourteen primary and secondary bile acids, either in the form of free salts or conjugated, were examined after 1, 2, and 3 hour exposures. The trihydroxy (3α, 7α, 12αOH) primary bile acid, cholate, at a concentration of 500 µM resulted in no greater loss of lymphoblastoid cell viability than that found under control conditions of incubation (Table 1). Its keto analogue, 3,7, 12 trioxocholanic acid (dehydrocholate), which is not know to be present in the bile of any species, similarly resulted in no decrease in viability under the same conditions. Neither taurocholate nor glycocholate showed decreased viability (Table 1). In striking contrast, the secondary bile acid deoxycholate (3α, 7αOH) at the same concentration, produced striking decreases in viability under these incubation conditions.

The cytotoxic effect of deoxycholate cellular viability was further examined with respect to increasing concentrations of deoxycholate (Figure 1). Lymphoblastoid cell viability decreased linearly within a concentration range of 15.6 to 500 µM. Addition to the incubation mixture of either glycine or taurine did not augment deoxycholate cytotoxicity. The conjugate, taurodeoxycholate possessed no cytotoxic effect on this cell line; glycodeoxcholate was not available for testing.

TABLE 1. Cytotoxic effect of free and conjugated cholate derivatives.

Compound	% Viable Cells		
	1hr	2hr	3hr
control (KRPG)	90	88	78
cholate (500 µM)	89	80	77
dehydrocholate (500 µM)	100	95	97
glycocholate (500 µM)	98	95	91
taurocholate (500 µM)	106	98	91
deoxycholate (500 µM)	48	37	15
taurodeoxycholate (500 µM)	95	83	77

TABLE 2. Cytotoxic effect of free and conjugated chenodeoxycholate derivatives.

Compound	% Viable Cells		
	1 h	2 h	3 h
chenodeoxycholate (500 µM)	15	5	1
glycochenodeoxycholate (500 µM)	81	72	63
taurochenodeoxycholate (500 µM)	87	81	75
ursodeoxycholate (500 µM)	109	ND	89

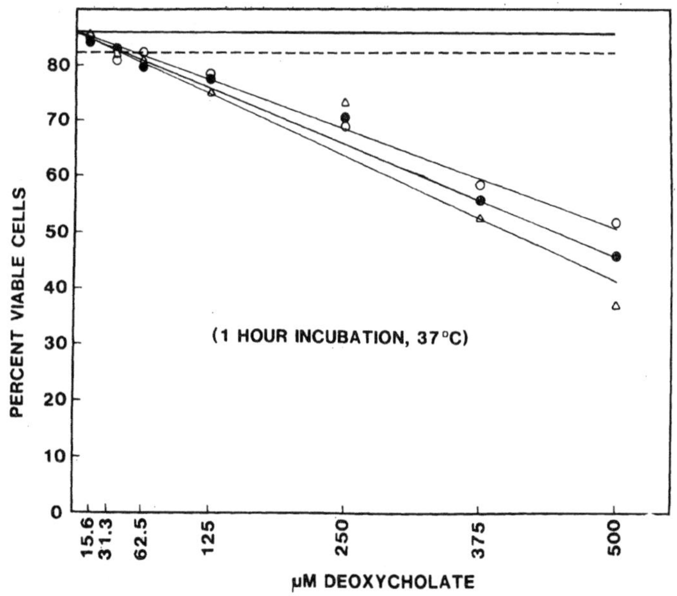

Figure 1. Human lymphoblastoid cell cytotoxicity with respect to increasing concentrations of deoxycholates. (●) deoxycholate, (○) deoxycholate plus 500 µM taurine, (△) deoxycholate plus 500 µM glycine, (——) control in buffer, (----) 500 µM taurodeoxycholate.

Chenodeoxycholate, the 3α, 7α OH form of cholanic acid, is also a primary acid in man. Under normal circumstances there is less of this dihydroxy bile acid than there is of the primary trihydroxy bile acid, cholate. This relationship may be reversed in certain disease states in man (1). It is of interest, therefore, that chenodeoxycholate results in strikingly decreased viability and that considerable protection is provided by its conjugation with glycine and, especially, with taurine (Table 2).

In this in vitro system the primary bile acid, chenodeoxycholate, is even more toxic than the secondary bile acid deoxycholate (Compare Table 1 with Table 2). However, when deoxycholate and chenodeoxycholate are conjugated with taurine, the loss in cellular viability is reversed. Interestingly, ursodeoxycholate (3α, 7 B OH), which is a primary bile acid in some species but is a secondary bile acid in others, results in little decrease in viability (Table 2).

Unlike cholate which results in a linear decrease in cellular viability with increasing concentrations, chenodeoxycholate toxicity shows a sigmoidal decrease (Figure 2). Between the concentrations of 15.6 and 125 μM chenodeoxycholate is relatively nontoxic. However, there is approximately a 50% decrease in viability between 125 and 250 μM which levels off with increasing concentrations. Although we now have no explanation for the shape of this dose response curve, it is known that high concentrations of bile acids exert a detergent effect on membranes (5). Such a mechanism could explain the toxicity at higher concentrations.

It is well known that abnormal accumulation of the secondary bile acid lithocholate in animals (1) results in hepatic and systemic toxici-

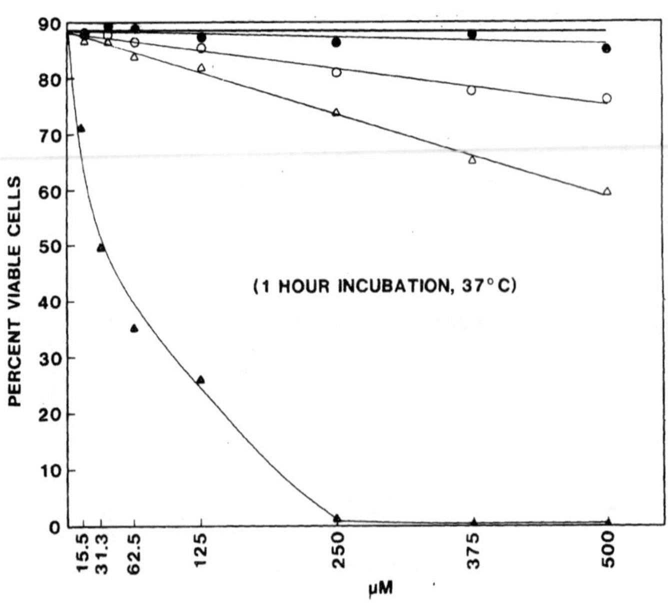

Figure 2. Human lymphoblastoid cell cytotoxicity with respect to increasing concentrations of lithocholate derivatives. (▲) lithocholate methyl ester, (△) Lithocholate, (○) lithocholate methyl ester bound to albumin, (●) lithocholate-3-sulfate, (——) control in buffer.

ty. Lithocholate, the monohydroxy (3 OH) cholanic acid, and other derivatives were examined in this cell culture system (Table 3). Lithocholate was found to be toxic at 500 µM; its methyl ester was even more toxic with complete cell death occurring at 250 µM (Figure 3). When sulfated at the 3 position, virtually complete protection was provided (Table 3). Conjugation with taurine or glycine alone was not as effective in this sytem as sulfation alone or sulfation in combination with conjugation with either taurine or glycine (Table 3). These results in an isolated cell system differ from those in vivo in the guinea pig (1) in which lithocholate sulfate itself resulted in toxicity, whereas conjugation with taurine, but not with glycine, alleviates the toxicity. The decrease in viability both of lithocholate and of its methyl ester was greatly reversed by their binding with albumin (Table 3, Figure 3). The spontaneous binding of lithocholates with proteins is a well known reaction (5) which decreases in vivo toxic effects.

Neither free taurine nor free glycine decreased the toxicity of the free secondary bile acid, lithocholate methyl ester, the free primary bile acid chenodeoxycholate, or the conjugated secondary bile acid, glycolithocholate (Table 4). Free taurine and perhaps free glycine may have decreased the toxicity of taurolithocholate, but this observation requires further study.

TABLE 3. Cytotoxicity of various lithocholate derivatives

Compound	% Viable Cells		
	1 h	2 h	3 h
lithocholate (500 µM)	55	32	13
lithocholate methyl ester (100 µM)	34	0	0
lithocholate-3-sulfate (500 µM)	ND	98	88
taurolithocholate (500 µM)	96	ND	69
glycolithocholate (500 µM)	87	ND	40
glycolithocholate-3-sulfate (500 µM)	ND	93	91
taurolithocholate-3-sulfate (500 µM)	ND	104	94
lithocholate methyl ester bound to albumin (100 µM)	87	91	ND
lithocholate bound to albumin (500 µM)	83	88	ND

TABLE 4. Effect of free taurine and free glycine on lithocholate cytotoxicity.

Compound	% Viable Cells		
	1 h	2 h	3 h
lithocholate methyl ester (100 μM)	34	0	0
+ glycine (100 μM)	42	3	0
+ taurine (100 μM)	30	4	1
chenodeoxycholate (300 μM)	43	32	13
+ glycine (300 μM)	36	32	16
+ taurine (300 μM)	37	30	12
glycolithocholate (300 μM)	90	ND	37
+ glycine (300 μM)	87	ND	40
+ taurine (300 μM)	89	ND	33
taurolithocholate (300 μM)	95	ND	76
+ glycine (300 μM)	98	ND	87
+ taurine (300 μM)	110	ND	98

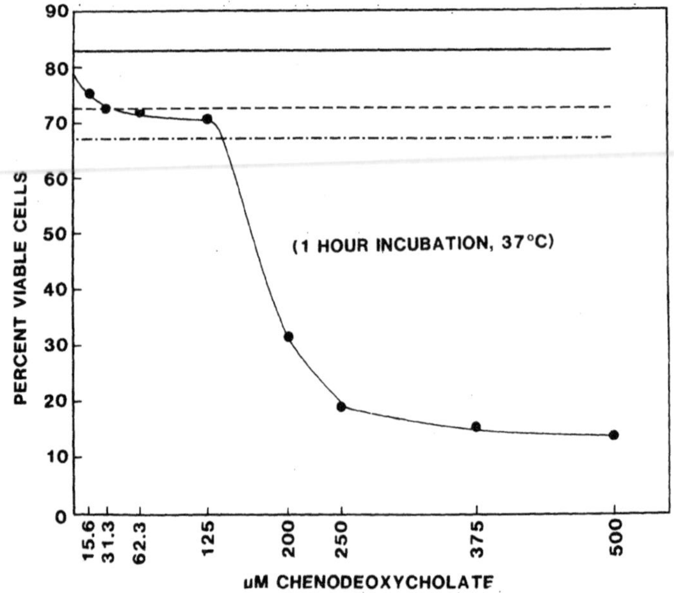

Figure 3. Human lymphoblastoid cell cytotoxicity with respect to increasing concentrations of chenodeoxycholates. (●) chenodeoxycholate, (—·—) glycochenodeoxycholate (- - -) taurochenodeoxycholate, (——) control in buffer.

SUMMARY

The primary trihydroxy bile acid, cholate, both free and conjugated, is nontoxic in cultured human lymphoblastoid cells incubated in vitro. The primary dihydroxy bile acid, chenodeoxycholate, was more toxic at high concentrations than at low concentrations, but conjugation with taurine reversed it. The free secondary bile acids, deoxycholate and lithocholate, corresponding respectively to cholate and chenodeoxycholate, are extremely toxic, the latter more than the former. Conjugation with taurine reversed the toxicity, as did glycine. Neither free taurine nor free glycine added to unconjugated bile acids decreased toxicity.

REFERENCES

1. Dowi, N.P., Yousef, I.M., Tuchseeber, B., and Roy, C.C., 1983, Taurine prevents cholestasis induced by lithocholic acid sulfate in guinea pigs, Am. J. Nutr. 37:221-222.
2. Emudianughe, T.S., Caldwell, J., and Smith, R.L., 1983, The utilization of endogenous taurine for the conjugation of xenobiotics in the ferret, Xenobiotica 13:133-137.
3. Gaull, G.E., Pasantes-Morales, H., and Wright, C.E., 1985, Taurine in human nutrition, in: S.S. Oja, L. Ahtee, P. Kontro, and M.K. Paasonen, eds., "Taurine: Biological Actions and Clinical Perspectives", Alan R. Liss, pp. 3-22.
4. Gottfries, A., Schersten, T., and Ekdahl, P.H., 1966, The capacity of human liver homogenates to synthesize taurocholic and glycocholic acid in vitro, Scand. J. Clin. Lab. Invest. 18;643-653.
5. Haselwood, G.A.D., 1962, Bile salts: structure, distribution and possible biological significance as a species character, in: M. Florkin, and H.S. Mason, eds., Comparative Biochemistry, Vol. IIIA, Academic Press, pp. 205-229.
6. Killenberg, P.G., and Jordan, J.T., 1978, Purification and characterization of bile acid-CoA:amino acid H-acyltransferase from rat liver, J. Biol. Chem. 253:1005-1010.
7. Pasantes-Morales, H., Wright, C.E., and Gaull, G.E., 1984, Protective effect of taurine, zinc and tocopherol on retinol-induced damage in human lymphoblastoid cell, J. Nutr. 114:2256-2261.
8. Pasantes-Morales, H., Wright, C.E., and Gaull, G.E., 1985, Taurine protection of lymphoblastoid cells from iron-ascorbate induced damage, Biochem. Pharmocol. 34:2205-2207.
9. Tallan, H.H., Jacobson, E., Wright, C.E., Schneidman, K., and Gaull, G.E., 1983, Taurine uptake by cultured human lymphoblastoid cells, Life Sci. 33:1853-1858.
10. Wright, C.E., Lin, T.T., Lin, Y.Y., Sturman, J.A., and Gaull, G.E., 1985, Taurine Scavenges Oxidized Chlorine in Biological Systems, in: S.S. Oja, L. Ahtee, P. Kontro, and M.K. Paasonen, eds., "Taurine: Biological Actions and Clinical Perspectives", Alan R. Liss, pp. 137-147.

DEVELOPMENT OF BIOSYNTHESIZING AND UPTAKE SYSTEMS FOR TAURINE IN CEREBRAL CORTICAL NEURONS IN PRIMARY CULTURE: ANALYSIS OF POSSIBLE FACTORS INVOLVED IN PERINATAL DECLINE OF CEREBRAL TAURINE

Kinya Kuriyama, Seitaro Ohkuma, Masataka Kishi and Misa Kimori

Department of Pharmacology, Kyoto Prefectural University of Medicine, Kawaramachi-Hirokoji, Kamikyo-Ku, Kyoto 602, Japan

It is well known that taurine (2-aminoethanesulfonic acid), one of most highly concentrated neuroactive amino acids in the mammalian CNS, exhibits a postnatal decline in the brain, which is different from the developmental elevation of the content of other neurotransmitters during the maturation of CNS (1,2,11,15). On the other hand, it has been reported that the activities of cysteine sulfinic acid decarboxylase (CSAD) and cysteine dioxygenase (CD), enzymes involved in taurine biosynthesis from cysteine, show an increase after birth (3,9,12).

Although studies using brain homogenate on the developmental changes in taurine content as well as the activities of enzymes catalyzing the formation of taurine from cysteine such as CSAD and CD have been reported, it is well known that the homogenate consists of different types of cells such as neurons, glial cells and endothelial cells. In addition, it has been reported that the proportion of neurons in brain tissue is only 40% (13). Accordingly, the developmental changes in taurine content found in brain homogenate does not necessarily reflect the changes in neurons during neuronal growth in vivo.

To analyze the developmental changes of the biosynthesizing and uptake systems for taurine in neurons, we have examined the contents of taurine, and cysteine sulfinic acid and cysteic acid (metabolic intermediates in taurine biosynthesis), the activities of enzymes involved in the formation of taurine from cysteine and the uptakes of taurine and cysteine in primary cultured neurons in comparison with those in the cerebral cortices obtained from age-matched fetal and neonatal mice. Additionally, immunohistochemical investigations using antitaurine antibody obtained from rabbit have been carried out for examining the developmental changes of intraneuronal localization of taurine.

MORPHOLOGICAL CHARACTERISTICS OF PRIMARY CULTURED NEURONS

Dissociated cerebral cortical neurons were obtained from the neopallium of fetus of the STD:ddy strain of mice by trypsin treatment and subjected to primary culture as previously described by Ohkuma et al. (10).

The primary cultured neurons used in this study showed bipolar or

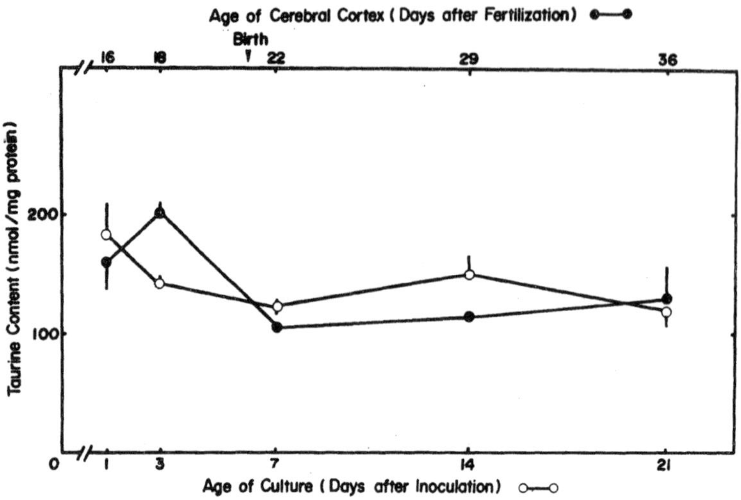

Fig. 1. Developmental pattern of taurine contents in primary cultured neurons and in cerebral cortices. Taurine was assayed by a HPLC equipped with a cation exchanger column following the extraction of taurine with 5% trichloroacetic acid. Each value represents mean ± SEM obtained from 4 to 5 separate experiments.

pyramidal shapes with neuronal processes which form fine networks and aggregated with each other during neuronal growth in vitro. Transmission electron microscopic studies revealed the formation of synaptic contacts which were completed 7 to 10 days after the inoculation, although incomplete synapses with vesicle-filled boutons with no asymmetric membrane thickening were also observed in neurons cultured for 3 days. These morphological investigations clearly indicated that cells cultured on poly-lysine coated surface possess morphological characteristics specific for neurons. In addition, it was found that more than 95% of cells used in this study had no immuno- reactivity to anti-glial fibrillary acidic protein (GFAP), a specific marker for astroglial cells (4). This result also indicates that the contamination of astroglial cells into the cultured neurons used in the present study is negligible.

DEVELOPMENTAL CHANGES OF THE CONTENTS OF TAURINE, CYSTEINE SULFINIC ACID (CSA) AND CYSTEIC ACID (CA) IN PRIMARY CULTURED NEURONS

The contents of taurine, CSA and CA in primary cultured neurons and cerebral cortices obtained from fetus and neonate were determined by the method of Ida and Kuriyama (6). As shown in Fig. 1, the content of taurine in neurons in vitro appeared to be reduced with the growth of neurons. Similarly, the concentration of taurine in the cerebral cortex obtained from fetus and neonate showed a remarkable reduction during the perinatal stage. The contents of CSA and CA in primary cultured neurons also showed a decrease during the first 7 days after the inoculation of cells and remained constant thereafter. Similar developmental patterns were noted in the cerebral cortex of age-matched fetus and neonate in vivo.

IMMUNOHISTOCHEMICAL CHANGES IN PRIMARY CULTURED NEURONS: ANALYSIS USING ANTITAURINE ANTIBODY

To clarify intraneuronal distribution of taurine, immunohistochemical investigations using rabbit antitaurine antibody were carried out. Antitaurine antibody was prepared according to the method previously described (18). For the immunohistochemical study, neurons were stained by the method previously reported (7) with a minor modification. Namely, neurons 1,3,7 and 14 days in culture were rinsed three times with ice-cold Krebs-Ringer bicarbonate buffer (pH 7.4) and then fixed with 0.1 M Dulbecco's phosphate buffer saline (PBS= pH 7.4) containing 4% paraformaldehyde, 0.3% glutaraldehyde and 0.2% picric acid at 4°C for 10 min. The fixed cells were fixed again with 4% paraformaldehyde and 0.2% picric acid dissolved in 0.1 M PBS at 4°C for 30 min followed by washing three times with the same solution used for refixation. After the rinse of fixed cells with 0.1 M PBS containing 15% sucrose, the cells were placed at 4°C for 2 days with 15% sucrose in 0.1 M PBS and then incubated with 0.1 M PBS containing 0.3% Triton X-100 at 4°C for 4 days. The cells were then subjected to treatment by the peroxidase-antiperoxidase (PAP) technique reported by Kimura et al. (7). The neurons fixed were incubated with antitaurine antibody (1/3000) obtained from rabbit at 4°C for 4 days, incubated with goat antirabbit IgG (1/200) for 2 h, incubated with PAP (rabbit, 1/200) for 1 h and finally reacted with 0.025% 3,3'-diaminobenzidine containing 0.005% H_2O_2 and 50 mM Tris-HCl buffer (pH 7.4) for 10 min. Following washing of the cells with Tris-HCl buffer, neurons were stained with hematoxylin and subjected to microscopic observations.

The reactivity to antitaurine antibody was localized in cell bodies and processes of neurons in primary culture, although several cells had no immunoreactivity. The immunoreactivity to antitaurine antibody was invariably found to be localized not only in cell bodies but also in neuronal processes of neurons cultivated in vitro for 1-7 days. Neurons cultured for 14 days possessed positive immunoreactive substances in cell bodies but were practically devoid of reactivity in neuronal processes. In addition, it was

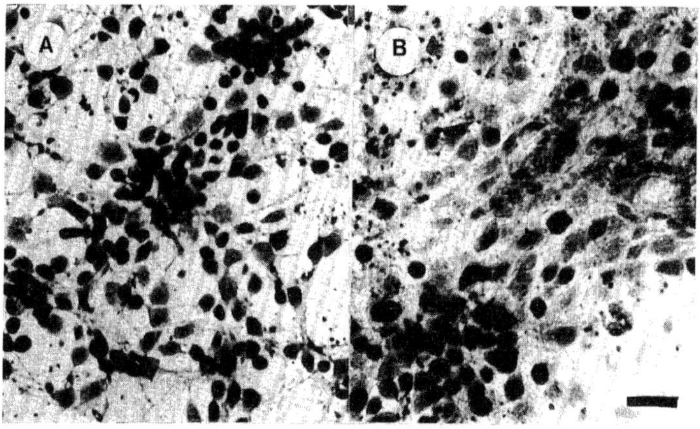

Fig. 2. Immunohistochemical staining of primary cultured neurons using antitaurine antibody. Neurons were fixed with 4% paraformaldehyde, 0.3% glutaraldehyde and 0.2% picric acid, and then placed with 0.3% Triton X-100 in 0.1 M phosphate buffer saline. After washing, cells were treated by the peroxidase-antiperoxidase (PAP) technique with antitaurine antibody (rabbit), goat antirabbit IgG and rabbit PAP. Finally, cells were reacted with 0.025% 3,3'-diaminobenzidine and 0.005% H_2O_2 followed by staining with hemotoxylin. (A) one day and (B) 14 days after the inoculation of neurons. Scale in (B) represents 50 M.

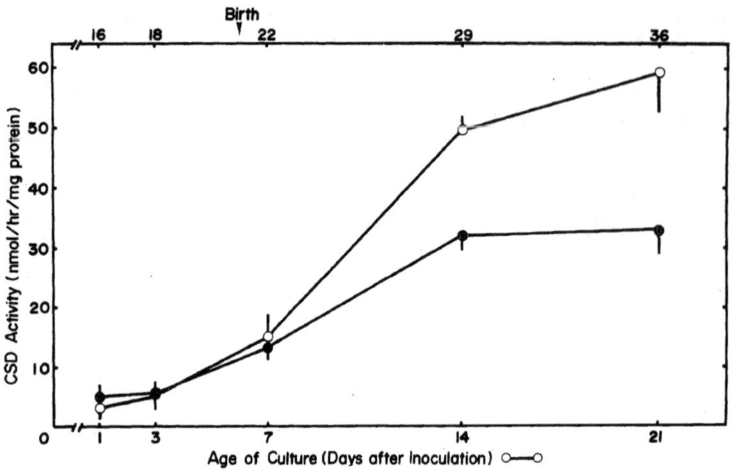

Fig. 3. Developmental pattern of cysteine sulfinic acid decarboxylase (CSAD) activities in primary cultured neurons and in cerebral cortices (days after fertilization ●-●). The activity of CSAD was determined using homogenates of primary cultured neurons and cerebral cortices with N_2-gassed distilled water. The homogenates were incubated with DL-[1-^{14}C]CSA and CSAD activity was assayed by measuring [^{14}C]CO_2 formed from labelled CSA. Each value represents mean ± SEM from 4 to 5 separate experiments.

found that the immunoreactivity in neurons cultured for 1 day was strongest in intensity, whereas that in neurons cultured for 14 days showed the weakest immunoreactivity (Fig. 2). These immunohistochemical changes observed in primary cultured neurons during neuronal growth in vitro correspond well with the changes of taurine content detected by biochemical procedures as shown in Fig. 1.

DEVELOPMENTAL CHANGES OF ACTIVITIES OF CSAD AND CD IN PRIMARY CULTURED NEURONS

The activities of CD (EC 1.13.11.20) and CSAD (EC 4.1.1.29) were measured by the methods of Misra and Olney (9) and Yoneda et al. (19), respectively.

The activity of CSAD in primary cultured neurons showed a significant increase with neuronal development. Especially, the magnitude of the increase in activity was remarkable between the 7th and 14th day after the seeding of cells, reaching a plateau on the 21st day in culture (Fig. 3). A similar developmental pattern of CSAD activity was observed in the cerebral cortices of fetus and neonate. The developmental pattern of CD activity in neurons grown in culture is shown in Fig. 4. Following 3 days of culture, the activity of CD in primary cultured cells began to be elevated, activity increasing almost linearly until 21 days. On the other hand, CD activity found in the cerebral cortices obtained from fetus 18 days after fertilization reached the level observed in the cerebral cortices of neonate immediately after delivery, and showed again a linear elevation thereafter.

These results clearly indicated that the activities of CD and CSAD, enzymes catalyzing the formation of CSA from cysteine and of hypotaurine from CSA and/or taurine from CA, respectively, have a significant increase not

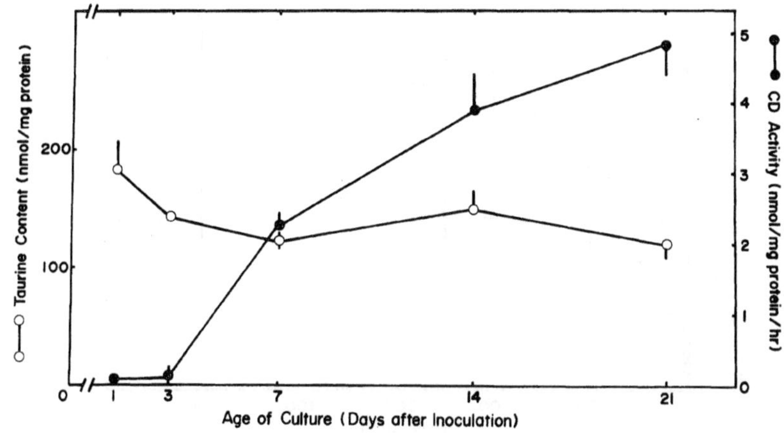

Fig. 4. Developmental patterns of taurine and CD activity in primary cultured neurons. The activity of CD was determined by the measurement of radioactivity of [^{14}C]CSA formed following the separation of [^{14}C]CSA from [^{14}C]cysteine by an anion exchange resin. Each value represents mean ± SEM from 4 to 5 separate experiments.

only in primary cultured neurons but also in cerebral cortices in vivo. The present results also strongly suggest that the observed decline of taurine during neuronal development in vitro (Figs. 1 and 2) may not be due to the decreased capacity of biosynthesizing system for taurine in these neurons.

DEVELOPMENTAL CHANGES OF THE UPTAKES OF TAURINE AND CYSTEINE BY PRIMARY CULTURED NEURONS

The uptakes of taurine and cysteine by primary cultured neurons were determined using [^{3}H]taurine and [^{35}S]cysteine. In brief, primary cultured neurons attached to culture dish were washed 3 times with ice-cold Krebs-Ringer bicarbonate buffer. After preincubating at 37°C for 20 min, the cells were incubated for 3-5 min with Krebs-Ringer bicarbonate buffer containing 10 µM [^{3}H]taurine or 100 µM [^{35}S]cysteine plus 1 mM dithiothreitol at 4°C or 37°C, respectively. Following aspiration of the medium, the cells attached to cultured dish were washed 5 times with ice-cold Krebs-Ringer bicarbonate buffer. The attached cells were then removed by a rubber policeman, transferred to a counting vial and finally subjected to the measurement of radioactivity using a liquid scintillation spectrometer.

The uptake of [^{3}H]taurine by primary cultured neurons was energy and Na^{+} dependent and inhibited competitively by β-alanine, hypotaurine and guanidinoethane sulfonate. Kinetic analysis using a Lineweaver-Burk plot of the uptake indicated that this uptake system consisted of a single component with a high affinity. From these results, it is concluded that the characteristics of taurine transport found in primary culture neurons are essentially identical to those found in various cerebral preparations obtained from mature brain (5). In addition, it has been found that the uptake of [^{3}H]taurine by primary cultured neurons is saturable and exhibits an increase with the development of neurons in vitro (Fig. 5). Furthermore, kinetic analyses of the uptake have indicated that this increase is due to an increase in Vmax value but not in that of Km value. For example, the Vmax value for neurons cultured for 3 days was found to be 0.485 ± 0.074 nmol/mg protein/min, while that for that for neurons cultured for 14 days was 0.764 ± 0.098 nmol/mg protein/min.

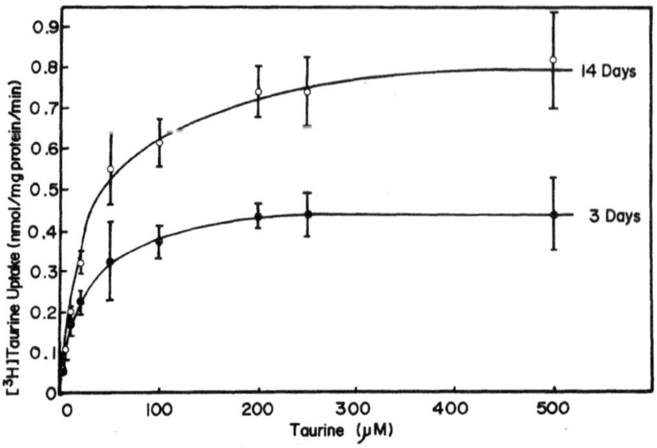

Fig. 5. Rate of [^{3}H]taurine uptake in mouse cerebral cortical neurons in primary culture as a function of external taurine concentration. Each value represents the mean ± SEM obtained from 3 to 4 separate experiments.

The uptake of [^{35}S]cysteine by primary cultured neurons was also found to be Na^+ and temperature dependent, saturable and to consist of a single component with a high affinity. It is also interesting to note that the uptake of [^{35}S]cysteine by these cells is competitively inhibited by CA and CSA but not by taurine, hypotaurine or isethionic acid. In contrast to a progressive increase of [^{3}H]taurine uptake with the time of culture, the uptake of [^{35}S]cysteine by primary cultured neurons showed a decrease with the development of neurons (Fig. 6). This decrease was found to be due to a decrease in Vmax value but not in Km value (Table 1).

The above results suggest that the observed decrease in cysteine uptake with the development of neurons, but not that of taurine, may have an important role in the establishment of the perinatal decline of taurine in the brain, since cysteine is a major precursor for cerebral taurine biosynthesis.

DISCUSSION

It has been reported that taurine content in the brain shows a significant reduction during development, especially at perinatal stages (1,2,11, 15). In the present study, it was also found that taurine content in the brain showed a significant reduction not only in developing primary cultured neurons but also in cerebral cortices obtained from age-matched fetus and neonate. In addition, the decrease in taurine content in neurons during in vitro development was also confirmed in the present immunohistochemical studies using antitaurine antibody. The results obtained in this study strongly suggest that the well known decrease in cerebral taurine during perinatal development (1,2,11,15) may be a reflection of its decrease in neurons, although the possibility that similar changes also occur in other types of cells such as glial cells and endothelial cells is not ruled out. The decrease of taurine content in the brain during developmental stages has attracted the interest of various investigators, and several explanations have been proposed. These hypotheses include greater capacity of immature brain to accumulate taurine as compared with mature brain (8), the increase

Table 1. Developmental changes of kinetic parameters of $[^{35}S]$cysteine uptake in mouse cerebral cortical neurons in primary culture

Period of culture	K_m (μM)	Vmax (nmol/mg protein/min)
3 days	189.73 ± 23.26	4.59 ± 0.22
14 days	151.00 ± 18.61	2.05 ± 0.27*

Each value represents the mean ± SEM obtained from 4 separate experiments. *Significantly different from 3 day value ($p < 0.001$).

of taurine diffusion with cerebral development (3), and the increase of metabolic turnover of taurine with development of the brain (14,16).

The present results have indicated that not only taurine but also the contents of CSA and CA show a significant reduction during developmental tages in both primary cultured neurons and cerebral cortices in vivo. These results also suggest that the activity of the biosynthesizing system for taurine in the brain may decrease during the perinatal stage. On the other hand, it has been found that the activities of CD and CSAD, enzymes catalyzing the formation of CSA from cysteine and of hypotaurine from CSA and/or taurine from CA, respectively, have a significant increase not only in primary cultured neurons but also in cerebral cortices in vivo.

Although the reasons for such a discrepancy are not clear at present, the results obtained in this study strongly suggest that a progressive decrease of the uptake of cysteine, the precursor of taurine biosynthesis, into neuronal cells during developmental stages may account for, at least in part, the perinatal decline of cerebral taurine. Furthermore, we have found in this study that the uptake of cysteine by developing cultured neurons is

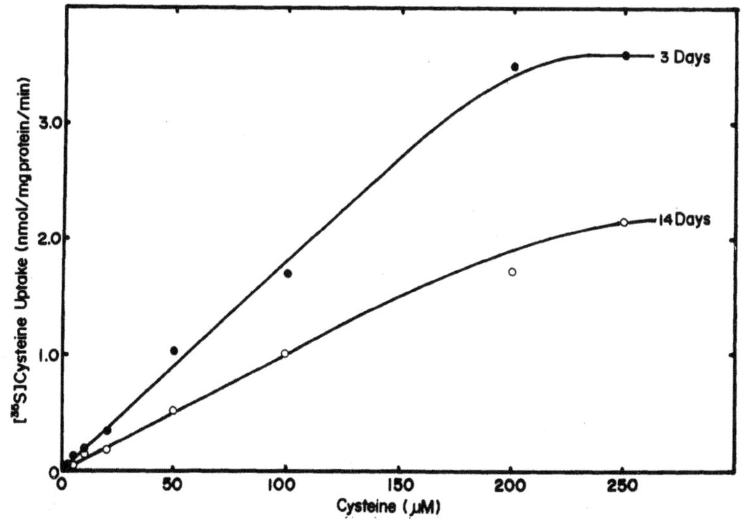

Fig. 6. Rate of $[^{35}S]$cysteine uptake in mouse cerebral cortical neurons in primary culture as a function of external cysteine concentration. Each value represents the mean obtained from 4 separate experiments.

competitively inhibited by CA and CSA, metabolic intermediates of taurine biosynthesis from cysteine. These facts also suggest that such a feedback control mechanism on cysteine uptake may play an important role in the occurrence of perinatal decline of taurine in the brain.

The other factors possibly involved in the perinatal decline of cerebral taurine content may be alterations in the passage of taurine through the blood-brain barrier and in the uptake of taurine by neuronal cells. Concerning the uptake of taurine, however, we have found that there is an increase during development. Therefore, an alteration in cellular transport is unlikely to be a factor in the postnatal decline in taurine content of the brain.

It has been well established that the blood-brain barrier for taurine is present in adult mammals, although it is not completed until 6-10 days after the birth in the case of rat brain. In fact, Sturman et al. (17) found that cerebral taurine content during prenatal and perinatal stages was well correlated with those in maternal blood and milk. These facts suggest that the establishment of a blood-brain barrier for taurine may be a factor related to the occurrence of perinatal decline of taurine.

CONCLUSION AND COMMENTS

Cerebral cortical neurons in primary culture have active biosynthesizing and uptake systems of taurine, and the developmental pattern of these systems in primary cultured neurons seems very similar to those found in age-matched fetal and neonatal brains in vivo. These results strongly suggest that these cultured cells may be a useful experimental model for studying the developmental biology of taurine-containing neurons. The present results also suggest that the well known decrease in cerebral taurine content in vivo during neonatal stages may be due to the decrease of taurine in neuronal cells. The biochemical mechanisms underlying the neonatal decline of taurine in the brain seem likely to involve a decreased capacity for cysteine uptake as well as the establishment of a blood-brain barrier for taurine passage into the brain. It is, however, obviously necessary to employ similar types of experiments on primary cultured and morphologically defined glial and endothelial cells in terms of the developmental patterns of the contents of taurine and related amino acids, biosynthesizing system of taurine and taurine transport, before drawing conclusions on the exact biochemical mechanisms underlying the perinatal decline of the taurine in the brain.

ACKNOWLEDGMENT

This work was supported in part by Grants-in-Aid (Nos 60126005 and 60770175, 1985) from the Ministry of Education and Culture, Japan.

REFERENCES

1. Agrawal, H.C., Davis, J.M., and Himwich, W.A., 1966, Postnatal changes in free amino acid pool of rat brain, J. Neurochem., 13:607-615.
2. Agrawal, H.C., Davis, J.M., and Himwich, W.A., 1968, Developmental changes in mouse brain: Weight, water content and free amino acids, J. Neurochem., 23:917-923.
3. Agrawal, H.C., Davison, A.N., and Kaczmarek, L.K., 1971, Subcellular distribution of taurine and cysteine sulphinate decarboxylase in developing rat brain, Biochem. J., 122:759-763.
4. Bignami, A., and Dahl, D., 1974, Astrocyte specific protein and neuroglial differentiation. An immunofluorescence study with antibodies

to the glial fibrillary acidic protein, J. Comp. Neurol., 153:27-38.
5. Hruska, R.E., Huxtable, R.J., an Yamamura, H.I., 1978, High affinity, temperature-sensitive, and sodium-dependent transport in rat brain, in: "Taurine and Neurological Disorders", A. Barbeau and R.J. Huxtable, eds., Raven Press, New York, pp. 109-177.
6. Ida, S., and Kuriyama, K., 1983, Simultaneous determination of cysteine sulfinic acid and cysteic acid in rat brain by high performance liquid chromatography, Anal. Biochem., 130:95-101.
7. Kimura, H., McGeer, P.L., Peng, J.H., and McGeer, E.G., 1981, The central cholinergic system studied by choline acetyltransferase immunohistochemistry in the cat, J. Comp. Neurol., 200:151-201.
8. Lahdesmaki, P., and Oja, S.S., 1972, Effect of electrical stimulation on the influx and efflux of taurine in brain slices of newborn and adult rats, Exp. Brain Res., 15:430-438.
9. Misra, C.H., and Olney, J.W., 1975, Cysteine oxidase in brain, Brain Res., 97:117-126.
10. Ohkuma, S., Tomono, S., Tanaka, Y., Kuriyama, K., and Mukainaka, T., 1986, Development of taurine biosynthesizing system in cerebral cortical neurons in primary culture, Int. J. Devl. Neuroscience, 4:383-395.
11. Oja, S.S., and Piha, R.S., 1966, Changes in the concentration of free amino acids in the rat brain during postnatal development, Life Sci., 5:865-870.
12. Pasantes-Morales H., Mapes, C., Tapia, R., and Mandel, P., 1976, Properties of soluble and particulate cysteine sulfinate decarboxylase of adult and the developing rat brain, Brain Res., 107:575-589.
13. Pope, A., 1978, Neuroglia: in: "Dynamic Properties of Glial Cells", E. Schoffeniels, G. Franck, L. Hertz, and D.B. Tower, eds., Pergamon Press, Oxford, pp. 13-20.
14. Sturman, J.A., 1973, Taurine pool sizes on the rat: Effects of vitamin B_6 deficiency and high taurine diet, J. Nutr., 103:1566-1580.
15. Sturman, J.A., and Gaull, G.E., 1975, Taurine in the brain and the liver of the developing human and monkey, J. Neurochem., 25:831-835.
16. Sturman, J.A., Rassin, D.K., and Gaull G.E., 1977, Taurine in developing rat brain: Maternal-fetal transfer of [^{35}S]taurine and its fate in the neonate, J. Neurochem., 28:31-39.
17. Sturman, J.A., Rassin, D.K., and Gaull, G.E., 1977, Taurine in developing rat brain: Transfer of [^{35}S]taurine to pups via the milk, Pediatr. Res., 11:28-33.
18. Tomida, Y., Ida, S., Kuriyama, K., and Kimura, H., 1986, Production and characterization of antibodies against taurine and its immunohistochemical application in the CNS, Neurochem. Res., in press.
19. Yoneda, Y., Takashima, S., Hirai, K., Kurihara, E., Yukawa, Y., Tokunaga, H., and Kuriyama, K., 1977, Microassay methods for taurine and cysteine sulfinate decarboxylase activity, Jap. J. Pharmacol., 27:881-888.

TAURINE EFFLUX FROM BRAIN SLICES: POTASSIUM-EVOKED RELEASE IS GREATER FROM IMMATURE THAN MATURE BRAIN TISSUE

Pirjo Kontro and S.S. Oja

Department of Biomedical Sciences
University of Tampere
Box 607
SF-33101 Tampere
Finland

Taurine is a free amino acid abounding in all brain regions, being enriched in synaptic structures (14). The concentration of taurine is very high in the developing brain, decreasing towards adulthood, a pattern in striking contrast to that of other neuroactive amino acids, whose concentrations tend to increase during development (26). It has therefore been suggested that taurine may have a specific regulatory role in brain development (24,32). In order to shed light on the functions of taurine in the developing brain, we compared the mechanisms of spontaneous and potassium-stimulated taurine release from slices prepared from adult and immature brain tissue.

MATERIALS AND METHODS

Materials

[1,2-^{3}H]Taurine (specific activity 1.07 PBq/mol) was obtained from Amersham International, U.K. 2-Guanidinoethanesulfonic acid (GES) was synthesized according to Huxtable et al. (8), taltrimide (2-phthalimido-ethanesulfon-N-isopropylamide) and MY-103 (2-phthalimidoethanesulfon-amide) according to Andersen et al. (1). 6-Aminomethyl-3-methyl-4H-1,2,4-benzothiadiazine-1,1-dioxide (TAG) was a gift from Dr. G. Yarbrough. Other reagents and drugs were from commercial sources. Adult and developing white NMRI mice of both sexes were used throughout.

Efflux experiments

Slices 0.4 mm thick weighing 15-20 mg were prepared with a Stadie-Riggs tissue slicer. The slices were first preloaded for 30 min with 10 µM [^{3}H]-taurine and then superfused at a rate of 0.25 ml/min for 50 min at 37°C in a system in which freely floating shaken slices were kept in a small volume of preoxygenated medium to preserve the viability of the preparations (18). The superfusion medium (Krebs-Ringer-Hepes-glucose) was pooled during the first 20 min and thereafter 2-min fractions (0.5 ml) were collected directly into scintillation vials with a fraction collector (Gilson TDC 220). The reversibility of stimulation was ascertained in a few experiments by re-exposing the slices again to standard superfusion medium from 40 min onwards. At 30 min the superfusion medium was changed to another containing 50 mM K^+ ions (stimulated efflux) or left unchanged (spontaneous efflux). At 30 min the medium was also supplemented with the analogues and drugs tested. After superfusion the slices were weighed, homogenized in ice-cold 5% (w/v) trichloracetic acid solution, centrifuged and the resulting clear supernatants measured for radioactivity.

Estimation of efflux rate constants

Desaturation curves of labelled taurine from the slices were plotted as a function of time on the basis of the radioactivities remaining in the slices after superfusion and recovered in the collected superfusate fractions (17). The efflux rate constants for the time intervals of 20 to 30 min (k_1) and 34 to 50 min (k_2) were computed as negative slopes for the regression lines of the logarithm of radioactivity permillage in the slices vs. superfusion time. The presence of statistically significant differences between the sample means was detected by variance analysis. Comparison of individual means was by Hartley's sequential method of testing (31).

Taurine determination

Taurine was determined by high performance liquid chromatography as described in Kontro and Oja (13).

RESULTS

Spontaneous efflux of preloaded [^{3}H]taurine from the slices was slow. For instance, about 74% of the original radioactivity remained after a 50-min superfusion in the cerebral cortex slices of adult mice (90-day-old) and 92% in the slices from 3-day-old mice. Potassium stimulation (50 mM K^+) clearly enhanced taurine release, the response being gradually reattenuated in low-potassium medium in both developing and adult brain (Fig. 1).

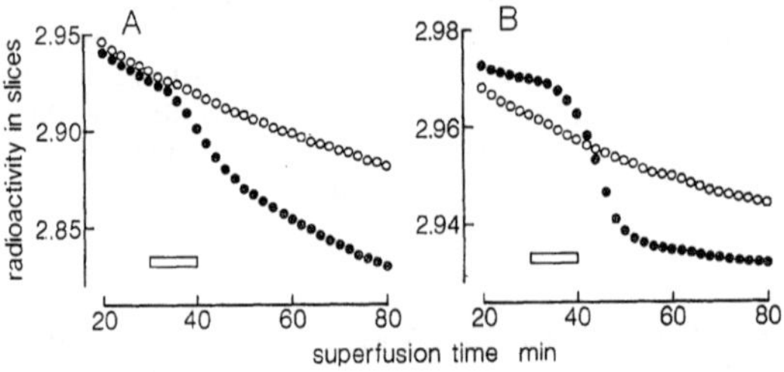

Fig. 1. Spontaneous (-o-) and K^+-stimulated (-•-) release of taurine from cerebral cortex slices of adult (A) and 3-day-old (B) mice. The logarithm of permillage of radioactivity remaining in the slices is plotted against superfusion time. The periods of superfusion with high-K medium is indicated by the horizontal bars. The negative slopes of the straight regression lines fitted to the experimental points for selected intervals represent efflux rate constants given in Tables 2 - 4. Single representative experiments are shown.

The response to K^+-stimulation was slower in both onset and offset but strikingly greater in developing than in adult animals in all brain areas studied, cerebral cortex, cerebellum, striatum and brain stem (Fig. 2). In 7-day-old mice the enhancement of taurine release was about 8-fold in the cerebral cortex, 6-fold in the cerebellum and 4- to 5-fold in the striatum and brain stem. On the other hand, there occurred only negligible K^+-

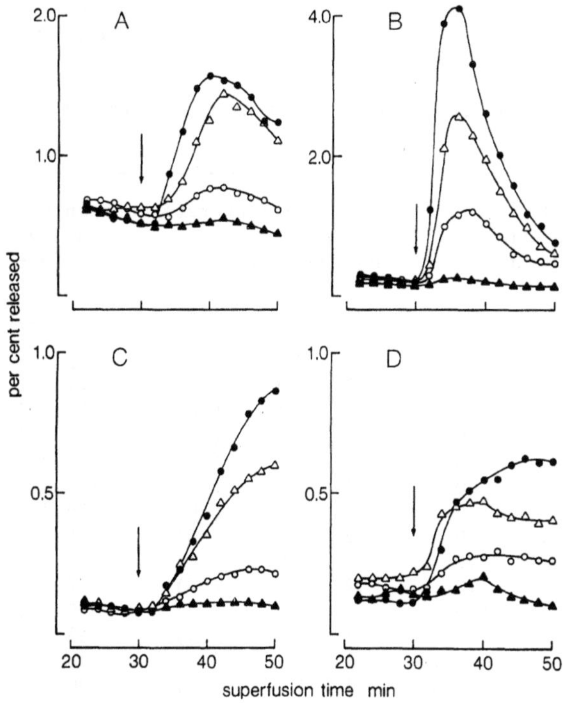

Fig. 2. K^+-stimulated taurine release from slices from adult, 90-day-old (-o-) and 7-day-old (-•-) mice. The high-K medium was superfused from 30 to 50 min (horizontal bars). (A) cerebral cortex, (B) cerebellum, (C) striatum, (D) brain stem. Single representative experiments are shown.

stimulated release from the slices of adult brain stem, 1.5-fold increment in the cerebellum and about 3-fold in the striatum and cerebral cortex. The spontaneous basal efflux from the slices during the interval of 20-30 min was considerably faster in the brain stem than elsewhere but invariably slower in all areas from slices of immature than mature brain (Fig. 2).

Table 1. Release of taurine from cerebral cortex slices of 3- and 90-day-old (adult) mice

Age of mice (days)	Taurine release $\mu mol.kg^{-1}.min^{-1}$ spontaneous		stimulated	
3	4.2 ± 0.6	(5)	46.6 ± 5.6	(6)
90	3.3 ± 0.6	(5)	11.2 ± 2.0**	(5)

The slices were preincubated for 30 min with 10 μM labelled taurine and superfused for 30 min with 5 mM potassium and then for 20 min with 50 mM potassium. The results, means ± SEM, are given per wet weight of incubated slices and refer to the superfusion period of 34 to 50 min. Number of experiments in parentheses. Significance of the difference: **p<0.01.

Table 2. Taurine efflux from cerebral cortex slices from adult and 3-day-old mice

Substance (mM) present in medium from 30 to 50 min		Efflux rate constants Adult k_2 (34-50 min) % of k_1		3-day-old k_2 (34-50 min) % of k_1	
None (control)		78 ± 2	(15)	78 ± 2	(9)
Taurine	1.0	148 ± 7**	(5)	177 ± 25**	(5)
GABA	1.0	160 ± 8**	(4)	153 ± 10**	(4)
Hypotaurine	1.0	191 ± 8**	(5)	155 ± 11**	(5)
β-Alanine	1.0	171 ± 8**	(5)	180 ± 9**	(5)
Homotaurine	1.0	129 ± 2**	(5)	119 ± 10**	(5)
GES	1.0	151 ±11**	(6)	139 ± 12**	(5)
Nipecotic acid	1.0	129 ± 4**	(5)	95 ± 10	(6)
Homocysteic acid	1.0	219 ±21**	(4)	528 ± 77**	(6)
Aspartic acid	1.0	184 ±13**	(7)	257 ± 32**	(4)
Kainic acid	1.0	164 ±16**	(5)	838 ±130**	(7)
Strychnine	0.1	74 ± 8	(4)	83 ± 3	(5)
TAG	0.001	76 ± 3	(6)	78 ± 3	(5)
Haloperidol	0.005	69 ± 6	(4)	90 ± 6	(4)

The slices were preloaded for 30 min in Krebs-Ringer-Hepes-glucose medium, pH 7.4, with 10 μM $[^3H]$taurine and then superfused for 50 min, with the indicated effectors added after 30 min. The results are given as percentages ± SEM of the rate constants (k_1, 20-30 min) for basal efflux from each slice. These constants were $(1.53 \pm 0.02) \times 10^{-3}$ min^{-1} (mean ± SEM, n=80) and $(0.41 \pm 0.02) \times 10^{-3}$ min^{-1} (n=75) for adult and 3-day-old mice, respectively. GES = guanidinoethanesulfonic acid, TAG = 6-aminomethyl-3-methyl-4H-1,2,4-benzothiadiazine-1,1-dioxide. Number of experiments in parentheses. Significance of differences from the corresponding controls: **p<0.01.

Table 1 shows the release of taurine from cerebral cortex slices from 3-and 90-day-old (adult) mice in molar terms calculated with the aid of the specific radioactivity of released taurine, the taurine content in the slices and the estimated efflux rate constants. Despite the severalfold higher taurine content in preincubated slices from 3-day-old than adult mice

Table 3. Potassium stimulated taurine efflux from cerebral cortex slices from adult and 3-day-old mice

Substance (mM) present in medium from 30 to 50 min		Efflux rate constants Adult k_2 (34-50 min) % of k_1		3-day-old k_2 (34-50 min) % of k_1	
Control (high K^+)		264 ± 9	(23)	860 ± 59	(12)
TAG	0.001	225 ± 18*	(6)	515 ± 45**	(5)
TAG	0.1	275 ± 14	(5)	1019 ± 73	(3)
Strychnine	0.1	210 ± 11*	(6)	560 ± 20**	(7)
Taltrimide	1.0	228 ± 17*	(11)	579 ± 37*	(5)
MY-103	1.0	232 ± 13*	(12)	636 ± 46*	(5)
Kainic acid	1.0	308 ± 19*	(7)	1128 ± 109**	(6)
Haloperidol	0.005	368 ± 34**	(6)	760 ± 23	(5)

The slices were preloaded as in Table 2 and then superfused for 50 min, from 30 min onwards with 50 mM K^+ together with the above effectors. The results are given as percentages ± SEM of the rate constants (k_1, 20-30 min) for basal efflux from each slice. These constants were $(1.53 \pm 0.02) \times 10^{-3}$ min^{-1} (mean ± SEM, n=76) and $(0.31 \pm 0.01) \times 10^{-3}$ min^{-1} (n=48) for adult and 3-day-old mice, respectively. TAG = 6-aminomethyl-3-methyl-4H-1,2,4-benzothiadiazine-1,1-dioxide, MY-103 = 2-phthalimidoethanesulfonamide. Number of experiments in parentheses. Significance of differences from the corresponding controls: *$p<0.05$, **$p<0.01$.

(15.5 ± 0.8 and 2.7 ± 0.4 mmol/kg, means ± SEM, n=6, respectively), the unstimulated taurine efflux was of the same order of magnitude in both age groups, only the stimulation-induced taurine release being more than fourfold in 3-day-old mice.

The efflux of taurine from the cerebral cortex slices of both adult and 3-day-old mice was markedly stimulated in the low-potassium (5 mM) medium by exogenous 1.0 mM taurine and several taurine analogs, including GABA, hypotaurine, β-alanine, homotaurine and GES (Table 2). The increment caused by nipecotic acid in mature brain was not discernible in immature brain. The excitatory aspartic, homocysteic and kainic acids enhanced the release more strongly from developing than from adult brain, kainic acid being particularly potent. Of the drugs tested, strychnine and haloperidol had no effect on spontaneous taurine efflux. The proposed taurine antagonist TAG was also without any effect. The K^+-stimulated taurine release was similarly accelerated by kainic acid in the cerebral cortex slices from both adult and 3-day-old mice, whereas haloperidol was stimulatory only in the adults (Table 3). TAG and the glycine antagonist strychnine inhibited the evoked release, as did also the anticonvulsant taurine derivatives taltrimide and MY-103.

The omission of sodium ions increased taurine efflux as much as potassium depolarization in the immature cerebral cortex, but the increment was nearly twofold in the mature cerebral cortex (Table 4). In both preparations 50 mM potassium proved ineffective in the absence of sodium. Veratridine (0.01 mM) caused a stimulation comparable to that evoked by 50 mM K^+ in the adults, whereas the response was relatively weaker in the 3-day-old animals (Table 4). In the presence of veratridine the potassium stimulation was absent in the adults but even greatly potentiated in the 3-day-old mice. An 0.1 mM concentration of verapamil did not significantly affect the spontaneous release, whereas the potassium-evoked release was

Table 4. Effects of cations on taurine efflux from cerebral cortex slices of adult and 3-day-old mice

Medium from 30 to 50 min	Efflux rate constants (± SEM) Adult			3-day-old		
	k_1 $\times 10^{-3}$ min^{-1}	k_2 % of k_1		k_1 $\times 10^{-3}$ min^{-1}	k_2 % of k_1	
Preincubation: Standard						
Na-free	1.54 ± 0.10	471 ± 50	(5)	0.34 ± 0.04	876 ± 85	(6)
Na-free, high K	1.56 ± 0.13	522 ± 50	(8)	0.47 ± 0.05	861 ± 83	(8)
Preincubation: Standard						
Veratridine	1.26 ± 0.09	263 ± 25	(5)	0.21 ± 0.03	309 ± 33	(6)
Veratridine, high K	1.78 ± 0.28	233 ± 2	(6)	0.66 ± 0.03	1287 ± 108**	(6)
Preincubation: Verapamil						
Verapamil	1.58 ± 0.17	81 ± 4	(4)	0.22 ± 0.01	81 ± 4	(4)
Verapamil, high K	1.84 ± 0.26	226 ± 21**	(6)	0.34 ± 0.02	500 ± 26**	(6)
Preincubation: Ca-free, EDTA						
High K	4.30 ± 0.09	117 ± 5	(6)	1.86 ± 0.26	234 ± 32	(4)
High K, high Ca	4.00 ± 0.20	138 ± 7	(7)	1.75 ± 0.31	200 ± 15	(6)
Preincubation: high Ca						
High Ca	1.38 ± 0.12	80 ± 5	(5)	0.27 ± 0.01	83 ± 2	(5)
High Ca, high K	1.01 ± 0.09	422 ± 30**	(5)	0.23 ± 0.03	700 ± 43**	(4)
Preincubation: high Mg						
High Mg	1.18 ± 0.14	78 ± 1	(5)	0.25 ± 0.02	77 ± 4	(5)
High Mg, high K	1.25 ± 0.09	202 ± 22**	(5)	0.26 ± 0.01	562 ± 36**	(5)

The slices were preincubated for 30 min with labelled 10 μM taurine and then superfused for 50 min with the medium changes as indicated. The cation omitted was substituted by equimolar choline. High K = 50 mM, high Ca = 2.4 mM, high Mg = 2.6 mM, EDTA = 0.5 mM, veratridine = 0.01 mM, verapamil = 0.1 mM. The constants k_1 and k_2 refer to superfusion periods of 20–30 min and 34–50 min, respectively. Number of independent experiments in parentheses. Statistical significance of the differences within each pair of experiments: $^{**}p<0.01$.

moderately attenuated in both age groups (Table 4). In the absence of calcium the potassium stimulation of taurine release was markedly reduced in both adult and 3-day-old cerebral cortex (Table 4). However, this was mainly due to an elevation of the basal prestimulation level of efflux in the presence of the calcium chelator EDTA (0.5 mM). A high Ca^{2+} concentration (2.4 mM) was then unable to enhance the release any further. In medium containing only high Ca^{2+}, K^{+}-stimulated taurine release was enhanced only in the mature cerebral cortex, while the high Mg^{2+} concentration (2.6 mM) in medium diminished the K^{+}-evoked release in both age groups (cf. Tables 2 and 4).

DISCUSSION

An enhanced taurine release from cerebral cortex slices could now be elicited by exogenous 1 mM taurine similarly in both adult and developing brain. Earlier a high taurine concentration (10 mM) had been seen to induce taurine release by homoexchange from whole brain synaptosomes of adult mice (9). The efflux was also facilitated by all studied structural analogs of taurine, the most potent being those substances that are also the most potent uptake inhibitors (10,11). Such a close correlation between inhibition of uptake and stimulation of efflux indicates that the efflux is at least partly mediated by the membrane carrier sites operating outwards and already functioning in early postnatal life.

The exact cellular origin of stimulation-released taurine remains open. The slow onset of the release upon stimulation may bespeak glial origin. The possibility of release from glial cells of the slices cannot be entirely excluded, though no potassium-stimulated taurine release has been found from astrocytes cultured from the cerebral cortex (6). The release of GABA induced by veratridine is considered to originate from neurons (22) while potassium-stimulated GABA release may come from glial cells as well (21). In adult mice potassium ions did not accentuate the veratridine-induced release of taurine, whereas in 3-day-old mice veratridine and potassium stimulations were additive. This might be taken to indicate that in mature brain the stimulated release of taurine is from neurons whereas in immature brain either a significant part of the release originates from glia, or immature neurons respond only poorly to veratridine. On the other hand, the distribution of taurine between glial cells and neurons in developing brain is not known. At least in cerebral astrocytes (7), the amount of taurine does not diminish in the same manner as in cerebral neurons (23) in primary culture and in the brain during development in vivo (16).

Strict calcium dependence is thought to be a prerequisite for vesicular but not for cytoplasmic release of synaptic transmitters (21,33). Evaluation as to whether or not the potassium-stimulated taurine release is dependent on calcium is hampered by the great enhancement of the basal taurine efflux in the presence of EDTA in calcium-free media. The calcium chelator, needed for total depletion of cytoplasmic calcium (33), destabilizes plasma membranes by chelating endogenous membrane-bound calcium ions which screen the negative surface charges (5). The permeability of membranes to sodium ions increases, sodium flows in, taurine out, and subsequent potassium stimulation fails to work (19). Similar is the behavior of GABA (3), whose depolarization-induced release is not completely calcium-dependent (2). The enhancement of stimulated taurine release by an elevated (2.4 mM) calcium concentration in adult but not in 3-day-old mice indicates at least a partial calcium dependence in mature brain tissue. On the other hand, a high magnesium concentration and verapamil, a blocker of the voltage-sensitive calcium channels (29), which prevents influx of extracellular calcium into the slices, diminished the stimulated taurine release similarly in both age groups.

Strychnine, TAG and all anticonvulsant drugs tested depressed similarly the potassium-stimulated taurine release in both age groups. The stimulation of taurine release by potassium ions from the cerebral cortex slices has already been shown to be strikingly slow in onset and in offset (20), though gradually fully reattenuated after the cessation of stimulation. The time course was here slow in all brain areas studied and even slower in 3-day-old than adult mice. Recently, the slow time course of potassium-stimulated taurine release has also been demonstrated with striatal neurons in primary culture (27). Could it be possible that taurine acts as a ubiquitous agent that switches off any prolonged excitation in neural networks of the brain? Such an assumption would explain the effectiveness of taurine (25) and several synthetic taurine derivatives (15) against epileptic seizures.

The stimulatory effects of both aspartate and kainate were more pronounced in 3-day-old than adult mice, showing that excitatory agents release taurine more readily from immature than mature brain tissue. In keeping, the combination of 50 mM potassium and 1 mM kainate caused a greater taurine release from the cerebral cortex slices of 3-day-old than adult mice, even though the potassium and kainate effects were directly additive in neither age group. Furthermore, depolarization of excitable membranes by potassium ions and veratridine and by omission of sodium ions from medium invariably evoked relatively more taurine release from the slices of immature brain than from those of mature brain. It is likely that the particularly prolonged and pronounced potassium-stimulated taurine release from immature brain signifies that taurine is an even more efficient regulator of electrical activity in developing than adult brain, as already suggested more than 20 years ago (24). Our present results seem to tally with the recording of inhibitory postsynaptic potentials of extraordinarily long duration in the cerebral cortex (28) shortly after synaptogenesis in situ (4), i.e. at a time when the other inhibitory mechanisms and transmitter systems (GABA and glycine) are not yet functionally matured (30) and cannot account for inhibition. To date, the potassium-stimulated release of both GABA (12) and glycine (P. Kontro and S.S. Oja, unpublished results) has been found rather negligible in all areas of the central nervous system during the early postnatal life of mice.

SUMMARY

The stimulated release of taurine was severalfold greater from slices of immature than mature brain. The release had a strikingly slow time course at all ages, the slower the younger were the animals. It is concluded that taurine is an important regulator of neuronal activity in the developing brain.

ACKNOWLEDGEMENTS

The skilful technical assistance of Miss Riitta Mero and Miss Paula Kosonen, assiduous typing of Mrs Tuija Mäkinen and financial support to Dr Pirjo Kontro from the Research Council for Medicine of the Academy of Finland and from the Scientific Fund of the University of Tampere are gratefully acknowledged.

REFERENCES

1. Andersen, L., Sundman, L.-O., Lindén, I.-B., Kontro, P., and Oja, S.S., 1984, Synthesis and anticonvulsant properties of some 2-aminoethanesulphonic acid (taurine) derivatives, J. Pharm. Sci., 73: 106-108.

2. Arias, C., and Tapia, R., 1986, Differential calcium dependence of γ-aminobutyric acid and acetylcholine release in mouse brain synaptosomes, J. Neurochem., 47: 396-404.
3. Arias, C., Sitges, M., and Tapia, R., 1984, Stimulation of [^{3}H]γ-aminobutyric acid release by calcium chelators in synaptosomes, J. Neurochem., 42: 1506-1514.
4. Crain, S.M., 1977, "Neurophysiologic Studies in Tissue Culture", Raven Press, New York.
5. Hille, B., 1968, Charges and potentials at the nerve surface. Divalent cations and pH. J. Gen. Physiol., 51: 221-236.
6. Holopainen, I., Kontro, P., and Oja, S.S., 1985, Release of preloaded taurine and hypotaurine from astrocytes in primary culture: stimulation by calcium-free media, Neurochem. Res., 10: 123-131.
7. Holopainen, I., Oja, S.S., Marnela, K.-M., and Kontro, P., 1986, Free amino acids of rat astrocytes in primary culture: changes during cell maturation, Int. J. Dev. Neurosci., 4: 493-496.
8. Huxtable, R.J., Laird, H.E., II, and Lippincott, S., 1979, The transport of taurine in the heart and the rapid depletion of tissue taurine content by guanidinoethyl sulfonate, J. Pharmac. Exp. Ther., 211: 465-471.
9. Kontro, P., 1979, Components of taurine efflux in rat brain synaptosomes, Neuroscience, 4: 1745-1749.
10. Kontro, P., 1984, Comparison of taurine, hypotaurine and β-alanine uptake in brain synaptosomal preparations from developing and adult mouse, Int. J. Dev. Neurosci., 2: 465-470.
11. Kontro, P., and Oja, S.S. 1981, Hypotaurine transport in brain slices: comparison with taurine and GABA, Neurochem. Res., 6: 1179-1191.
12. Kontro, P., and Oja, S.S., 1986, Taurine and GABA release from developing mouse brain: effects of cations, Int. J. Dev. Neurosci., 4: S46-S47.
13. Kontro, P., and Oja, S.S., 1987, Taurine and GABA binding in mouse brain: effects of freezing, washing and Triton X-100 treatment on membranes, Int. J. Neurosci., in press.
14. Kontro, P., Marnela, K.-M., and Oja, S.S., 1980, Free amino acids in the synaptosome and synaptic vesicle fractions of different bovine brain areas, Brain Res., 184: 129-141.
15. Kontro, P., Lindén, I.-B., Gothóni, G., and Oja, S.S., 1983, Novel anticonvulsant taurine derivatives, in: "Sulfur Amino Acids: Biochemical and Clinical Aspects", H. Iwata, R.J. Huxtable, and K. Kuriyama, eds., Alan R. Liss, New York, pp. 211-220.
16. Kontro, P., Marnela, K.-M., and Oja, S.S., 1984, GABA, taurine and hypotaurine in developing mouse brain, Acta Physiol. Scand., Suppl. 537: 71-74.
17. Korpi, E.R., and Oja, S.S., 1979, Efflux of phenylalanine from rat cerebral cortex slices as influenced by extra- and intracellular amino acids, J. Neurochem., 32: 789-796.
18. Korpi, E.R., and Oja, S.S., 1984, Comparison of two superfusion systems for study of neurotransmitter release from rat cerebral cortex slices, J. Neurochem., 43: 236-242.
19. Korpi, E.R., and Oja, S.S., 1984, Calcium chelators enhance the efflux of taurine from brain slices, Neuropharmacology, 23: 377-380.
20. Korpi, E.R., Kontro, P., Nieminen, K., Marnela, K.-M., and Oja, S.S., 1981, Spontaneous and depolarization-induced efflux of hypotaurine from mouse cerebral cortex slices: comparison with taurine and GABA, Life Sci., 29: 811-816.
21. Levi, G., 1984, Release of putative transmitter amino acids, in: "Handbook of Neurochemistry, 2nd edn, Vol. 6", A. Lajtha, ed., Plenum Press, New York, pp. 463-509.
22. Levi, G., Gallo, V., and Raiteri, M. 1980, A reevaluation of veratridine as a tool for studying the depolarization-induced release of neurotransmitters from nerve endings, Neurochem. Res., 5: 281-295.

23. Ohkuma, S., Tomono, S., Tanaka, Y., Kuriyama, K., and Mukainaka, T., 1986, Development of taurine biosynthesising system in cerebral cortical neurons in primary culture, Int. J. Dev. Neurosci., 4: 383-395.
24. Oja, S.S., 1966, Postnatal changes in the concentration of nucleic acids, nucleotides and amino acids in the rat brain, Ann. Acad. Sci. Fenn. A5, 125: 1-69.
25. Oja, S.S., and Kontro, P., 1983, Free amino acids in epilepsy: possible role of taurine, Acta Neurol. Scand. 67, Suppl. 93: 5-20.
26. Oja, S.S., Uusitalo, A.J., Vahvelainen, M.-L., and Piha, R.S., 1968, Changes in cerebral and hepatic amino acids in the rat and guinea pig during development, Brain Res., 11: 655-661.
27. Pin, J.-P., Weiss, S., Sebben, M., Kemp, D.E., and Bockaert, J., 1986, Release of endogenous amino acids from striatal neurons in primary culture, J. Neurochem., 47: 594-603.
28. Purpura, D.P., Shofer, R.J., and Scarff, T., 1965, Properties of synaptic activities and spike potentials of neurons in immature neocortex, J. Neurophysiol., 28: 925-942.
29. Schramm, M., and Towart, R., 1985, Modulation of calcium channel function by drugs, Life Sci., 37: 1843-1860.
30. Sedlacek, J., 1978, The development of supraspinal control of spontaneous motility in chick embryo, in: "Progress in Brain Research, Vol. 48, Maturation of the Nervous System", M.A. Corner, R.E. Baker, N.E. van de Poll, D.F. Swaab, and H.B.M. Uylings, eds., Elsevier, Amsterdam, pp. 367-384.
31. Snedecor, G.W., 1956, "Statistical Methods", 5th edn, The Iowa State College Press, Ames, pp. 351-353.
32. Sturman, J., and Gaull, G.E., 1975, Taurine in the brain and liver of the developing human and monkey, J. Neurochem., 25: 831-835.
33. Vizi, E.S., 1984, Physiological role of cytoplasmic and non-synaptic release of transmitter, Neurochem. Int., 6: 435-440.

TAURINE AND PHOTORECEPTOR STRUCTURE: BIOCHEMICAL AND ELECTROPHYSIOLOGICAL STUDIES

H. Pasantes-Morales, O. Quesada, A. Picones and R.López-Escalera

Instituto de Fisiología Celular, Universidad Nacional Autoñoma de México Apartado Postal 70-600, 04510 México, D.F., México

The retina has proved to be one of the most useful experimental models for studies on the physiological role of taurine in animal cells. Some features of the retina, including its large content of taurine, the association of taurine with photoreceptors and particularly, photoreceptor vulnerability to taurine deficiency, have importantly contributed to the advancement in the search for taurine function.

There is now little doubt that taurine is involved in the mechanisms preserving the structural and functional integrity of photoreceptor cells in the vertebrate retina. The supportive evidence comes from 1) in vivo experiments, in taurine deficient animals, in which structural and functional disturbances occur in the retina following a decrease in taurine retinal levels (5,6,11,16) and 2) in vitro experiments, in preparations of isolated photoreceptor outer segments (OS) where taurine affords protection against different experimental conditions causing a disruption of the highly organized structure of the OS (10,12). Although in none of these situations has the mechanism underlying the protective action of taurine been clarified, recent morphological, physiological and biochemical studies have provided further information contributing to our understanding of the role of taurine in photoreceptors.

In this review we present our own contributions on this aspect of taurine research together with a survey of recent progress in the area. In the first part, new results on taurine-deficient animals will be described. In the second part, recent observations of taurine effects on isolated OS will be presented and discussed.

IN VIVO STUDIES: TAURINE DEFICIENT ANIMALS

It is now firmly established that a decrease of taurine retinal levels beyond a critical point results in a retinal degeneration characterized by an alteration of the structural and functional integrity of photoreceptors and other retinal cells. The bioelectrial response, the electroretinogram (ERG), is markedly affected with reduced amplitudes in both the a-wave and the b-wave. The structural organization of photoreceptors is also severely disturbed. Swelling, disorganization and vesiculation of membranes in photoreceptors characterize the retinal degeneration subsequent to taurine retinal depletion. Eventually, cell death occurs resulting in blindness. The degenerative pattern associated with taurine deficiency has been described

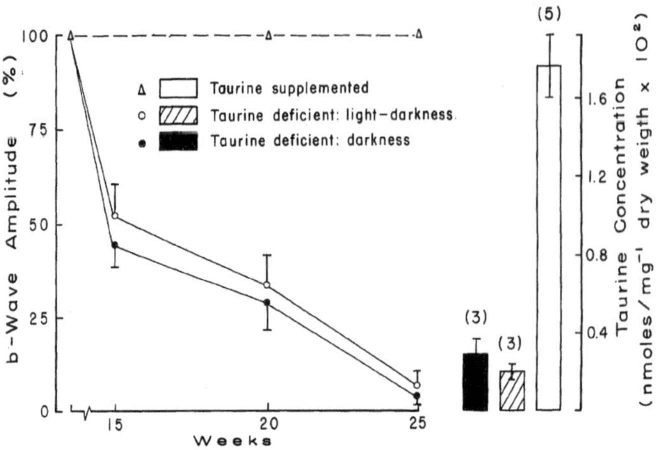

Fig. 1. The effect of light deprivation on taurine levels and on the b-wave amplitude in taurine deficient cats. Electroretinograms were recorded in the dark as described in Pasantes-Morales et al., (Exp. Eye Res., 43,55, 1986). Amplitudes of the b-wave are expressed as percentage of values in taurine-supplemented cats, exposed to a light-dark cycle. The b-wave amplitude in these cats averaged 0.15 ± 0.03 mV. Taurine concentration was determined by high performance liquid chromatography of the fluorescent o-phthalaldehyde ethanol derivative, prepared in borate buffer. Results are the means ± SEM of the number of determinations indicated in parenthesis.

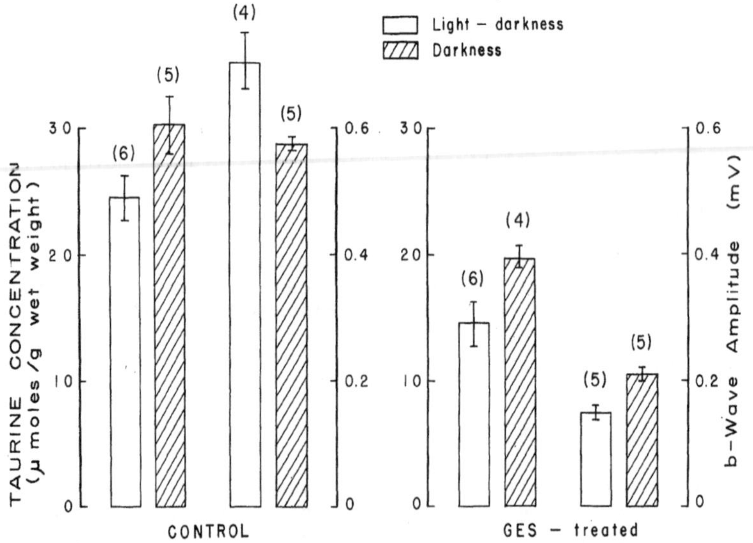

Fig. 2. The effect of light deprivation on taurine levels and on the b-wave amplitude in taurine-deficient rats. Rats were made taurine deficient by administration of GES and exposed to total darkness or to a 12 h light-dark cycle. The ERG was recorded as a function of light intensity. Bars in the figure represent the response at the highest light intensity utilized (10^5 lux). Taurine retinal levels were determined as described in Fig. 1. Results are the means ± SEM of the number of animals indicated in parenthesis.

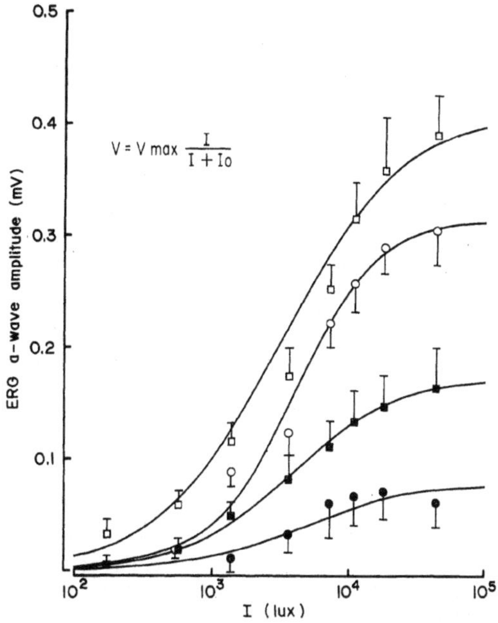

Fig. 3. The effect of light deprivation on the amplitude of the a-wave of the ERG as a function of stimulus intensity in taurine-deficient rats. The experimental conditions are as described in Fig. 2. Normal control, dark (□); control, light-dark (○); GES-treated, dark (■); GES-treated, light-dark (●). The points are the means + SEM of 4 animals. The curves follow the equation shown in inset.

in several species, regardless of the experimental approach used to decrease taurine levels in the retina. Retinal degeneration is observed in cats fed a taurine-free diet (6), in monkeys fed a taurine deficient milk formula (16), in humans receiving long term total parenteral nutrition without taurine (5) and in rats treated with a taurine transport antagonist (11). This latter approach is useful in species in which, even though having a high ability to synthesize taurine, the endogenous pools of taurine in some organs or cells are maintained via an exogenous supply of taurine reaching the cell through active transport mechanisms.

The Effect of Light Deprivation

The mechanisms responsible for the retinal degeneration associated with taurine deficiency are still unclear. We have addressed this question by the study of the influence of illumination on the degenerative changes occuring in taurine-deficient cats and rats. By this approach it is possible to investigate whether the damage is associated with a functional state of the retina related to physiological levels of illumination or if it occurs even when the retina is unstimulated by light. The experiments were carried out in cats fed a taurine-free diet and in rats treated with the taurine antagonist, guanidinoethane sulfonate (GES). To examine the effect of light, animals were kept in a dark room provided with a double door, ventilated and with controlled temperature, with free access to food and water. A dim red light was turned on for about 10 min every day for feeding and cleaning the animals. Another group of animals was maintained under standardized lighting conditions of 12 h light and 12 h dark.

Light deprivation in cats did not modify the decrease of taurine in tissues produced by the dietary restriction. Taurine levels in the retina decreased more than 75% in both light-exposed and light-deprived cats. The amplitude of the b-wave of the ERG also decreased with an identical time course in the two groups (Fig. 1). Also, the structural damage in photoreceptors of taurine-deficient animals was the same, regardless of the lighting condition.

In rats treated with GES and exposed to different illumination conditions, it was observed that taurine levels of animals kept in the dark were higher in both untreated and GES-treated animals as compared with their respective controls exposed to light. After six weeks of light-deprivation, taurine levels in the retina were significantly higher than those in rats exposed to light (Fig. 2). GES treatment produced a similar decrease of about 40% in taurine levels in the two groups but the taurine content in animals kept in the dark were still higher than in GES-treated, light-exposed animals (Fig. 2). The amplitude of the b-wave of the ERG was markedly reduced in GES-treated rats, with similar values in the animals kept in the dark and in those exposed to light (Fig. 2).

Fig. 3 shows the amplitude of the a-wave at different light intensities. GES treatment markedly decreased the response amplitude, particularly at the high intensity stimulus. A close relationship between taurine levels in the retina and the a-wave amplitude was observed for the different groups. The best response was observed in the groups containing the highest taurine concentration, whereas the groups with lower taurine levels showed the poorest response. This was observed regardless of the lighting conditions of the different groups. The points in the V-log I graph fit a curve defined by the relation shown in Fig. 3, which characterizes the vertebrate photoreceptor response (1,8). V represents the response amplitude, V_{max} is the maximal response, I is the stimulus intensity and I_o is a constant representing the half-saturating intensity. I_o values obtained from the curves for the different groups of animals were similar for all groups, suggesting that the sensitivity of the registered response is unaffected by illumination conditions or by GES treatment. In contrast, V_{max} values were markedly lower in GES treated animals for both light-exposed and dark-reared groups of rats (Table 1). Since V_{max} is the parameter affected by light as well as by GES treatment, while I_o remains unchanged, it may be concluded that the number of intact photoreceptors contributing to the response is lower in rats exposed to light and in the taurine-deficient animals but that the response of individual photoreceptors is unaffected. The correlation observed between the functional response and the concentration of the taurine may result from the effect of taurine protecting the structural integrity of photoreceptor cells. This interpretation agrees with the well known structural damage of photoreceptors observed in taurine-deficient animals. From these observations it may be concluded that taurine participates in the generation of a normal functional response by maintaining the structural integrity of photoreceptors, but that the decrease in taurine levels is not affecting the normal functioning of those photoreceptors with an intact morphology. These results also indicate that the presence of taurine in the retina is required for preserving the cell structure and function in the dark as well as during the physiological stimulation by light.

The b-wave is most likely generated at the glial cells (Müller cells). A close relationship between taurine retinal levels and the b-wave amplitude has been reported by Schmidt et al. (15) and is confirmed in our studies in rats. Taurine effects on Muller cells are not documented but a protective action of the amino acid on the cell membrane may explain its requirement for the generation of a normal bioelectrical response.

Table 1. Numerical values for the equation parameters of the theoretical V-log I curve for the a-wave

		V_{max} (uV)	Io (lux)
LIGHT	Control	315	3890
	GES	80	5012
DARK	Control	410	3548
	GES	175	3981

Calculated from least squares fit to the equation shown in Fig. 3.

The Effect of Antioxidants and Membrane Stabilizers

Another approach that may help to clarify the mechanism underlying the requirement for taurine in photoreceptors is the investigation of the effect of conditions or compounds able to counteract the adverse effects of taurine deficiency. The conditions examined in this study include mainly the effect of compounds acting as antioxidants and of substances with a known effect as membrane stabilizers. A protective effect of taurine on membranes damaged by peroxidation processes is now well documented (9,12,23), suggesting that the requirements of taurine may be linked to processes protecting the retina against membrane lipid peroxidation. The strikingly high levels of polyunsaturated fatty acids which characterizes the photoreceptor membrane (18) makes it particularly susceptible to lipid peroxidation. This feature, together with the highly aerobic metabolism of the retina, probably requires the existence of additional protection mechanisms in the retina against the risk of membrane lipid peroxidation. In this study, we examined the effect of antioxidants including cysteine, chromium and selenium on the retinal degeneration associated with taurine deficiency. These compounds were added as supplements to the taurine free diet during the whole treatment of about 25 weeks. The structure of photoreceptors appears somewhat well preserved in these cats as compared to the alteration observed in unsupplemented cats. The question of a possible protection by these antioxidants could not be definitely solved, since taurine levels in these cats, although significantly diminished, were never as low as those in unsupplemented cats. This observation confirms an early report of Berson and coworkers (2) on the presence of higher taurine levels in the retina of cats supplemented with cysteine. An inducible activation of the cysteine sulfinate decarboxylase may account for these observed protection of cysteine against tissue taurine depletion.

The possibility of a protective effect of zinc and tocopherol on the retinal dysfunction associated with taurine deficiency was considered with the described effects of a deficiency in these compounds on the photoreceptor structure as well as on their known actions as membrane stabilizers in a variety of biological preparations (3). Zinc and tocopherol deficiency in rats produce a degeneration of the photoreceptor structure that has many similarities with that developed by taurine deficiency (7,14). Disturbance of the discs membranes, disorientation and vesiculation are observed in zinc- and tocopherol-deficient rats. Tocopherol was administered to taurine-deficient cats from the beginning of the treatment by subcutaneous injection twice a week, whereas zinc was added in the drinking water at a concentration of 200 μM. At the end of the treatment, the ERG was measured. As shown in Fig. 4, zinc and tocopherol provided a partial protection of the a-wave amplitude, whereas no effect was observed on the b-wave amplitude.

The effects of GABA and glycine, which are neutral amino acids occurring naturally in the retina and that are structurally similar to taurine, were examined in terms of ability to protect the retinal structure in taurine-deficient cats. The amino acids were administered in the drinking water at a concentration of 1%. The rationale for this experiment is that the Müller cells, where the b-wave originates, possess active mechanisms for accumulating GABA and glycine (17). In conditions of taurine deficiency and of a surplus of GABA and glycine, the cell might be able to replace taurine by increasing the accumulation of these amino acids. Results in Fig. 4 show that GABA and glycine provide only partial protection Altogether, the results of these experiments, although preliminary, indicate a specific role for taurine on membrane functioning, which cannot be completely replaced by antioxidants, membrane stabilizers or structurally close amino acids.

ISOLATED ROD OUTER SEGMENTS

Another experimental approach which has provided strong support for an involvement of taurine in the mechanisms preserving photoreceptor structure is the study of the protective effect of taurine on isolated frog rod outer segments. This structure is easily detached from the photoreceptor by mechanical disruption. The isolated OS have a relatively simple subcellular structure, mainly constituted by the lamellar discs which are highly organized. The discs are formed from the cell membrane by invaginations, so that the outer side of the cell membrane is identical to the inner side of the discs, and the inner side of the cell membrane is the same as the outer side of the disc membranes. This observation is relevant for the interpretation of the results described below.

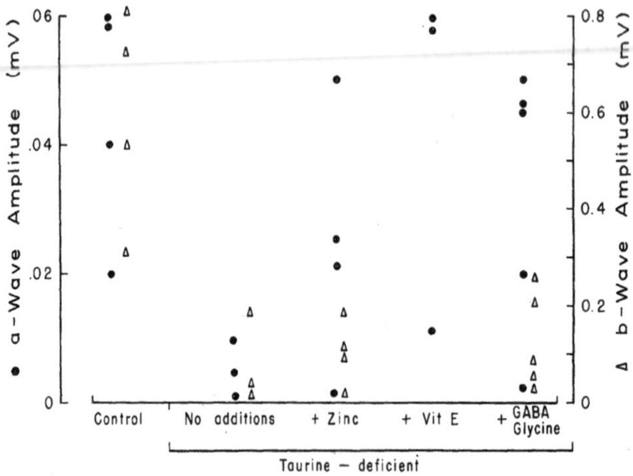

Figure 4. The effect of a taurine-deficient diet supplemented with zinc, GABA and glycine, and tocopherol on the ERG of taurine-deficient cats. Zinc (200 uM) and amino acids (1%) were added to the drinking water. Tocopherol was injected subcutaneously twice a week. Treatment duration was 25 weeks. The ERG recording was as described in Fig. 1. Each point represents the response of a different cat.

Table 2. Total fatty acids in phospholipids of rod outer segment membranes exposed to illumination and in the presence of taurine, hypotaurine and albumin

Fatty Acids (umol/mg protein)	Dark	Light	+Taurine	+ Hypotaurine	+ Albumin
14:0	3 ± 2	3 ± 2	2 ± 2	3 ± 2	2 ± 2
16:0	127 ± 15	116 ± 36	117 ± 23	119 ± 11	112 ± 30
18:0	116 ± 20	110 ± 35	112 ± 27	122 ± 40	99 ± 12
18:1	69 ± 17	63 ± 16	63 ± 12	60 ± 11	54 ± 24
18.2	12 ± 2	10 ± 3	11 ± 3	10 ± 2	11 ± 4
20:4	33 ± 6	33 ± 7	33 ± 7	34 ± 5	35 ± 13
22:5	17 ± 3	20 ± 6	3 ± 8	23 ± 8	19 ± 5
22:6	267 ± 56	310 ± 94	319 ± 95	336 ± 119	287 ± 55
Total	661 ± 91	679 ± 189	690 ± 163	713 ± 148	638 ± 107

Lipids were extracted and purified by the procedure of Folch et al. (J. Biol. Chem., 226, 496, 1957) and free fatty acids and phospholipids were separated by gradient thickness thin-layer chromatography, according to Bazán and Bazán (Res. Methods in Neurochemistry, Plenum Press, New York pp. 309, 1985) and quantitated by gas liquid chromatography. Results are the means ± SD of 6 separate experiments.

Isolated OS, incubated in a Krebs-bicarbonate medium in the dark, maintain their normal size and appearance during several hours. When the incubation occurs in the light (2500 lux), a disturbance of the OS structure is observed. In previous reports we have described that the presence of taurine in the incubation medium at concentrations of 5-25 mM affords a complete protection of the OS structure disturbed by light (10). Swelling of the OS and protrusion of the discs characterize the effect of light. Now we are attempting to identify the processes responsible for the disruption produced by light and in a subsequent step, to investigate the mechanisms underlying the protective effect of taurine. An increase in lipid peroxidation in the OS membranes observed after illumination was effectively counteracted by taurine (12). However on the basis of results in a number of other models of cell damage associated with lipid peroxidation, where taurine protects the structure without antagonizing the peroxidative process (13), it was concluded that the effect of taurine is primarily located at a step previous to the membrane lipid peroxidation. The hypothesis proposing that a release of free fatty acids from the membrane damaged by light might be the trigger for the chain of events leading to membrane lipid peroxidation was supported by investigations reporting increased membrane permeability and swelling in cells exposed to free fatty acids (19,20). This hypothesis proved to be wrong since the analysis of fatty acid composition in OS exposed to light and that in OS kept in the dark showed no significant differences (Table 2).

Investigation on the specificity of taurine effects, and the characteristics of other compounds also able to protect the OS structure, may provide information about the mechanism of taurine action. A number of compounds were tested for their ability to protect the OS structure from the damage caused by light. Table 3 shows that compounds with a marked protective effect comparable to that of taurine include hypotaurine, cysteine, and albumin. Compounds providing partial protection include sucrose, GABA, glycine, β-alanine, mercaptoethanol and mannitol. Substances without effect include amino acids, such as glutamic and aspartic acids and proline, the taurine analogue GES, strychnine, and bicuculline. Among the compounds

Table 3. The effect of amino acids, antioxidants, ion transport inhibitors and other compounds on the structure of OS disturbed by light

Compound	mM	disrupted outer segment (%)
None	25	71
Taurine	25	23
GABA	25	39
Glycine	25	34
Hypotaurine	25	20
β-Alanine	25	51
GES	25	68
Cysteine	25	32
Mercaptoethanol	25	69
Tocopherol	1	58
*Urea	2	80
Sucrose	30	62
Mannitol	30	57
Mersalyl	2	65
*N-Ethylmaleimide	2	100
*Furosemide	1	100
*DIDS	1	72
*PCMBS	1	98

Incubation in Krebs-bicarbonate medium, during 2 h. Light intensity, 2500 lux. The percentage of outer segments with disturbed structure in darkness was 24.6. The asterisk denotes compounds which also produce disruption of outer segments structure in darkness. PCMBS= p-chloromercuribenzene-sulfonate; DIDS= 4,41-diisothiocyanatostilbene-2,2'-disulfonic acid.

examined, some of them caused cell damage by themselves, in the dark. These include urea, the anion-transport inhibitor DIDS, n-ethyl-maleimide, p-chloromercuribenzensulfonate, and furosemide (Table 3). From these results it is concluded that the effect of taurine on protecting the outer segments is quite specific but not exclusive, and that compounds having a structural similarity to taurine are also effective. Thus, β-alanine and hypotaurine worked better than GABA and glycine. Albumin and cysteine may be acting through an effect on SH groups, although the inability of mercaptoethanol to protect the cell structure militates against this possibility (Table 3).

The approach used for the identification of the ionic currents responsible for the swelling and disruption of the OS structure after illumination was to examine the effect of replacement of the major ions in the physiological medium on the alteration caused by light. Table 4 shows that omission of sodium, replacing sodium chloride by choline chloride, has a beneficial effect preventing disruption. The same effect is observed when chloride is replaced in the medium by sodium gluconate. Omission of potassium in the presence of sodium and chloride has no effect. These results suggest that the increase in membrane permeability, which probably subserves the structural damage, requires the simultaneous presence of sodium and chloride. The significance of this observation will be discussed later.

Removal of the divalent cations calcium and magnesium from the incubation medium does not prevent the damage caused by light. On the contrary, OS kept in the dark showed a disruption pattern similar to that of illuminated OS in a complete medium. Swelling and discs protrusion were also

Table 4. Effect of ion replacement on the structure of rod outer segment disrupted by light and by omission of calcium-magnesium

Conditions	Disrupted ROS (%) Light	Ca^{++}-Mg^{++}free
Krebs-bicarbonate*	(9) 71.0 ± 2.6	(4) 67.0 ± 4.2
Krebs-HEPES	(3) 24.8 ± 1.0	(4) 32.3 ± 4.7
Sodium-free (choline chloride)	(3) 18.5 ± 2.1	(4) 27.1 ± 2.3
Chloride-free (sodium gluconate)	(4) 23.6 ± 1.9	(4) 28.2 ± 3.1

Isolated rod outer segments were resuspended in Krebs-bicarbonate medium and exposed to illumination (2500 lux) for 2 h, or to calcium-free magnesium-free medium, containing 100 uM EGTA, for 1 h in darkness. After incubation, about 200 ROS were examined in a light-microscope, and the number of ROS with altered structure was counted. Results are the means ± SEM of the number of experiments indicated in parentheses.
* Composition of Krebs-bicarbonate buffer: 118 mM NaCl, 4.7 mM KCl, 1.2 mM KH_2PO_4, 2.5 mM $CaCl_2$, 1.17 mM $MgSO_4$, 25 mM $NaHCO_3$, 5.6 mM glucose, pH 7.0.

characteristics of the OS exposed to divalent cation-free medium. After one hour of incubation in these conditions, 75% of the OS show an altered structure. Again, taurine affords almost complete protection, maintaining intact the OS structure (Fig. 5). In order to compare the ionic requirements of this new model of structural damage with those of illumination, sodium chloride was replaced by sodium gluconate (chloride-free medium) or by choline chloride (sodium-free medium), and potassium was omitted. Table 4 shows that ionic replacements produced about the same effects on the OS structure as those observed after illumination. The presence of sodium, and chloride is necessary for the cell damage to occur. These observations suggest that swelling observed under the experimental conditions investigated requires the transport of sodium and chloride in a concerted way, most probably accompanied by water entry.

When chloride is replaced by anions more permeant than gluconate, the OS swells very rapidly. Isethionate, acetate, sulfate and nitrate all induce a marked disruption of the OS structure. This observation confirms a previous report of Cohen et al. (4), where a study on the effect of a series of anions and cations on the structure of OS in the rat retina was carried out. The addition of taurine to the incubation medium partially protected the OS structure against the disruption produced by permeant anions. The protective effect of taurine against the anions examined showed differences in efficacy. The highest protection was observed with sodium isethionate, followed by sodium acetate and sodium nitrate. Taurine did not protect the cell structure in the presence of sulfate.

Altogether, these observations confirm an interaction of taurine with the OS membrane, resulting in a regulation of ionic fluxes and/or water entry. The mechanism of this action of taurine is still unclarified. Among the possibilities to be explored, we are considering 1) a modulatory action of taurine on ion cotransport systems, which may be either activated or inhibited by light and in the absence of divalent ions, and 2) a nonspecific interaction of taurine with membranes, probably at the polar heads of phospholipids, creating a barrier for water influx, carried or uncarried by ions. Experiments deviced for providing evidence supporting one or another of these possibilities are now in progress.

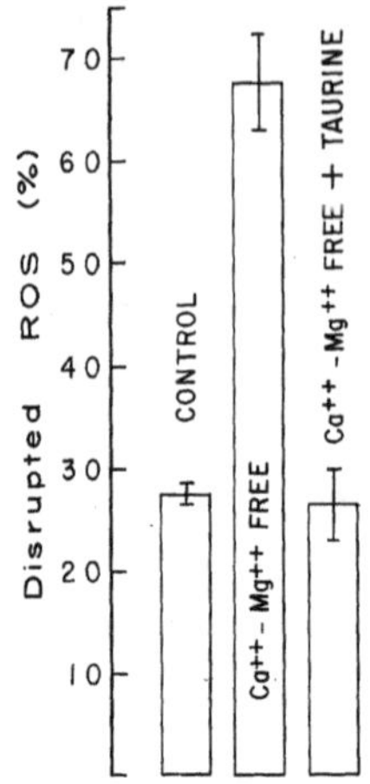

Figure 5. Protective effect of taurine on the disruption of isolated rod outer segments produced by remotion of divalent cations. Frog outer segments were incubated in a Krebs-bicarbonate medium (control) or in a medium without calcium and magnesium, in the presence or absence of taurine (25 mM). After 60 min of incubation in darkness, the number of outer segments intact or with disrupted structure were counted at the light microscope. Results are the means ± SEM of 8 separate experiments.

The protective effect of taurine on OS structure occurs through an interaction of the outer side of the membrane, whereas taurine is found concentrated intracellularly. In preliminary experiments, we have observed that all the experimental conditions producing disruption of the OS structure induce the release of taurine from retinal cells. From these observations we propose that these disrupting conditions generate an intracellular signal - ions, water or a volume change - which in turn induces taurine release. This response may have a dual beneficial effect for the cell: on one hand, the decrease of intracellular taurine contributes to restore the isosmotic condition and on the other hand the released taurine, acting extracellularly, contributes to regulate the membrane permeability, thus terminating the detrimental effects of ion overload or water entry.

ACKNOWLEDGEMENTS

This work was supported in part by grants No. EY0254-07 from NIH and PCSABNA 03775 from CONACYT. We Thank Ms. Y. Diaz de Castro for typing the manuscript.

REFERENCES

1. Baylor, D.A., and Fourtes, M.G., 1970, Electrical responses of single cones in the retina of the turtle, J. Physiol. Lond., 207:77-92.
2. Berson, E.L., Hayes, R.C., Robin, A.R., Schmidt, S.Y., Watson, G., 1976, Retinal degeneration in cats fed casein II. Supplementation with methionine, cysteine and taurine, Invest. Opthalm. Vis. Sci., 15:52-58.
3. Bettger, W.J., and O'Dell, B.C., 1982, A critical physiological role of zinc in the structure and function of biomembranes, Life Sci., 28:1425-1438.

4. Cohen, A.I., 1971, Electron microscope observations on form changes in photoreceptor outer segments and their saccules in response to osmotic stress, J. Cell Biol., 48:547-565.
5. Geggel, H.S., Ament, M.E., Heckenlively, J.R., Martin, D.A., Martin, D.S., and Kopple, J.D., 1985, Nutritional requirements for taurine in patients receiving long term parenteral nutrition, New Eng. J. Med., 312:142-146.
6. Hayes, K.C., Carey, R.E., and Schmidt, S.Y., 1975, Retinal degeneration associated with taurine deficiency in the cat, Science, 188:949-951.
7. Laure-Dupree, A.E., and Bridges, D.B., 1982, Changes in retinal morphology and vitamin A metabolism as a consequence of decreased zinc availability, Retina, 2:294-304.
8. Naka, K., and Rushton, W.A., 1966, S-potentials from luminosity units in the retina fish (Cyprinidae), J. Physiol. Lond., 185:587-599.
9. Nakashima, T., Takino, T., and Kuriyama, K., 1983, Therapeutic and prophylactic effects of taurine administration on experimental liver injury, Prog. Clin. Biol. Res., 125:449.
10. Pasantes-Morales, H., Ademe, R.M., and Quesada, O., 1981, Protective effect of taurine on the light induced disruption of isolated frog rod outer segments, J. Neurosci. Res., 6:337-346.
11. Pasantes-Morales, H., Quesada, O., Carabez, A., and Huxtable, R.J., 1983, Effect of taurine transport antagonists, guanidino ethanesulfonate and β-alanine on the morphology of the rat retina, J. Neurosci. Res., 9:135-146.
12. Pasantes-Morales, J., and Cruz C., 1985, Taurine and hypotaurine inhibit light-induced lipid peroxidation and protect rod outer segment structure, Brain Res., 330:154-157.
13. Pasantes-Morales, H., Wright, C.E., and Gaull, G.E., 1985, Taurine protection to lymphoblastoid cells from iron ascorbate induced damage, Biochem. Pharmacol., 34:2205-2207.
14. Robinson, W.G., Kuwabara, T., and Bieri, J.G., 1982, The roles of vitamin E and unsaturated fatty acids in the visual process, Retina, 2:263-281.
15. Schmidt, S.Y., Berson, E.L., Watson, G., and Huang, C., 1977, Retinal degeneration in cats fed casein. III. Taurine deficiencey and ERG amplitudes, Invest. Ophthalm., 16:673-684.
16. Sturman, J.A., Wen, G.Y., Wisniewski, H.M., and Neuringer, M., 1984, Retinal degeneration in primates raised on a synthetic human infant formula, Int. J. Devel. Neurosci., 2:121-128.
17. Voaden, M.J., 1978, The localization and metabolism of neuroactive aminoacids in the retina, in: "Amino Acids as Chemical Transmitters", F. Fonnun, Ed., Plenum Press, New York, pp. 257-274.
18. Wheeler, T.G., Benolken, R.M., and Anderson, R.E., 1975, Visual membrane: specificity of fatty acid precursors for the electrical response to illumination of the rat retina, Science, 188:131-134.
19. Wiegand, R.D., Giusto, N.M., Rapp, L.M., and Anderson, R.E., 1983, Evidence of rod outer segment lipid peroxidation following constant illumination of the rat retina, Invest. Ophthalm. Vis. Sci., 10:1433-1439.
20. Willmore, L.J., and Rubin, J.J., 1982, Formation of malonaldehyde and focal edema induced by subpial injection of $FeCl_2$ into rat isocortex, Brain Res., 246:113-119.

TAURINE AVAILABILITY AND FUNCTION IN NEUROGENETIC RETINOPATHIES

Charles E. Wright, Harris H. Tallan, Tena R. Wright
Bruce Gillam, Gerald E. Gaull[1] and John A. Sturman

New York State Institute for Basic Research in
Developmental Disabilities, Staten Island, NY 10314 USA

Several neurogenetic diseases manifest abnormal retinal electrophysiology and pathology; such changes in both retinal integrity and function resemble those found in the retina of individuals with retinitis pigmentosa. Some of these inherited neurodegenerative disease include: ceroid lipofuscinosis, Bardet-Beidl-Lawrence-Moon syndrome and Refsum disease, (36). The etiologies of the retinal degeneration observed in these neurogenetic diseases and retinitis pigmentosa remain unknown. Preliminary results, however, from this laboratory (68) and published reports from other investigators (16,21) suggest altered transport of the amino acid taurine in several neurodegenerative diseases.

Many lines of evidence suggest a relationship between retinal tissue integrity and function and the sulfur-containing amino acid taurine. Taurine is exceptionally abundant as the major free amino acid in the vertebrate retina (15,42). Radiochemical incorporation methods show that taurine concentrates in the retinal pigment epithelium and photoreceptor cell layer, with maximal concentrations (80 mM) present in the inner segments and outer nuclear layer (41,43).

Taurine may participate in the mechanism of electromagnetic wave excitation, in that photostimulation of dark-adapted retina (43), and X-ray irradiation of retina induce a rapid release of taurine (38). This release of taurine is reminiscent of the general taurinuria caused by whole body X-ray irradiation (17). Electrophysiologically, taurine reversibly abolishes the b-wave of the electroretinogram (11,35,60). In several animal species, including cat (27), monkey (29,52), rat (44), and man (23,24,48), taurine depletion results in abnormal retinal electrophysiology. Except for man, all of the aforementioned species can develop retinal pathology (27,29,44,52). Recently, cysteine sulfinic acid decarboxylase deficiency has been observed in an individual with bilateral retinitis pigmentosa (45).

Some of the electrophysiological and pathological changes that occur in the retina of taurine-deficient animals resemble those found in the retina of individuals with retinitis pigmentosa and several inherited

[1]Present Address: Department of Nutritional Sciences, The NutraSweet Company, Skokie, Illinois 60076 U.S.A.

neurodegenerative disorders, suggesting that these groups of human eye diseases might be associated with a defect in taurine availability or utilization (9,10,28,61). Published evidence in support of this idea has not been forthcoming. The concentration of taurine in the plasma of individuals with retinitis pigmentosa has been found to be normal (3,10). It has been reported (59) that among several affected individuals, some had normal plasma taurine concentrations but others had a mean plasma taurine concentration only 4% of the normal mean. The reason for this difference is not evident, but may involve variations in type of inheritance or in diet. Measurement of whole blood free amino acids (7) showed normal taurine concentrations in a large group of affected individuals. The taurine values thus obtained probably represented the contents of the platelets and leukocytes, which together contain 90% of the taurine present in whole blood (49). Finally, direct analysis of platelets from individuals affected with retinitis pigmentosa also revealed normal taurine concentrations (3). Thus, it is evident that retinitis pigmentosa occurs in persons with normal concentrations of taurine in plasma, platelets and probably leukocytes.

The possibility remains that the transport of taurine into the retina is defective in persons with retinitis pigmentosa and similar retinal degenerations associated with inherited neurodegeneration. Since retinal tissue is not readily available, uptake measurements have been made with isolated human platelets. Platelets possess multiple active uptake systems (1,12,25,37), including a system similar to that in retina (46). In one study (2), platelet transport of taurine was found to be decreased, while in another investigation (62), platelets from affected patients had normal or increased uptake. Thus, there is no consistent evidence from platelet transport studies that altered taurine uptake mechanisms are involved in the etiology of retinitis pigmentosa. As an alternative, taurine uptake (53) and efflux (54) have been examined in human lymphocyte-derived cultured lymphoblastoid cell lines, cultured human fibroblasts cell strains and cultured human retinal pigment epithelial cells (Wright, unpublished). All three of these types of cultured cells possess a high affinity, active uptake system for taurine similar to the transport systems described in platelets (1,12,25,37), retina (2,46) and retinal pigment epithelium (61). All of these taurine transport systems differ in detail, but demonstrate similarities to one another with regard to kinetic parameters and to the inhibition by various substances. Use of these cultured human cells provides important advantages over primary tissue samples such as platelets and retina. Any inherited defect in taurine transport can be propagated indefinitely, thereby providing an abundant and accessible source of material to study. In addition such cultures can be manipulated in vitro, thus providing a means of examining the biochemical nature of the defect. This experimental strategy has been routinely employed to study certain inborn errors of metabolism; for example gyrate atrophy (36).

RESULTS

Cultured human lymphoblastoid cells acquire taurine from the medium by two processes: a temperature-dependent, sodium-dependent, saturable "active" transport system and a sodium-independent, nonsaturable transport system ("diffusion"). The active transport has properties similar to those reported for taurine transport by other tissues, including retina. The apparent Km of 21 control cell lines is about 15 μM and the apparent Vmax is about 11 pmol/min/10^6 cells; saturation occurs at 100 μM taurine. Sodium-independent transport is proportional to taurine concentrations up to 20 mM. However, at the concentration of taurine in human plasma, 40-100 μM, active transport would provide 90% of the taurine present in these cells.

Similar to lymphoblastoid cells, cultured human fibroblasts derived from individuals with no known genetic defects also acquire taurine from the medium by an active and diffusional processes. For five fibroblast strains so far examined, the apparent Km is similar (range 15-18 uM) cells). The observed increase in Vmax reflects the larger surface area However, the Vmax is significantly greater (range 45-57 pmol/min/10^6 of fibroblast cells, since they average almost eight times the surface areas of lymphoblastoid cells.

In order to verify the in vitro taurine transport results in nonocular cells, the kinetic parameters of taurine uptake have been examined in two cultures of retinal pigment epithelial cells. The mean apparent Km for these cultures were 13.6 ± 1.2 and 16.4 ± 0.8 µM with a mean Vmax of 36.7 ± 3.9 and 40.2 ± 5.6 pmol/min/10^6 cells respectively. These observed values are similar to those found for normal lymphoblastoid cells and cultured fibroblasts. Differences in Vmax among these different types of cells reflect differences in surface area. When calculated with respect to unit surface area the velocity of taurine transport in normal retinal pigment epithelial and normal lymphoblastoid cells are identical, 3.7-3.8 pmol/min/ mm^2. This value is well within the range recently reported by Hussain and Voaden for postmortem retinal pigment epithelium of baboon (30). Thus cultured lymphoblastoid cells and cultured retinal pigment epithelial cells possess similar taurine transport systems with regard to the kinetic parameters, Km and Vmax.

Preliminary studies (Fig. 1) have shown that lymphoblastoid cell lines derived from individuals with either autosomal recessive, autosomal dominant, X-linked retinitis pigmentosa or Usher syndrome a related reti-

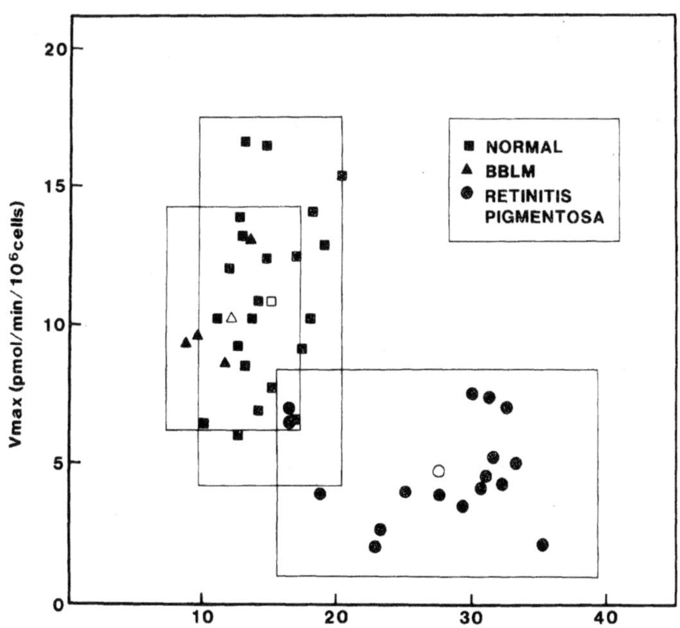

Figure 1. Taurine uptake in lymphoblastoid cells derived from individuals with retinitis pigmentosa. See insert for legends. Open symbols represent the mean values of Km and Vmax. The boxed in area represents $\pm$ two standard deviations.

nitis pigmentosa possessed a decreased affinity for transporting taurine (Km), a decreased rate of taurine uptake (Vmax), or both, as compared to cell lines from clinically unaffected individuals. The mean Vmax for 17 affected cell lines (4.8 ± 2.0 pmol/min/10^6 cells) is lower ($P < 0.005$; Student t test) than the mean Vmax of 21 control cell lines (11.1 ± 3.2 pmol/min/10^6 cells), whereas the mean apparent Km of the affected cell lines (27.5 ± 3.8 µM) is greater ($P < 0.001$) than that for control cell lines (14.9 ± 2.5 µM). These findings indicate that an altered taurine uptake could be a contributory factor in the manifestation of the constellation of retinitis pigmentosa disorders. In order to determine whether altered taurine transport is specific to retinitis pigmentosa, glutamate transport has been examined in lymphoblastoid cell lines derived from three retinitis pigmentosa affected individuals and two normal individuals. No differences were observed in the kinetic parameters, Km and Vmax, for glutamate uptake in normal or affected cells.

In order to determine whether the kinetic parameters of taurine transport could be used as a relatively noninvasive method of preclinical detection of inherited retinopathy, several double blind studies were performed. The first study examined a family presenting a case of autosomal recessive inheritance, with an unaffected son, an affected son and an affected daughter (Fig. 2). Both affected siblings demonstrated a decrease in Vmax when compared to their parents and unaffected sib. In control studies, two normal pedigrees were examined. Neithert family showed alterations in the uptake of taurine when compared to the clinically unaffected population. In a second double blind study (Table 1), two affected males with X-linked retinitis pigmentosa were identified from a group of lymphoblastoid cells derived from an extensive pedigree containing X-linked retinitis pigmentosa. In a third and recent double blind study 72.2% (13/18) of the unkown cell cultures were correctly assigned from a mixed group of six normal and twelve individuals with retinitis pigmentosa (Table 2). Of the five individuals that could not be correctly assigned, three were of the autosomal dominant, Type I form; one was a normal individual assigned as being affected and one individual could not be assigned to either the normal or affected classification.

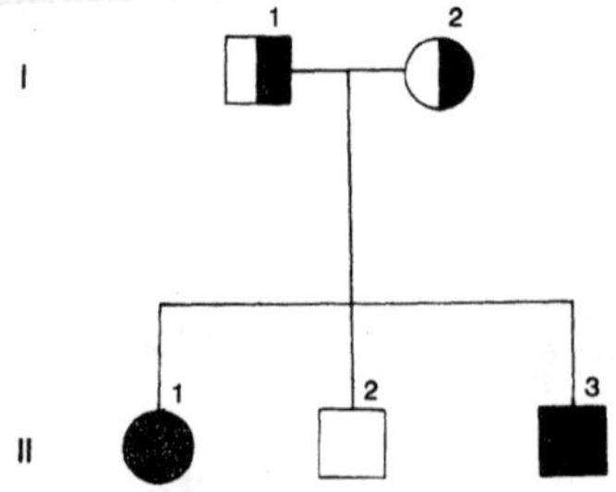

	Vmax (pmols/min/10^6 cells)	Km (µM Taurine)	Vd (nmols/min/10^6 cells)
I-1	11.9	11.4	14.8
I-2	12.3	14.8	9.1
II-1	7.7	16.5	12.1
II-2	16.5	14.7	11.1
II-3	6.5	16.7	12.2

Figure 2. Taurine uptake in a pedigree presenting autosomal receiving retinitis pigmentosa.

TABLE 1. Taurine uptake in sex-linked retinitis pigmentosa

Status	Vmax	Km
normal[a]	11.41	15.73
affected ♂	2.03	17.84
affected ♂	3.91	13.72
heterozygote ♀	8.53	13.14
heterozygote ♀	9.22	9.69
normal ♂	6.84	14.33
normal ♂	7.03	12.94

[a] average values from 35 unaffected individuals

TABLE 2. Assignment of individuals in a double blind study.

Individual	Condition[c]	Normal	Affected
1[a]	normal		+
2	normal	+	
3	AD, Type II		+
4	XL		+
5	XL		+
6	XL		+
7	normal	+	
8	normal	+	
9	AD, Type II		+
10	XL		+
11	XL		+
12[a]	AD, Type I	+	
13[a]	AD, Type I	+	
14[a]	AD, Type I	+	
15	normal	+	
16	XL		+
17[b]	XL		
18	normal	+	

[a] incorrectly assigned; [b] could not be assigned to either group
[c] AD, autosomal dominant; XL, sex-linked

Bardet-Beidl-Lawerence-Moon syndrome is an inherited neurogenetic disease characterized by mental retardation, retinitis pigmentosa and, in some cases, obesity (36). Several cultured lymphoblastoid cell lines derived from individuals with Bardet-Beidl-Lawrence-Moon syndrome have been examined for alterations in taurine transport. In all studies, taurine uptake parameters (Km and Vmax) were within the range of the clinically unaffected population (cf. Fig. 1). These data suggest that the retinal degeneration present in this neurogenetic syndrome may not be associated with taurine availability. However, more individuals should be examined.

Several forms of ceroid lipofuscinosis manifest retinal degeneration (36). Three fibroblast cell strains derived from juvenile ceroid lipofuscinosis and one from the late infantile have been examined for taurine uptake (Fig 3). In four cell strains no changes were observed in the apparent Km (normal, 15-18 µM; juvenile, 14-20 µM; late infantile, 16 µM). However, an eight to ten fold decrease in Vmax was observed in the affected cell strains (normals 45-57 pmol/min/10^6 cells; juvenile 6-8 pmol/min/10^6 cells; late infantile, 6 pmol/min/10^6 cells). These results suggest that an altered taurine transport system may be a contributing factor in the retinal degeneration common to these neurogenerative genetic diseases. These results agree with earlier taurine accumulation studies reported by Dawson (16) for fibroblasts derived from an individual with the infantile form of ceroid lipofuscinosis. In addition, four lymphoblastoid cell cultures established from infantile ceroid lipofuscinosis have been examined with respect to taurine transport. Similar to fibroblast cells from such individuals there was no change in apparent Km (16.3 - 19.4 µM). However, the apparent Vmax was significantly reduced (2.6 - 4.3 pmol/min/10^6 cells). This pattern of reduced Vmax is in accordance to that observed for ceroid lipofuscinosis fibroblast cells and retinitis pigmentosa lymphoblastoid cells.

Refsum disease is an inherited neurodegenerative disease that also manifests retinal degeneration (36). In this disease a defect in phytanic acid β-hydroxylase results in the accumulation of phytanic acid (3,7,11,15-tetramethylhexadecanoic acid) in blood and tissues of affected individuals (51). Two lymphoblastoid cell lines have been examined for uptake of taurine. In neither case did the transport of taurine in the Refsum lymphoblastoid cells differ from control cell lines. The apparent Km was 13.3 µM and the Vmax was 13.1 pmol/min/10^6 cells for one cell line and the apparent Km was 15.8 µm and the Vmax was 17.2 pmol/ min/10^6 cells for the other cell line (see Table 3 for comparison).

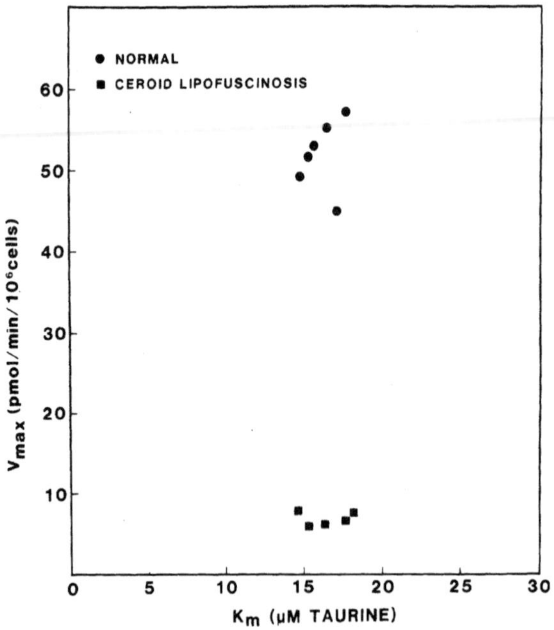

Figure 3. Taurine uptake in fibroblasts derived from individuals with various forms of ceroid lipofuscinosis. See insert for legends.

TABLE 3. Summary of taurine transport in cultured cells derived from inherited retinal degenerations.

Disease[a]	Cell Type	Apparent Km (μM)	Vmax (pmol/min/10^6)
Normal(21)	Lymphoblastoid	15	11
RP(17)	Lymphoblastoid	28	5
BBLM(3)	Lymphoblastoid	17	13
Refsum(2)	Lymphoblastoid	13	13
CL(4)	Lymphoblastoid	17	3
Normal(5)	Fibroblast	16	51
RP(1)	Fibroblast	26	20
CL(3)	Fibroblast	16	6
Normal(2)	Retinal pigment epithelium	15	38

[a] RP, Retinitis Pigmentosa; BBLM, Bardet-Biedl-Lawrence-Moon; CL, Ceroid Lipofuscinosis

Although taurine transport appears to be normal in lymphoblastoid cells from individuals with Refsum disease, a possibility exists that phytanic acid could be inhibiting taurine transport. Phytanic acid accumulates in the plasma and tissues of individuals affected with Refsum disease (51). Preliminary studies indicate that phytanic acid noncompetitively inhibits taurine uptake in cultured human lymphoblastoid cells. The ID50 concentration (the concentration of compound giving 50% inhibition of taurine uptake for 50 μM taurine) of this phytol derivative is approximately 250 μM. At this concentration phytanic acid is nontoxic and is as effective as chloroquine in inhibiting the uptake of taurine (53). The ID50 concentration is at least one log lower than the amounts of phytanic acid present in the plasma of individuals with Refsum disease. This initial experiment suggests that phytanic acid could effectively alter taurine transport, leading to retinal abnormalities present in this neurogenetic disease. In additional studies, the phytanic acid precursor, phytol, had no effect on taurine transport.

DISCUSSION AND SUMMARY

Ocular tissues are excellent targets for damage by oxidants formed by radiation and metabolism. The deleterious effects of light on the eye have been firmly established (26,34,39,50). The vertebrate eye possesses a battery of low molecular weight scavengers and enzyme mediated protective mechanisms against oxidants (5,6,20), yet failure in any one of these mechanisms might result in ocular tissue pathogenesis, due to an inability to cope with the continual production of reduced oxygen species and their cascade of toxic by-products.

Retinal pigment epithelial cells eliminate aged and shed photoreceptor rod outer segments by a mediator-induced phagocytic process, (4, 19,20,40). Phagocytic dysfunction, as observed in the RCS strain of rats (18), results in retinal degeneration. Retinal pigment epithelial cells have many of the same chemical properties and enzyme systems present in other phagocytic cells such as neutrophils, eosinophils and monocytes

(6,8,14,18,19,20), all of which possess high concentrations of taurine. In addition to certain low molecular weight oxidant scavengers, all of these phagocytic cells possess peroxidase which reduce hydrogen peroxide (13,14,33,55-58,63-65). These specific peroxide catalyze the formation of hypochlorous acid (HOCl) from hydrogen peroxide and chloride anion (Cl^-). Hypochlorous acid is a potent oxidizing agent, and its toxic properties are well known. It can oxidize a variety of biologically significant substances such as carbohydrates, nucleic acids, peptides, amino acids and lipids. In addition, HOCl can form secondary chlorinating agents, such as molecular chlorine N-chloroamides, and N-chloroamines (47,55,66,69,70). Most alpha amino acid chloroamines are unstable and spontaneously deaminate, decarboxylate, and dechlorinate to produce a corresponding aldehyde which can form toxic Schiff bases (55-58,70,71). However, amino acids that possess amino groups at other carbon positions, react with HOCl to form relatively stable N-chloroamines (66,67,69,70). Taurine is a beta amino acid which can be chlorinated by the myeloperoxidase/hydrogen peroxide/Cl^- system, thus forming a stable taurine chloramine (55-58,67,69,70).

Many of the biological and chemical properties of taurine make it is an excellent candidate as an intracellular scavenger of hypochlorous acid. Taurine is one of the most abundant free amino acids present in vertebrate tissues (31), including the retina (15,42) lymphocytes (22) and neutrophils (22). Since taurine is a sulfonic acid, it is not normally incorporated into the structure of peptides or proteins. Earlier studies (70) have shown that taurine is a competitive inhibitor of the decarboxylation and deamination of amino acids by myeloperoxidase. Due to its lower oxidizing potential taurine chloramine participates in few reactions, oxidizing reduced sulfhydryls and thioether bonds (55-58,63, 71). Taurine chloramine has also been shown to inhibit myeloperoxidase-catalyzed reactions (58,70).

The presence of a peroxidase in the pigment epithelium of the retina (5,6) a tissue that actively acquires and maintains mM amounts of taurine (15,32,41,42) and has phagocytic function (4,18,20,40), warrants further investigation. To our knowledge, formation of hypochlorous acid by a peroxidase present in retinal tissue and the ability of taurine to scavenge this cytotoxic oxidant has never been examined. A function of taurine in the retina may be to attenuate HOCl produced during the normal phagocytic maintenance of the photoreceptors. Should deprivation of taurine occur, pathological changes in the retina could be initiated or aggravated, for that tissue is particularly vulnerable to taurine deficiency.

REFERENCES

1. Ahtee, L., Boullin, D.J., and Paasonen, M.K., 1974, Transport of taurine by normal human blood platelets, Br. J. Pharmacol. 52: 245-251.
2. Airaksinen, E.M., Sihvola, P., Airaksinen, M.M., Sihvola, M., and Tuovinen, E., 1979, Uptake of taurine by platelets in retinitis pigmentosa, Lancet i:474-475.
3. Airaksinen, E.M., Oja, S.S., Marnela, K.M., and Sihvola, P., 1980, Taurine and other amino acids of platelets and plasma in retinitis pigmentosa, Ann. Clin. Res. 12:52-54.
4. Anderson, D.H., Fisher, S.K., Erickson, P.A., and Tabor, G.A., 1980, Rod and cone disc shedding in the rhesus monkey retina, Exp. Eye Res. 30:559-563.
5. Armstrong, D., Connole, E., Feeney, L., and Berman, E.R., 1978, Peroxidase in the neural retina and pigment epithelium, J. Neurochem. 31:761-769.

6. Armstrong, D., Santangelo, G., and Connole, E., 1981, The distribution of peroxide regulating enzymes in the canine eye, Cur. Eye Res. 1:225-242.
7. Arshinoff, S.A., McCulloch, J.C., Macrae, W., Stein, A.N., and Marliss, E.B., 1981, Amino acids in retinitis pigmentosa, Br. J. Ophthalmol. 65:626-630.
8. Badwey, J.A., and Karnovsky, M.L., 1980, Active oxygen species and the functions of phagocytic leukocytes, Ann. Rev. Biochem. 49: 695-726.
9. Berson, E.L., Hayes, K.C., Rabin, A.R., Schmidt, S.Y., and Watson, G., 1976, Retinal degeneration in cats fed casein: II Supplementation with methionine, cysteine or taurine, Invest. Ophthalmol. 15: 52-58.
10. Berson, E.L., Schmidt, S.Y., and Rabin, A.R., 1976, Plasma amino acids in hereditary retinal disease: Ornithine, lysine and taurine, Br. J. Ophthalmol. 60:142-147.
11. Bolscher, B.G.J.M., Plat, H., and Wever, R., 1984, Some properties of human eosinophil peroxidase: A comparison with other peroxidases, Biochim. Biophys. Acta 784:177-186.
12. Bonaventure, N., Wioland, N., and Mandel, P., 1974, Antagonists of the putative inhibitory transmitter effects of taurine and GABA in the retina, Brain Res. 80:281-289.
13. Boullin, D.J., Airaksinen, E.M., and Paasonen, M.K., 1975, Platelet taurine in Down's Syndrome, Med. Biol. 53:184-186.
14. Buys, J., Wever, R., and Ruitenberg, E.J., 1984, Myeloperoxidase is more efficient than eosinophil peroxidase in the in vitro killing of newborn larvae of Trichinella spirosis, Immunology 51:601-607.
15. Cohen, A.I., McDaniel, M., and Orr, H.T., 1973, Absolute levels of some free amino acids in normal and biologically fractionated retinas, Invest. Ophthalmol. 12:686-693.
16. Dawson, G., 1982, Approaches to the detection of neuronal ceroid lipofuscinosis in cultured skin fibroblasts, in: "Ceroid-lipofuscinosis (Batten's Disease)", D. Armstrong, N. Koppang, J. A. Rider, eds., Elsevier Biomedical Press, pp. 229-240.
17. Dilley, J.V., 1972, The origin of urinary taurine excretion during chronic radiation injury, Radiat. Res. 50:191-196.
18. Edwards, R.B., and Szamier, R.B., 1977, Defective phagocytosis of isolated rod outer segments by RCS rat retinal pigment epithelium in culture, Science 197:1001-1003.
19. Edwards, R.B,. and Bakshian, S., 1980, Phagocytosis of outer segments by cultured rat pigment epithelium. Reduction by cyclic AMP and phosphodiesterase inhibitors, Invest. Ophthalmol. Vis. Sci. 19:1184-1188.
20. Feeney, L., and Berman, E.R., 1976, Oxygen toxicity: Membrane damage by free radicals, Invest. Ophthalmol. 15:789-792.
21. Filla, A., Butterworth, R.F., and Barbeau, A., 1979, Pilot studies on membranes and some transport mechanisms in Friedreich's ataxia, Can. J. Neurol. Sci. 6:285-289.
22. Fukada, K., Hirai, Y., Yoshida, H., Nakajima, T., and Usu, T., 1982, Free amino acid content of lymphocytes and granulocytes compared, Clin. Chem. 29:1758-1761.
23. Geggel, H.S., Ament, M.E., Heckenlively, J.R., and Kopple, J.D., 1982, Evidence that taurine is an essential amino acid in children receiving total parenteral nutrition, Clin. Res. 30:486A.
24. Geggel, H.S., Ament, M.E., Heckenlively, J.R., Martin, D.A., and Kopple, J.D., 1985, Nutritional requirement for taurine in patients receiving long-term parenteral nutrition, N. Eng. J. Med. 312:142-146.
25. Grant, Z.N., and Nauss, C.B., 1976, Uptake of taurine by human blood platelets: A possible model for brain, in: "Taurine", R. Huxtable, A. Barbeau eds., Raven Press, pp. 91-98.

26. Ham, W.T., Jr., Ruffulo, J.J., Jr., Mueller, H.A., Clark, A.M., and Moon, M.E., 1978, Histologic analysis of photochemical lesions produced in rhesus retina by short-wave-length light, Invest. Ophthalmol. Vis. Sci. 17:1029-1035.
27. Hayes, K.C., Carey, R.E., and Schmidt, S.Y., 1975, Retinal degeneration associated with taurine deficiency in the cat, Science 188: 949-951.
28. Hayes, K.C., Rabin, A.R., and Berson, E.L., 1975, All ultrastructural study of nutritionally induced and reversed retinal degeneration in cats, Am.'J. Pathol. 78:505-524.
29. Hayes, K.C., Stephan, Z.F., and Sturman, J.A., 1980, Growth depression in taurine-depleted infant monkeys, J. Nutr. 110:2058-2064.
30. Hussain, A.A., and Voaden, M.J., 1985, Postenucleation survival of taurine uptake by pigment epithelium and chroid of the baboon eye, Exp. Eye Res. 40:643-646.
31. Jacobsen, J.G., and Smith, L.N., Jr., 1968, Biochemistry and physiology of taurine and taurine derivaties, Physiol. Rev. 48:424-511.
32. Kennedy, A.J., and Voaden, M.J., 1974, Free amino acids in the photoreceptor cells of the frog retina, J. Neurochem. 23:1093-1095.
33. Klebanoff, S.J., 1967, Myeloperoxidase-halide-hydrogen peroxide antibacterial system, J. Bacteriol. 95:2131-2138.
34. Kuwabara, T., and Gorn, R.A., 1968, Retinal damage by visible light: An electron m;icroscope study, Arch. Ophthalmol. 79:69-78.
35. Mandel, P., Pasantes-Morales, H., and Urban, P.F., 1976, Taurine, a putative neurotransmitter in retina, in: "Transmitters in the Visual Process", S.L. Bonting, ed., Pergamon Press, pp. 89-105.
36. McKusick, V.A., 1983, "Mendelian Inheritance in Man", Johns Hopkins Univ. Press.
37. Nauss-Karol, C., and VanderWende, C., 1981, High affinity taurine uptake in human blood platelets, in: "The Effects of Taurine on Excitable Tissues", S.W. Schaffer, S.I. Baskin, J.J. Kocsis, eds., SP Medical and Scientific Books, pp. 81-91.
38. Nishimura, C., Ida, S., and Kuriyama, K., 1983, Biosynthesis of Taurine in Frog Retina-differential Effects of Physiological Light and x-ray Irradiation, in: "Sulfur Amino Acids Biochemical and Clinical Aspects", K. Kuriyama, R.J. Huxtable, H. Iwata, eds., Alan R. Liss, pp. 233-250.
39. Noell, W.K., Walker, V.S., Kang, B.S., and Berman, S., 1966, Retinal damage by light in rats, Invest. Ophthalmol. 5:450-473.
40. Ogino, N., Matsumura, M., Shirakawa, H., and Tsukahara, I., 1983, Phagocytic activity of cultured retinal pigment epithelial cells from chick embryo, Ophthalmic Res. 15:72-89.
41. Orr, H.T., Cohen, A.I., and Lowry, O.H., 1976, The distribution of taurine in the vertebrate retina, J. Neurochem. 26:609-611.
42. Pasantes-Morales, H., Klethi, J., Ledig, M., and Mandel, P., 1972, Free amino acids of chicken and rat retina, Brain Res. 41:494-497.
43. Pasantes-Morales, H., Urban, P.F., Klethi, J., and Mandel, P., 1973, Light stimulated release of [^{35}S]taurine from chicken retina, Brain Res. 51:375-378.
44. Pasantes-Morales, H., Quesada, O., Carabez, A., and Huxtable, R.J., 1983, Effects of the taurine transport antagonists, guanidinoethane sulfate, and beta-alanine on the morphology of rat retina, J. Neurosci. Res. 9:135-143.
45. Perry, T.L., Norman, M.G., Yong, V.W., Whiting, S., Crichton, J.U., Hansen, S., and Kish, S.J., 1985, Hallervorden-Spatz Disease: Cysteine accumulation and cysteine dioxygenase deficiency in the globus pallidus, Ann. Neurol. 18:482-489.
46. Schmidt, S.Y., and Berson, E.L., 1980, Postmortem metabolic capacity of photoreceptor cells in human and rat retinas, Invest. Ophthalmol. Vis. Sci. 19:1274-1280.

47. Scully, F.E., Jr, and Bempong, M.A., 1982, Organic N-chloramines: Chemistry and toxicology, Environ. Health Perspect. 46:111-116.
48. Sheikh, K. 1981, Taurine deficiency and retinal defects associated with small intestine bacterial growth, Gastroenterology, 80:1363.
49. Soupart, P., 1962, Free Amino Acids of Blood and Urine in the Human, in: "Amino Acid Pools", J.T. Holden, ed., Elsevier, pp. 220-262.
50. Sperling, H.G., 1980, Are ophthalmologists exposing their patiens to dangerous light levels?, Invest. Ophthalmol. Vis. Sci. 19:989-990.
51. Stanbury, J.B., Wyngaarden, J.B., Fredrickson, D.S., Goldstein, J.L., and Brown, M.S., eds., 1983, "The Metabolic Basis of Inherited Diseases", 5th Ed., McGraw-Hill.
52. Sturman, J.A., Wen, G.Y., Wisniewski, H.M., and Neuringer, M.D., 1984, Retinal degeneration in primates raised on synthetic human infant formula, Int. J. Devel. Neurosci. 2:121-130.
53. Tallan, H.H., Jacobson, E., Wright, C.E., Schneidman, K. and Gaull, G.E., 1983, Taurine uptake by cultured lymphoblastoid cells, Life Sci. 33:1853-1860.
54. Tallan, H.H., and Schneidman, K., 1984, Taurine release from cultured human lymphoblastoid cells, Fed. Proc. 43:1779.
55. Thomas, E.L., 1979, Myeloperoxidase-hydrogen peroxide-chloride antimicrobial system: Nitrogen-chlorine derivatives of bacterial components in bactericidal action against Escherichia coli, Infect. Immun. 23:522-532.
56. Thomas, E.L., 1979, Myeloperoxidase-hydrogen peroxide-chloride antimicrobial system: Effect of exogenous amines on antibacterial action against Escherichia coli, Infect. Immun. 25:110-116.
57. Thomas, E.L., Jefferson, M.M., and Grisham, M.B., 1982, Myeloperoxidase-catalyzed incorporation of amines into protein: Role of hypochlorous acid and dichloramines, Biochemstry 21:6299-2308.
58. Thomas, E.L., Grisham, M.B., and Jefferson, M.M., 1983, Myeloperoxidase-dependent effect of amines on functions of isolated neutrophils, J. Clin. Invest. 72:441-454.
59. Uma, S.M., Satapathy, M., and Sitaramayya, A., 1983, Decreased plasma taurine levels in retinitis pigmentosa, Biochem. Med. 30: 49-52.
60. Urban, P.F., Dreyfus, H., and Mandel, P., 1976,Influence of various amino acids on the bioelectrical response to light stimulation of a superfused frog retina, Life Sci. 18:473-480.
61. Voaden, M.J., 1982, Taurine and Retinitis Pigmentosa, in: "Problems of Normal and Genetically Abnormal Retinas", R.M. Clayton, J. Haywood, H.W. Reading, A. Wright, eds., Academic Press, pp. 353-362.
62. Voaden, M.J., Hussain, A.A., and Chan, I.P.R., 1982, Studies on retinitis pigmentosa in man: I. Taurine and blood platelets, Br. J. Ophthalmol. 66:771-775.
63. Weiss, S.J., Klein, R., Slivka, A., and Wei, M., 1982, Chlorination of taurine by human neutrophils: Evidence for hypochlorous acid generation, J. Clin. Invest. 70:598-607.
64. Weiss, S.J., Lampert, M.B., and Test, S.T., 1983, Long-lived oxidants generated by human neutrophils: Characterization and bioactivity, Science 222:625-628.
65. Wever, R., Plat, H., and Hamers, M.Y., 1981, Human eosinophil peroxidase: A novel isolation procedure, spectral properties and chlorinating activity, FEBS Lett. 123:327-331.
66. Wright, C.E., Lin, T., Lin, Y.Y., and Sturman, J.A., 1984, Taurine reacts with HOCl in cultured cells, Fed. Proc. 43:616.
67. Wright, C.E., Lin, T., Lin, Y.Y., Sturman, J.A., and Gaull, G.E., 1985, Taurine scavenges oxidized chlorine in biological systems, in: "Taurine: Biological Actions and Clinical Perspectives", S.S. Oja, L. Ahtee, M.K. Paasomem, ed., Alan R Liss, pp. 113-147.

68. Wright, C.E., Tallan, H.H., and Gaull, G.E., 1985, Altered taurine uptake in cultured cell lines derived from patients with retinitis pigmentosa, Invest. Ophthalmol. Vis. Sci. 26:Suppl. 132.
69. Wright, C.E., Tallan, H.H., Lin, Y.Y., and Gaull, G.E., 1986, Taurine: Biological update, Ann. Rev. Biochem. 55:427-453.
70. Zgliczynski, J.M., Stelmaszymska, T., Dumanski, J., and Ostrowski, W., 1971, Chloramines as intermediates of oxidation reaction of amino acids by myeloperoxidase, Biochim. Biophys. Acta 235:419-424.
71. Zgliczynski, J.M., Selvaraj, R.J., Paul, B.B., Stelmaszynska, T., Poskitt, P.K.F., and Sbarra, A.J., 1977, Chlorination by the myeloperoxidase-H_2O_2-Cl antimicrobial system at acid and neutral pH, Proc. Soc. Expt. Biol. Med. 154:418-422.

NUTRITIONAL TAURINE DEFICIENCY AND FELINE PREGNANCY AND OUTCOME

J.A. Sturman[1], T. Palackal[1], H. Imaki[1], R.C. Moretz[2], J. French[2], and H.M. Wisniewski[2]

[1]Department of Developmental Biochemistry
[2]Department of Pathological Neurobiology
New York State Office of Mental Retardation and Developmental Disabilities
Institute for Basic Research
1050 Forest Hill Road
Staten Island, New York 10314

The high concentrations of taurine and the function of taurine in excitable tissues have been the subject of speculation for many years. In recent years it has become possible to examine the functions of taurine further by perturbing the concentration in tissues; in the cat by dietary restriction and in rodents by administering the taurine transport inhibitor guanidinoethane sulfonate. Both of these strategies result in reduced tissue taurine concentrations and retinal degeneration. There has been additional interest in the possible functions of taurine in the developing nervous system because taurine concentrations are especially high, taurine generally being the free amino acid present is the greatest concentration in late fetal and early postnatal brain (4). We have explored the role of taurine in development by mating and breeding female cats that had been acclimated to synthetic diets containing various amounts of taurine for long periods of time (more than 6 months). It soon became clear that although taurine-deficient females apparently came into estrus normally and generally appeared to conceive normally, they experienced difficulty in successfully completing pregnancies (5). Reproductive losses included fetal resorption, abortion, stillbirth, and low birthweights (mean birth weight 70 g) of live kittens at term (Fig. 1). Live kittens had a poor survival rate, and grew slowly. Females fed the same diet supplemented with 0.05% taurine did not experience such difficulties (mean birthweight 104 g). Table 1 details the outcome of pregnancies to date.

To date, there has been a maximum of two live kittens born to a taurine-deficient female, and only on one occasion have two kittens survived to weaning. Some taurine-deficient females have never produced a live kitten. In contrast, females consuming the diet containing 0.05% taurine average almost four live kittens per litter. We have observed severe hydrocephalus in two aborted fetuses from pregnant taurine-deficient females and anencephaly in one stillborn kitten (Fig. 2). These severe manifestations may be more frequent because aborted fetuses and stillborn kittens are usually eaten by their mothers, and may account for some of the excessive reproductive losses. The brain taurine concentration at birth in live kittens from taurine-deficient females is 30% of that in brain of kittens from taurine-supplemented females and that in brain of stillborn kittens is even lower at 15% (Table 2). Surviving kittens exhibit abnormal hind leg development and a peculiar gait characterized by excessive abduction and paresis. The hind limbs have diminished bulk and tone in comparison to the fore limbs, or to the hind limbs of age-matched control kittens (Fig. 3). The patellar and achilles tendon jerks are reduced. Maturational reflexes are delayed and their spontaneous activity has a tremulous quality. X-Rays revealed obvious

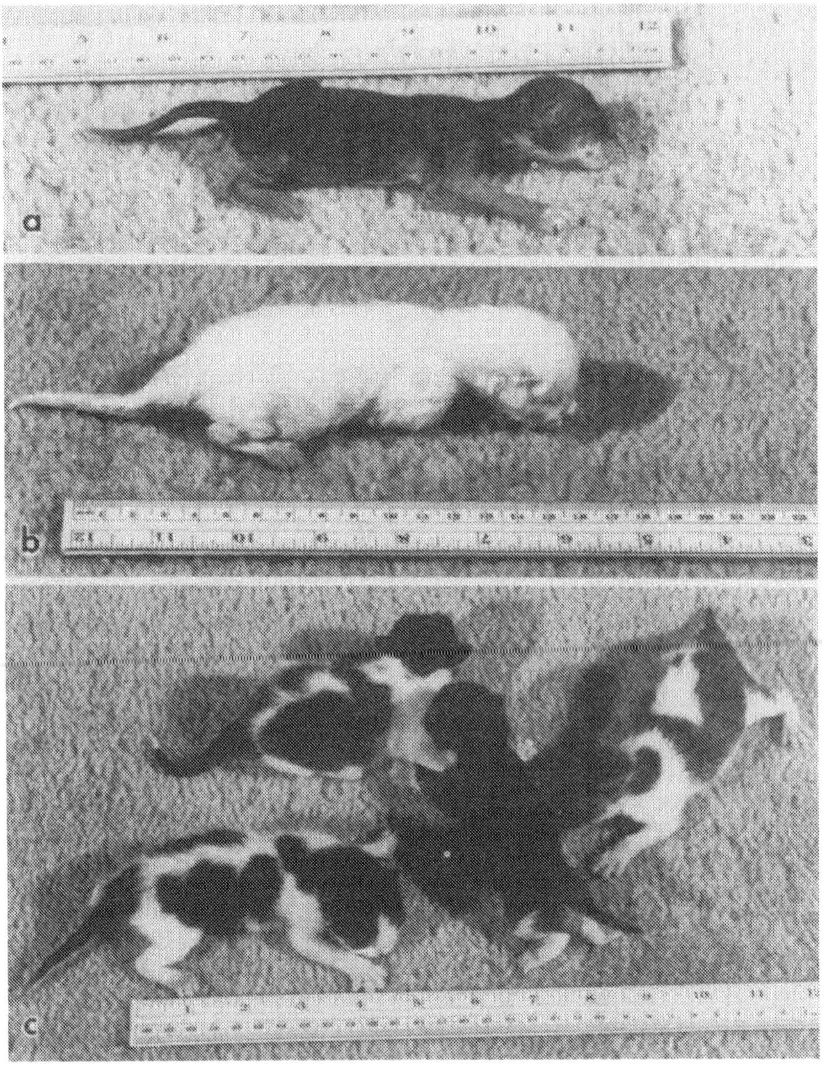

Figure 1. Newborn kittens from taurine-deficient mothers (a) and (b) and from a mother supplemented with 0.05% taurine (c). The taurine-deficient mothers each had only one live kitten compared with the average of 4 live kittens for taurine-supplemented mothers. In addition, the taurine-deficient kittens are smaller and developmentally delayed.

thoracic kyphosis. Morphological examination of the brains of 8-week-old kittens from taurine-deficient females revealed a persistence of cells in the cerebellar external granule cell layer and the presence of numerous mitotic figures in these cells indicating continuing cell division. These abnormalities are accompanied by greatly reduced cerebellar taurine concentrations. In normal kittens, cell division and migration is completed by 3 weeks after birth.

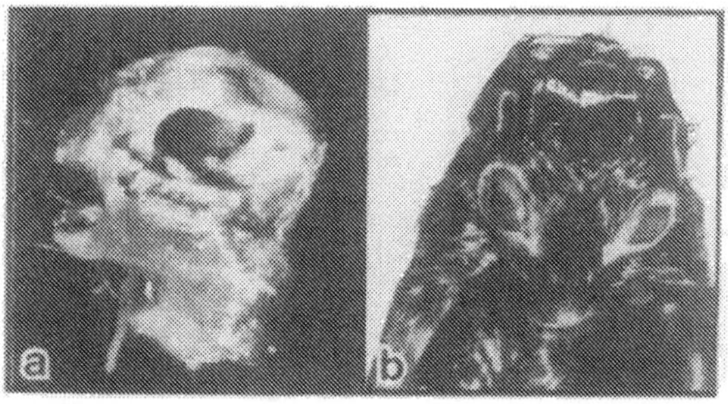

Figure 2. Fetus aborted by a taurine-deficient mother showing extreme hydrocephalus (a) and a full term stillborn kitten from a taurine-deficient mother with anencephaly (b).

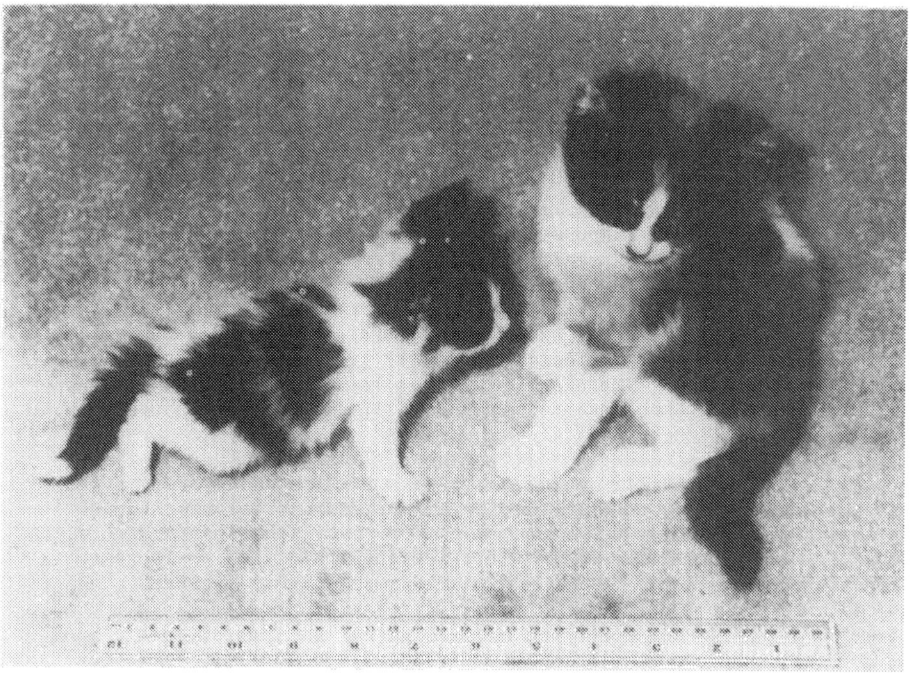

Figure 3. 8-week-old kittens from a taurine-deficient mother (left) and a taurine-supplemented mother (right). Note the difference in size and the abnormal back leg development of the taurine-deficient kitten.

Table 1. Outcome of pregnancies from females fed a taurine-free diet alone or supplemented with 0.05% taurine

Maternal Diet	Pregnancies	To Term	Kittens Stillborn	Kittens live	Survived to weaning
Taurine-supplemented (0.05% taurine)	26	25	0	97	93
Taurine-free	31	15	14	32	14

Examination of the visual cortex of these same 8-week-old kittens, and of newborn kittens, by the rapid Golgi staining technique revealed even more striking abnormalities (Fig. 4) (2). This result was not unexpected because the proliferation of neuroblasts and neuroglioblasts in the visual cortex takes place prenatally. In newborn kittens from taurine-deficient mothers neuroblasts are aggregated both at the ventricular and pial zones, having failed to migrate and differentiate normally. Numerous neuritic fibers oriented in all directions and making extensive interconnections have formed around the cellular plaques suggesting that they are unable to migrate freely. The visual cortex of newborn kittens from mothers supplemented with 0.05% taurine shows normal development with immature cellular plaques at the ventricular and pial regions and more mature cellular plaques showing isolation and differentiation in the middle region. By weaning at 8 weeks after birth the neuroblasts and neuroglioblasts that migrated from the germinative zone have differentiated into neurons and neuroglia and the majority of the embryonic radial glial cells have differentiated into protoplasmic astrocytes that direct and distribute the differentiating neurons to their final destination. Differentiating pyramidal neurons have developed dendritic processes with extensive arborization giving the cortex a fibrous appearance. In contrast, in the kittens from taurine-deficient mothers, some astrocytes remain undeveloped throughout the cortical parenchyma and fibrous astrocytes are scattered among different layers of the cortex indicating that axons have diverged and become randomly distributed before reaching the pial surface. Only a few of the normally expected pyramidal and nonpyramidal neurons are found. Those present have heavily spined dendritic processes with little arborization and low density of synaptic connections which give the cortex a nonfibrous appearance. The changes in the visual cortex of kittens from taurine-deficient mothers are accompanied by substantially reduced taurine concentrations.

It is clear from our measurements of taurine concentrations in numerous brain regions that the whole nervous system becomes taurine-depleted with olfactory bulb and optic nerve being the least affected (Table 3). We have examined the morphology of some other brain regions (frontal lobe, superior colliculus, thalamus) and have found that the abnormalities accompanying taurine deficiency are widespread. In the spinal cord, the dorsal roots are not aligned in the correct fashion and the proportion of white matter is noticeably reduced (Fig. 5). Our preliminary results suggest that this abnormality is most severe in the cervical region of the spinal cord.

Ophthalmoscopic examination of kittens from taurine-deficient mothers revealed no obvious abnormalities, at least up to weaning at 8 weeks after birth, but ultrastructural examination revealed a number of abnormalities in the retina and tapetum (1). The most common alteration in the retina was a reduction in the thickness of the outer segments of the photoreceptors and a derangement of the regular columnar configuration of the outer segments (Fig. 6a). More severe forms of degeneration of the outer segments also were observed, but they were less widespread than those found in long-term taurine-deficient adults (Fig. 6b).

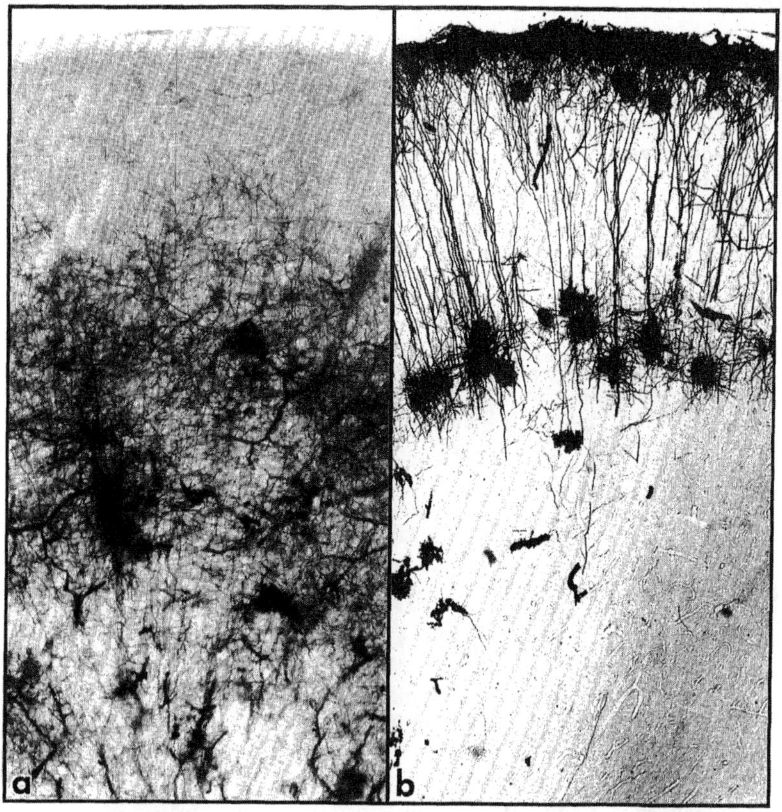

Figure 4. 80 µM-thick sections of silver-impregnated visual cortex of 8-week-old kittens using the rapid Golgi method. (a) From a taurine-supplemented mother showing extensive arborization of axonal and dendritic processes which obscure the pyramidal neurons. Protoplasmic astrocytes with elaborate radial processes are scattered throughout. (b) From a taurine-deficient mother showing few pyramidal neurons which have thick, spiny dendritic processes directed towards the pial surface and little arborization. Magnification X 96.

The tapetum of taurine-deficient kittens was found to include many cells containing, in addition to the normal tapetal rods, electron-dense droplets, not only between the melanocytes but also in the outer and mid levels (Fig. 7). Cells with large numbers of these droplets had a less regular shape. Other changes noted were tapetal rods with greatly dilated limiting membranes, irregularly-shaped vesicles filled with granular and membranous materials (Fig. 8). Some tapetal cells contained densely-packed rods within a markedly dense cytoplasmic matrix accompanied by other degenerative signs. These changes suggest an abnormal development as well as degeneration of tapetal cells in the kittens.

Table 2. Taurine concentration in newborn kitten brain

Source of kittens	Occipital cortex	Cerebellum
	(µmole/g wet weight)	
Taurine-supplemented mothers (12)	6.7 ± 1.0	8.0 ± 0.5
Taurine-deprived mothers (14) (live births)	1.6 ± 1.0	2.4 ± 0.3
Taurine-deprived mothers (6) (stillborn)	1.1 ± 0.3	1.1 ± 0.4

Each value represents the mean ± SEM from the number of kittens in parentheses.

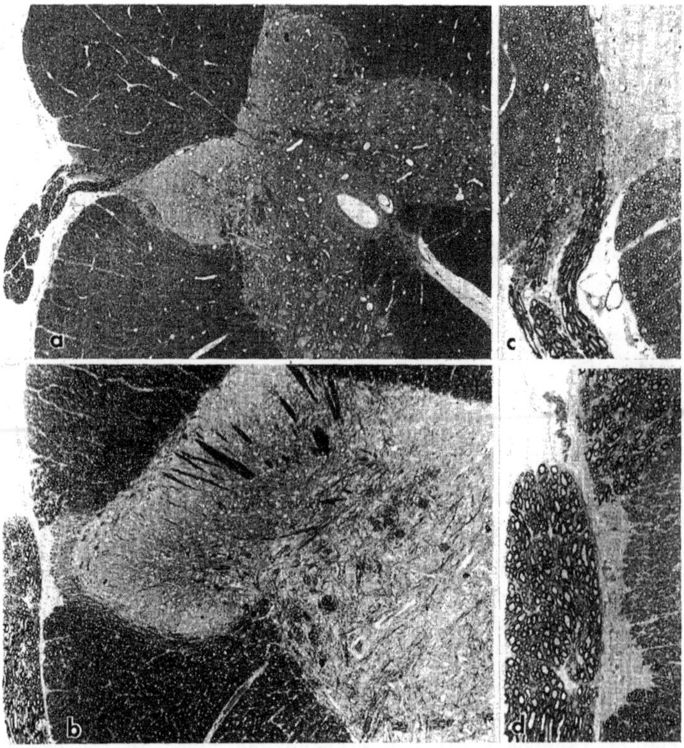

Figure 5. µM sections stained with toluidine blue of the T8 region of spinal cord from 8-week-old kittens. (a) From a taurine-supplemented mother showing normal entry of the dorsal root into the spinal cord. (b) From a taurine-deficient mother showing failure of the dorsal root to enter the spinal cord. This observation was confirmed in serial sections through the whole thickness of the root. The greatly reduced proportion of white matter is also evident. Magnification X 16. (c) and (d) The same sections showing the root and spinal cord in greater detail. Magnification X 24.

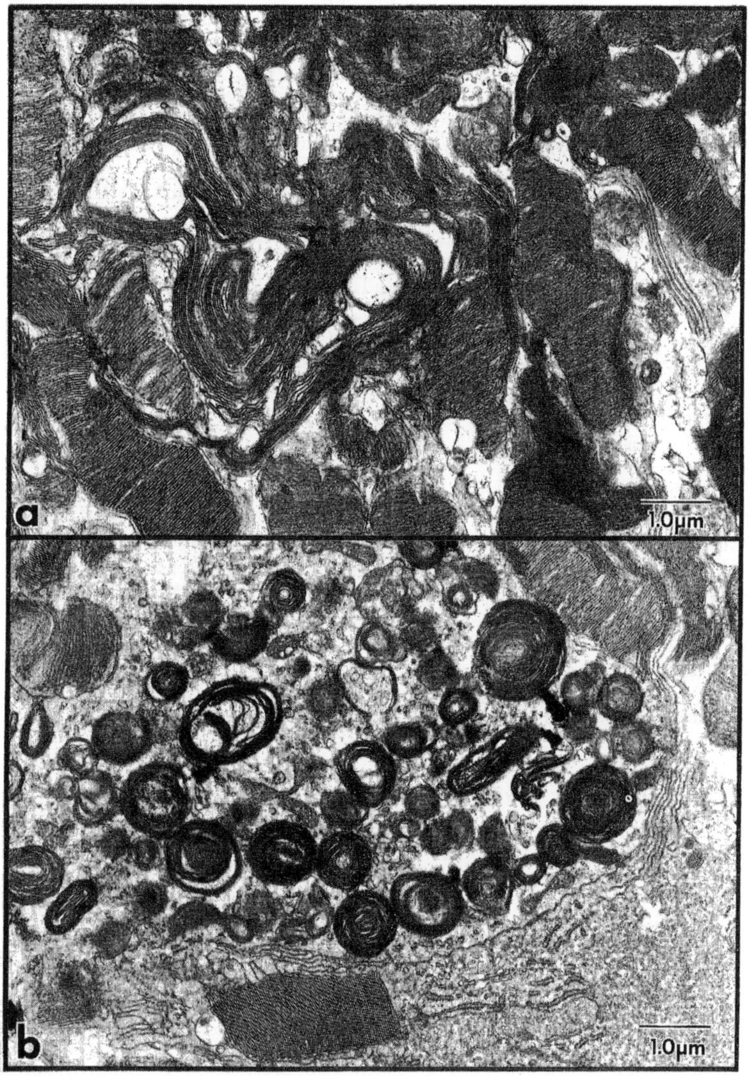

Figure 6. Electron micrographs of transverse sections through the outer retina of 8-week-old taurine-deficient kittens. (a) Showing moderately severe degeneration in which the outer segments are misaligned and many of the disc membranes are fragmented and vesiculated. (b) Advanced degeneration and other cell debris in the process of being phagocytosed.

Finally, we examined a number of biochemical parameters in the mother's milk and in the tissues of the kittens (3). In milk, there was no difference in free amino acid concentration (other than taurine), total protein (Table 4), or amino acid composition of the protein. In 8-week-old kitten tissues, aside from a 40-80% higher concentration of cystathionine in nervous tissue of taurine-deficient kittens, there were no differences in free amino acids. There were no significant differences in the activities of several enzymes in liver and brain involved in the biosynthesis of taurine (Table 5) suggesting that kittens are not capable of compensating metabolically for a dietary deficiency of taurine.

Table 3. Concentration of taurine in nervous tissue of 8-week-old kittens from mothers fed a synthetic diet alone or supplemented with 0.05% taurine

Tissue	Taurine concentration Synthetic diet + 0.05% taurine	Synthetic diet alone
	(µmole/g wet weight)	
Occipital lobe	5.33 ± 0.87	0.96 ± 0.28
Frontal lobe	4.85 ± 0.45	0.76 ± 0.41
Temporal lobe	5.54 ± 0.82	1.08 ± 0.42
Parietal lobe	4.99 ± 0.47	1.23 ± 0.63
Cerebellum	5.36 ± 1.00	1.48 ± 0.65
Superior colliculus	4.47 ± 1.20	0.68 ± 0.17
Inferior colliculus	3.44 ± 0.52	0.69 ± 0.23
Hippocampus	5.29 ± 0.93	0.89 ± 0.24
Corpus callosum	5.52 ± 0.78	0.98 ± 0.42
Thalamus	3.75 ± 1.08	0.63 ± 0.22
Pons	2.71 ± 0.76	0.61 ± 0.23
Medulla	2.67 ± 0.22	0.70 ± 0.31
Olfactory bulb	8.61 ± 1.14	4.11 ± 1.06
Lateral geniculate nucleus	4.97 ± 0.90	0.56 ± 0.10
Optic tract	4.23 ± 1.18	0.98 ± 0.26
Optic nerve	6.62 ± 1.02	2.47 ± 0.21
Spinal cord	2.34 ± 0.81	0.38 ± 0.11
Sciatic nerve	3.32 ± 0.58	0.65 ± 0.26

Each value represents the mean ± SD from 6 to 12 kittens.

Table 4. Total protein content of milk

Measure and analysis	Synthetic diet + 0.05% taurine	Synthetic diet alone
Protein content, mg/ml	53.7 ± 14.4 (136)	56.1 ± 20.4 (70)
Slope of best line of fit	0.39	0.22
Correlation coefficient	0.44	0.17
Value at birth (0 d)	43.3	49.9
Value at weaning (8 wk)	64.8	61.6

These data were obtained from the number of samples in parentheses by using a standard computer program for linear regression. These samples were evenly distributed between birth (0 d) and weaning (8 wk). The mean value of all samples is presented ± SD.

Our present studies include measurement of all of the above-mentioned parameters in kittens from mothers fed diets supplemented with 0.005%, 0.01% and 0.02% taurine in order to determine the effects of marginal taurine deficiency. Results to date indicate that female cats fed the diet supplemented with 0.005% taurine are taurine-deficient, suffer from retinal degeneration, and experience the same difficulties with pregnancy and outcome as females fed the taurine-free diet. Recently we were able to salvage three of four fetuses aborted by a female on this dietary regimen at approximately 50 days gestation (Fig. 9). The great variation in body size and weight indicates that normal in utero development does not take place with insufficient taurine, and probably is the reason that taurine-deficient mothers have a high incidence of resorption and abortion, and usually can give birth to only one or two live kittens. The concentration of taurine in these fetuses was extremely low (less than 1 μmole/g for all tissues).

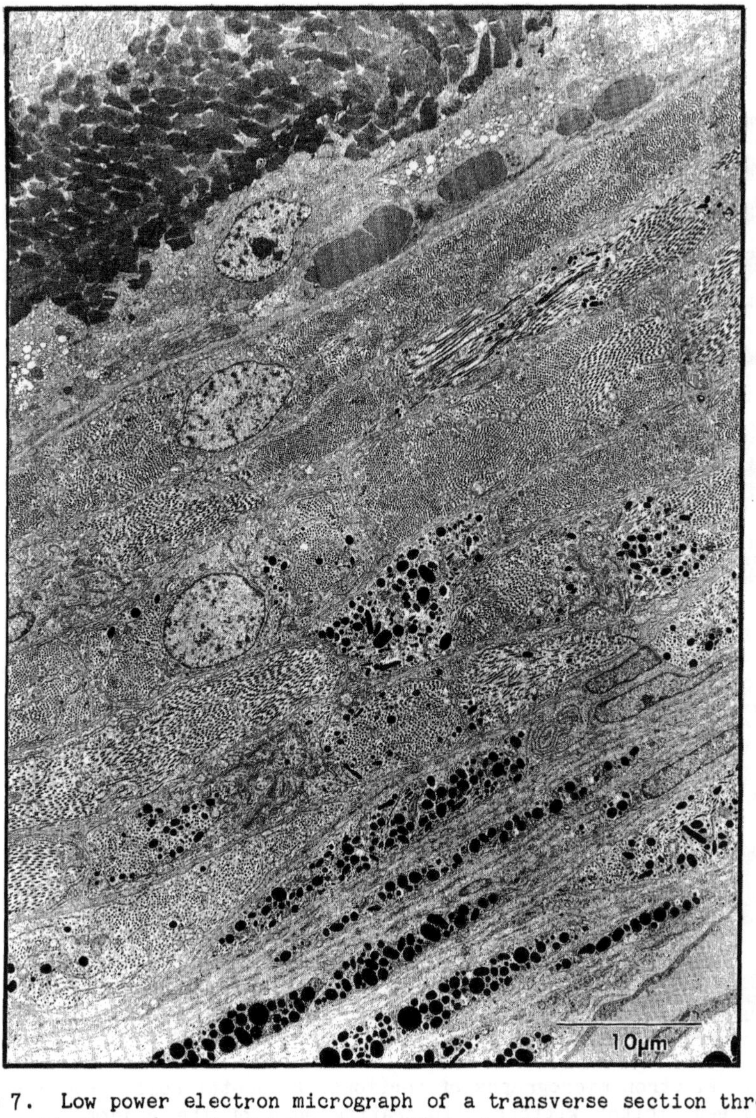

Figure 7. Low power electron micrograph of a transverse section through the choroid of an 8-week-old taurine-deficient kitten. There are fewer layers of tapetal cells than normal, and many tapetal cells contain electron-dense globules which are formed from degenerating tapetal rods.

Finally, two female cats in our colony that had been consuming the taurine-free diet for approximately 3 yr and had each had at least six pregnancies, none of which resulted in live-born kittens, were switched to the same diet supplemented with 0.05% taurine. After 6 mo on this diet, the cats were returned to the usual breeding schedule, and, to date, each has produced two litters of live, apparently normal, kittens. These results provide convincing evidence that a dietary deficiency of taurine alone is sufficient to interrupt normal feline pregnancy dramatically and suggest further that the process is reversible by dietary intervention.

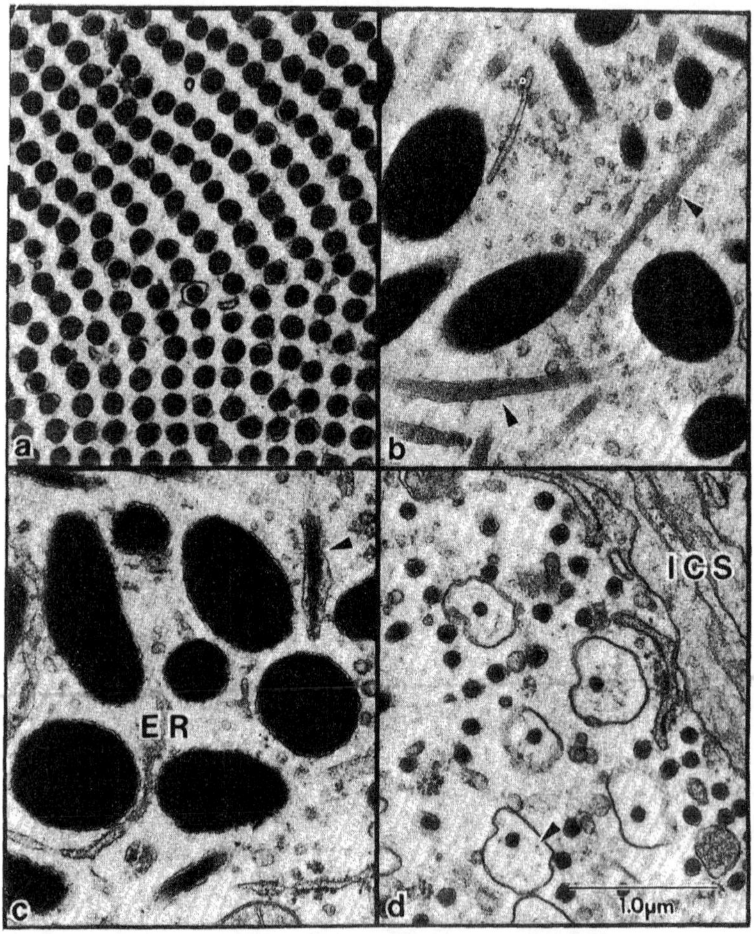

Figure 8. Electron micrographs of portions of tapetal cells from 8-week-old taurine-deficient kittens illustrating some of the abnormalities observed. (a) A cell with the normal regular array of electron-dense cores but showing some with dilated limiting membranes. (b) A cell without the normal regular array of rods containing large electron-dense globular bodies and a few immature tapetal rods with a periodic cross striation (arrowheads). (c) A similar cell to (b) illustrating the slightly mottled inner structure of the globular bodies and showing them clearly bounded by a limiting membrane as is the more normal tapetal rod (arrowhead). ER- endoplasmic reticulum. (d) A cell with irregular distribution of apparently normal electron-dense tapetal rods, some surrounded by greatly dilated limiting membranes.

Table 5. Activities of some enzymes in kitten liver and brain

Enzyme and tissue	Synthetic diet + 0.05% taurine	Synthetic diet alone
	nmol product (mg protein. h^{-1})	
Cystathionine synthase		
Liver	180 ± 19	189 ± 18
Brain	18.7 ± 6.3	25.8 ± 6.3
Cystathionase		
Liver	362 ± 47	268 ± 44
Cysteine dioxygenase		
Liver	123 ± 26	97 ± 22
Brain	336 ± 92	233 ± 32
Cysteinesulfinic acid decarboxylase		
Liver	11.8 ± 1.9	13.7 ± 3.1
Brain	81.0 ± 18.7	70.9 ± 9.7
Brain + 100 mM L-glutamic acid	11.1 ± 1.8	11.1 ± 1.9

Each value represents the mean ± SD of determinations from 6 to 12 kittens. There is no significant difference between the activities of enzymes from kittens suckling mothers fed the diet supplemented with taurine and those suckling mothers fed the taurine-free diet as determined by Student's t-test.

Figure 9. Fetuses aborted at approximately 50 days gestation by a female fed the taurine-free diet supplemented with 0.005% taurine. They weighed 36, 49, and 9 g, respectively.

REFERENCES

1. Imaki, H., Moretz, R.C., Wisniewski, H.M. and Sturman, J.A., 1986, Feline maternal taurine deficiency: Effects on retina and tapetum of the offspring, Dev. Neurosci., 8:160-181.

2. Palackal, T., Moretz, R., Wisniewski, H. and Sturman, J.A., 1986, Abnormal visual cortex development in the kitten associated with maternal dietary taurine deprivation, J. Neurosci. Res., 15:223-239.

3. Sturman, J.A., Gargano, A.D., Messing, J.M. and Imaki, H., 1986, Feline maternal taurine deficiency: Effect on mother and offspring, J. Nutr., 116:655-667.

4. Sturman, J.A. and Hayes, K.C., 1980, The biology of taurine in nutrition and development, Adv. Nutr. Res., 3:231-299.

5. Sturman, J.A., Moretz, R.C., French, J.H. and Wisniewski, H.M., 1985. Taurine deficiency in the developing cat: Persistence of the cerebellar external granule cell layer, J. Neurosci. Res., 13:405-416.

ABNORMAL VISUAL ACUITY AND RETINAL MORPHOLOGY IN RHESUS MONKEYS FED A TAURINE-FREE DIET DURING THE FIRST THREE POSTNATAL MONTHS

Martha Neuringer, Humi Imaki, John A. Sturman, Roger Moretz and Henryk M. Wisniewski.

Oregon Regional Primate Research Center, Beaverton, Oregon 97006; and Institute for Basic Research in Developmental Disabilities, Staten Island, New York 10314, USA

Over ten years ago, taurine was identified as an essential nutrient for cats (7). Without a dietary source of this amino acid, such as their natural taurine-rich diet of meat and fish, both kittens and adult cats undergo a decline in plasma and tissue taurine levels, reduced electroretinogram amplitudes with prolonged peak latencies, and progressive retinal degeneration which eventually results in blindness (8,24,26). As Dr. Sturman describes elsewhere in this volume, taurine deprivation of pregnant cats and their kittens also produces more widespread nervous system disturbances, including abnormal development of the cerebellum and visual cortex (19,27). Although cats appear to be unique in the severity of the effects produced by dietary taurine deprivation, it is now clear that they are not the only species affected. Recent studies of both human children and infant monkeys show that changes in retinal structure and function are associated with taurine deprivation.

Two lines of evidence provoke concern for possible adverse effects of taurine deprivation in human infants and children. First, recent studies by Geggel and his colleagues (6) showed that total parenteral nutrition, using taurine-free amino acid preparations, led to reduced plasma taurine levels, ophthalmoscopic retinal changes, and electroretinographic abnormalities. These changes were reversed in a small number of documented cases by the addition of taurine to the infusate. Although most amino acid mixtures now in use are supplemented with taurine, taurine-free amino acid mixtures for intravenous feeding were standard until a few years ago. Such mixtures were commonly used for the intravenous feeding of premature infants, who might be expected to be particularly vulnerable to taurine deprivation (23). The second line of evidence comes from the earlier studies of Gaull, Järvenpää, and colleagues (5,12) which showed that human infants, when fed taurine-poor infant formulas based on cow's milk, underwent a fall in taurine levels in plasma and urine which was not seen in infants fed taurine-rich human milk. The reductions in plasma taurine levels were comparable to those seen in the studies of children fed by total parenteral nutrition. If formula-fed infants suffer similar functional abnormalities, the implications are of course even greater. However, despite the importance of this question, it has not been addressed directly. Because taurine supplementation of infant formula has been instituted during the last two years by most manufacturers, studies of taurine-deprived infants will now be difficult to justify, despite the fact that this issue remains unresolved.

The next best approach, in our view, is to examine the effects of taurine deprivation on nonhuman primate infants, which are very similar to human infants in both their nutritional requirements and their visual system structure and function. At the last taurine symposium two years ago, we reported the results of our first study of dietary taurine deprivation in rhesus monkeys (18,28). In that study, in which we fed infant monkeys a low-taurine human infant formula, we found a reduction in the amplitude of the cone electroretinogram and consistent degenerative changes in the outer segments of cone photoreceptors. However, because retinal morphology was not examined until two years of age, we did not know the time course of the changes induced by taurine deprivation. In particular, we did not know whether they arose within the first few months of life, during the period when a human infant might exclusively consume such a formula. We also did not answer directly the most important question regarding the effects of taurine deprivation: whether the functional visual capabilities of the monkeys were impaired. Indeed, no previous studies in any species have directly measured visual capacities in taurine-deprived animals. Previous studies have monitored only the electroretinogram, which is limited in its ability to assess visual impairment. We are now conducting a study designed to address these questions, and I will present the results of the first part of this new study, including data obtained up to three months of age.

Our previous study showed specific degenerative changes in cone outer segments but not those of rods. Cone photoreceptors underlie daylight vision, color vision, and high levels of visual acuity. Therefore, in the present study, we chose visual acuity as a measure of visual function which is sensitive to loss of integrity of cones. In addition, we chose visual acuity because it can be measured in very young infants by methods which have recently been developed.

METHODS

Five rhesus monkeys were fed a taurine-free, soy-protein-based human infant formula (Isomil^R, Ross) from the day of birth. A second group of six were fed the same formula supplemented with taurine (70 µM/100 ml, similar to the amount in rhesus monkey milk). The infants' formulas were coded so that those involved in their care and experimental testing did not know whether they were taurine-deprived or taurine-supplemented.

Visual acuity was tested at ages 3-4, 7-8, and 11-12 weeks. We used the preferential looking method, a technique recently developed for measuring visual acuity in both human and monkey infants (14,29). This method is based on infants' strong inherent tendency to attend and orient to patterned stimuli such as stripes, checkerboards or faces. The pattern in this test was an acuity target, consisting of a 10 x 10 cm card covered with high-contrast black and white stripes. The acuity target was paired with an unpatterned gray card of the same overall luminance, and both stimuli were presented within a gray background. An infant was held 35 cm from the stimuli while an observer, who did not know the position of the striped card, watched its eye movements and fixations. Based on the infant's looking behavior, the observer judged whether the striped card was on the right or left. Several sizes of stripes were presented, each for at least 20 trials. In general, the striped stimulus attracts an infant's gaze if it is visible; if the stripes are unresolvably small, they appear gray, like the gray card and the background, and fail to elicit preferential looking. Typically, the observer's judgements are correct on 85-100% of trials when the size of the stripes is above the infant's resolution threshold, and drop to 50% or chance levels when the stripes are too small to be distinguished from gray. Here the acuity threshold was defined as the size at which judgements were correct 75% of the time.

At 3 months of age, the monkeys were killed for biochemical and morphological studies. The retinas were processed for electron microscopy according to standard procedures (11).

RESULTS

Plasma levels of taurine fell to 35% of control levels in the taurine-deprived monkeys by 8 weeks of age and were approximately 50% of control levels at 12 weeks (Fig. 1). These results are very similar to those obtained in human infants receiving low-taurine, cow's-milk-based formulas versus human milk or taurine-supplemented formulas (5,12). Retinal levels of taurine were reduced by 22% at 3 months of age.

Monkeys receiving the taurine-supplemented formula showed normal visual acuity development, compared to macaque infants tested in other laboratories (14). In the taurine-deprived group, however, visual acuity was poorer by almost a factor of two at all three ages (Fig. 2). Thus, even as early as four weeks of age, dietary taurine deprivation had a significant impact on visual acuity.

This acuity loss was not correctable; that is, it was not due to refractive errors. Refractions were determined at least once within the first 12 weeks by slit lamp retinoscopy under cycloplegia. None of the monkeys had significant myopia, hyperopia, or astigmatism which would impair visual acuity, and there were no differences in these measurements between the groups. Therefore, the acuity loss is attributable to retinal rather than optical factors.

Retinal morphology was examined at 3 months of age. In the taurine-deprived group, photoreceptor outer segments were swollen and irregularly oriented, particularly in the foveal region. Ultrastructurally, cone outer

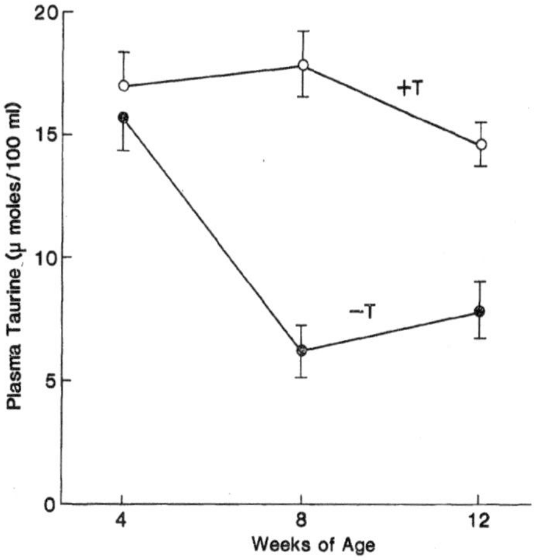

Fig. 1. Taurine concentrations in the plasma of infant rhesus monkeys fed taurine-supplemented (+T) or taurine-free (−T) infant formulas.

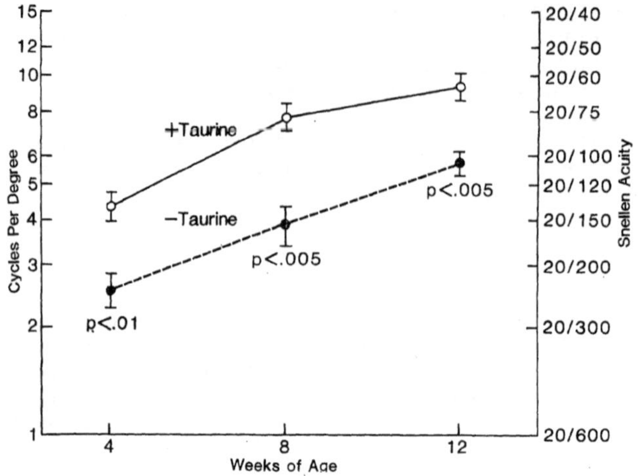

Fig. 2. Visual acuity thresholds (mean ± SEM) as determined by the preferential looking method for taurine-deprived and taurine-supplemented infant rhesus moneys. Thresholds are expressed in cycles per degree of visual angle (left ordinate) and in the equivalent Snellen acuity values (right ordinate). N = 5 for the taurine-deprived group; for the taurine-supplemented group, N = 6 at 4 and 8 weeks and 3 at 12 weeks.

segments exhibited extensive vesiculation and disorientation of disk membranes and disruption of outer plasma membranes (Fig. 3). Rod outer segments were less affected but also showed some disorganization, with expanded spaces between disk membranes (Fig. 4). The inner segments of both rods and cones also showed several abnormalities, including dense membrane whorls within the mitochondria of ellipsoids and markedly dilated cisternae of the endoplasmic reticulum in the cytoplasm of myoids and perikarya. The synaptic endings of both cones and rods contained swollen mitochondria and large vacuoles. In pigment epithelial cells, large vacuoles and spherical droplets were frequently observed in the perinuclear cytoplasm. Such abnormalities were rarely seen in the retinas of taurine-supplemented monkeys.

DISCUSSION

This study strengthens the evidence that taurine is an essential nutrient for normal visual development in primate infants. First, it demonstrates that significant retinal degenerative changes occur within the first three months after birth in infants fed a taurine-free formula. Because this is a period during which human infants often receive infant formulas as their only food source, these results increase the concern for retinal damage in infants fed low-taurine formulas. Secondly, this study provides the first direct evidence of functional visual impairment resulting from these retinal changes. Previous anatomical and electrophysiological data strongly implied a loss of visual capacity, but the exact extent and type of loss could not be inferred. A significant effect on visual acuity is consistent with the observed disorganization of cone photoreceptors. More subtle changes were also seen in rod photoreceptors at three months of age. Further studies will be needed to determine if these changes are associated with abnormalities in

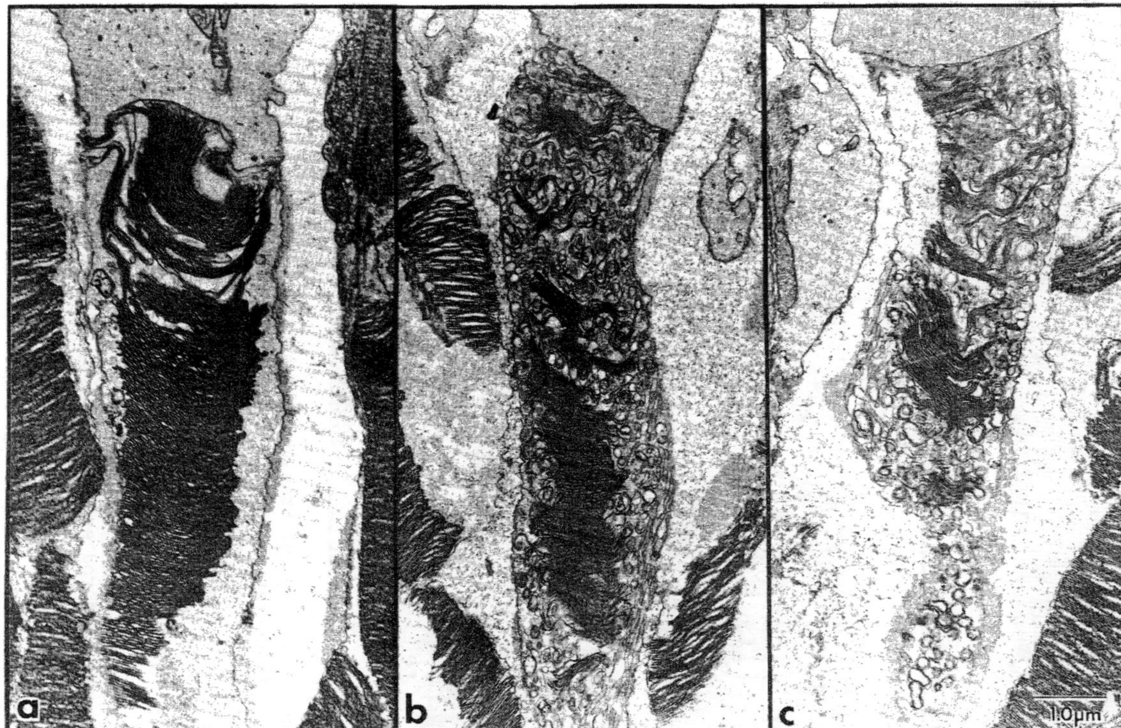

Fig. 3. Electron micrographs of transversely-sectioned retina in the foveal region illustrating cone outer segments from 3-month-old rhesus monkeys. A. Cone outer segment from the retina of a taurine-supplemented rhesus monkey, showing relatively regular stacking of disks, absence of vesiculation, and intact outer plasma membrane. B and C. Cone outer segments from taurine-deprived rhesus monkeys, showing two representative degrees of disorganization, including disk disorientation, vesiculation, and disruption of outer plasma membranes.

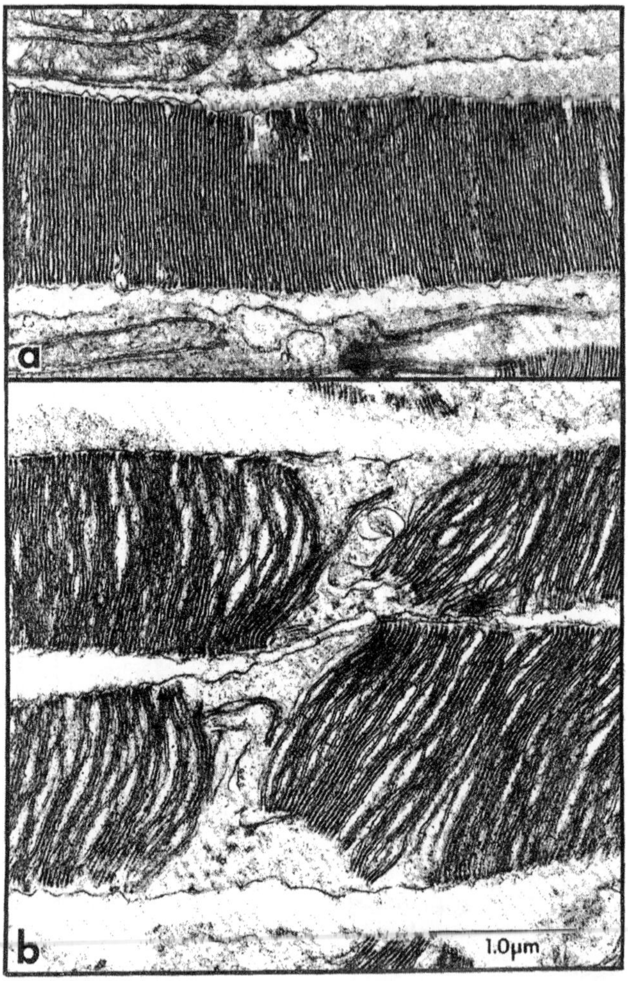

Fig. 4. Electron micrographs of transversely-sectioned retina in the parafovea illustrating rod outer segments from 3-month-old rhesus monkeys. A. Normal rod outer segment from taurine-supplemented monkey. B. Rod outer segment from taurine-deprived monkey, showing ruffled and loosened appearance due to distention of intradisk spaces.

psychophysical measures of rod system function, such as changes in absolute visual thresholds and dark adaptation.

The anatomical and functional changes reported here were not accompanied by significant changes in the electroretinogram at 3 months of age. It is interesting that a direct measure of vision was more sensitive than the electroretinogram and a better correlate of the morphological changes. This may be due to the nature of the electroretinogram itself as a somewhat indirect measure of the visual process. It may also be due to the anatomical distribution of morphological changes within the retina, with fovea and parafovea showing the greatest ultrastructural changes. This small retinal area is critical for normal acuity, but does not make a significant contribution to the full-field electroretinogram. In addition, photoreceptor swelling and disorientation would seriously disrupt the tight packing and regular spacing of cones and their precise alignment with the path of light, all of which are necessary for optimum spatial resolution but may have less effect on generation of the evoked potential to a flash stimulus.

As always, several critical questions remain to be answered. Some of these are being addressed by subsequent phases of this ongoing study. First, we are following the time course of functional and morphological changes at later ages. Secondly, we are assessing the reversibility of these changes. We previously showed that adult rhesus monkeys appear not to depend on dietary sources of taurine, and can maintain both normal plasma and tissue taurine levels and normal visual function despite years of complete taurine deprivation (18). We do not know at what age rhesus monkeys lose their dependence on dietary taurine, although our earlier studies suggested that this change begins within the second year. Therefore, it is possible that the effects of taurine deprivation are self-reversing. The evidence from cats suggests that morphological and electroretinographic changes in rod photoreceptors are almost entirely reversible upon rehabilitation with taurine-containing diets, but that cones show some persistent alterations. However, even if retinal morphological changes prove to be completely reversible in primates, there is still cause for considerable concern regarding long-term functional consquences. A large literature on the effects of early visual deprivation has shown that alterations in input from the retina can produce profound and persistent effects on the organization of the visual cortex (2,3,15). These studies have indicated the need for aggressive early treatment of infants and children with such ocular abnormalities as congenital cataracts and astigmatism if permanent visual loss is to be prevented (4). With these studies in mind, we are now examining the visual cortex of taurine-deprived monkeys, and preliminary results indicate a retardation in the development of visual cortical organization. It remains to be determined whether this effect on the visual cortex is a secondary result of retinal damage or represents a more general effect of taurine deprivation on primate nervous system development.

We do not know whether human infants fed taurine-poor formulas show the same abnormalities observed in our infant monkeys. However, the two species are closely similar in nutritional requirements, retinal structure, and visual abilities, and their plasma taurine levels are similarly affected by dietary taurine deprivation. Therefore, serious consideration must be given to the possibility that similar changes occur in formula-fed human infants and have gone undetected because they produce no grossly observable symptoms. Indeed, human infants might be more vulnerable because the retina and brain are less developed at birth than in rhesus monkeys (1,9,10). Although major manufacturers have begun to supplement their infant formulas with taurine, a number of smaller companies, at least in the United States, have not, and no requirement for taurine has been established.

The function of taurine in the retina remains mysterious, although significant progress in understanding has been achieved by the contributors to several chapters of this book. It seems clear that taurine acts as a stabilizer of membranes, particularly those physiologically active, excitable membranes which have a high content of polyunsaturated fatty acids. For several reasons, retinal outer segment membranes are perhaps the most inherently unstable of this class of membranes, and therefore the most vulnerable to insults such as taurine deprivation. First, their structural phospholipids have the highest content of polyunsaturated fatty acids of any tissue. The predominant fatty acid is docosahexaenoic acid which, with six double bonds, is the most unsaturated major fatty acid in living tissue (17); it is therefore particularly prone to peroxidation and is selectively reduced in light-damaged retinas (31). Levels of taurine closely parallel the levels of this fatty acid in various tissues. Diet-induced retinal depletion of docosahexaenoic acid is also associated with acuity loss in infant monkeys, together with electroretinogram abnormalities (16). In addition to their high level of polyunsaturated fatty acids, outer segment membranes are bathed in very high levels of oxygen, are exposed to light, and contain high levels of retinol, bound to opsin to form the visual pigment. All of these factors promote oxidative damage and other forms of membrane destabilization. Therefore, it is not surprising that the retina contains multiple antioxidant and membrane-stabilizing systems, including vitamin E, vitamin C, selenium and glutathione, superoxide dismutase, xanthophylls, zinc, and taurine.

The capacity of taurine to protect outer segment membranes from the effects of light exposure and oxidants has been demonstrated in vitro by Dr. Pasantes-Morales and her coworkers (20). Little information is available in vivo on the interaction of taurine with oxidative stressors. However, in vivo studies of taurine and light exposure have produced contradictory results. Rats depleted of taurine by treatment with guanidinoethane sulfonate (GES) were protected from retinal degeneration when kept in the dark (13). On the other hand, cats with dietary taurine deficiency were not protected by light deprivation (21). Furthermore, oral taurine administration did not protect rats from the damaging effects of continuous illumination (30), and GES-treated rats showed no increased vulnerability to continuous light (22). The hypothesis that taurine protects the retina from light-induced damage is also inconsistent with our finding that cone photoreceptors, rather than the more light-sensitive rods, are more affected by taurine deprivation. Thus, retinal degeneration in taurine-deprived animals probably cannot be attributed to increased sensitivity to the damaging effects of light alone, but may result from increased vulnerability to a variety of other destabilizing influences on membrane structure.

ACKNOWLEDGEMENTS

These studies were supported by grants from the National Institutes of Health (HD-18678 to John Sturman and RR-00163 to the Oregon Regional Primate Research Center). Coded taurine-free and taurine-supplemented formulas were provided by Ross Laboratories. This is publication number 1483 of the Oregon Regional Primate Research Center.

REFERENCES

1. Abramov, I., Gordon, J., Hendrickson, A., Hainline, L., Dobson, V., and LaBossiere, E., 1982, The retina of the newborn human infant, Science, 217:265-266.
2. Barlow, H.B., 1975, Visual experience and cortical development, Nature, 258:199-124.
3. Blakemore, C., and Vital-Durand, F., 1981, Postnatal development of the monkey's visual system, in: "The Fetus and Independent Life", Ciba Foundation Symposium 86, Pitman, London, pp. 152-171.
4. Boothe, R.G., Dobson, V., and Teller, D.Y., 1985, Postnatal development of vision in human and nonhuman primates, Ann. Rev. Neurosci., 8:495-545.

5. Gaull, G.E., Rassin, D.K., Raiha, N.C.R., and Heinonen, K., 1977, Milk protein quantity and quality in low-birth-weight infants. III. Effects on sulfur amino acids in plasma and urine, J. Pediat., 90:348-355.
6. Geggel, H.S., Ament, M.E., Heckenlively, J.R., Martin, D.A., and Kopple, J.D., 1985, Nutritional requirement for taurine in patients receiving long-term parenteral nutrition, New Engl. J. Med., 312:142-146.
7. Hayes, K.C., Carey, S.Y., and Schmidt, S.Y., 1975, Retinal degeneration associated with taurine deficiency in the cat, Science, 188:949-951.
8. Hayes, K.C., Rabin, A.R., and Berson, E.L., 1975, An ultrastructural study of nutritionally induced and reversed retinal degeneration in cats, Am. J. Pathol., 78:505-516.
9. Hendrickson, A., and Kupfer, C., 1976, The histogenesis of the fovea in the macaque monkey, Invest. Ophthalmol., 15:746-756.
10. Hendrickson, A., and Yuodelis, C., 1984, The morphological development of the human fovea, Ophthalmology, 91:603-612.
11. Imaki, H., Moretz, R.C., Wisniewski, H.M., and Sturman, J.A., 1986, Feline maternal taurine deficiency: Effects on retina and tapetum of the offspring, Devel. Neurosci., 8:160-181.
12. Järvenpää, A.-L., Rassin, D.K., Räihä, N.C.R., and Gaull, G.E., 1982, Milk protein quantity and quality in the term infant. II. Effects on acidic and neutral amino acids, Pediatrics, 70:221-230.
13. Lake, N., and Cocker, S., 1986, The influence of lighting on the effects of taurine depletion in the rat retina, Invest. Ophthalmol. Vis. Sci., 27(Suppl. 3):57.
14. Lee, C., and Boothe, R.G., 1981, Visual acuity development in infant monkeys (Macaca nemestrina) having known gestational ages, Vision Res., 21:805-809.
15. Movshon, J.A., and Van Sluyters, R.C., 1981, Visual neural development, Ann. Rev. Psychol., 32:477-522.
16. Neuringer, M., Connor, W.E., Lin, D.S., Barstad, L., and Luck, S., 1986, Biochemical and functional effects of prenatal and postnatal omega-3 fatty acid deficiency on retina and brain in rhesus monkeys, Proc. Natl. Acad. Sci. USA, 83:4021-4025.
17. Neuringer, M., and Connor, W.E., 1986, Omega-3 fatty acids in the retina and brain: Evidence for their essentiality, Nutr. Rev., 44:285-294.
18. Neuringer, M., Sturman, J.A., Wen, G.Y., and Wisniewski, H.M., 1985, Dietary taurine is necessary for normal retinal development in monkeys, in: "Taurine: Biological Actions and Clinical Perspectives", S.S. Oja, L. Ahtee, P. Kontro and M.K. Paasonen, eds., Alan R. Liss, New York, pp. 53-62.
19. Palackal, T., Moretz, R., Wisniewski, H.M., and Sturman, J., 1986, Abnormal visual cortex development in the kitten associated with maternal dietary taurine deprivation, J. Neurosci. Res., 15:223-239.
20. Pasantes-Morales, H., Ademe, R.M., and Quesada, O., 1981, Protective effect of taurine on the light-induced disruption of isolated frog rod outer segments, J. Neurosci. Res., 6:337-348.
21. Pasantes-Morales, H., Dominguez, L., Campomanes, M.A., and Pacheco, P., 1986, Retinal degeneration induced by taurine deficiency in light-deprived cats, Exp. Eye Res., 43:55-60.
22. Rapp, L.M., Thum, L.A., Tarver, A.P., and Wiegand, R.D., 1986, Vitamin E and taurine deficiency: Effects on the retina in cyclic light-maintained and light-damaged rats, Invest. Ophthalmol. Vis. Sci., 27(Suppl.3):131.
23. Rigo, J., and Senterre, J., 1977, Is taurine essential for the neonate?, Biol. Neonate, 32:73-76.

24. Schmidt, S.Y., Berson, E.L., and Hayes, K.C., 1976, Retinal degeneration in cats fed casein. I. Taurine deficiency, Invest. Ophthalmol., 15:45-52.
25. Schmidt, S.Y., Berson, E.L., and Hayes, K.C., 1976, Retinal degeneration in the taurine-deficient cat, Trans. Amer. Acad. Ophthalmol., 81:687-693.
26. Schmidt, S.Y., Berson, E.L., Watson, G., and Huang, G., 1977, Retinal degeneration in cats fed casein. III. Taurine deficiency and ERG amplitudes, Invest. Ophthalmol. Vis. Sci., 16:673-678.
27. Sturman, J.A., Moretz, R.C., French, J.H., and Wisniewski, H.M., 1985, Postnatal taurine deficiency in the kitten results in a persistence of the cerebellar external granule cell layer: Correction by taurine feeding, J. Neurosci. Res., 13:521-528.
28. Sturman, J.A., Wen, G.Y., Wisniewski, H.M., and Neuringer, M., 1984, Retinal degeneration in primates raised on a synthetic human infant formula, Int. J. Devel. Neurosci., 2:121-129.
29. Teller, D.Y., Morse, R., Borton, R., and Regal, D.M., 1974, Visual acuity for vertical and diagonal gratings in human infants, Vis. Res., 14:1433-1439.
30. Voaden, M.J., Hussain, A.A., and Lalji, K., 1984, Photochemical damage in the albino rat retina: Oral taurine has no effect on DNA and protein loss from severely damaged photoreceptor cells, J. Neurochem., 42:582-583.
31. Wiegand, R.D., Giusto, N.M., Rapp, L.M., and Anderson, R.E., 1983, Evidence for rod outer segment lipid peroxidation following constant illumination of the rat retina, Invest. Ophthalmol. Vis. Sci., 24:1433-1435.

TAURINE CONCENTRATIONS IN HUMAN HEART - A RETROSPECTIVE ANALYSIS

H. Böhles[1], D. Michalk[1], and E. Wagner-Thiessen[2]

Departments of [1]Pediatrics and [2]Pathology, University of Erlangen, F.R. Germany

Taurine is present in human heart in high concentrations and there is evidence that it is involved in the regulation of myocardial performance. However, there exists only scarce information about the physiological functions of taurine. Taurine analysis in human myocardium was performed in the present study to examine for a correlation with previous severe disease and to get further insight into its functional role.

TISSUE SAMPLING AND METHODS

For a retrospective analysis the myocardial taurine concentrations were determined in left ventricular muscle. A tissue sample of about 100 mg was taken during 49 consecutive autopsies within 24 h of death. The age ranged from 12 to 85 years. 26 were male and 23 female. The muscle tissue was homogenized in 5% sulfosalicylic acid. The taurine concentrations were determined in the acid supernatant on an amino acid analyzer. Myocardial Zn and Ca concentrations were measured in the same substrate by atomic absorption photometry. All concentrations were expressed per gram wet weight. The statistical signifcance of the data was tested with the nonparametric U-test according to Mann and Whitney.

RESULTS

The myocardial taurine concentrations ranged from 3.56 to 10.44 with a mean of 6.57 $\pm$ 1.41 µmol/g wet weight. There was no relationship to either age or gender (male: 6.33 $\pm$ 1.47; female: 6.92 $\pm$ 1.05; n.s.). The autopsied patients were grouped retrospectively into those suffering and dying exclusively from severe heart disease (group 1; n = 17); additional cardiac problems to underlying severe disease (group 2; n = 11) and those without obvious cardiac problems (group 3; n = 21). Among the patients with exclusive severe heart disease, 14 had advanced coronary sclerosis and signs of acute or old infarction. The other three in this group had heart failure as a consequence of congenital cardiac malformation (Fig. 1). There was no difference between myocardial taurine concentrations (µmol/g wet weight) in group 1 (mean 6.65 $\pm$ 1.45), group 2 (mean 6.30 $\pm$ 1.69) and group 3 (mean 6.55 $\pm$ 1.06). Also no difference was found when reported diabetics (n = 9; 6.94 $\pm$ 0.85) were compared with reported nondiabetics (n = 26; $\bar{x}$:6.44 $\pm$ 1.51). Patients with renal failure that had not been treated with hemodialysis (n = 4) presented myocardial taurine concentrations in the upper range (mean 7.23 $\pm$ 0.21). Heart taurine concentrations above 8 umol/g wet weight were exclusively found in those with reported extreme congestive heart failure. The patients with concentrations above 9 µmol/g wet weight had been reported to be dehydrated.

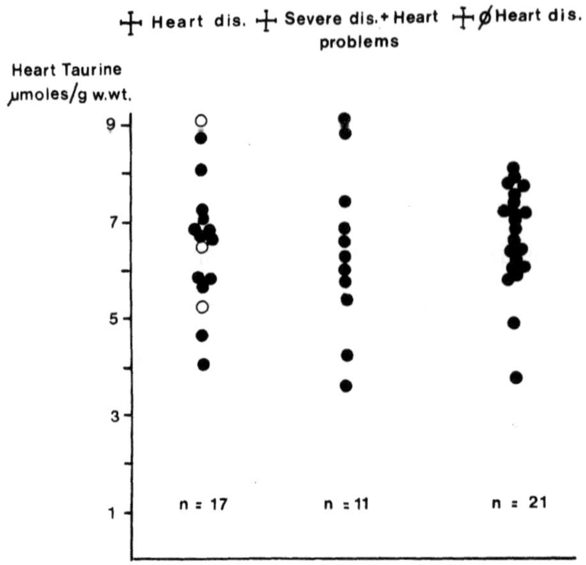

Fig. 1. Myocardial taurine concentrations in patients with cardiac and noncardiac cause of death. O Heart failure as a consequence of congenital cardiac malformation.

Taurine concentrations below 5 µmol/g wet weight were found among patients with (n = 4) and without (n = 2) heart disease. However taurine concentrations were significantly higher ($p < 0.05$) in the heart muscle of patients with hypertension (n = 15; mean 7.66 ± 1.08) compared to those without reported high blood pressure (n = 26; mean 5.85 ± 1.08) (Fig. 2). The previously mentioned myocardial taurine concentrations above 8 µmol/g wet weight were exclusively found in those patients who had suffered from severe heart disease plus hypertension.

The myocardial zinc and calcium concentrations are presented in Table 1. Concentrations of both calcium and zinc tended to be higher in samples from patients with heart disease, especially if there was high blood pressure. However there was no correlation between taurine and calcium and zinc concentrations.

DISCUSSION

The present data give some indirect evidence that taurine may interfer with myocardial performance. Peterson et al. demonstrated an increase in myocardial taurine concentration in experimental right sided congestive heart failure in dogs (7). Huxtable and Bressler showed a similar increase of taurine levels in the left ventricles of patients dying of chronic congestive heart failure (4,5). The myocardial taurine concentrations in our autopsy specimens correspond to those determined by Huxtable and Bressler (5.6 ± 0.5 µmol/g wet weight) (4). We could not prove a relationship of ventricular taurine concentrations to the severity of heart disease. However there was a clear tendency to higher taurine concentrations when the clinical data indicated an extremely severe cardiac condition. The discrepancy may be due to the crude way of defining an eventually underlying cardiac problem. Our results concerning higher taurine concentrations in the heart of patients with high blood pressure are in accordance to values reported by Huxtable and Bressler (5). They found a positive correlation between taurine

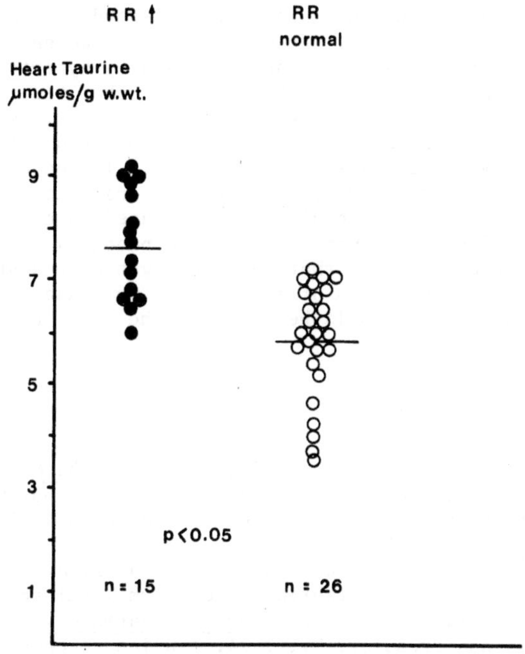

Fig. 2. Increased myocardial taurine concentraions in patients with reported arterial hypertension (●) and patients with normal pressures (○).

concentration and blood pressure. Patients with an average systolic blood pressure of 145 mm Hg had ventricular taurine levels of 8.5 µmol/g wet weight, whereas patients averaging 108 mm Hg systolic pressure had levels of 5.0 µmol/g wet weight.

Table 1. Myocardial zinc and calcium concentrations (per g wet weight). The number of patients in each group is given in parentheses. Groups are not mutually exclusive.

Group		Zinc µg/g	Calcium µmole/g
Death from severe heart disease	(17)	6.8 ± 1.5	0.5 ± 0.2
Death from severe disease with additional heart disease	(11)	5.0 ± 2.2	0.7 ± 0.2
Death from severe disease without heart problems	(21)	5.5 ± 1.9	0.3 ± 0.2
With high blood pressure	(15)	6.5 ± 1.3	0.5 ± 0.2
With normal blood pressure	(26)	4.2 ± 1.8	0.4 ± 0.2
Severe heart disease with additional high blood pressure	(8)	6.3 ± 1.2	0.6 ± 0.2

The effect of taurine on blood pressure as tested in different animal experiments is under debate (2,6). However there is increasing evidence that it is interrelated to the metabolic changes caused by hypertension. Franconi et al., using guinea pig ventricular strips, found that taurine was positively inotropic especially at low calcium concentrations (0.5 mM), whereas at high calcium concentrations (2.7 mM) the action of taurine was reversed and a negative inotropy was produced (3). Increased myocardial taurine concentrat- ions may therefore represent a regulatory increase of a "natural calcium antagonist". The slightly increased calcium concentrations in tissue specimens of our patients with cardiac problems fit into this conception. It has been presumed that taurine might have a stimulatory effect on Na,K-ATPase by chelation of zinc in the membrane. However no changes in tissue zinc levels could be found in the heart of patients dying of heart failure (1).

REFERENCES

1. Barbeau, A., 1974, Zinc, taurine and epilepsy, Arch. Neurol., 30:52-58.
2. Cession-Fossion, A., Leconte, J., and Bacq, Z.M., 1963, Comparision chez le rat des effects generaux de la taurine avec ceux de la cysteamine, C. R. Soc. Biol. (Paris), 157:1833-1835.
3. Franconi, F., Martini, F., Stendardi, I., Matucci, R., Zilletti, L., and Giotti, A., 1982, Effect of taurine on calcium level and contractility in guinea pig ventricular strips, Biochem. Pharmacol., 31:3181-3186.
4. Huxtable, R.J., and Bressler, R., 1974, Taurine concentration in congestive heart failure, Science, 184:1187-1188.
5. Huxtable, R.J., and Bressler, R., 1974, Evaluation of taurine in human congestive heart failure, Life Sci., 14:1353-1359.
6. Nara, Y., Yamori, Y., and Lovenberg, W., 1978, Effects of dietary taurine on blood pressures in spontanously hypertensive rats, Biochem. Pharm., 27:2689-2692.
7. Peterson, M.B., Mead, R.J., and Welty, J.D., 1973, Free amino acids in congestive heart failure, J. Mol. Cell. Cardiol., 5:139-147.

THE DEVELOPMENT OF HEART AND BRAIN FUNCTION IN LOW-BIRTH-WEIGHT INFANTS FED WITH TAURINE-SUPPLEMENTED FORMULA

D.V. Michalk, F. Tittor, R. Ringeisen, K.H. Deeg and H. Bohles

University Children's Hospital, D-8520 Erlangen, F.R.G.

Taurine is important for the development of the nervous (7,18) and cardiovascular system (1,8). The biosynthesis of taurine seems to be reduced in preterm and term infants (3,18). The taurine content of human breast milk is high compared to infant formula or bovine milk, from which such formulas are prepared (2,16). The taurine concentrations of plasma and urine were found to be lower in formula-fed infants than in those fed human milk or taurine-supplemented formula (3,11,17), but there were no differences in growth of the infants. In contrast, infant monkeys raised from birth on an infant formula showed growth depression, which could be prevented by taurine supplementation (5). According to these studies taurine was thought to be essential for human infants, and was added to infant formulas. But clinical data supporting this hypothesis are still lacking, especially those regarding the need of supplemental taurine for the development of the nervous and cardiovascular system. We, therefore, investigated the effect of dietary taurine deficiency and supplementation on heart and brain function in low-birth-weight infants.

METHODS

Eighteen low-birth-weight infants were assigned randomly to be fed an adapted formula (Aptamil, Fa. Milupa, Friedrichsdorf, FRG, Table 1) to which supplemental taurine (0.48 mmol/l) either was or was not added. Full enteral feeding was tolerated from the first day of life in all but three infants, who additionally received glucose and electrolytes intravenously for 2 to 5 days. Intake was maintained at 178 ml/kg/day until a body weight of 2500 g was reached; thereafter it was reduced to 167 ml/kg/day (Table 2). The formulas were fed for 16 weeks as the only protein source, fruit juices were allowed to be added in the third month.

Table 1. Composition of formula (100 ml)

Protein (g)	1.9	Fat (g)	3.5
Whey:casein ratio	42:58	Minerals (g)	0.3
Carbohydrates (g)	8.9	Osmolarity (mosM)	270
Lactose (g)	7.0	Calories (kcal)	75
Starch (g)	1.9	Taurine (µmol)	3.8*

* adapted from Erbersdobler et al. (2)
Supplemental taurine concentration: 48 µmoles/100ml

Table 2. Clinical data of infants

	Control infants	Taurine-supplemented infant
Number	10	8
Birth weight (g)	2043 ± 357 (1400-2420)	2122 ± 190 (1890-2370)
Gestational age	37.1 ± 1.5 (35-40)	6.9 ± 2.0 (33-39)
Daily food intake (ml/kg/d)	(167 - 178)	(167 - 179)
Daily energy intake (kcal/kg/d)	(125 - 133)	(124 - 134)
Daily taurine intake (μmol/kg/d)	(6.3 - 6.7)	(86.4 - 92.6)

data are shown ± SD and/or as ranges (in parentheses)

In the second week and after 16 weeks of life the following parameters were measured: 1. growth parameters: weight, crown-heel length, head circumference, 2. cardiologic parameters: electrocardiography, heart ultrasonography in real-time and TM-mode technic for the calculation of the percent systolic shortening fraction as a measure of contraction force (21), 3. neurologic parameters: electroencephalography, brain ultrasonography (real-time) and clinical neurologic examination (sensory and motor functions, infant reflexes, performed always by the pediatric neurologist), 4. chemical analysis: venous blood and urine samples were obtained 2 h after feeding for the determination of BUN, creatinine, protein, electyrolytes and taurine. The routine analysis were done at once (Autoanalyzer), blood and urine specimens for taurine determination were stored at -20°C until column chromatography. After centrifugation of the blood only the upper part of plasma was taken for taurine determination to avoid contamination with platelets.

Differences between both feeding regimens were evaluated with Wilcoxon's test for unpaired samples. Parental information consent was obtained for each infant.

RESULTS

Growth

Food intake was similar in both groups (Table 2); the formula was well tolerated (no vomiting, no meteorism, normal stool frequency and consistency). There were no significant differences in any growth variable between the groups of infants (Table 3).

Cardiologic Parameters

A normal heart configuration and cardiac action was seen in all infants studied by ultrasonography (Table 4). Systolic shortening fraction was within the normal range in all infants at the beginning and the end of the study, and there was no difference between the feeding groups. ECG recordings showed disturbances of repolarization (T-wave inversion) in 4 control and in 3 taurine-supplemented infants after the first week of life. These abnormalities disappeared after 16 weeks in all but 2 infants of the control group. Interestingly, these 2 infants were identical twins, suggesting a constitutional abnormality.

Table 3. Growth in control and taurine-supplemented infants

	Control infants	Taurine suppl. infants
Weight (g)		
at birth	2043 ± 357	2123 ± 190
16 weeks	5265 ± 717	5324 ± 588
increase (g/wk)	202 ± 29	199 ± 34
Crown-heel length (cm)		
at birth	44.3 ± 2.1	45.0 ± 1.8
16 weeks	56.2 ± 3.4	58.3 ± 3.4
growth rate (cm/wk)	0.73 ± 0.12	0.83 ± 0.20
Head circumference		
at birth	31.6 ± 1.5	31.3 ± 0.9
16 weeks	39.6 ± 1.7	39.8 ± 1.1
growth rate (cm/wk)	0.50 ± 0.05	0.52 ± 0.05

Neurologic Parameters

Ultrasonography of brain revealed a normal parenchymal structure and ventricular system in all infants throughout the study. EEG recordings showed normal age related electrical activity in all low birth-weight-infants 1 week after birth and a regular maturation of 17 infants. One infant exhibited hypersynchronic (spike-wave) activity after 16 weeks, but clinical seizures never occurred. This small-for-date infant was born from a heavy-smoking mother after prolonged delivery, presenting with intrauterine deceleration of the heart sounds and postnatal respiratory disorder for one day. Neurologic development was found to be slightly retarded in 3 infants of each feeding group (increased muscular hypertonicity), which could be cured by medical gymnastic only. The abnormalities were not related to the dietary taurine supply. The other infants had a normal maturation of their psychomotor activity. No infant had signs of impaired vision.

Chemical analysis

The taurine concentrations of plasma and urine are shown on Table 5. Mean plasma levels of taurine were similar in control and taurine-supplemented infants, and did not change with age. In contrast, urinary taurine concentration, either expressed as µmol/100 ml or related to mg creatinine/

Table 4. Cardiologic parameters in control and taurine-suplemented infants

	Week	Control infants (10)	Taurine-suppl. infants (8)
Heart sonography	1	normal	normal
	16	normal	normal
Shortening fraction(%)[a]	1	38.6 + 4.6	33.2 + 2.8
	16	37.1 + 5.8	39.2 + 3.7
ECG	1	abnormal: 4[b]	abnormal: 3[b]
	16	abnormal: 2	abnormal: 0

[a] normal range 29-42; [b] disturbance of repolarization

Table 5. Plasma and urine taurine in control and taurine-supplemented infants

	Week	Control infants	Taurine-suppl. infants
Plasma taurine (μM)	1	87 ± 30	104 ± 32
	16	97 ± 36	95 ± 45
Urine taurine (μM)	1	55 ± 48	659 ± 435
	16	96 ± 104	1118 ± 204*
Urine taurine / Urine creatinine (umol/mg)	1	0.57 ± 0.44	2.61 ± 1.16
	16	0.72 ± 0.41	4.02 ± 1.15*

*$p < 0.05$ compared to control

100 ml were higher in the taurine-supplemented infants. There was a marked increase in the taurine excretion with age in both goups; amounting to 357% in the control and 458% in the taurine-supplemented infants, relative to creatinine excretion. The standard deviation, however, was high in the taurine-deprived infants. Other blood parameters (haemoglobin, WBC, BUN, creatinine, protein and electrolytes) were within the normal range under both feeding regimens.

DISCUSSION

Taurine supplementation of the formula did not affect growth in our low-birth-weight infants. This is in accordance with previous studies, in which either breast milk (10,15,16,20) or taurine-supplemented formula (14, 16) were compared to formulas without taurine.

There was also no difference between the control and the taurine-supplemented infants in respect to heart and brain function, measured with clinical methods. With these methods, however, it cannot be ruled out that there were changes on the molecular level, e.g. lower cardiac and cerebral taurine concentrations in the controls; obviously these data cannot be obtained from living infants. But depletion of the heart and brain taurine content seemed to be improbable, since taurine plasma levels were normal in both feeding groups. In rats, decreased taurine tissue concentrations have only been observed in connection with markedly reduced plasma taurine levels (6,12,13).

Whereas in the early studies from Gaull et al. (3), Jarvenpaa et al. (11), and Rassin et al. (15) decreased plasma taurine levels were found in formula-fed term and low-birth-weight compared to breast milk or taurine-supplemented formula-fed infants, more recent studies (14,20) including the present one were unable to detect a correlation between taurine plasma levels and dietary taurine supply. The reason for this discrepancy is obscure. For direct comparison the data of the different studies are listed on Table 6. Unfortunately, in some studies the values are given in means ± SD, in the others in means ± SEM or without any variance. The mean taurine levels after 1 week were similar in all studies except that of Volz et al. (20), who found generally higher values. In contrast to all other studies, Rassin et al. (15) observed lower taurine plasma levels already after one week of life in formula-fed infants, which remained constant with increasing age, suggesting, that the taurine depletion in these infants was rather due to an insufficient prenatal taurine accumulation than to the taurine deficient food. The data of Gaull et al. (3), which are only presented on figures without SD or SEM, show a slight decrease of plasma taurine after 8 weeks feeding of a whey-predominant formula with 1.5 g protein/100 ml. The increase of the protein

Table 6. Plasma concentrations (μmol/dl) of taurine in infants fed with human milk or formulas with different taurine content

Study/subject	N	Food	taurine suppl. (μmol/dl)	Age - weeks 1	8	16
Järvenpää et al.(11)	22	Human Milk		8.2 + 0.6(SD)	9.1 + 0.4	11.2 + 1.1
term infants	10	1.5(60/40)[b]		7.8 + 0.6	5.4 + 0.5*	6.1 + 0.5* 12 wks
Volz et al.(20)	14	Human Milk		16.1 + 1.9(SEM)	13.2 + 1.3	14.8 + 1.6
term infants	15	1.5(60/40)[b]		16.3 + 1.7	14.4 + 1.4	12.6 + 1.4
Gaull et al.(3)	22	Human Milk		6.4	5.8	
LBW[a](1200-1900 g)	21	1.5(60/40)[b]		8.1	5.0*	
	11	3.0(60/40)[b]		7.1	7.2	
Rassin et al.(17)	10	Human Milk		8.5 + 0.7(SEM)	8.0 + 0.8	9.4 + 1.0
LBW[a](1400-2200 g)	12	1.5(60/40)[b]		6.2 + 0.7*	5.2 + 0.6* 7 wks	5.8 + 0,8*
	11	1.5(60/40)+T	30	8.4 + 0.8	9.9 + 1.2	10.6 + 1.0
Okamoto et al.(14)	5	1.5(60/40)[b]		9.8 + 4.6(SD)	no change in 5 weeks	
LBW[a](1300-1700 g)	5	1.5(60/40)[b]+T	25	6.8 + 3.0	" " " "	
Michalk et al.	10	1.9(42/58)[b]		8.7 + 3.0(SD)		9.7 + 3.6
LBW[a](1400-2400 g)	8	1.9(42/58)[b]+T	48	10.4 + 3.2		9.5 + 4.5

*$p < 0.05$ compared to human milk; [a]low birth weight infant; [b]g protein/100ml (whey:casin ratio)

content to 3.0 g/100 ml resulted in higher plasma taurine level than that obtained in breast milk fed infants, although the taurine concentration in that formula has been far below the levels present in human milk. The only study which undoubtedly found a slight decrease of plasma taurine in formula fed infants is that of Jarvenpaa et al. (11), but the number of the investigated infants were quite different in each group, and the standard deviations were smaller than in the other studies.

The only effect of taurine supplementation in our infants was an increase in taurine excretion, which was in accordance with previous observations (3,11,14,15). The amount of taurine excretion was higher than in the other studies. This is not suprising, since our infants received a formula with a higher taurine content. A low urinary taurine concentration in the presence of normal plasma taurine levels does not mean taurine deficiency, but rather reflects the ability of the kidney to conserve taurine in the case of low intake.

Since for ethical reasons (general anesthesia) electroretinagrams were not recorded in our infants, retinal abnormalities due to taurine deficiency could not be excluded. But again the normal plasma taurine levels argue against this possibility, since changes of the ERG were only observed at low plasma taurine concentrations in infants with short bowel syndrome and long-term parenteral nutrition (4), and in monkeys raised from birth for 26 month on a casein hydrolysate (19).

The major argument for a possible requirement of taurine in infant nutrition is its presence in human milk, and an assumed insufficient biosynthetic rate of taurine in humans (3). CSAD activity, expressed by the rate-limiting enzyme of taurine biosynthesis, has been found to be low in the liver of human fetuses and adults (0.3 nmol/mg protein/kg) compared to other mammals (3,18). However, taking this low activity as a basis for the calculation of the daily endogenous taurine production, an infant weighing 3000 g could be able to synthesize 216 umoles of taurine per day, which is far more than the daily taurine intake of 150 umoles present in 500 ml of breast milk. This amount would be sufficient ot replenish the expanding taurine body pool of the growing child, and to compensate the daily taurine losses in bile and urine (calculation: liver weight = 150 g, protein content (20% wt/wt) = 30,000 mg; 30,000 mg x 0.3 nmol x 24 h = 216,000 nmol/day). Direct evidence for a significant ability to synthesize taurine in man was obtained recently from $^{18}O_2$ inhalation studies (9).

Inview of the available data, there is no real evidence that taurine is essential in the nutrition of infants under normal conditions; i.e. sufficient cysteine supply, no serious complications with taurine loss (short bowel syndrome), addition of meat to the diets in the fifth month of life. Therefore, taurine supplementation of formulas with high whey content (enough cysteine) seems not to be necessary at present.

REFERENCES

1. Chesney, R.W., 1985, Taurine: its biological role and clinical implications, in: "Advances in Pediatrics Vol. 32", Z.A. Barnes ed., Year book Med Publishers, Inc., pp. 1-43.
2. Erbersdobler, H.F., Trautwein, E., and Greulich, H.-G., 1984, Determinations of taurine in milk and infant formula diets. Eur. J. Pediatr., 142:133-134.
3. Gaull, G.E., Rassin, D.K., Räihä, N.C.R., and Heinonen, K., 1977, Milk protein quantity and quality in low-birth-weight infants III. Effects on sulfur amino acids in plasma and urine. J. Pediatrics, 90:348-355.
4. Geggel, H.S., Amnet, M.E., Heckenlively, J.R., Martin, D.A., and Kopple,

J.D., 1985, Nutritional requirement for taurine in patients receiving long-term parenteral nutrition N. Engl. J. Med., 312:142-146.
5. Hayes, K.C., Stephan, Z.F., and Sturman, J.A., 1980, Growth depression in taurine-depleted infant monkeys. J. Nutr., 110:2058-2064.
6. Huxtable, R.J., Chubb, J., and Azari, J., 1980, Physiological and experimen- tal regulation of taurine content in the heart, Fed. Proc.,39:2685-2690.
7. Huxtable, R.J., 1981, Insights on function: Metabolism and pharmacology of taurine in the brain, in: "The Role of Peptides and Amino Acids as Neurotransmitters", J.B. Lombardini and A.D. Kenny eds., Alan R. Liss, New York, pp. 53-97.
8. Huxtable, R.J., and Sebring, L.A., 1983, Cardiovascular actions of taurine, in: "Sulfur Amino Acids: Biochemical and Clinical Aspects", K. Kuriyama, R.J. Huxtable and H. Iwata eds., Alan R. Liss, New York, pp. 5-37.
9. Irving Ch.S., Marks, L., Klein, P.D., Fister, N., Gadde, P.L., Chase, T.N., and Samuel, D., 1986, New evidence for taurine biosynthesis in man obtained from 18-O_2 inhalation studies. Life Sciences, 38:491-495.
10. Järvenpää, A.-L., Räihä, N.C.R., Rassin, D.K., and Gaull, G.E., 1982, Milk protein quantity and quality in the term infant. I. Metabolic responses and effects on growth, Pediatrics, 70:221-230.
11. Järvenpää, A.-L., Rassin, D.K., Räihä, N.C.R., and Gaull, G.E., 1982, Milk protein quantity and quality in the term infant II. Effect on acidic and neutral amino acids, Pediatrics, 70:221-230.
12. Lombardini, J.B., and Medina, E.V., 1978, Effects of dietary inorganic sulfate, taurine, and methionine on tissue levels of taurine in the growing rat, J. Nutr., 108:428-433.
13. Michalk, D.V., Bohles, H.J., and Scharer, K., 1983, Growth, taurine, and calcium metabolism in chronic renal failure: effect of taurine and methionine deficiency. in: "Sulfur Amino Acids: Biochemical and Clinical Aspects", K. Kuriyama, R. J. Huxtable, and H. Iwata eds., Alan R. Liss, New York, pp. 305-328.
14. Okamoto, E., Rassin, D.K., Zucker, Ch.L., Salen, G.S., and Heird, W., 1984, Role of taurine in feeding the low-birth-weight infant, J. Pediatr., 104:936-940.
15. Räihä, N.C.R., Heinonen, K., Rassin, D.K., and Gaull, G.E., 1976, Milk protein quantity and quality in low-birthweight infants: I. Metabolic responses and effect on growth, Pediatrics, 67:659-674.
16. Rassin, D.K., Sturman, J.A., and Gaull, G.E., 1978, Taurine and other free amino acids in milk of man and other mammals, Early Human Dev., 2:1-13.
17. Rassin, D.K., Gaull, G.E., Järvenpää, A.-L., and Räihä, N.C.R., 1983, Feeding the low-birth-weight infant: II. Effect of taurine and cholesterol supplementation on amino acids and cholesterol, Pediatrics, 71:179-186.
18. Sturman, J.A., and Hayes, K.C., 1980, The biology of taurine in nutrition and development, in: "Advances in Nutritional Research," Vol.3, H.H. Draper ed., Plenum Publishing Corp., New York, pp. 231-299.
19. Sturman, J.A., Wen, G.Y., Wisniewsky, H.M., and Neuringer, M.D., 1984, Retinal degeneration in primates raised on a synthetic human infant formula, Int. J. Devel. Neuroscience, 2:121-129.
20. Volz, V.R., Book, L., and Churella, H.R., 1983, Growth and plasma amino acids concentrations in term infants fed either whey-predominant formula or human milk, J. Pediatr., 102:27-31.
21. Weitzel, D., Dinkel, E., Dittrich, M., and Peters, H., 1984, Padiatrische Ultraschalldiagnostik, Springer-Verlag, Berlin, Heidelberg, New York, pp. 64-73.

SECTION III: PHARMACOLOGY OF EXCITED TISSUE

INTRODUCTION: PHARMACOLOGY OF EXCITABLE TISSUES

Flavia Franconi [1] and Ryan J. Huxtable [2]

[1] Cattedra di Saggi e Dosaggi Farmacologici, Universita di Sassari, Sassari, ITALY
[2] Department of Pharmacology, College of Medicine, University of Arizona, Tucson, Arizona 85724

This section is concerned with the pharmacological action of taurine in the central nervous system, heart, skeletal and smooth muscle. In addition it contains attempts to identify the mechanism whereby phenomena affected by taurine modify physiological responses. For clarity we wish to subdivide this part into four subsections.

The first subsection discusses the cardioprotective action of taurine and the positive inotropic effects of two taurine-related compounds. Schaffer et al. show that taurine-depleted rats were more susceptible to heart damage induced by ischemia than hearts obtained from non-depleted rats. Although not a definitive explanation of the mechanism, these data together with the fact that taurine is lost during ischemia, suggest that taurine can control ischemic damage. Azuma et al. show clearly that taurine prevents cardiotoxicity of isoproterenol and adriamycin. These two substances increase cardiac calcium levels and taurine prevents the increase as it does in many other experimental conditions such as calcium paradox, hamster cardiomyopathy and hypoxia followed by reoxygenation. Thus this article describes the modulating role played by taurine in a situation of calcium overload without specifying the etiological agents. The Kordenat paper is involved with propionylcarnitine taurinamide in acute myocardial infarction. The last chapter of this section deals with the positive inotropic effect of two taurine related compounds, L-cysteic and orthanilic acids. The conclusion of the paper is that, although these compounds mimic some taurine actions, they are not equal to taurine.

The second subsection is concerned with smooth muscle. One paper is focussed on vascular smooth muscle and the other on bronchial smooth muscle. In fact, Kurachi et al. present the first observation on the bronchodilating effect of taurine on PAF-induced bronchospasm. This effect is not direct but mediated by a decrease in serotonin release due to a decrease of platelet aggregation. This observation opens the way to new fields in taurine research. Manzini's paper presents an analysis in which the calcium pool is involved in the vasodilating effect of taurine; based on indirect response this paper shows that taurine mainly interferes with extracellular pools.

The third subsection deals with the action of taurine on skeletal muscle. The first paper, from Iwata et al., takes into account the control of taurine and uptake in two different kinds of muscle: fast-twitch and slow-twitch muscle. The bulk of the data suggests that taurine transport and levels are controlled by muscular and neuronal activity. On the other hand, the paper of Camerino et al. shows that taurine increases chloride conductance in a manner distinct from that of GABA. These data point out that taurine could have a pivotal role in the control of excitability. The works of Maturo and Kulakowski show that taurine controls metabolic activity. Therefore at a skeletal muscle level taurine plays a role in the control of metabolic and electrical activity.

Izumi's work shows that taurine reduces epileptic activity induced by opioid peptides in the hippothalamus, not only interfering with epileptic activity but reducing wet-dog shakes. In their paper, Fariello et al. reach the conclusion that taurine is not important in the explored activities related to control of cortical neuronal excitability. The paper of Palmi and Sgaragli, which studies the effect of taurine on body temperature, correlated this effect with calcium levels in samples collected from cisterna magna.

POTENTIATION OF MYOCARDIAL ISCHEMIC INJURY BY DRUG-INDUCED TAURINE DEPLETION

S.W. Schaffer, S. Allo and M. Mozaffari

University of South Alabama, School of Medicine, Department of Pharmacology, Mobile, Alabama

The discovery that taurine treatment reverses arrhythmias associated with epinephrine- and digoxin-induced toxicity stimulated research into the myocardial actions of taurine and eventually led to the hypothesis that taurine must have an important physiological function in the heart (10,14, 25,28). Two subsequent lines of evidence have provided the most compelling support for this hypothesis. First, cardiac taurine levels are extremely high and are maintained by an energy-dependent transport process (14,29). Second, exposure of the heart to a medium containing taurine results in a significant change in myocardial energy and calcium metabolism (5-7,15, 18,20,28).

In spite of the attractiveness of the hypothesis, a clearly defined role for the amino acid in the heart still remains to be established. To date, the search for a function has been hampered by the proliferation of effects attributed to taurine and the likelihood that most of these are pharmacological rather than physiological (14,20). Therefore, the recent discovery of a series of drugs which specifically decrease intracellular taurine content has introduced a powerful tool to test for possible physiological effects of the amino acid (13,32). The aim of this report is to describe some of the effects of drug-induced taurine depletion on myocardial function and metabolism.

METHODS

Male Wistar rats (250-300 g) fed Purina rat chow ad lib were maintained on tap water (control) or tap water containing 3% (w/v) β-alanine (taurine-depleted). The β-alanine treated rats lost over 50% of their cardiac taurine content after 3-4 weeks of treatment; heart taurine content fell from 85 $\pm$ 4 mol/g dry wt for controls to 41 $\pm$ 3 mol/g dry wt in β-alanine treated rats.

Hearts from control and taurine-depleted rats were perfused on a working heart apparatus within 45 sec of decapitation (34). All hearts were paced at 300 beats/min and stabilized for 20 minute prior to the experimental protocol. The perfusate used in the ischemic studies was standard Krebs-Henseleit buffer containing 5 mM glucose, 2.5 U/l insulin and 2.5 mm Ca^{2+}, while in the contractility-relaxation studies Krebs-Henseleit buffer at varying calcium concentrations and supplemented with 11 mM glucose was employed. Preload and afterload were fixed at 15 cm and 100 cm H_2O respectively during nonischemic perfusion. In the global ischemia studies, coronary flow was reduced from 15 to approximately 1.0 ml/min using the protocol described by Vary et al. (34). After the appropriate period of ischemia, some of the hearts were frozen for analysis of metabolic intermediates while others were reperfused for 15 minutes under normal conditions and used in measuring recovery of mechanical function or metabolic intermediate content. As an indicator of mechanical function in the reperfused hearts, recovery of cardiac work was calculated according to Vary et al. (34). In all studies, aortic pressure was monitored by a 23Gb Statham pressure transducer placed above the aorta while ventricular pressure, +dP/dt were measured with a transducer by inserting a 22 gauge needle through the ventricle wall.

Hearts for metabolic analysis were rapidly frozen following the specific period of controlled perfusion, ischemia or reperfusion using a Wollenberger clamp precooled in liquid nitrogen. The samples were freeze-dried and a known weight of ventricular tissue was homogenized in 3.0 ml of 6% perchloric acid. After the extraction procedure, the levels of key metabolic intermediates were assayed according to the methods described by Vary et al. (34).

RESULTS

Previous studies have revealed that mechanical function of the isolated heart perfused with buffer containing subnormal levels of calcium improves following introduction of pharmacological levels of taurine into the medium (5). These results and a series of similar findings led to the hypothesis that one putative role of taurine is to regulate calcium homeostasis (28). As a further test of this hypothesis, the effect of drug-induced taurine depletion on cardiac contractility was examined. It was found that in contrast to taurine treatment, taurine depletion had no effect on the contractility parameter +dP/dt over a perfusate calcium concentration range of 0.4 to 1.6 mM (Table 1). On the other hand, the relaxation parameter -dP/dt was depressed in taurine-depleted hearts perfused with medium containing higher calcium concentrations, implying that the rate of relaxation but not contractility is affected by intracellular taurine content.

One possible consequence of the taurine-linked defect in myocardial relaxation relates to its ability to resist stress-induced damage. Table 2 reveals that taurine-depleted hearts recover considerably less mechanical function than hearts containing normal tissue taurine levels when subjected to a period of ischemia followed by reperfusion. This effect is independent of the length of ischemia and the degree of mechanical impairment. Moreover, a change in the substrate composition of the buffer from glucose to glucose plus acetate dues not prevent the response to taurine depletion.

In order to define more completely the effect of taurine depletion on the ischemic heart, tissue adenine nucleotide content was also determined. Since this parameter is a measure of the degree to which adenine nucleotides are degraded to nucleosides, it serves as a good marker of impaired energy metabolism. As shown in Table 3, the size of the adenine nucleotide pool was found to be lower in ischemic hearts relative to the control. Ischemic hearts also exhibited a large decrease in the tissue ATP/ADP ratio, indicating a change in the oxidative phosphorylation state as a result of oxygen deficiency. Interestingly, taurine-depleted hearts contained lower ATP and AMP levels following ischemia than control hearts subjected to a similar ischemic insult. However, the ATP/ADP ratio was unaffected by taurine depletion, indicating that the differences in adenine nucleotide content are not caused by varying degrees of oxygen deficiency or disruption of oxidative phosphorylation.

The importance of certain glycolytic intermediates in development of ischemic damage has recently received considerable attention (4,22,26). While the mechanism underlying the harmful action of intermediates such as lactate remains unclear, excellent negative correlations between the

Table 1. Mechanical performance of taurine-depleted hearts

Buffer Ca^{2+} Concentration	+dP/dt (cm H_2O/s) Control	+dP/dt (cm H_2O/s) Taurine-depleted	-dP/dt (cm H_2O/s) Control	-dP/dt (cm H_2O/s) Taurine-depleted
0.4 mM	1180 ± 90	1290 ± 130	930 ± 50	1120 ± 150
0.8 mM	2300 ± 140	2230 ± 43	2040 ± 120	2070 ± 69
1.2 mM	2430 ± 40	2440 ± 77	2440 ± 120	2230 ± 63
1.6 mM	2590 ± 67	2580 ± 48	2530 ± 100	2260 ± 61*

Hearts from taurine depleted and control rats were perfused with Krebs-Henseleit buffer containing 11 mM glucose and the indicated concentration of calcium. Following stabilization +dP/dt and -dP/dt were measured with a P23Gb Statham pressure transducer by inserting a 22 gauge needle through the ventricle wall. Values shown represent means ± SEM of 4 to 5 hearts. *Denotes significant difference from control ($p < 0.05$).

accumulation of lactate by the ischemic heart and recovery of mechanical function after reperfusion of the ischemic heart have been observed (22). Since taurine depletion is known to influence anaerobic metabolism(20), it was thought that an elevation in the amount of lactate accumulated during ischemia could explain the effect of taurine depletion on ischemic damage. However, this proved not to be the case since lactate levels were identical in taurine-depleted and control, ischemic hearts (Table 3).

In accordance with previous studies (34), reperfusion of the ischemic heart leads to complete restoration of tissue creatine phosphate content but only partial recovery of ATP levels (Table 4). The failure to restore ATP does not appear to be related to mitochondrial dysfunction since the ATP/ADP ratio in both groups of hearts increases more than two fold upon reperfusion to levels approaching the control (Table 4). Instead, the results in Table 4 suggest that ATP loss may be linked to the depletion of the adenine nucleotide pool, which is 25% less in the reperfused heart than the control perfused heart. While differences in tissue adenine nucleotide content of ischemic hearts from taurine-depleted and control rats were observed, no such effect of taurine depletion could be detected in the reperfused hearts. Rather, the only significant difference between the reperfused hearts from taurine-depleted and untreated control animals was their tissue lactate content, which was lower in the former. This effect on lactate content appears to be unrelated to ischemia since a similar difference exists between

Table 2. Effect of taurine depletion on myocardial ischemic injury

Period of ischemia	% Recovery Control	% Recovery Taurine-depleted
20 min	87.2 ± 2.5	70.0 ± 5.5*
25 min	67.8 ± 3.5	50.2 ± 7.6*
28 min	57.6 ± 6,0	46.0 ± 3.0

Hearts from control and taurine-depleted rats were perfused with Krebs-Henseleit buffer containing 5 mM glucose, 2.5 U/l insulin and 2.5 mM Ca^{2+} for a stabilizing period of 20 minutes. This was followed by the indicated period of severe ischemia and then 15 minutes of reperfusion. Cardiac work was determined before ischemia and after the reperfusion period and the data were expressed as % recovery of cardiac work according to the method of Vary et al. (34). Values shown are means ± SEM of six to nine hearts. *Denotes significant difference from control ($p < 0.05$).

Table 3. Effect of taurine depletion on tissue high energy phosphate content of ischemic myocardium

Intermediate	Nonischemic Control	Nonischemic Taurine-depleted	Ischemic Control	Ischemic Taurine-depleted
	Metabolic intermediates (μmol/g dry wt)			
ATP	21.9 ± 0.4	23.8 ± 0.3	10.3 ± 0.5	7.8 ± 0.5*
ADP	4.2 ± 0.4	4.2 ± 0.2	5.9 ± 0.3	5.5 ± 0.5
AMP	0.8 ± 0.2	0.5 ± 0.1	4.8 ± 0.5	2.8 ± 0.3*
ATP/ADP	5.3	5.7	1.6	1.5
Total adenine nucleotide	26.6 ± 1.0	28.5 ± 0.2	20.7 ± 0.4	15.4 ± 0.9*
Phosphocreatine	31.4 ± 1.4	29.9 ± 0.9	6.1 ± 0.9	5.5 ± 0.5
Lactate	1.0 ± 0.2	0.6 ± 0.1	139 ± 12	137 ± 10

Hearts from control and taurine-depleted rats were perfused with Krebs-Henseleit buffer containing 5 mM glucose, 2.5 U/l insulin and 2.5 mM Ca^{2+} for a stabilizing period of 20 minutes followed by 25 minutes of severe global ischemia. The preparation was then frozen with Wollenberger clamps precooled in liquid nitrogen. Tissue intermediates were assayed according to standard procedures. Values shown represent means ± SEM of 5 to 8 hearts. *Denotes significant difference from ischemic control ($p<0.05$).

taurine-depleted and control hearts perfused for 20 minutes under normal conditions (Table 3).

DISCUSSION

Taurine is now considered a nutritional requirement for cats and certain primates (9,11,30). In the absence of a dietary source of the amino acid, tissue taurine levels fall and the animals develop retinal lesions (30). Offspring of these animals exhibit an even wider range of abnormalities, including increased mortality, abnormal hind-limb development, stunted growth and neurological abnormalities (33). Many of these effects can be duplicated by treatment with taurine-depleting drugs such as guanidinoethyl sulfonate or β-alanine (17,23).

Very few studies have focused on the effect of taurine depletion on the heart. In 1982 Franconi et al. (7) depleted taurine from guinea pig ventricular strips through superfusion of the strips with taurine-free medium. The taurine-depleted strips exhibited reduced contractility, which could be restored by inclusion of 10 mM taurine in the buffer. Although Franconi et al. (7) proposed that the improvement in contractility was related to normalization of intracellular taurine content, other means of altering myocardial taurine content do not affect cardiac contractility (12,20). Thus, these differences in contractile response could arise from the nature of the preparations of effects related to the method of taurine depletion.

In contrast to myocardial contractility, the drug-induced, taurine-depleted hearts exhibit reduced rates of myocardial relaxation, an effect often linked to abnormalities in calcium uptake by the sarcoplasmic reticulum (28). Yet, in the taurine-depleted hearts an equally attractive explanation for the defect is that abnormal relaxation arises from changes in ion flux associated with repolarization. While further studies are required to clarify this point, it is noteworthy that the reduction in ERG amplitude in

Table 4. Effect of taurine depletion on the reperfused myocardium

Intermediate	Metabolic intermediates (μmol/g dry wt) Reperfused Control	Reperfused Taurine-depleted
ATP	16.1 ± 0.3	15.7 ± 0.5
ADP	3.6 ± 0.1	3.5 ± 0.1
AMP	0.5 ± 0.1	0.5 ± 0.1
ATP/ADP	4.6	4.6
Total adenine nucleotides	20.2 ± 0.5	20.1 ± 1.1
Phosphocreatine	30.4 ± 1.1	33.2 ± 1.3
Lactate	1.43 ± 0.2	0.6 ± 0.1*

Hearts from control and taurine-depleted rats were perfused with Krebs-Henseleit buffer containing 5 mM glucose, 2.5 U/1 insulin and 2.5 mM calcium for a stabilizing period of 20 minutes of reperfusion, the preparation was then frozen with Wollenberger clamps precooled in liquid nitrogen. Tissue intermediates were assayed according to standard procedure. Values shown represent means ± SEM of 7 to 9 hearts. *Denotes significant differences from control ($p<0.05$).

taurine-deficient retina is probably caused by taurine-related effects on sodium and potassium flux (30).

The other major finding of this study is that taurine-deficient hearts are more susceptible to ischemia-mediated tissue damage than normal hearts. The idea that taurine has cardioprotective activity is not new. In 1980 Azari et al. (1) reported that taurine administered decreases lesion severity in hearts of cardiomyopathic hamsters, an effect believed at the time to be caused by taurine-mediated reductions in tissue calcium content (18). In support of this hypothesis, subsequent work revealed that the amino acid was protective against such diverse calcium-overload models as the calcium paradox (16), isoproterenol toxicity (35) and hypoxic injury (8,27). Although the mechanism of action has not been established, taurine treatment also improves the status of congestive heart failure induced by aortic regurgitation (3). Based on a double-blind randomized crossover clinical trial, Azuma et al. (2) concluded that taurine therapy is beneficial in patients with congestive heart failure. Thus, the uniqueness of the present finding is not that taurine can affect the heart, but that a reduction in the size of a physiological, intracellular pool of taurine adversely alters myocardial function.

The observed cardioprotective effects of taurine are consistent with many of its other actions. First, and perhaps foremost, are the membrane stabilizing effects of taurine. Chovan et al. (5) and Sebring and Huxtable (31) have reported that taurine interacts with membrane components, leading to changes in the membrane which affect ion transport. It is now believed that by reducing calcium uptake by the heart, taurine diminished and some causes prevents calcium overload (7,18,35). This would have a major impact on the ischemic myocardium since the accumulation of calcium appears to contribute to the injury process (21).

Another mechanism by which taurine could influence the ischemic myocardium is through regulation of myocardial metabolism. Data from Table 3 reveal that ATP content of the taurine-depleted, ischemic heart is statistically lower than that of the nontaurine-depleted, ischemic heart. Previous studies have shown that taurine depletion leads to a stimulation in anaerobic metabolism (20). Yet, since anaerobic metabolism is the major source of ATP during ischemia, this effect alone could not account for the less favorable energy state of the taurine-depleted, ischemic heart. The only possible mechanism by which stimulation of anaerobic metabolism could contribute to the loss of tissue ATP would be if taurine depletion led to the accumulation of a toxic intermediate such as lactate (22). However, this possibility does not seem to be viable since lactate levels, which also reflect the activity of glycolysis, are not affected by taurine depletion. Thus, the basis for differences in ATP content of the two groups of ischemic hearts remains unclear. The possibility that the difference is related to alterations in energy utilization rather than production deserves further consideration.

Recently, several investigators have concluded that taurine possesses antioxidant activity (24). In light of growing evidence that free radicals play a central role in ischemic injury (19), this property of taurine could easily account for the cardioprotective action of the amino acid. Further studies are required to explore this possibility.

REFERENCES

1. Azari, J., Brumbaugh, P., and Huxtable, R.J., 1980, Prophylaxis by taurine in the hearts of cardiomyopathic hamsters, J. Mol. Cell. Cardiol., 12:1353-1366.
2. Azuma, J., Sawamura, A., Awata, N., Hasegawa, H., Ogura, K., Harada, H., Ohata, H., Yamauchi, K., Kishimoto, S., Yamagami, T., Ueda, E., and Ishiyama, T., 1983, Double blind randomized crossover trial of taurine in congestive heart failure, Curr. Therap. Res., 34:543-557.
3. Azuma, J., Takihara, K., Awata, N., Ohata, H., Sawamura, A., Harada, H., and Kishimoto, S., 1984, Benefician effect of taurine on congestive heart failure induced by chronic aortic regurgitation in rabbits, Res. Commun. Chem. Pathol. Pharmacol., 45:261-270.
4. Bunger, R., Swindall, B., Brodie, D., Zdunek, D., Stiegler, H., and Walter, G., 1986, Pyruvate attenuation of hypoxia damage in isolated working guinea pig heart, J. Mol. Cell. Cardiol., 18:423-438.
5. Chovan, J.P., Kulakowski, E.C., Sheakowski, S., and Schaffer, S.W., 1980, Calcium regulation by the low-affinity taurine binding sites of cardiac sarcolemma, Mol. Pharmacol., 17:295-300.
6. Dolara, P., Agresti, A., Giotti, A., and Pasquini, G., 1973, Effect of taurine on calcium kinetics of guinea pig heart, Eur. J. Pharmacol., 24:352-358.
7. Franconi, F., Martini, F., Stendardi, I., Matucci, R., Zilletti, L., and Giotti, A., 1982, Effect of taurine on calcium levels and contractility in guinea pig ventricular strips, Biochem. Pharmacol., 31:3181-3185.
8. Franconi, F., Stendardi, I., Matucci, R., Bennardini, F., Baccaro, C., and Giotti, A., A protective effect of taurine on hypoxic and reoxygenated guinea pig heart, in: "Taurine: Biological Actions and Clinical Perspectives", S.S. Oja, L. Ahtee, P. Kontro, and M.K. Paasonen, eds., Alan R. Liss, Inc., New York (1985), pp. 177-182.
9. Gaull, G.E., 1986, Taurine as a conditionally essential nutrient in man, J. Am. Coll. Nutr., 5:121-125.
10. Grosso, D.S., and Bressler, R., 1976, Taurine and cardiac physiology, Biochem. Pharmacol., 25:2227-2232.
11. Hayes, K.C., Stephen, Z.F., and Sturman, J.A., 1980, growth depression in taurine-depleted infant monkeys, J. Nutr., 110:2058-2064.
12. Huxtable, R.J., Regulation of taurine in the heart, in: "Taurine and Neurological Disorders", A. Barbeau and R.J. Huxtable, eds., Raven Press, New York (1985), pp. 5-17.
13. Huxtable, R.J., Laird, H.E.II, and Lippincott, S.E., 1979, The transport of taurine in the heart and the rapid depletion of tissue taurine content by guanidinoethyl sulfonate, J. Pharmacol. Exp. Therap., 211:465-471.
14. Huxtable, R.J., and Sebring, L.A., Cardiovascular actions of taurine, in:

"Sulfur Amino Acids: Biochemical and Clinical Aspects", K. Kuriyama, R.J. Huxtable and H. Iwata, Eds., Alan R. Liss, Inc., New York (1983), pp. 5-37.

15. Iwata, H., and Fujimoto, S., Potentiation by taurine of the inotropic effect of ouabain and the content of intracellular Ca^{2+} and taurine in the heart, Experientia, 32:1559-1561.
16. Kramer, J.H., Chovan, J.P., and Schaffer, S.W., 1981, Effect of taurine on calcium paradox and ischemic heart failure, Am. J. Physiol., 240:H238-H246.
17. Lake, N., 1983, taurine depletion of lactating rats: effects on developing pups, Neurochem. Res., 8:881-887.
18. McBroom, M.J., and Welty, J.D., 1977, Effects of taurine on heart calcium in the cardiomyopathic hamster, J. Mol. Cell. Cardiol., 9:853-858.
19. Meerson, F.Z., Kagan, V.E., Kozlov, Y.P., Belkina, L.M., and Arkhipenko, Y.V., 1982, The role of lipid peroxidation in pathogenesis of ischemic damage and the antioxidant protection of the heart, Basic Res. Cardiol., 77:465-485.
20. Mozaffari, M.S., Tan, B.H., Lucia, M.A., and Schaffer, S.W., 1986, Effect of drug-induced taurine depletion on cardiac contractility and metabolism, Biochem. Pharmacol. 35:985-989.
21. Nayler, W.G., Poole-Wilson, P.A., and Williams, A., 1979, Hypoxia and calcium, J. Mol. Cell. Cardiol., 11:683-706.
22. Neely, J.R., and Grotyohann, L.W., 1984, Role of glycolytic products in damage to ischemic myocardium, Circ. Res., 55:816-824.
23. Pasantes-Morales, H., Quesada, O., Carabez, A., and Huxtable, R.J., 1983, Effects of the taurine transport antagonist, guanidinoethane sulfonate, and β-alanine on the morphology of rat retina, J. Neurosci. Res., 9:135-143.
24. Pasantes-Morales, H., Wright, C.E., and Gaull, G.E., 1985, Taurine protests lymphoblastoid cells from iron-ascorbate induced damage, Biochem. Pharmacol., 34:2205-2207.
25. Read, W.O., and Welty, J.D., 1963, Effect of taurine on epinephrine and digoxin induced irregularities of the dog heart, J. Pharmacol. Exp. Therap., 139:283-289.
26. Rovetto, M.J., Lamberton, W.F., and Neely, J.R.., 1975, Mechanisms of glycolytic inhibition in ischemic rat hearts, Circ. Res., 37:742-751.
27. Sawamura, A., Sperelakis, N., and Azuma, J., 1986, Protective effect of taurine against decline of cardiac slow action potentials during hypoxia, Eur. J., Pharmacol., 120:235-239.
28. Schaffer, S.W., Kramer, J., and Chovan, J.P., 1980, Regulation of calcium homeostasis in the heart by taurine,Fed. Proc., 39:2691-2694.
29. Schaffer, S.W., Kulakowski, E.C., and Kramer, J.H., Taurine transport by reconstituted membrane vesicles, in: "Taurine in Nutrition and Neurology", R.J. Huxtable and H. Pasantes-Morales, eds., Plenum Press, New York (1982), pp. 143-160.
30. Schmidt, S.Y., Taurine in retinas of taurine-deficient cats and RCS rats, in: "The Effects of Taurine on Excitable Tissues", S.W. Schaffer, S.I. Baskin and J.J. Kocsis, Ed., Spectrum Publications, New York (1981), pp. 177-185.
31. Sebring, L.A., and Huxtable, R.J., 1985, Binding of taurine to cardiac sarcolemma and phospholiposomes, Fed. Proc., 44:2052.
32. Shaffer, J.E., and Kocsis, J.J., Methods of reducing tissue taurine levels, in: "The Effects of Taurine on Excitable Tissues", S.W. Schaffer, S.I. Baskin and J.J. Kocsis, Eds., Spectrum Publications, Inc., New York (1981), pp. 219-229.
33. Sturman, J.A., Moretz, R.C., French, J.H., and Wisniewski, H.M.., 1985, Postnatal taurine deficiency in the kitten results in a persistence of the cerebellular external granule cell layer: Correction by taurine feeding, J. Neurosci. Res., 13:521-528.
34. Vary, T.C., Angelaskos, E.T., and Schaffer, S.W., 1979, Relationship between adenine nucleotide metabolism and irreversible ischemic tissue

damage in isolated perfused rat heart, Circ. Res., 45:218-225.
35. Welty, M.C., Welty, J.D., and McBroom, M.J., 1982, Effect of isoproterenol and taurine on heart calcium in normal and cardiomyopathic hamsters, J. Mol. Cell. Cardiol., 14:353-357.

INOTROPIC ACTIVITY OF ORTHANILIC AND L-CYSTEIC ACID ON ISOLATED GUINEA-PIG VENTRICULAR STRIPS

F. Franconi[1], I. Stendardi[2], P. Failli[2], A. Fazzini[2] and A. Giotti[2]

[1]Cattedra di Saggi e Dosaggi Farmacologici, Università di Sassari, Sassari, Italy. [2]Dipartimento di Farmacologia Preclinica e Clinica "Mario Aiazzi- Mancini", Università di Firenze, Firenze, Italy

The positive inotropic effect of taurine in low calcium medium has been demonstrated extensively in guinea-pig (1,2,4,6,9), and chick heart (11). This effect of taurine is dose- and calcium-concentration dependent (2,6,8), and is not suppressed by the addition of propranolol (4,11), cimetidine or indomethacin (8). Furthermore, taurine antagonizes the negative inotropic effect of D-600 (7) in guinea-pig ventricular strips. More recently it has been demonstrated that two taurine-related compounds, orthanilic acid (2-aminobenzenesulfonic acid) and L-cysteic acid (Fig. 1), exert a positive inotropic effect on guinea-pig ventricular strips incubated in a medium containing 0.9 mM $CaCl_2$ (4). This effect is dose-dependent and occurs over the same concentration range as the inotropic action of taurine (5-20 mM). It is also similar with respect to the time course (4). Both compounds are characterized by primary amino and sulfonic groups bound to adjacent carbons (Fig. 1). Chain length, sulfonic and primary amino groups are a structural requirement for taurine-like inotropic activity (4). The present study investigates other properties of these two compounds in guinea-pig ventricular strips.

$H_2N-CH_2-CH_2-SO_3H$

Taurine

$H_2N-CH(COOH)-CH_2-SO_3H$

L-Cysteic Acid

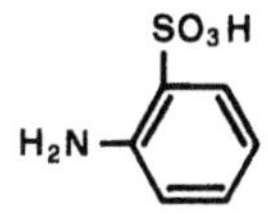

Orthanilic Acid

Fig. 1. Structures of taurine and its analogs.

METHODS

Guinea pigs weighing 250-300 g were stunned by a blow on the head. The hearts were rapidly removed after exsanguination and rinsed with ice-cold oxygenated saline solution. Right ventricular strips were suspended vertically in two 20 ml organ baths and superfused at a constant rate of 6 $ml.min^{-1}$ with aerated (95% O_2 - 5% CO_2) gas mixture Krebs-Henseleit bicarbonate buffer (pH = 7.4) containing (mM): NaCl 115, KCl 4.7, $MgSO_4$ 1.2, $CaCl_2$ 1.8, KH_2PO_4 1.2, $NaHCO_3$ 25, glucose 10 mM at 30°C. Orthanilic acid and L-cysteic acid increased the acidity of the solution; therefore the pH was adjusted with NaOH. The NaCl concentration was modified to keep the Na^+ concentration equal to that of the control solution. Preparations were electrically stimulated at 2 Hz with square wave pulse of 0.2 ms duration. The intensity was twice the usual threshold current of 0.3-0.4 mA using a platinum-point stimulation electrode. Isometric force of contraction was recorded with a force displacement transducer and a MARB polygraph recorder with a resting tension adjusted to 1 g. An equilibration period of 60 min with stimulation was allowed. Immediately after this equilibration and prior to the addition of the drugs and change of $CaCl_2$ concentration, mean tension magnitudes were recorded and considered as the absolute initial controls. They were taken as 100% compared with experimental values. In a set of experiments before the addition of the compound under examination the strips were incubated for 20 min with cimetidine and indomethacin. Where used, propranolol was added 20 min before orthanilic acid and L-cysteic acid. In the case of taurine, after equilibration, strips were incubated for 40 min with 20 mM taurine before the addition of propranolol because propranolol is an inhibitor of taurine transport (10).

In order to test the effect of phenylephrine, experiment were performed at 2.2 mM $CaCl_2$. After 40 min superfusion with taurine, orthanilic acid and

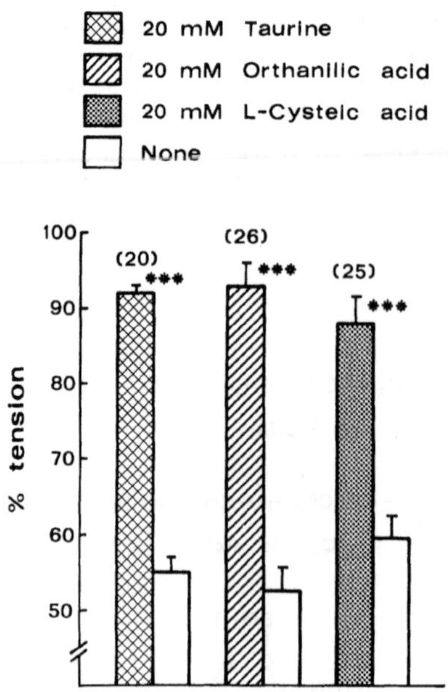

Fig. 2. Effect of 20 mM taurine, orthanilic acid and L-cysteic acid on the decrease in tension due to low calcium medium on guinea pig ventricular strips. Values are means ± SEM, with the number of experiments shown in parentheses. ***$P < 0.001$.

L-cysteic acid, both strips were admitted to a medium containing an effective β-blocking concentration of propranolol. Propranolol was allowed to act 10 min before administration of phenylephrine dissolved in saline solution containing ascorbic acid (1 $mg.ml^{-1}$). In order to equalize the osmotic effect of the compounds control strips were exposed to equimolar concentrations of sucrose.

RESULTS AND DISCUSSION

Tension in guinea pig ventricular strips was linearly linked to $CaCl_2$ at least between 0.45-3.6 mM $CaCl_2$ (6). The change of $CaCl_2$ from 1.8 to 0.9 mM in the medium produced a decrease in developed tension of about 50%. The superfusion of strips with 20 mM taurine, 20 mM orthanilic acid or 20 mM L-cysteic acid prevented statistically the decrease in tension induced by low calcium concentration as shown in Fig. 2. Typical time courses of the positive inotropic activity of these compounds are shown in Fig. 3.

The time course of inotropy produced by orthanilic and L-cysteic acids was like that of taurine. The positive inotropic effect of these compounds started after 10 min of superfusion and reached the maximum after 40 min. The activity of these compounds was studied at three $CaCl_2$ concentrations and as shown in Fig. 4 it is affected by the calcium concentration. Inotropy is more pronounced at 0.9 mM $CaCl_2$, decreases at 1.8 mM $CaCl_2$ and disappears at 2.2 mM $CaCl_2$.

More detailed studies have been performed at 0.9 mM $CaCl_2$. At 0.9 mM $CaCl_2$ the effect of a single concentration of propranolol (10^{-6} M), indomethacin (10^{-6} M) and cimetidine (10^{-6} M) was evaluated on the positive inotropic effect induced by taurine, orthanilic acid and L-cysteic acid. The effect of taurine and taurine analogs was not influenced by the beta adrenergic or histamine antagonists or indomethacin (Table 1).

Table 1. Effect of cimetidine, propranolol and indomethacin on the positive inotropic effect of 20 mM taurine, orthanilic acid and L-cysteic acid measured at 0.9 mM $CaCl_2$ in guinea pig ventricular strips.

	% OF CONTROL	TREATMENT
		10^{-6} M cimetidine
Taurine	107.9±0.8	105.3±3.6
Orthanilic acid	89.4±7.9	93.8±7.1
L-Cysteic acid	80.3±8.7	92.5±7.0
		10^{-6} M propranolol
Taurine	85.5±4.9	86.0±4.3
Orthanilic acid	95.0±5.6	88.5±5.1
L-Cysteic acid	94.7±7.5	86.2±5.9
		10^{-6} M indomethacin
Taurine	95.0±0.7	88.8±3.2
Orthanilic acid	107.4±9.3	100.2±9.7
L-Cysteic acid	89.2±5.1	94.4±9.8

Values are means ± SEM of at least 4 experiments. Control experiments were run at 1.8 mM $CaCl_2$.

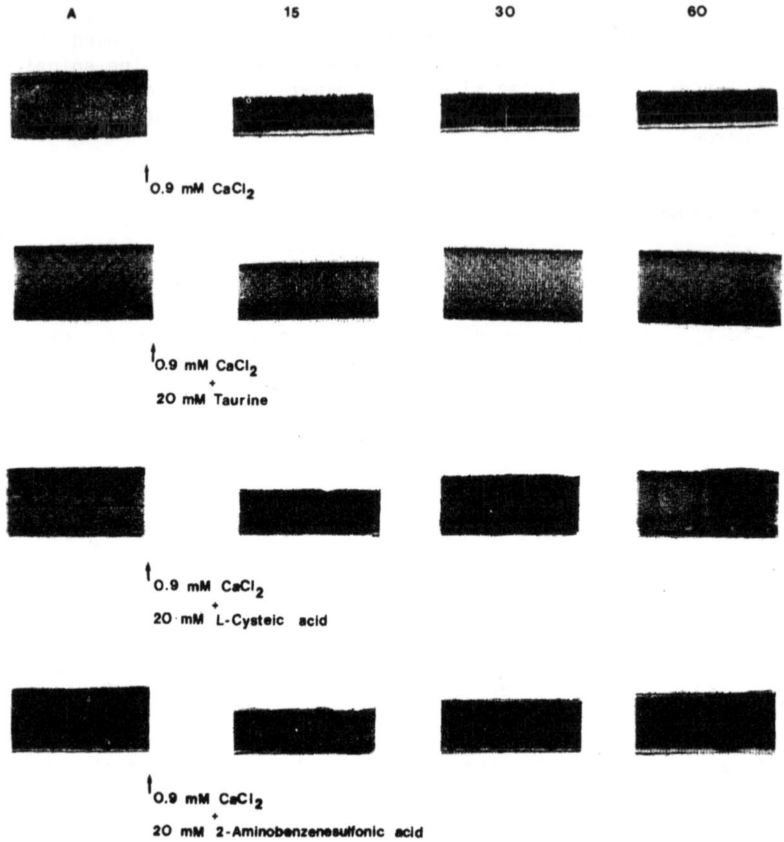

Fig. 3. Time-course of the positive inotropic effect of 20 mM taurine, L-cysteic acid and orthanilic acid (aminobenzene sulfonic acid) when guinea pig ventricular strips were exposed to low calcium medium.

The superfusion of guinea-pig ventricular strips with a medium containing 10 mM β-alanine, an inhibitor of taurine transport, (5) plus 10 mM orthanilic acid and or L-cysteic acid showed that the positive inotropic effect of these two compounds was not modified by β-alanine (Table 2), while the activity of taurine was reduced (8).

Another set of experiments was performed at 1.8 mM $CaCl_2$ to determine whether orthanilic acid and L-cysteic acid were able to counteract the negative inotropic effects of 0.1 µM D-600, a so-called "calcium channel antagonist". In this model 20 mM orthanilic acid reduced the negative inotropic effect of D-600, as does taurine, while 20 mM L-cysteic acid is ineffective (Table 3).

We also evaluated if orthanilic acid and L-cysteic acid were able to interfere with the positive inotropic effect mediated by stimulation of alpha-adrenoceptors through the superfusion of 3 µM phenylephrine (a selective alpha agonist) in the presence of 1 µM propranolol (experimental procedure has been described in ref. 3). In this test both orthanilic acid and L-cysteic acid (20 mM) were unable to interfere with this alpha-mediated

Table 2. Effect of 10 mM beta-alanine on the positive inotropic effect of 10 mM L-cysteic acid or orthanilic acid measured at 0.9 mM $CaCl_2$ on guinea pig ventricular strips.

		% of Control	
None	51.7±3.2	Beta-Alanine	53.0±6.7
L-Cysteic acid	85.1±7.1	L-Cysteic acid + Beta-Alanine	81.3±6.5**
Orthanilic acid	83.5±7.4	Orthanilic acid + Beta-Alanine	76.9±3.5*

Values are means ± SEM of at least 5 experiments. Control experiments were run at 1.8 mM Calcium.
* 0.05>P>0.01 vs beta-alanine ** 0.01>P>0.001 vs beta-alanine

positive inotropic effect, while taurine at the same dose statistically reduced it (Table 4).

The inotropic effect of orthanilic acid and L-cysteic acid (20 mM) was not influenced by the histamine H_2-receptor antagonist cimetidine (added to the bath before the addition of these compounds), the non-selective beta-blocker propranolol (added before orthanilic acid and L-cysteic acid and 40 min after taurine administration because propranolol is an inhibitor of taurine transport (10), or indomethacin (added before the compounds). These lack of responses are also found with taurine (8).These concentrations of cimetidine and propranolol were sufficient to block the effect of histamine and isoproterenol, respectively. Thus, the positive inotropic effect induced by orthanilic acid and L-cysteic acid is not elicited by activation of the histamine H_2 receptor, $beta_1$-$beta_2$-adrenoceptors or through the prostaglandin system.

The effect of orthanilic acid and L-cysteic acid is calcium-dependent, being more marked at low calcium concentrations while at 2.2 mM $CaCl_2$ it is absent. The effect of these compounds is not equal to that of taurine because orthanilic acid and L-cysteic acid do not prevent the positive

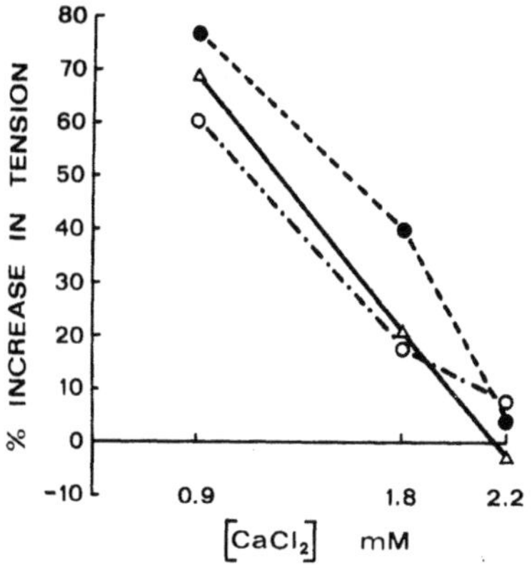

Fig. 4. Effect of 20 mM taurine (●), orthanilic acid (△) and L-cysteic acid (○) on tension measured at different calcium concentrations. Values are shown as percent increase over tension in the absence of the test substance, and are the means of at least 6 experiments.

Table 3. Effect of taurine, L-cysteic and orthanilic acid on the negative inotropic effect induced by 0.1 uM D-600

Control	20 mM Taurine	20 mM L-Cysteic acid	20 mM Orthanilic acid
43.8±4.0	74.2±5.7***	--	--
45.0±3.5	--	58.5±4.5	--
53.4±4.1	--	--	84.0±9.3**

Values are expressed as % of the control and are the means ± SEM of at least 9 experiments. ** 0.01>P>0.001, ***P<0.001.

inotropic effect induced by phenylephrine as taurine does (3). Furthermore, their activity is not prevented by beta-alanine; this fact suggests that beta-alanine does not interfer with the mechanism of action of either compound whereas the taurine effect is prevented by beta-alanine in the incubation medium (8).

Orthanilic acid has been measured in cardiac tissue after 30 min of washing, suggesting that it is linked to same cardiac structure (Franconi et al., unpublished results). Neither orthanilic acid and nor L-cysteic acid prevent the loss of taurine from ventricular tissue during superfusion (4). The mechanism of action of L-cysteic acid seems to be different from that of orthanilic acid because it does not prevent the negative inotropic effect of D-600. Our results demonstrate that two taurine analogs exert a positive inotropic effect but that their mechanism of action could be different.

Table 4. Effect of taurine, L-cysteic acid and orthanilic acid on the positive inotropic effect induced by 3 μM phenylephrine in the presence of 1 μM propranolol on guinea-pig ventricular strips at 2.2 mM $CaCl_2$.

% of Control	
3 μM Phe	3 μM Phe + 20 mM taurine
140.0±5.6	110.0±6.5***
	3 μM Phe + 20 mM L-Cysteic acid
135.1±6.6	124.7±9.5
	3 μM Phe + 20 mM Orthanilic acid
128.8±6.6	128.6±7.5

Values are means ± SEM of at least 7 experiments. Phe: phenylephrine. ***P<0.001.

ACKNOWLEDGEMENTS

This work was supported by a grant from M.P.I. We would like to thank Mrs. Mary Forrest for English revision of the manuscript and Dr. Anna Mancini for skillful technical assistance.

REFERENCES

1. Dietrich, J., and Diacono, J., 1971, Comparison between ouabain and taurine effects on isolated rat and guinea-pig hearts in low calcium medium, Life Sci., 10:499-507.
2. Dolara, P., Ledda, F., Mugelli, A., Mantelli, L., Zilletti, L., Franconi, F., and Giotti, A., 1978, Effect of taurine on calcium, inotropism and electrical activity of the heart, in: "Taurine and Neurological Disorders", A. Barbeau and R.J. Huxtable, eds., Raven Press, New York, pp. 151-159.
3. Franconi, F., Bennardini, F., Matucci, R., Stendardi, I., Failli, P., Manzini, S., and Giotti, A., 1986, Functional and binding evidence of taurine inhibition of alpha-adrenoceptor effects on guinea-pig ventricle, J. Mol. Cell. Cardiol., 18:461-468.
4. Franconi, F., Failli, P., Stendardi, I., Matucci, R., Bennardini, F., Baccaro, C., and Giotti, A., 1986, Positive inotropic effect of some taurine-related compounds on guinea-pig ventricular strips superfused with low calcium medium, Eur. J. Pharmacol., 124:129-133.
5. Franconi, F., Martini, F., Manghi, N., Galli, A., Bennardini, F., and Giotti, A., 1981, Uptake of ^{3}H-taurine into myocardial membranes, Biochem. Pharmacol., 30:77-80.
6. Franconi, F., Martini, F., Stendardi, I., Matucci, R., Zilletti, L., and Giotti, A., 1982, Effect of taurine on calcium level and contractility in guinea-pig ventricular strips, Biochem. Pharmacol., 31:3181-3185.
7. Franconi, F., Stendardi, I., Martini, F., Zilletti, L., and Giotti, A., 1982, Interaction between organic calcium-channel blockers and taurine in vitro and in vivo, J. Pharm. Pharmacol., 34:329-330.
8. Franconi, F., Stendardi, I., Matucci, R., Failli, P., Bennardini, F., Antonini, A., and Giotti, A., 1984, Inotropic effect of taurine in guinea-pig ventricular strips, Eur. J. Pharmacol., 102:511-514.
9. Guidotti, A., and Giotti, A., 1969, Digitalis inotropic effect on the auricular myocardium of taurine treated guinea-pigs, in: "Mèdicaments et mètabolisme du myocarde et du muscle strié", M. Lamarche and R. Royer eds., Nancy, pp. 487-491.
10. Huxtable, R.J., and Chubb, J., 1977, Adrenergic stimulation of taurine transport by the heart, Science, 198:409-411.
11. Sawamura, A., Azuma, J., Harada, H., Hasegawa, H., Okura, K., Sperelakis, N., and Kishimoto, S., 1983, Protection by oral pretreatment with taurine against the negative inotropic effects of low-calcium medium on isolated perfused chick heart, Cardiovasc. Res., 17:620-626.

CALCIUM OVERLOAD-INDUCED MYOCARDIAL DAMAGE CAUSED BY ISOPROTERENOL AND BY ADRIAMYCIN: POSSIBLE ROLE OF TAURINE IN ITS PREVENTION

Junichi Azuma, Tomoyuki Hamaguchi, Hideyuki Ohta, Keiko Takihara, Nobuhisa Awata, Akihiko Sawamura, Hisato Harada, Yuya Tanaka and Susumu Kishimoto

The Third Department of Internal Medicine, Osaka University Medical School, 1-1-50 Fukushima, Osaka 533, Japan

Calcium ion (Ca^{2+}) is essential for excitation-contraction coupling and for maintenance of cell integrity in the myocardium. On the other hand, it is clear that cytosolic Ca^{2+} loading may be the first event leading to cell death in certain forms of myocytic injury, such as Ca^{2+} paradox and isoproterenol (ISO) toxicity. The Ca^{2+} overload injury is characterized by an exhaustion of tissue high-energy phosphate, massive release of enzymes and extensive ultrastructural damage, as well as excessive influx of Ca^{2+} into the myocardial cells.

It is well known that a toxic dose of the synthetic catecholamine, ISO, causes massive heart cell necrosis (27). Therefore, ISO-treated animals have been used extensively as a model of cardiomyopathy (26). Several hypotheses have been proposed to explain the pathogenic mechanism of ISO-induced cardiotoxicity. These include an intracellular Ca^{2+}-overload (9) and depletion of high-energy phosphate reserves (37). It was recently proposed that free radical generation and lipid peroxidation could be involved in the process of ISO-induced cardiac damage (32).

Adriamycin (ADR), an anthracycline antibiotic derived from Streptomyces var. calcius is one of the most potent antineoplasmic agents for the treatment of various hematological malignancies and solid tumors. Unfortunately, its clinical usefulness is limited by its propensity to produce a severe, and sometimes life-threatening, cardiomyopathy in patients treated with cumulative doses greater than 550 mg/m^2 (20). Several proposals regarding the pathogenesis of ADR-induced cardiotoxicity have been made, including an increase in intracellular Ca^{2+} (4), free radical formation (21), effects on nucleic acid synthesis (29), effects on mitochondrial respiration (16) and release of cardio- or vaso-active substances (7).

Taurine (2-aminoethanesulfonic acid) is an amino acid that constitutes 50% of the free amino acid pool in the myocardium (13). Its myocardial level is altered in various cardiac conditions, suggesting that taurine has a possible pathophysiologic role in the heart. It has been reported that taurine modulates Ca^{2+} movement through the sarcolemma, and consequently exerts beneficial effects in modifying susceptibility to heart cell injury due to Ca^{2+}-overload in certain animal models (2,19,34). Alvarez and Storey (1) have suggested that in rabbit spermatozoa taurine acts as a superoxide scavenger and reduces free radical formation.

Ca^{2+} movement and free radical metabolism are considered as possible common pathways both in the mechanism of ISO- and ADR-induced cardiotoxicity, and in the physiological role of taurine. Therefore, it is possible that taurine might prevent these cell damage. We designed the present study to determine whether administration of taurine could exert any protective action against deterioration of the myocardium which occurs in response to ISO or ADR treatment in animals.

MATERIALS AND METHODS

Isoproterenol-induced myocardial damage

Three-day-old male chicks were randomly divided into four groups as follows: (i) Group 1, control; (ii) Group 2, taurine-treated; (iii) Group 3, ISO-treated; and (iv) Group 4, ISO and taurine-treated. Control and ISO-treated animals (Group 1 and 3) received forced oral administration of 1 ml of 7.45% mannitol (iso-osmolar with the taurine solution) for 7 days by means of a cannula inserted into the esophagus. Taurine-treated animals (Group 2 and 4) received 1 ml of 5% taurine for 7 days by the same procedure used in the control animals. ISO was administered subcutaneously (80 or 240 mg/kg) twice a day for 4 days following 3 days of mannitol or taurine treatment (Group 3 and 4). Chicks in Group 1 and 2 were given subcutaneous injection of 0.9% saline instead of ISO. The chicks were weighted and killed by decapitation 17 h after the last treatment of ISO or saline. Tissue from the apex region of the heart was rapidly excised, washed in 0.9% ice-cooled saline, blotted and weighed.

Adriamycin-induced myocardial damage

Male ddY mice (15-20 g body weight) at 4 weeks of age were randomly divided into three experimental groups as follows: (i) control group, (ii) ADR-treated group, (iii) ADR plus taurine-treated group. Taurine (0.5 ml of 5% solution) was administered ip once a day for 6 consecutive days to the mice in the ADR+taurine group, while an equal volume of sterile 0.9% saline was given to the two other groups. ADR (15 μg/kg body weight), freshly prepared with sterile 0.9% saline, was administered as a single ip injection 24 h after the last injection of taurine in the ADR+taurine group or of saline in the ADR group. Sterile 0.9% saline was injected ip (instead of ADR) in the control group. Throughout the study, 3% taurine in tap water was also given ad libitum as drinking fluid to animals in the ADR+taurine group. The mice were killed by a blow on the head at various times after the injection of ADR or saline. The hearts were excised and immediately washed in iced 0.9% saline. The ventricular muscles were blotted and rapidly frozen.

Biochemical determination

Protein and Taurine: The tissue was homogenized with 1 ml of H_2O, and the homogenate was used for the assay of protein and taurine. Protein was assayed by the method of Lowry (18). For the assay of taurine, the homogenate was mixed with 10 volumes of 10% sulfosalicylic acid, and centrifuged at 3,000 rpm for 10 min at 4°C. Taurine content was determined with a high-performance amino acid analyzer.

Ca^{2+} and Mg^{2+}: The tissue was dissolved with nitric acid and perchloric acid, then evaporated to dryness, and ashed. The residue was resuspended in 1% lanthanum chloride containing 0.1 N HCl. Analyses were performed by an atomic absorption spectrophotometer.

CPK, GOT and LDH: Myocardial CPK, GOT and LDH activities were assayed by modification of the methods of Spath et al. (33) and Rosalki et al (28). The homogenate in 0.25 M sucrose was centrifuged at 16,000 g for 30 min at 0°C. After dilution with a buffer containing 0.2% bovine serum albumin and trihydromethylaminomethane, these enzyme activities were assayed using a TBA 360 auto-analyser.

High-energy phosphates: After overnight lyophilization of tissue, the high-energy phosphates were extracted in 0.5 N perchloric acid and neutralized with KOH. The supernatant was assayed for adenosine nucleotides with a high-performance liquid chromatography apparatus and UV monitor, at a wavelength of 260 nm. The samples were chromatographed on an anion exchange column.

Malondialdehyde: The determination of lipid peroxide in muscles was carried out by measuring the formation of the 2-thiobarbituric acid-reacting substance, malondialdehyde. This was done by fluorophotometric assay (35).

Glutathione (GSH): The tissue was homogenized on ice in 0.1 M phosphate buffer (pH 8.0) containing 5 mM EDTA and 5% meta-phosphoric acid. The homogenate was centrifuged at 100,000 g for 30 min, and the supernatant was used for the assay. GSH was analyzed with o-phthalaldehyde using a fluorometric method (12).

Glutathione peroxidase (GSH-Px): Ventricular homogenates were analysed for GSH-Px activity by Hafeman's procedure (11). This method is based on the 5,5'-dithio-bis-2-nitrobenzoic acid reaction with sulfhydryls. An enzyme unit of activity was defined as a decrease in the log[GSH] of 0.001/min.

Phospholipid: Lipid was extracted from the tissue with a solution of chloroform-methanol-H_2O according to the method of Bligh and Dyer (6). Total phospholipid was determined by the measurement of inorganic phosphorus release upon hydrolysis of the lipid extract. The myocardial phospholipid profile was determined by two-dimensional thin-layer chromatography, using a TLC Silica Gel G glass plate. After development in the first dimension in a saturated chamber containing chloroform-methanol-ammonia-water, the plate was dried for 1 h in a chamber connected to a vacuum pump. The run in the second dimension was performed in a saturated chamber containing a mixture of chloroform-methanol-acetone-acetic acid-water. Each phospholipid spot visualized by inodine vapours was scrapped off for assay. Phospholipid was determined by inorganic phosphorus release, and each phospholipid was identified by use of authentic standards.

Statistical analysis

The results are presented as the mean + SEM of n experiments. Test of significance was calculated by Student's *t*-test, or analysis of variance (Bonferroni's method was used to compare individual data when a significant F value was obtained), depending on the design of the experiments. Differences were considered significant when the calculated p value was less than 0.05.

RESULTS

Isoproterenol-induced myocardial damage

During the 7-day experimental period, the overall mortality was 16% in the ISO-treated group and 5% in the ISO+taurine group; no death occurred in chicks treated with taurine or in the control chicks.

Table 1. Effects of taurine on changes in myocardial contents of protein, taurine, electrolytes, CPK, and adenine nucleotides induced by isoproterenol (80 mg/kg for adenine nucleotides, 240 mg/kg for protein, taurine, electrolytes and CPK).

	Control	Taurine treated	Isoproterenol treated	Isoproterenol + taurine treated
Protein content (mg/g dry weight)	(5) 436 ± 23	(5) 454 ± 12	(9) 484 ± 8	(7) 486 ± 19
Taurine content (g/100g protein)	(7) 5.20 ± 0.23	(8) 6.12 ± 0.18[a]	(9) 5.11 ± 0.23	(7) 6.07 ± 0.25[a,b]
Calcium content (μmol/g dry weight)	(5) 7.7 ± 0.9	(5) 8.3 ± 0.3	(16) 14.6 ± 1.1[a]	(15) 10.5 ± 1.3[a,b]
Magnesium content (μmol/g dry weight)	(5) 48.3 ± 2.4	(5) 52.6 ± 2.9	(9) 47.4 ± 1.2	(6) 50.5 ± 1.3
CPK activity (at 5 h) (U/100mg wet weight)	(9) 123.3 ± 3.2	(5) 124.6 ± 1.3	(12) 99.2 ± 7.3[a]	(7) 107.2 ± 6.6[a]
CPK activity (U/100mg wet weight)	N.D.	N.D.	(12) 90.3 ± 3.0[a]	(8) 110.9 ± 2.2[a,b]
ATP content (μmol/g dry weight)	(10) 25.0 ± 1.0	N.D.	(8) 14.2 ± 0.8[a]	(9) 18.6 ± 1.0[a,b]
Total adenine nucleotides content (μmol/g dry weight)	(10) 31.6 ± 1.0	N.D.	(8) 23.3 ± 1.2[a]	(8) 27.7 ± 1.1[a,b]

Data given are means ± SEM. Numbers in the parentheses indicate the number of hearts. Hearts were excised 17 h after the last treatment of isoproterenol or saline, except for CPK activity which was 5 h. [a]Significantly different from the control group ($p<0.05$). [b]Significantly different from the isoproterenol group ($p<0.05$). Abbreviation: ATP, adenine triphosphate; CPK, creatine phosphokinase; N.D., not determined.

Protein: There was no significant change in myocardial protein level in any of the treatment groups (Table 1).

Taurine: A significant increase of myocardial taurine level after taurine administration was observed, whereas ISO administration did not alter taurine content (Table 1).

Ca^{2+} and Mg^{2+}: In the ISO group, the myocardial Ca^{2+} content was about double the control value ($p<0.05$ vs. control). Administration of taurine significantly reduced the ISO-induced elevation of tissue Ca^{2+} level ($p<0.05$ vs. ISO). Taurine administration alone did not alter myocardial Ca^{2+} content. There were no significant changes in myocardial Mg^{2+} level in any of the groups (Table 1).

CPK: Myocardial CPK activity, 5 h after the last injection, was significantly reduced by ISO ($p<0.05$ vs. control). CPK activity of hearts excised 17 h after the last injection of ISO was higher than those excised after 5 h. Taurine treatment partially protected against the decrease in CPK activity following ISO treatment ($p<0.05$ vs. ISO) (Table 1).

High-energy phosphates: ATP content in the myocardium declined to below 60% of the control value when ISO (80 mg/kg) was administered ($p<0.05$ vs. control). Total adenine nucleotide content was also significantly decreased by ISO ($p<0.05$ vs. control). Taurine partially protected against this ISO-induced depletion of tissue ATP and total adenine nucleotide level ($p<0.05$ vs.ISO) (Table 1).

Malondialdehyde: Following ISO treatment, myocardial malondialdehyde levels increased ($p<0.05$ vs control). Taurine treatment protected against the increase in malondialdehyde produced by ISO ($p<0.05$). Administration of taurine alone did not affect the malondialdehyde content (Table 2).

GSH and GSH-Px: ISO substantially decreased myocardial GSH content to approximately 70% of the control level ($p<0.05$ vs. control). Taurine treatment protected against the decrease in GSH following ISO treatment ($p<0.05$). Administration of taurine alone did not affect the GSH content. There were no significant changes in myocardial GSH-Px activity in any of the treatment groups (Table 2).

Phospholipid: Myocardial total phospholipid content diminished significantly following ISO treatment ($p<0.05$ vs. control). Taurine treatment protected against the ISO-induced degradation of tissue phospholipid content ($p<0.05$). The major phospholipids in the control myocardium were; phosphatidylcholine (38%), phosphatidylethanolamine (38%), and cardiolipin (17%), with minor amounts of phosphatidylserine, phosphatidylinositol, and sphingomyelin. Neither lysophosphatidylcholine nor lysophosphatidylethanolamine was detectable. ISO treatment decreased substantially tissue phosphatidylcholine, phosphatidylethanolamine, cardiolipin and phosphatidylserine ($p<0.05$). However, taurine treatment significantly prevented these decreases following ISO treatment ($p<0.05$) (Table 3).

Adriamycin-induced myocardial damage

Over an experimental period of 144 h after the injection of ADR, the overall survival rate was 17% (2/12 animals) in the ADR group and 58% (7/12 animals) in the ADR+taurine group. Thus, taurine significantly improved the survival rate of the ADR-treated mice ($p<0.05$). This result suggests that taurine protects the mice against the acute toxicity of ADR.

Taurine: ADR significantly elevated myocardial taurine level ($p<0.05$) (Table 4).

Table 2. Effects of taurine on changes in myocardial content of malondialdehyde, GSH, and GSH-Px induced by isoproterenol (240 mg/kg).

	Control	Taurine treated	Isoproterenol treated	Isoproterenol + taurine treated
Malondialdehyde content (nmol/g wet weight)	(9) 166.6 ± 4.9	(9) 172.4 ± 6.6	(12) 216.6 ± 13.0[a]	(12) 178.6 ± 8.2[b]
GSH content (μg/g wet weight)	(9) 214.8 ± 7.6	(5) 211.9 ± 15.9	(9) 151.1 ± 9.1[a]	(10) 195.7 ± 7.7[b]
GSH-Px activity (U/g wet weight)	(10) 435.5 ± 38.6	(6) 447.2 ± 40.4	(11) 381.8 ± 21.3	(9) 449.7 ± 26.7

Data given are means ± SEM. Numbers in the parentheses indicate the number of hearts. [a]Significantly different from the control group ($p<0.05$). [b]Significantly different from the isoproterenol group ($p<0.05$).

Table 3. Effects of taurine on changes in myocardial phospholipid profile induced by isoproterenol (240 mg/kg)

	Control (4)	Isoproterenol treated (5)	Isoproterenol + taurine treated (4)
Unknown	0.42 ± 0.06	0.82 ± 0.21[a]	0.51 ± 0.12[b]
Origin	0.34 ± 0.12	0.09 ± 0.03[a]	0.21 ± 0.03[a,b]
Phosphatidylserine	0.32 ± 0.04	0.15 ± 0.05[a]	0.31 ± 0.04[a,b]
Phosphatidylinositol	0.29 ± 0.10	0.20 ± 0.04	0.45 ± 0.05[b]
Sphigomyelin	1.05 ± 0.09	0.68 ± 0.14	0.57 ± 0.03[a]
Cardiolipin	5.51 ± 0.29	2.68 ± 0.44[a]	3.90 ± 0.23[a,b]
Phosphatidylethanolamine	12.11 ± 0.16	5.81 ± 0.55[a]	9.29 ± 0.65[a,b]
Phosphatidylcholine	12.12 ± 0.09	6.42 ± 0.96[a]	10.80 ± 0.71[b]
Total	31.99 ± 0.70	16.97 ± 1.93[a]	26.03 ± 1.21[a,b]

Data given are means ± SEM in μmol/g wet weight. Numbers in parentheses indicate the number of hearts. [a]Significantly different from the control group ($p<0.05$). [b]Significantly different from the isoproterenol group ($p<0.05$).

Table 4. Effects of taurine on changes in myocardial contents of taurine, calcium, CPK, LDH, GOT, malondialdehyde and GSH-Px induced by adriamycin (15 mg/kg).

	Control	Adriamycin treated	Adriamycin + taurine treated
Taurine content (μmol/g dry weight)	(11) 180 ± 4	(9) 194 ± 4[a]	N.D.
Calcium content (μmol/g dry weight)	(9) 4.20 ± 0.10	(10) 4.98 ± 0.18[c]	(10) 4.36 ± 0.17[b]
CPK activity (IU/100 mg wet weight)	(15) 63.2 ± 3.1	(15) 52.0 ± 1.6[a]	(15) 59.5 ± 2.2[b]
LDH activity (IU/100 mg wet weight)	(15) 53.0 ± 1.2	(15) 46.4 ± 1.5[c]	(15) 52.6 ± 1.0[d]
GOT activity (IU/100 mg wet weight)	(15) 16.1 ± 0.3	(14) 14.5 ± 0.5[a]	(15) 16.0 ± 0.3[b]
Malondialdehyde content (nmol/g wet weight)	(11) 112 ± 9	(13) 185 ± 18[c]	(15) 139 ± 10[b]
GSH-Px activity (U/g wet weight)	(19) 891 ± 62	(21) 645 ± 70[c]	(20) 685 ± 42[c]

Data given are means ± SEM. Numbers in the parentheses indicate the number of hearts. Hearts were excised at 24 h for GSH-Px activity, at 48 h for CPK, LDH, and GOT activities, and at 72 h for the others after the single treatment of adriamycin or saline. [a,c]Significantly different from the control group, [a]: $p<0.05$; [c]: $p<0.01$. [b,d]Significantly different from the adriamycin group, [b]: $p<0.05$; [d]: $p<0.01$.

Ca^{2+}: In the ADR group, Ca^{2+} content was significantly raised over the control value ($p<0.01$). Combined administration of taurine significantly reduced the ADR-induced elevation of Ca^{2+} in the myocardium ($p<0.05$ vs. ADR group) (Table 4).

CPK, GOT and LDH: Myocardial CPK activity was significantly decreased by ADR treatment ($p<0.05$). Combined administration of taurine caused a significant retention of CPK activity ($p<0.05$ vs. ADR group). Myocardial GOT and LDH activities were significantly decreased by ADR ($p<0.05$). Taurine produced a significant retention of GOT and LDH activities ($p<0.05$ and $p<0.01$, respectively vs. ADR group). Thus, administration of taurine protected against the decrease in these enzyme activities upon exposure to ADR (Table 4).

Malondialdehyde: Following ADR treatment, myocardial malondialdehyde level markedly increased ($P<0.01$). Taurine treatment protected against the elevation of malondialdehyde produced by ADR ($P<0.05$ vs. ADR group) (Table 4).

GSH-Px: There was a significant drop in myocardial GSH-Px activity by ADR treatment ($P<0.01$). Despite administration of taurine, this activity remained significantly low ($P<0.01$ vs. control group) (Table 4).

DISCUSSION

The present results demonstrate for the first time that taurine can protect against the cardiac cell injury induced by ISO and by ADR, both of which are in part caused by intracellular Ca^{2+} overload. We observed that administration of taurine lowered the elevated cytosolic Ca^{2+} level. Therefore, it seems likely that the beneficial effect of taurine in the present animal models was mediated through the reduction of myocardial Ca^{2+} overload.

The present study confirmed the well-known cytotoxicity of exogenous catecholamines on the myocardium, including a variety of biochemical changes. In our previous study (24), we observed mild edema of the interstitium in the hearts treated with 80 mg/kg ISO. At 240 mg/kg, there were marked eosinophilia, degeneration of myocytes, interstitial edema, and migratory cell infiltration. However, in the ISO+taurine group, the severity of the lesions observed was milder compared to the ISO group.

In the present study, a single high dose (15 mg/kg) of ADR was used. Although this dose might cause acute enteric or bone marrow lesions, death was probably due to its cardiotoxicity. Several investigators have demonstrated severe ultrastructural damage to the myocardium after administration of ADR at a dose similar to that used in the present experiments (21). In addition, we also showed various biochemical perturbations in the myocardium after ADR administration. Taurine significantly improved the survival rate of ADR-treated mice. This beneficial effect of taurine might result from the protection against myocardial injury.

One hypothesis for the mechanism of the ISO- and ADR- induced cardiotoxic reaction is an increase in intracellular Ca^{2+} content. We have proposed that the enhanced Ca^{2+} inflow (via stimulation of slow channels) could be a factor in the Ca^{2+}-overload associated with adriamycin-induced cardiomyopathy (4). Activation of myocardial $_1$-adrenergic receptors, in ISO-treated animals, with resultant elevation of cyclic AMP, increases Ca^{2+} influx and slow inward current. The inhibition of phosphodiesterase activity by ADR (17) may be responsible for raised cyclic AMP levels. A raised cytosolic Ca^{2+} may hasten myocardial cell death by various events (23). Mitochondrial Ca^{2+} overloading impairs its function, followed by failure of

ATP production by this organelle. Depletion of the tissue stores of ATP results in a localized accumulation of Ca^{2+} at the inner surface of the cell membrane, which may activate the various membrane-located phospholipases and proteases, resulting in the partial destruction of the cell membrane and loss of its selective permeability. Activation of phospholipases stimulates peroxidation of membrane lipids. Damage to the cell membrane produces release of intracellular enzymes. The present study showed that taurine antagonized myocardial Ca^{2+} accumulation and protected against depletion of intracellular enzyme activity. Huxtable suggested that the protective effect of taurine against the ISO-induced cardiomyopathy involved an interaction of taurine with the cell membrane (15). We have recently shown that taurine transiently (0.5-2.0 min) decreased ISO-induced Ca^{2+} dependent slow action potentials in perfused heart preparation (30).

The partial protective effect of taurine against cell necrosis has been demonstrated in several types of Ca^{2+}-overload cardiotoxicity. For example, Azari et al. reported that, in cardiomyopathic Syrian hamster, taurine could relieve the intracellular Ca^{2+} accumulation and retard the development of cardiomyopathy (2). Welty et al. noted that taurine exerted a protective effect against ISO-induced Ca^{2+} accumulation in the Syrian hamster (36). Fleckenstein and coworkers demonstrated that an ISO-induced loss of high-energy phosphates and myocardial cellular necrosis can be prevented by an attenuation of the myocardial Ca^{2+}-overload (9). They considered that Ca^{2+}-induced high-energy phosphate exhaustion is a crucial point in the etiology of ISO-induced myocardial fiber necrosis. We showed that taurine partially protected against the ISO-induced depletion of the high-energy phosphate level in the myocardium.

We demonstrated that ISO and ADR substantially increased malondialdehyde content in the myocardium. A mechanism involving free radical formation seems to be the most prevalent opinion on ISO- and ADR-induced cardiotoxicity in recent years (5, 31, 32, 35). Free radicals are toxic to tissues, and their dismutation to peroxide initiates a radical chain reaction, resulting in conversion of unsaturated fatty acids to lipid peroxides and in increase in membrane permeability. As the result, the sarcolemma is destroyed, followed by leakage of intracellular enzymes from the cardiomyocyte. Taurine significantly reduced the elevation of malondialdehyde content, and depletion of intracellular enzyme activity in the myocardium. Taurine may exert these protective effects through some of the following actions: (i) an action as a superoxide scavenger, (ii) an indirect effect through normalization of myocardial Ca^{2+} content, (iii) a direct prevention of sarcolemmal phospholipids from attack by oxygen radical. In the present study, parenteral administration of taurine suppressed lipoperoxide formation, inhibited phospholipid degradation and diminished intracellular Ca^{2+} accumulation. Alvarez and Storey (1) suggested in rabbit spermatozoa that taurine acts as a superoxide scavenger. It was also demonstrated that taurine suppresses lipid peroxidation in the liver of CCl_4-intoxicated rats (22).

A decrease of myocardial GSH content was observed in ISO-treated animals in the present study. Scavenger enzymes, such GSH-Px, catalase and superoxide dismutase, directly catalyze the free radicals and inhibit the peroxidation of membrane phospholipids (25). However, superoxide dismutase and catalase activities in heart are very low compared to other organs (10). Therefore, in hearts, the glutathione/GSH-Px system is considered to be the major enzymatic pathway for detoxification of free radicals (8). We suppose that if free radicals were formed, GSH would interact readily with these oxidative products and thereby would become depleted. In addition, leakage of GSH may occur due to sarcolemmal disruption. Prolonged depletion of myocardial GSH content might, in turn, lead to the elevation of lipoperoxide content. We showed a higher level of myocardial GSH content in ISO+taurine-treated animals compared with animals treated with ISO alone. Therefore, taurine may reduce the free radical formation and decrease GSH release (as well as CPK) via the myocardial sarcolemma.

In the present study, we demonstrated that the taurine content of the hearts from the ADR-treated mice was elevated. This observation may be related to the increase of myocardial cyclic AMP level by phosphodiesterase inhibition (17) and/or, by β-adrenergic stimulation. Taurine transport into the myocardium is stimulated by β-adrenergic agonists (14) and by other conditions that increase cellular cyclic AMP level (3).

In conclusion, we demonstrated that taurine partially protected against myocardial damage induced by ISO and by ADR. The beneficial effect of taurine may be due in part to inhibition of lipoperoxide formation and to protection against the deterioration of the cell membrane, resulting in reduction of intracellular massive Ca^{2+} accumulation and overload. However, it might be preferable to rationalize its action in terms of taurine primarily being a calcium modulator.

ACKNOWLEDGEMENT

We thank Dr. T. Sadatome for his contribution and Miss Y. Ikeda for her assistance in preparing the manuscript.

REFERENCES

1. Alvarez, J.G., and Storey, B.T., 1983, Taurine, hypotaurine, epinephrine and albumin inhibit lipid peroxidation in rabbit spermatozoa and protect against loss of motility, Biol. Reprod., 29:548-555.
2. Azari, J., Brumbaugh, P., and Huxtable, R., 1980, Prophylaxis by taurine in the hearts of cardiomyopathic hamsters, J. Mol. Cell. Cardiol., 12:1353-1366.
3. Azari, J., and Huxtable, R.J., 1980, The mechanism of the adrenergic stimulation of taurine influx in the heart, Eur. J. Pharmacol., 61:217-223.
4. Azuma, J., Sperelakis, N., Hasegawa, H., Tanimoto, T., Vogel, S., Ogura, K., Awata, N., Sawamura, A., Harada, H., Ishiyama, T., Morita, Y., and Yamamura, Y., 1981, Adriamycin cardiotoxicity: possible pathogenic mechanisms, J. Mol. Cell. Cardiol., 13:381-397.
5. Bechur, N.R., Gordon, S.L., and Gee, M.B., 1977, Anthracycline antibiotic augmentation of microsomal electron transport and free radical formation, Mol. Pharmacol., 13:911-917.
6. Bligh, E.G., and Dyer, M.J., 1959, A rapid method of total lipid extraction and purification, Can. J. Biochem. Physiol., 13:911-917.
7. Bristow, M.R., Kantrowitz, N.E., Harrison, S.D., Minobe, W.A., Sageman, W.S., and Billingham, M.E., 1983, Mediation of subacute anthracycline cardiotoxicity in rabbits by cardiac histamine release, J. Cardiovasc. Pharmacol., 5:913-919.
8. Doroshow, J.H., Locker, G.Y., and Myers, C.E., 1980, Enzymatic defenses of the mouse heart against reactive oxygen metabolites, Clin. Invest., 65:128-135.
9. Fleckenstein, A., Janke, J., Doring, H., and Leder, O., 1974, Myocardial fiber necrosis due to intracellular Ca overload - a new principle in cardiac pathophysiology, in: "Recent Advances in Studies on Cardiac Structure and Metabolism, Myocardial Biology, vol. 4", N.S. Dhalla, ed., University Park Press, Baltimore, pp. 563-587.
10. Fridovich, I., 1978, The biology of oxygen radicals, Science (Wash., D.C.), 201:875-880.
11. Hafeman, D.G., Sunde, R.A., and Hoekstra, W.G., 1974, Effect of dietary selenium on erythrocyte and liver glutathione peroxidase in the rat, J. Nutr., 104:580-587.

12. Hissin, P.J., and Hilf, R., 1976, A fluorometric method for determination of oxidized and reduced glutathione in tissues, Anal. Biochem., 74:214-226.
13. Huxtable, R.J., 1976, Metabolism and function of taurine in the heart, in: "Taurine", Raven Press, New York, pp. 99-120.
14. Huxtable, R.J., and Chubb, J., 1977, Adrenergic stimulation of taurine transport by the heart, Science, 198:409-411.
15. Huxtable, R.J., and Sebring, L.A., 1983, Cardiovascular actions of taurine, Prog. Clin. Biol. Res., 125: 5-37.
16. Kishi, T., Watanabe, T., and Folkers, K., 1976, Bioenergics in Clinical Medicine: Prevention by forms of conenzyme Q_{10}-enzymes in mitochondira of the myocardium, Proc. Natl. Acad., 73:4653-4656.
17. Lau, J., Fumagalli, A., Schein, P.S., and Rahman, A., 1985, Selective inhibition of cardiac cyclic nucleotide phosphodiestereases by doxorubicin and daunorubicin, Life Sci., 35:589-599.
18. Lowry, O.H., Rosenbrough, H.J., Farr, A.L., and Randall, R.J., 1959, Possible measurement with Folin phenol reagent, J. Biol. Chem., 193: 254-275.
19. McBroom, M.J., and Welty, J.D., Effects of taurine on heart calcium in the cardiomyopathic hamster, J. Mol. Cell. Cardiol., 9:853-858.
20. Minow, R.A., Benjamin, R.S., and Gottlieh, J.A., 1975, Adriamycin (NSC-123127) cardiomyopathy - an overview with determination of risk factors, Cancer Chemother. Rep., Part 3, 6:195-201.
21. Myers, C.E., McGuire, W.P., Liss, H.L., Ifrim, I., Grutzinger, K., and Young, R.C., 1979, Adriamycin: the role of lipid peroxidation in cardiac toxicity and tumor response, Science, 197:165-167.
22. Nakashima, T., Takino, T., and Kuriyama, K., 1983, Therapeutic and prophylactic effects of taurine administration on experimental liver injury, Prog. Clin. Biol. Res., 179:449-459.
23. Nayler, W.G., 1981, The role of calcium in the ischemic myocardium, Am. J. Pathol., 102:262-270.
24. Ohta, H., Azuma, J., Ohnishi, S., Awata, N., Takihara, K., and Kishimoto, S., 1986, Protective effect of taurine against isoprenaline-induced myocardial damage, Basic. Res. Cardiol., 81:473-481.
25. Reed, J.D., and Beatty, P.W., Biosynthesis and regulation of glutathione: toxicological implication, Rev. Biochem. Toxicol., 2:213-241.
26. Rona, G., Catecholamine cardiotoxicity, J. Mol. Cell Cardiol., 17:292-306.
27. Rona, G., Chappel, C.L., Balazs, T., and Gaudry, R., An infarct-like myocardial lesion and other toxic manifestations produced by isoproterenol in the rat, Arch. Pathol., 67:443-455.
28. Rosalki, S.B., 1967, An improved procedure for serum creatine phosphokinase determination, J. Lab. Clin. Med., 69:696-705.
29. Rosenoff, S.H., Brooks, E., Bostick, F., and Young, R.C., 1971, Alterations in DNA synthesis in cardiac tissue induced by adriamycin in vivo-relationship to fatal toxicity, Biochem. Pharmacol., 24:1898-1901.
30. Sawamura, A., Sperelakis, N., Azuma, J., and Kishimoto, S., 1986, Effects of taurine on the electrical and mechanical activities of embryonic chick heart, Can. J. Physiol. Pharmacol., 64:649-655.
31. Singal, P.K., Beamish, R.E., and Dhalla, N.S., 1983, Potential oxidative pathways of catecholamines in the formation of lipid peroxides and genesis of heart disease, Adv. Exp. Med. Biol., 161:391-401.
32. Singal, P.K., Kapur, N., Dhillon, K.S., Beamish, R.E., and Phalla, N.S., 1982, Role of free radicals in catecholamine-induced cardiomyopathy, Can. J. Physiol. Pharmacol., 60: 1390-1397.
33. Spath, J.A., Gee, M.H., and Gwirts, P.A., 1979, Normalization of the measurement of cardiac creatine phosphokinase activity, Cardiology, 64:222-230.
34. Takihara, K., Azuma, J., Awata, N., Ohta, H., Sawamura, A., Kishimoto, S., and Sperelakis, N., 1985, Taurine's possible protective role in

age-dependent response to calcium paradox, Life Sci., 37:1705-1710.
35. Tanizawa, H., Sazuka, Y., and Takino, Y., 1981, Microdetermination of lipoperoxide in the mouse myocardium by thiobarbituric acid fluorophotometory, Chem. Pharm. Bull., 29:2910-2914.
36. Welty, M.C., Welty, J.D., and McBroom, M.J., 1982, Effect of isoproterenol and taurine on heart calcium in normal and cardiomyopathic hamsters, J. Mol. Cell Cardiol., 14:353-357.
37. Zimmer, H.G., and Gerlach, E., 1976, Influence of isoproterenol, propranolol, and D600 on the de novo synthesis of adenine nucleotides in rat hearts, in: "Recent Advances in Studies on Cardiac Structure and Metabolism, Biochemistry and Pharmacology of Myocardial Hypertrophy, Hopoxia, and Infarction, vol 7.", P. Harris, R.J. Bing, and A. Fleckenstein, eds., University Park Press, Baltimore, pp. 131-136.

AN ANALYSIS OF THE EFFECTS OF TAURINE ON CELLULAR AND EXTRACELLULAR CALCIUM MOBILIZATION IN RABBIT EAR ARTERY

Stefano Manzini[1], Adriana Castellucci, Flavia Franconi[2], and Alberto Giotti[3]

[1] Malesci S.p.A. - Instituto Farmacobiologico , Pharmacology Department, Via Porpora, 22 Firenze, Italy
[2] University of Sassari, Faculty of Pharmacy, Via Muroni, 23 Sassari, Italy
[3] University of Florence, Pharmacology Department, Viale Morgagni, 65 Firenze, Italy

Tonic vasoconstriction of rabbit ear artery depends mainly upon influx of extracellular Ca^{++} through plasmalemmal calcium channels activated by depolarization (voltage-operated channel) or by stimulation of specific receptors (receptor operated channel) (4,5,16,17,19). Taurine has been shown to be an effective antagonist of depolarization- but not norepinephrine-induced tonic vasoconstrictions in rabbit ear artery (8,9), suggesting a selective interference on the availability of the voltage operated channel. However, in vascular smooth muscle other sources of calcium are involved in excitation-contraction coupling; for example release of calcium from internal store such as sarcoplasmic reticulum (4).

A mobilization of cellular calcium stores could be achieved both with autacoids such as norepinephrine (3,24) and serotonin (15,20) and also with caffeine (6,22). In view of this it appeared worthwhile to investigate in a more comprehensive way the overall effect of taurine on excitation-contraction coupling by studying its potential interference with norepinephrine-induced cellular and extracellular Ca^{++}-dependent vasoconstriction. In these studies a comparison with the effects of a well-known calcium entry blocker (nifedipine) (13) and of an inhibitor of cellular Ca^{++} mobilization (procaine) (12,7) was performed. In addition, functional experiments were carried out to assess the potential effects of taurine on kinetics of depletion by norepinephrine of cellular Ca^{++} store.

MATERIALS AND METHODS

New Zeland male albino rabbits (2.5-3.5 kg) were heparinized (1000 IU iv) and then killed by a blow to the back of the head. A 3 cm segment of central ear artery was dissected free from adhering tissues, cannulated at both ends with polyethylene tubing, and transferred to a 7 ml organ bath thermostatically controlled at 37°C. The volume of the fluid was kept constant by means of an overflow. The arterial segments was perfused intraluminally by means of a SOCREL five channel peristaltic pump at a rate of 5 ml/min, while extraluminal perfusion at a rate of 8 ml/min was done by means of a Mariotte bottle. Both intraluminal and extraluminal perfusion fluids were gassed with 95% O_2 and 5% CO_2 and heated to 37°C.

Changes in intraluminal perfusion fluid were performed by changing the reservoir supplying the peristaltic pump. All drugs were added intraluminally. Changes in intraluminal perfusion pressure recorded by means of a pressure transducer (Bentley Trantec) attached to the inlet side of the perfusion system, were taken as an indirect measure of increase of the arterial contractile tone. Intraluminal pressure was continuously recorded by means of a polygraph (Basile Unirecord 7050).

Solutions used had the following composition (mM): Physiological salt solution: NaCl 119, KCl 4.7, $MgSO_4$ 1.5, KH_2PO_4 1.2, $CaCl_2$ 2.5, $NaHCO_3$ 25, glucose 11; high-K^+ solution: NaCl 69, KCl 54.7, $MgSO_4$ 1.5, KH_2PO_4 1.2, $CaCl_2$ 2.5, $NaHCO_3$ 25, glucose 11; high-K^+ Ca^{++}-free solution: NaCl 69, KCl 54.7, $MgSO_4$ 1.5, KH_2PO_4 1.2, $NaHCO_3$ 25, glucose 11, EDTA 0.77; Ca^{++}-free solution NaCl 117, KCl 4.7, $MgSO_4$ 41.5, KH_2PO_4 1.2, $NaHCO_3$ 25, glucose 11, EDTA 0.77.

All experiments commenced after an equilibration period of at least 60 min. In a first set of experiments, preparations were contracted by perfusing with norepinephrine (5 μM) or high-K^+ (54 mM) solution. When the tonic component of the vasoconstriction was stable the drug under study was administered intraluminally and the effect observed for 15-20 min.

In other experiments the effects of taurine, nifedipine and procaine were investigated on norepinephrine-induced cellular and extracellular Ca^{++} -dependent vasoconstriction according to the method of Manzini et al. (17) (see fig. 1 left panel). Briefly, after the stabilization period the extraluminal and intraluminal physiological salt solution was replaced with high-K^+ Ca^{++}-free solution which produced a rapid vasoconstriction with gradual return to resting values. Five minutes later the intraluminal perfusion fluid was substituted for 3 min with high-K^+ Ca^{++}-free medium containing norepinephrine (5 μM) which elicited a transient vasoconstriction followed by return to resting values. Such a response has been demonstrated to be purely dependent on mobilization of cellular calcium (16,17,18).

Subsequent intra- and extraluminal perfusion with physiological salt solution resulted in a further rapid and transient vasoconstriction whose amplitude was assumed to be a functional measure of the ability of norepinephrine to induce extracellular Ca^{++}-dependent responses (16,17,18,19).

The effect of the test drug on these responses was assessed by administering the drug at the stated concentration 2 min after the start of perfusion with high-K^+ Ca^{++}-free solution and then throughout the experimental procedure described above.

Caffeine-induced vasoconstriction was elicited after 5 min of intra and extraluminal perfusion with a Ca^{++}-free medium. Caffeine administration lasted for 3 min and then the preparation was reperfused intra- and extraluminally with physiological salt solution. Preliminary experiments indicated that reproducible responses could be elicited at 15 min interval. In the same preparation, caffeine-induced vasoconstriction was elicited in the absence and in the presence of the stated concentration of taurine, nifedipine or procaine.

These drugs were added to the intraluminal Ca^{++}-free medium five minutes before and during the challenge with caffeine. Only one drug was tested in each preparation.

In another experimental set, the rate of norepinephrine-induced depletion of cellular calcium stores was assessed by using the following protocol: after equilibration in physiological salt solution, preparations

Table 1. Effect of taurine (40 mM) on amplitude of norepinephrine and high-K^+-induced tonic vasoconstrictions in isolated perfused rabbit ear artery.

Vasoconstrictor agent	Control (mm Hg)	Taurine (40 mM) (mm Hg)	Beta-alanine (40 mM) (mm Hg)
Norepinephrine (5 μM)	187 ± 13	162 ± 15 (6)	N.T.
High-K^+ (54 mM)	153 ± 11	83 ± 13* (7)	180 ± 20 (4)

Each value is the mean ± SEM for the number of experiments shown in parentheses; * $p<0.01$ as compared to control response; N.T. = Not Tested

were perfused intra and extraluminally with a Ca^{++} free medium for 15 min and then challenged for 3 min with norepinephrine (1 μM) intraluminally perfused (always in a Ca^{++} free medium). This series of 15 min in Ca^{++} free medium and 3 min of challenge with norepinephrine was carried out four times and then the preparation were reperfused with physiological salt solution.

One hour later the whole protocol was repeated under control conditions or in the presence of taurine or beta-alanine. Each response was calculated as mm Hg and as a % of the first challenge. In all experiments, in order to avoid any effect of change in osmolarity in determining the actions of taurine, solutions used for control responses contained sucrose at the same concentration used for taurine.

Statistical Analysis

All results in the text are given as mean ± SEM. Statistical evaluation of results was done using the paired or unpaired Student's t test, as appropriate.

Drugs

Drugs used were: taurine (Merck), beta-alanine (Merck), L-norepinephrine hydrochloride (Fluka), nifedipine (Sigma), procaine hydrochloride (Gianni), and caffeine (Sigma). A stock solution of 1 mM nifedipine in absolute ethanol was prepared and from this the appropriate dilutions in physiological salt solution, in high-K^+ Ca^{++} free solution or in Ca^{++}-free solution were made just before the start of the experiments. Care was taken to avoid exposure of the norepinephrine and nifedipine solution to the light. Ascorbic acid (0.05 mg/ml) was added to the solutions containing norepinephrine.

RESULTS

Effect of taurine on norepinephrine- and depolarization-induced tonic vasoconstrictions in rabbit ear artery

Perfusion of rabbit ear artery with intraluminal norepinephrine (5 μM) or intra- and extraluminal high-K^+ solution (54 mM) resulted in a prompt vasoconstriction which reached a steady state within 10-15 min and lasted for at least one hour. In such conditions, intraluminal administration of taurine (40 mM) resulted in a sharp and significant relaxation of high-potassium- but not norepinephrine-induced tonic vasoconstriction (Table 1).

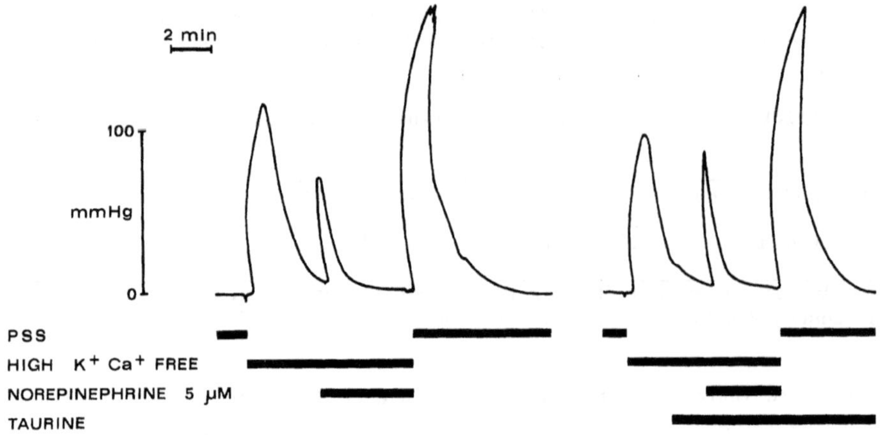

Figure 1. Typical tracings (redrawing from the originals) showing the experimental protocol for the study of cellular (obtained in high-K^+,Ca^{++}-free solution) and extracellular Ca^{++}-dependent (obtained after reperfusion with physiological salt solution) norepinephrine-induced vasoconstrictions in rabbit ear artery. The left tracing shows the effect of norepinephrine in control conditions while the right tracing shows the responses in presence of taurine (40 mM). Control and treated responses were obtained in the same specimen. PSS: physiological saline.

Mimickry of such activity could not be obtained with beta-alanine (Table 1).

Effect of taurine, nifedipine and procaine on norepinephrine-induced cellular and extracellular Ca^{++}-dependent vasoconstriction

Norepinephrine (5 µM) induced transient vasoconstrictions in high-K^+ Ca^{++}-free medium, with amplitudes of 75 ± 7 mm Hg (n=20). These vasoconstrictions were unaffected by the calcium entry blocker nifedipine (0.3 µM) but were markedly reduced by an inhibitor of Ca^{++} release from sarcoplasmic reticulum, procaine (1 mM) (See Table 2). This further suggesting the dependence of vasoconstriction upon mobilization of calcium from cellular stores. Taurine (40 mM) consistently produced a very slight,

Table 2. Effect of nifedipine, procaine and taurine on amplitude of cellular and extracellular Ca^{++}-dependent norepinephrine (5 µM)-induced vasoconstriction in rabbit ear artery.

	Cellular Ca^{++}-dependent (mm Hg)	Extracellular Ca^{++}-dependent (mm Hg)
Control	68 ± 5	170 ± 7
+ nifedipine (0.3 uM)	70 ± 4	49 ± 6**
Control	65 ± 3	179 ± 20
+ procaine (1 mM)	35 ± 2**	220 ± 22
Control	72 ± 3	200 ± 14
+ taurine (40 mM)	88 ± 9*	234 ± 11

Each value is the mean ± SEM of at least five experiments.
* $P < 0.05$, ** $P < 0.01$

albeit significant, enhancement of these cellular Ca^{++}-dependent responses (See Table 2 and Fig. 1).

When, after the three minutes challenge with norepinephrine (5 μM), the arteries were perfused with physiological salt solution, a further vasoconstriction was obtained, which was demonstrated to be strictly dependent upon extracellular Ca^{++} concentration (17).

Dependence of this response on influx of extracellular Ca^{++} through receptor operated channels was confirmed by the observation that it was markedly reduced by nifedipine (0.3 μM) and virtually unaffected by procaine (1 mM) (Table 2).

Taurine (40 mM) did not significantly modify these responses (Table 2 and Fig. 1) confirming previous data obtained in other experimental conditions (8,9).

<u>Effect of taurine, nifedipine and procaine on caffeine-induced vasoconstriction</u>

An aim of this set of experiments was to investigate whether or not taurine might be able to affect cellular Ca^{++}-dependent vasoconstriction

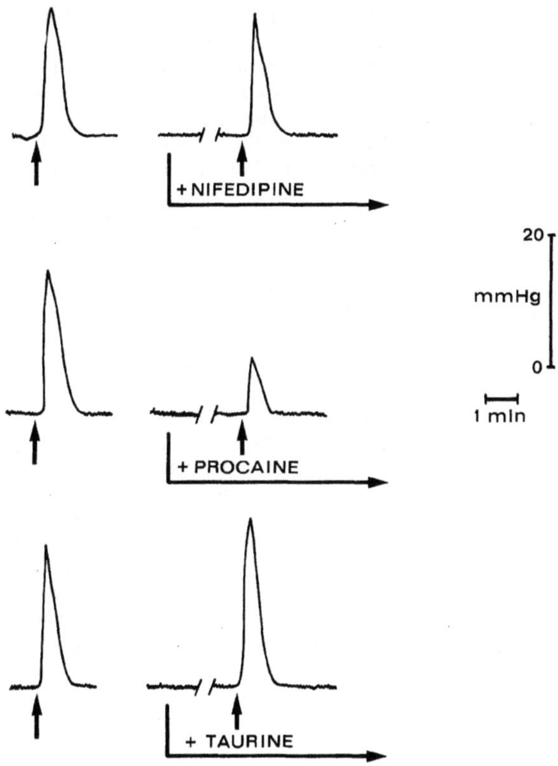

Figure 2. Typical tracing (redrawn from the originals) showing the effect of nifedipine (0.3 μM), procaine (1 mM) and taurine (40 mM) on caffeine (15 mM)-induced cellular Ca^{++}-dependent vasoconstriction. Caffeine was administered at the arrow, the other drugs 5 min before and during the challenge. Vasoconstrictions were obtained in a Ca^{++}-free medium containing EDTA (0.77 mM).

Table 3. Vasoconstrictor response in a Ca^{++}-free medium to successive challenges with norepinephrine 1 μM in control conditions and in presence of taurine or beta-alanine.

CHALLENGE	Control (mm Hg)	Taurine (40 mM) (mm Hg)	Beta-alanine (40 mM) (mm Hg)
1	69.6 ± 10.6	71.0 ± 7.2	69.7 ± 6.2
2	4.0 ± 0.6	1.5 ± 0.4*	1.7 ± 0.3*
3	2.1 ± 0.3	0.8 ± 0.3**	1.4 ± 0.3
4	1.9 ± 0.2	1.0 ± 0.1**	0.9 ± 0.2**

Each value is the mean ± SEM of at least five experiments. * $P < 0.05$; ** $P < 0.02$

elicited by a nonadrenergic calcium-mobilizing agent, such as caffeine (7, 12). Intraluminal administration of caffeine (15 mM) in arteries perfused with a Ca^{++}-free medium produced a rapidly-rising short-lived vasoconstriction (see typical tracing in Fig. 2). The amplitude of such vasoconstriction was 18 ± 2 mm Hg (n = 18). As shown in the typical tracing of Fig. 2, these responses were virtually unaffected by nifedipine (0.3 μM) being 17.5 ± 2 and 17 ± 3 mm Hg in the presence and absence of the calcium entry blocker, respectively (n = 6; n.s.). On the other hand, procaine (1 mM) reduced the height of these vasoconstrictions from 18.5 ± 3 to 9.5 ± 1 mm Hg (n = 6: $p < 0.01$; Fig. 2). In the presence of taurine (40 mM) an slight but definite increase in the amplitude of caffeine-induced vasoconstriction was always observed (from 19 ± 3 to 26 ± 4 mm Hg; n = 6; $p < 0.02$; Fig. 2).

Effect of taurine on rate of depletion of cellular calcium stores by norepinephrine

Since the above experiments indicated that taurine enhanced cellular Ca^{++}-dependent responses elicited by norepinephrine and caffeine, it appeared worthwhile to evaluate the possible interference of taurine on the rate of emptying by norepinephrine of cellular calcium stores.

Administration of norepinephrine in Ca^{++}-free medium produced a transient vasoconstriction whose amplitude was 67.0 ± 5.0 mmHg (n = 24). A marked decrease in the amplitude of transient vasoconstrictions induced by repeated challenges with norepinephrine was observed; a reduction to 5.2 ± 1.0 mm Hg (n = 17) at the second, 2.2 ± 0.4 mm Hg (n = 19) at the third and finally 1.7 ± 0.3 mm Hg (n = 19) at the fourth challenge. Repeating such a protocol after one hour of perfusion in physiological salt solution gave a similar result (see first column of Table 3), indicating the reproducibility of such responses. On the other hand, when the protocol was repeated in presence of taurine (40 mM) or beta-alanine (40 mM) a significant reduction in these responses was observed (Table 3).

DISCUSSION

Taurine has been proposed and utilized as therapy in vascular diseases such as intermittent claudication (2) and hypertension (23). Taurine administration also lowered blood pressure and impaired development of hypertension in spontaneously hypertensive rats (21). Such experimental and clinical data could be indicative of possible interferences of this amino acid with excitation-contraction coupling in vascular smooth muscle.

Present and previous studies strongly suggest that in rabbit ear artery taurine selectively inhibits depolarization mediated vasoconstrictions while responses mediated by receptor operated channel are virtually unaffected (8,9).

It has been recently reported that taurine reduces 4-aminopyridine- but not acetylcholine-induced contraction in guinea pig ileum (1). Besides prejunctional effects, 4-aminopyridine enhances electrical excitability of smooth muscle through a blockade of potassium conductance and consequent depolarization, increased calcium influx and contraction (14). On the other hand contractile response of intestinal preparations to cholinergic stimulation is at least partially mediated by opening of receptor operated channel (4).

It is thus possible that in smooth muscle taurine relaxing activity is mainly,if not exclusively, evident on contraction sustained by depolarization of the sarcolemma. This could be explained on the basis of a selective interference on the availability of voltage-operated calcium channels and/or an hyperpolarizing effect. It has been reported that taurine, by interacting with potassium fluxes, increased rates of membrane repolarization in frog and rat skeletal muscle (11).

Further electrophysiological studies in smooth muscle preparations are needed to identify the mechanism(s) involved. In addition, present findings indicate that taurine (40 mM) produces also some small, but significant, actions on the availability of calcium from cellular sources. In fact, an enhancement of both norepinephrine- and caffeine-induced cellular calcium-dependent vasoconstriction and an increase in rate of depletion of cellular calcium stores releasable by norepinephrine was observed. Although these data might be suggestive of an unspecific potentiation of calcium release from cellular storage sites followed by a depletion of calcium in the sarcoplasmic reticulum, the small magnitude of these effects, the high concentration of taurine required, and the observation that at least part of these actions are mimicked by beta-alanine indicate that such an action probably play a minor role, if any, in the overall vascular actions of taurine.

ACKNOWLEDGMENTS

We thank Miss Anna Giannini for typing the manuscript and for her skillful secretarial help. This research was partially supported by a grant from the Public Education Ministry, Italy.

REFERENCERS

1. Arzate, M.E., Ponce, H., and Pasantes-Morales, H., 1984, Antagonistic effects of taurine and 4-aminopyridine on guinea pig ileum, J. Neurosci. Res., 11:271-280.
2. Becattini, U., 1967, Prospettive di un nuovo farmaco nella terapia delle vasculopatie: la taurina, Sett. Med., 55:1255-1265.
3. Bevan, J.A., Garstka, W., Su, C., and Su, M.O., 1973, The biomodal basis of the contractile response of the rabbit ear artery to norepinephrine and other agonists, Eur. J. Pharmacol., 22:47-53.
4. Bolton, T.B., 1979, Mechanisms of action of transmitters and other substances on smooth muscle, Physiol. Rev., 59:606.
5. Casteel, R., 1980, Electro- and Pharmacomechanical coupling in vascular smooth muscle, Chest, 78:150-156.
6. Deth, R.C., and Lynch, C.J., 1981, Mobilization of a common source of smooth muscle calcium ions by NE and methylxanthines, Am. J. Physiol., 240 (Cell Physiol. 9): C239-C247.

7. Flaim, S.F., and Di Pette, D.J., 1979, Digoxin-norepinephrine response and calcium blocker effects on vascular smooth muscle, Am. J. Physiol., 236:H613-H619.
8. Franconi, F., Giotti, A., Manzini, S., Martini, F., Stendardi, I., and Zilletti, L., 1982, The effect of taurine on high potassium- and noradrenaline-induced contraction in rabbit ear artery, Br. J. Pharmac., 75:605-612.
9. Franconi, F., Manzini, S., Stendardi, I., Bennardini, F., Antonini, G., Failli, P., Matucci, R., and Giotti, A., 1983, Differential inhibitory effect of taurine on contractile responses to potassium an noadrenaline in rabbit ear artery, J. Pharm. Pharmacol., 35:46-48.
10. Glossmann, H., Ferry, D.R., Lubbecke, F., Mewes, R., and Hofmann, F., 1983, Identification of voltage operated calcium channels by binding studies: differentiation of subclasses of calcium antagonist drugs with ^{3}H-nimodipine radioligand binding, J. Receptor Res., 3(1&2):177-190.
11. Gruener, R., Markovitz, D., Huxtable, R., and Bressler, R., 1975, Excitability modulation by taurine: transmembrane measurement of neuromuscular transmission, J. Neurol. Sci., 24:351-360.
12. Hudgins, P.M., and Weiss, G.B., 1968, Differential effects of Ca^{2+} removal upon vascular smooth muscle contraction induced by norepinephrine histamine and potassium, J. Pharmacol. Expt. Ther., 159:91-97.
13. Kolhardt, M., and Flecknstein, A., 1977, Inhibition of the slow inward current by nifedipine in mammalian ventricular myocardium, Naunyn-Schmed. Arch. Pharmacol., 298:267-272.
14. Leander, S., Arner, A., and Johansson, B., 1977, Effects of 4-aminopyridine on mechanical activity and noradrnaline release in the rat portal vein in vitro, Eur. J. Pharmacol., 46:351-361.
15. Maggi, C.A., Manzini, S., and Meli, A., 1983, Contribution of cellular and extracellular Ca^{2+} during 5-hydroxytryptamine-induced contractions of rabbit ear artery, Eur. J. Pharmacol., 94:251-260.
16. Maggi, C.A., Manzini, S., and Meli, A., 1984, Differential effects of vasodilators on the mobilization of calcium pools during contractions of rabbit ear artery induced by norepinephrine and high potassium, J. Pharm. Pharmacol., 36:27-30.
17. Manzini, S., Maggi, C.A., and Meli, A., 1982, A simplex procedure for assessing norepinephrine induced cellular and extracellular calcium ions mobilization in rabbit ear artery, J. Pharmacol. Methods, 8:47-57.
18. Manzini, S., Maggi, C.A., and Meli, A., 1982, Aminophylline-induced contractions of rabbit ear artery in high-K^+ Ca^{2+}-free medium, J. Pharm. Pharmacol., 34:195-196.
19. Manzini, S., Maggi, C.A., and Meli, A., 1983, Alpha-adrenoceptors subtypes and Ca^{2+}-mobilization in the rabbit ear artery, J. Pharm. Pharmacol., 35:584-589.
20. Manzini, S., Maggi, C.A., and Meli, A., 1986, 5-hydroxytryptamine delays relaxation time of norepinephrine-induced vasoconstriction, Am. J. Physiol., 250(Heart Circ. Physiol. 19):H121-H130.
21. Nara, Y., Yamori, Y., and Lovenberg, W., 1978, Effect of dietary taurine on blood pressure in spontaneously hypertensive rats, Biochem. Pharmac., 77:2689-2692.
22. Saida, K., and Van Breemen, C., 1984, Characteristics of the norepinephrine sensitive Ca^{2+} stores in vascular smooth muscle, Blood Vessels, 21:43-52.
23. Schaffer, S.W., and Kocsis, J.J., 1979, Taurine: research surges after 150 years, Am. Pharm., 19:652-654.
24. Steinsland, O.S., Furchgott, R.F., and Kirpekar, S.M., 1973, Biphasic vasoconstriction of the rabbit ear artery, Circ. Res., 32:49-58.

SUPPRESSION OF BRONCHIAL RESPONSE TO PLATELET ACTIVATING FACTOR FOLLOWING TAURINE ADMINISTRATION

Michio Kurachi, Kayoko Hongoh, Akiko Watanabe, and Hironaka Aihara

Research Center, Taisho Pharmaceutical Co., Ltd., Yoshiono-cho 1-403, Ohmiya, Saitama 330, Japan

Platelets contain a high level of taurine (10), and accumulate this neutral amino acid by an energy-dependent process (7). Taurine inhibits the aggregation induced by ADP (1), and hence is thought to modulate the platelet function.

Platelet activating factor (PAF), one of the most potent aggregators of platelets, is biosynthetized by various cells, including basophils, macrophages, neutrophils and platelets (17). The molecular structure of PAF is 1-0-alkyl-2-acetyl-sn-glyceryl-3-phosphorylcholine (5). PAF appears to be a putative mediator of asthma (4,6), inflammation (8) and thrombosis (3,15,18). In vitro, PAF has been shown to promote calcium-dependent platelet aggregation, serotonin release and thromboxane A2 production (12,16). When administered to guinea pigs intravenously, PAF induces constriction of the airway smooth muscle (19). This constriction has been shown to depend on platelets since it is inhibited by pretreatment with antiplatelet antibody or by the administration of PGI2 (2,19). The relationship between the actions of PAF and taurine has not been clarified. In our studies we found that taurine inhibits the bronchospasm induced by PAF in guinea pigs. The mechanism of this inhibitory action of taurine was investigated herein.

MATERIALS AND METHODS

Animals

Male Hartley guinea pigs weighing 500-650 g, male New Zealand White rabbits weighing 2-3 kg, male Wistar rats weighing 250-350 g and male ICR mice weighing 21-27 g were used for the experiment.

Antiserum

Rabbits were sensitized with an intramuscular injection of an emulsion of 0.5 ml of physiological saline containing 10 mg of egg albumin and 0.5 ml of Freund's complete adjuvant, 4 times at intervals of 1 week. The antisera were prepared one week after the final sensitization.

Measurement of bronchoconstriction

Guinea pigs were anesthetized with 30 mg/kg of sodium pentobarbital given intraperitoneally. The trachea was cannulated and the animal ventilated with a respiratory pump with a stroke volume of 5 ml at a rate of 70 stroke/min. These animals were immobilized by giving an intravenous injection of decamethonium bromide. Airway resistance was measured by the overflow technique of Konzett and Roessler (11). The sidearm of the tracheal cannula was connected to a nonreturn water valve set at a pressure of 10 cm water. The air overflow from the water valve was measured with a pneumotachograph and recorded on a multichannel electronic recorder. During the course of experiments, ssytemic blood pressure was continuously monitored and recorded on a multichannel electronic recorder. The bronchoconstriction was expressed as a % of the maximum overflow, obtained by clamping the trachea, at the end of the experiment.

Platelet aggregation

Arterial blood, anticoagulated with 0.32% sodium citrate, was obtained from guinea pigs anesthetized with ether. The blood was centrifuged at 150 g for 10 min to obtain platelet-rich plasma (PRP). The red cell layer was centrifuged at 1500g for 10 min to obtain platelet-poor plasma (PPP). Platelet aggregation was monitored in an aggregometer at 37°C, with PPP as the 100% standard. A sample without taurine was preincubated 5 min at 37°C. Taurine was added 2 min prior to PAF.

Contraction of lung parenchymal strips

Lung parenchymal strips from guinea pigs were suspended in an organ bath at 37°C. The contraction of lung parenchyma induced by spasmogen was isometrically recorded on a multichannel electronic recorder.

Blood pressure of rats

Rats were anesthetized with 40 mg/kg of sodium pentobarbital given intraperitoneally. Blood pressure was recorded on a polygraph via an electromanomenter from a polyethylene cannula inserted into the carotid artery.

PAF-induced death in mice

PAF and taurine were simultaneously given iv to a lateral tail vein and the mice were observed for at least 6 h.

Drugs

Drugs used were taurine (Tokyo Kasei), 1-0-octadecyl-2-acetyl-sn-glyceryl-3-phosphocholine (PAF, Wako Pure Chem.), leukotriene D4 (Funakoshi), egg albumin (Biochem. Indust.), serotonin creatinine sulfate (Tokyo Kasei), pyrilamine and indomethacin (Sigma).

RESULTS

PAF-induced bronchospasm in guinea pigs

Intravenous injection of PAF into guinea pigs in a dose of 0.1 ug/kg induced an immediate bronchospasm (51.0+10.4%, n=13), the peak of which was reached within 1 min (Fig. 1). An initial decrease then increase, followed by a long-lasting decrease of blood pressure was observed after the administration of PAF (Fig. 1). The bronchoconstriction was reduced gradually, but the tone did not completely revert to the control level. The second administration of PAF revealed a different response, consistent with the findings of

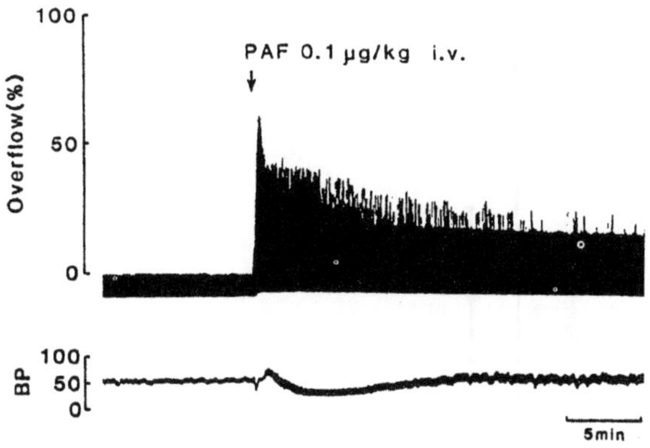

Fig. 1. Typical recording of airway contraction induced by the injection of PAF into guinea pigs (0.1 μg/kg).

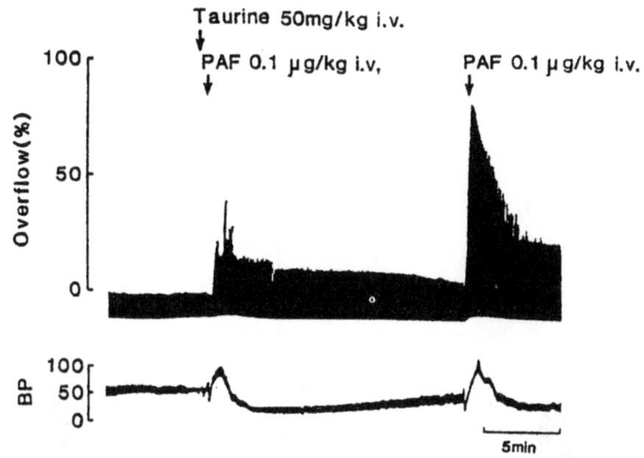

Fig. 2. Typical record of the inhibitory effect of taurine on the bronchoconstriction induced by the intravenous injection of PAF (0.1 μg/kg) in the guinea pig. Upper trace shows the ventilation overflow, and lower trace the blood pressure. Note that PAF-induced bronchoconstriction was reduced by the injection of taurine.

other workers (12,19). Accordingly, the response to PAF in animals pretreated with taurine was compared with that in animals not given taurine. The administration of taurine 30 s prior to the injection of PAF markedly inhibited the bronchospasm induced by PAF (Fig. 2) and this inhibitory action was dose dependent (Fig. 3). To ascertain whether the animal was sensitive to PAF, PAF was administered 30 min after the administration of taurine (Fig. 2).

Antigen-induced bronchospasm in sensitized guinea pig pretreated with pyrilamine and indomethacin

When an antigen was administered to a sensitized guinea pig pretreated with pyrilamine, an antihistaminic, and indomethacin, a cyclooxygenase inhibitor, the bronchospasm occured more than 1 min after the administration of the antigen, although a sensitized animal pretreated with no drug responded immediately to the antigen. Taurine in a dose of 300 mg/kg iv did not modulate the bronchial response to the antigen (Fig. 4).

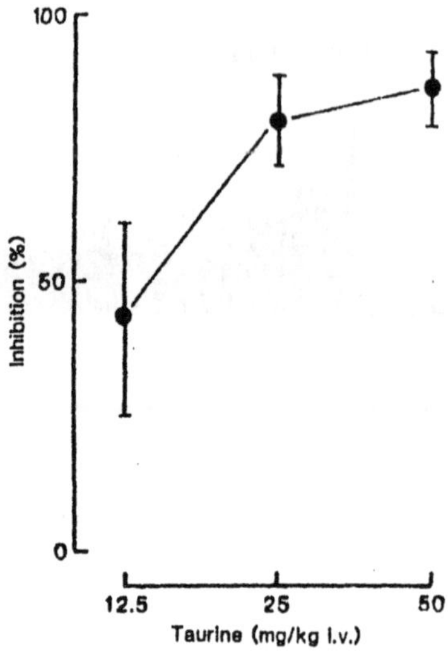

Fig. 3. Inhibitory effect of taurine on the bronchoconstriction induced by PAF (0.1 µg/kg iv) in the guinea pig. Each point represents the mean ± SEM of 5 experiments.

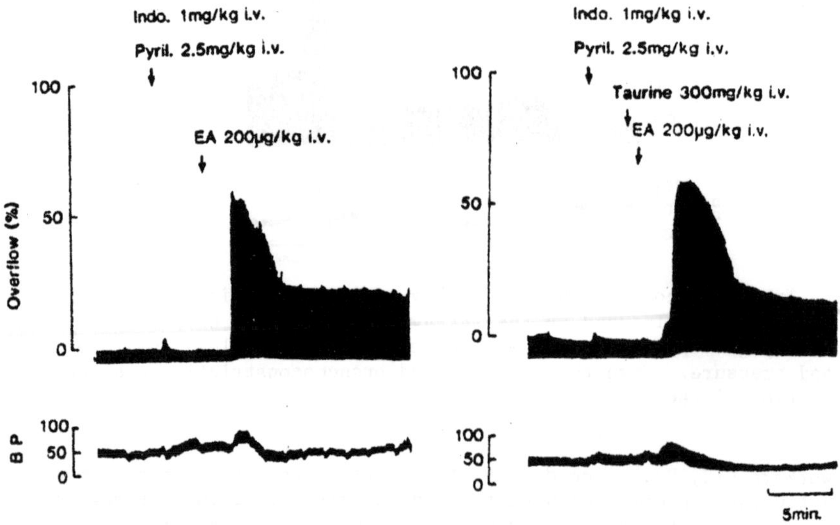

Fig. 4. Typical record showing the lack of effect of taurine on the bronchoconstriction evoked by an antigen, egg albumin (EA), in the guinea pig pretreated by indomethacin (Indo) and pyrilamine (Pyril).

LTD4-induced bronchospasm in guinea pig

A rapid bronchial contraction was induced following the administration of LTD4 in a dose of 0.3 µg/kg iv (Fig. 5). The bronchial response then was gradually reduced and the bronchial tone reverted to the control level. The response to LTD4 was unchanged following the administration of taurine, even at a high dose of 300 mg/kg iv (Fig. 5).

PAF-induced aggregation of guinea pig platelet

The aggregation of guinea pig platelets induced by PAF at a final concentration of 3 nM was inhibited slightly with 20 mM of taurine and markedly with 40 mM of this compound (Fig. 6).

PAF- and LTD4-induced contractions of guinea pig lung parenchymal strips

The lung parenchymal strips showed a various sensitivity to PAF(1 μM). However, the responses obtained in lung parenchymal strips prepared from the same animal were much the same. The contractile forces induced by PAF (1 μM), were in the range of 0 to 20% of the force induced by histamine (100 μM). Peak times of the responses were 4 to 5 min. PAF-induced contracture of lung parenchymal preparation, which was more than 10% of the contractile force induced by histamine (100 μM), was not modified when taurine (40 mM) was added to the bath fluid (Fig. 7A).

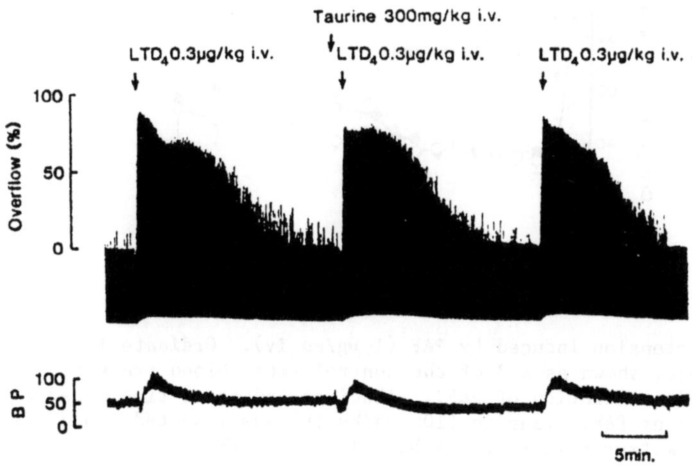

Fig. 5. Experimental record showing the lack of effect of taurine on the bronchospasm induced by LTD4.

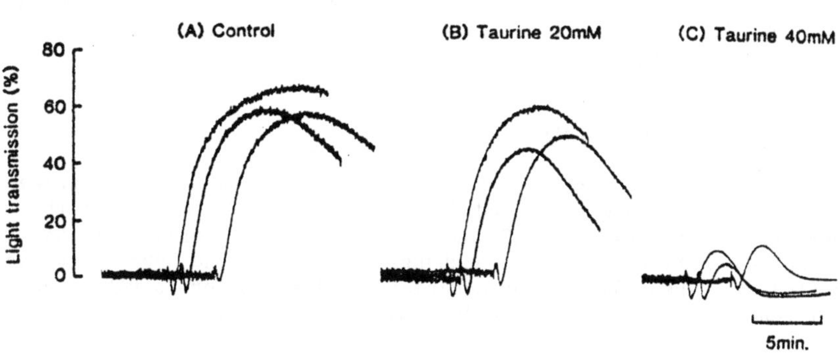

Fig. 6. Effect of taurine on the platelet aggregation induced by PAF (final concentration 3nM) in vitro.

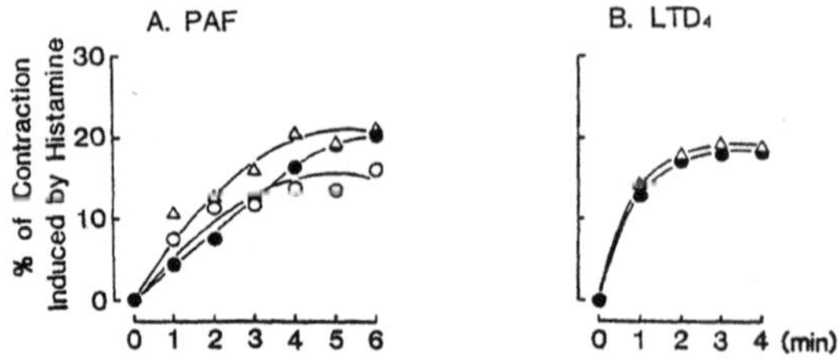

Fig. 7. Effects of taurine on contraction induced by (A) PAF and (B) LTD4. Ordinate indicates contractile force which is shown as a % of the contraction induced by histamine (100 μM). Abscissa indicates time after the addition of PAF or LTD4. ●; control, ○;taurine 20 mM, △; taurine 40 mM.

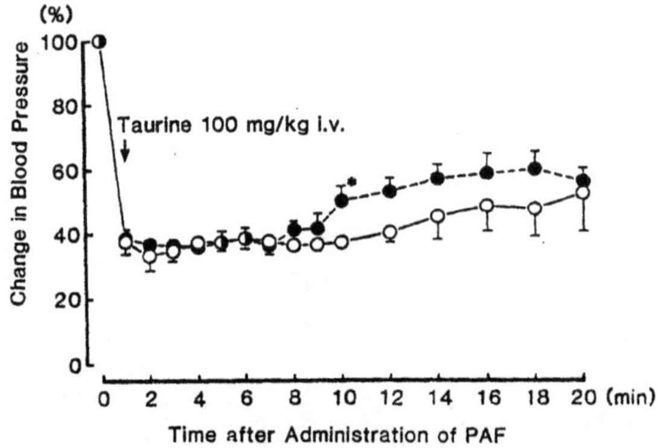

Fig. 8. Hypotension induced by PAF (1 μg/kg iv). Ordinate indicates mean blood pressure, shown as a % of the control (mean blood pressure at a time before the administration of PAF). Abscissa indicates time after the administraion of PAF. Taurine (100 mg/kg iv) was injected 1 min after PAF. ○;control, ●;taurine. Vertical bars indicate SEM.

Contraction of the lung parenchyma was induced by adding LTD4 (10 nm) to the bath fluid. The response to LTD4 reached a peak at about 3 min after the application of LTD4 and was about 20% of the response to histamine (100 μM). Taurine 40 mM) did not inhibit the LTD4-induced contraction of lung parenchymal strips (Fig. 7B).

PAF-induced hypotension in rats

Blood pressure in the rat carotid artery was lowered by PAF (1 μg/kg iv) and this hypotensive state persisted for over 30 min. One min after the administration of PAF, tauriné (100 mg/kg iv) was administered, but the hypotension evoked by PAF remained unchanged (Fig. 8). The hypotension was not modified even when taurine was administered 1 min prior to PAF. The bronchial tone recorded simultaneously remained constant.

PAF-induced death in mice

PAF (31 μg/kg iv) produced shock in mice and 7/10 mice died within 60 min. The mortality in mice pretreated simultaneously with PAF and taurine (100 mg/kg iv) was 80%.

Serotonin-induced bronchospasm in guinea pig

Bronchial tone was increased immediately after the treatment with serotonin given at a dose of 3 µg/kd iv. The duration of bronchospasm was short and occured within 5 min. The responses to serotonin were stable even in case of administration at intervals of 15 min. Taurine in a dose of 100 mg/kg iv diminished the bronchial spasm induced by serotonin (Fig. 9).

Fig. 10 shows that taurine dose-dependently inhibited the bronchoconstriction induced by serotonin.

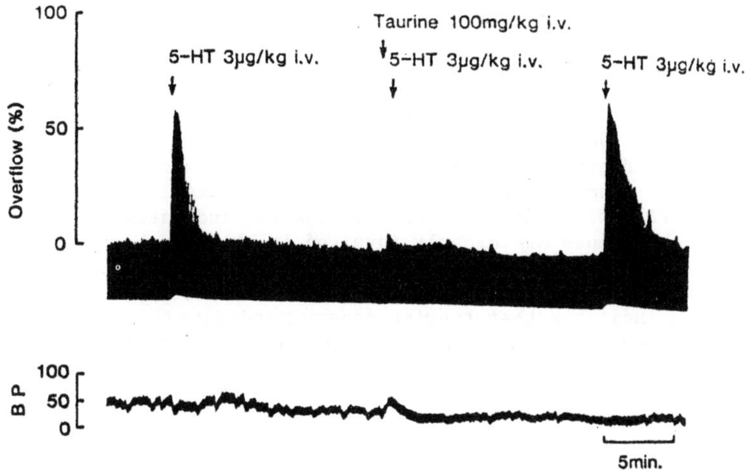

Fig. 9. Typical record indicating the effect of taurine on the serotonin (5-HT)-induced bronchoconstriction.

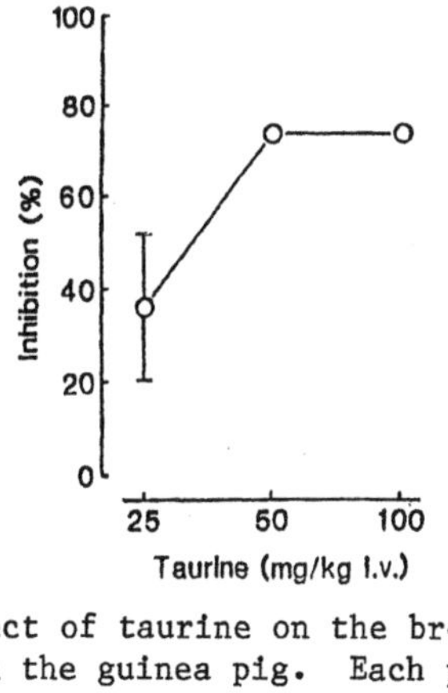

Fig. 10. Inhibitory effect of taurine on the bronchospasm induced by serotonin (3 µg/kg iv) in the guinea pig. Each point represents the mean ± SEM of 4 experiments.

DISCUSSION

Suppression of the PAF-induced bronchoconstriction in guinea pigs was observed following the administration of taurine. Taurine (40 mM) also inhibited the PAF-induced aggregation of guinea pig platelets. The bronchoconstriction induced by PAF is known to depend on platelet, since the response to PAF is inhibited by pretreatment with antiplatelet antibody (19) or PGI2 (2). Our results suggest that the inhibitory action of taurine on PAF-induced bronchospasm is due to inhibition of the aggregation of platelets.

It is presumed that the PAF-induced bronchospasm involves leukotrienes since the spasm is reduced by lipoxygenase inhibitors (2) or the leukotriene antagonist, FPL-55712 (data not shown). To determine whether taurine modulates the production or biological action of leukotrienes, taurine was administered to sensitized guinea pigs pretreated with pyrilamine and indomethacin. In these treated animals, an antigen caused a slowly progressive bronchoconstriction, relative to that in sensitized guinea pigs given no pretreatment. Mediators of anaphylactic bronchospasm in sensitized guinea pigs are thought to be mainly histamine, leukotrienes, PAF and prostaglandins (4). Pretreatment with pyrilamine and indomethacin, an antihistaminic and cyclooxygenase inhibitor, respectively, will eliminate the response due to endogenous histamine and prostaglandins released by challenge with the antigen. The antigen-evoked response was makedly inhibited by treatment with FPL-55712 (data not shown), thereby indicating that under these conditions, leukotrienes are closely involved in the bronchospasm induced by the antigen. Taurine (100 mg/kg iv) did not inhibit the bronchospasm induced by the antigen in sensitized guinea pigs pretreated with pyrilamine and indomethacin. This indicates that taurine does not modulate the production or the action of leukotrienes. Moreover, in lung parenchymal strips, the contraction induced by LTD4 was not altered when taurine was added to the bath fluid. Taurine did not inhibit the PAF-induced death of mice. Lipoxygenase inhibitors or leukotriene antagonist prevents the death (20). From these results, it is suggested that suppressive acion of taurine on the bronchospasm induced by PAF is not relevant to leukotrienes.

Taurine had neither a prophylactic effect nor a reverse action on the PAF-induced hypotension in rats. The PAF-induced hypotension in rats is not associated with thrombocytopenia (14). Therefore, our results indicate that platelets are required for the inhibitory effect of taurine on PAF-induced bronchospasm.

Platelets contain serotonin which is released by the application of PAF (9) and serotonin provokes contraction of the airway smooth muscle. Therefore, it is possible that the PAF-induced bronchoconstriction is partly linked to serotonin. We next examined whether taurine would affect the serotonin-induced bronchospasm. Taurine reduced the response to serotonin, in a dose-dependent manner, thereby indicating that the inhibitory action of taurine on the PAF-induced bronchospasm is in part explained by the suppressive action of taurine on the serotonin-induced contraction.

In conclusion, taurine inhibited the bronchospasm induced by PAF in guinea pig. This inhibitory action of taurine seems to be due to the suppression of PAF-induced platelet aggregation and serotonin-induced bronchospasm. Our results also indicate that the inhibitory action of taurine on the PAF-induced bronchoconstriction is not due to the antagonism of PAF, or to the production or the antagonism of leukotriene.

REFERENCES

1. Almazov, V.A., Gurerich, V.S., Mikhailova, I.A., and Streltsova, E.N., 1985, The influence of taurine on Ca^{2+}, Mg^{2+}, ATPase activity and aggregation of human platelet (Russian), Byull. Eksp. Biol. Med., 100:398-400.
2. Bennett, J., Thibaudeau, D., and Bessin, P., 1983, Dependency of the PAF-acether induced bronchospasm on the lipoxygenase pathway in the guinea pig, Prostaglandins, 26:457-466.
3. Chignard, M., LeCouedic, J.P., Tence, M., Vargaftig, B.B., and Benveniste, J., 1979, The role of platelet-activating factor in platelet aggregation, Nature, 279:799-800.
4. Darius, H., Lefer, D.J., Smith, J.B., and Lefer, A.M., 1986, Role of platelet-activating factor-acether in mediating guinea pig anaphylaxis, Science, 232:58-60.
5. Demopoulos, C.A., Pinckard, R.N., and Hanahan,D.J., 1979, Platelet activating factor. Evidence for 1-0-alkyl-2-acetyl-sn-glyceryl-3-phosphorylcholine as the active component (a new class of lipid chemical mediators), J. Biol. Chem., 254:9355-9358.
6. Fitzgerald, M.F., Moncada, S., and Parente, L., 1986, The anaphylactic release of platelet-activating factor from perfused guinea pigs, Br. J. Pharmac., 88:149-153.
7. Gaut, Z.N., and Nauss, C.B., 1976, Uptake of taurine by human blood platelets: A plssible model for brain, in: "Taurine", R. J. Huxtable and A. Barbeau eds., Raven Press, New York, pp. 91-98.
8. Hwang, S., Lam, M., Li, C., and Shen, T., 1986, Release of platelet activating factor and its involvement in the first phase of carrageenin-induced rat foot edema, Eur. J. Pharmac., 120:33-41.
9. Inarrea, P., Gomez-Cambronero, J., Nieto, M., and Crespo, M.S., 1984, Characteristics of the binding of platelet-activating factor to platelets of different animal species, Eur. J. Pharmac., 105:309-315.
10. Jacobsen, J.G., and Smith, L.H., 1968, Biochemistry and physiology of taurine and taurine derivatives, Physiol. Rev., 48:424-511.
11. Konzett, H., and Rossler, R., 1940, Verschsanordnung zu Utersüchungen an der Bronchialmuskulatur, Naunyn-Schmiedebergs Arch. Exp. Path. Pharmak., 195:71.
12. Lee, T-C., Mlone, B., Blank, M.L., and Snyder, F., 1981, 1-Alkyl 2-acetyl-sn-glycero-3-phosphocholine (platelet-activating factor) stimulates calcium influx in rabbit platelet, Biochem. Biophys. Res. Commun., 102:1262-1268.
13. Maridonneau-Parini, I., Lagente, V., Lefort, J., Russo-Marie, F., and Vargaftig, B.B., 1985, Desensitization to PAF-induced bronchoconstriction and to activation of th alveolar macrophages by repeated inhalations of PAF in the guinea pig, Biochem. Biophys. Res. Commun., 131:42-49.
14. Namm, D.H., Tadepalli, A.S., and High, J.A., 1982, Species specificity of the platelet responses to 1-0-alkyl-2-acetyl-sn-glycero-3-phosphocholine, Thromb. Res., 25:341-350.
15. Robertson, D.N., and Smith, G.M., 1986, Cv88 inhibits in vivo platelet aggregation induced by PAF-acether and collagen, Eur. J. Pharmac., 123:91-97.
16. Shaw, J.O., Klusick, S.J., and Hanahan, D.J., 1981, Activation of rabbit platelet phospholipase and thromboxane synthesis by 1-0-hexadecyl/-octadecyl-2-acetyl-2-sn-glyceryl-3-phosphorylcholine (platelet activating factor), Biochem. Biophys. Acta, 663:222-229.
17. Vargaftig, B.B., and Benveniste, J., 1983, Platelet-activating factor today, Trends Pharmacol. Sci., 4:341-344.

18. Vargaftig, B.B., Chignard, M., Benveniste, J., Lefort, J., and Wal, F., 1981, Background and present status of research on platelet-activating factor (PAF-acether), Ann. N. Y. Acad. Sci., 370:119-137.
19. Vargaftig, B.B., Lefort, J., Chignard, M., and Benveniste, J., 1980, Platelet-activating factor induces a platelet-dependent bronchoconstriction unrelated to the formation of prostaglandin derivatives, Eur. J. Pharmac., 65:185-192.
20. Young, J.M., Maloney, P.J., Jobb, S.N., and Clark, J.S., 1985, Pharmacological investigation of the mechanisms of platelet-activating factor induced mortality in the mouse, Prostaglandins, 30:545-551.

NEURAL REGULATION OF TAURINE TRANSPORT IN SKELETAL MUSCLE

Heitaroh Iwata, Bong Ki Kim, Yoshihiro Fukui and Akemichi Baba

Department of Pharmacology, Faculty of Pharmaceutical Sciences, Osaka University, 1-6 Yamada-oka, Suita, Osaka 656, Japan

Taurine is present at high concentration in skeletal muscles, and affects various electrophysiological and biochemical parameters of skeletal muscle; it hyperpolarized muscle fibers and decreased the duration of the action potential (3), and it enhanced the capacity for calcium uptake by sarcoplasmic reticulum (5). Response of isolated skeletal muscles to cholinergic stimulation is attenuated by taurine (9). With respect to the possible involvement of taurine in skeletal muscle pathology, it has been known that the skeletal muscle of dystrophic animals had a significantly higher amount of taurine (18). Experiments with myotonic muscles have shown that taurine prevents or reduces the electrical signs of muscular hyper-excitability and favors potassium accumulation into the muscle fibers (2,11).

Mammalian skeletal muscles are physiologically distinguishable into fast-twitch and slow-twitch muscles. Extensor digitorum longus (EDL) and soleus muscles are representative fast-twitch and slow-twitch muscles, respectively. Denervation of motor neurons causes typical biochemical or physiological changes in skeletal muscles which are dependent on the muscle fiber phenotypes; that is, slow-twitch and fast-twitch muscles (1,10). Recent studies from our laboratory showed that denervation of skeletal muscles increased the transport and the content of taurine in fast-twitch muscle, but not in slow-twitch. In addition, the steady state level of the content and the transport of taurine are much higher in slow-twitch muscles than in the fast-twitch (7). The functional and biochemical expression of different types of skeletal muscles are largely determined by information from the innervating motor neurons. This information might be nerve impulses or neurotrophic substances released from nerves to muscles. Our previous studies suggest that the transport of taurine in skeletal muscles is under neural regulation of which the direction is dependent on the muscular phenotypes. It is well known that chronic nerve stimulation of fast-twitch muscles at low frequency can transform various cellular parameters of the muscles to those which resemble slow-twitch muscles (4,12,15,16). In this communication, we report the effects of acute and chronic nerve stimulation on taurine transport in rat skeletal muscles.

MATERIALS AND METHODS

Surgical procedures

Sprague-Dawley rats aged 4 weeks were used throughout the experiments. In chronic experiments, the electrode was implanted lateral to the left sciatic nerve of the animals and passed subcutaneously to the back under sodium pentobarbital anesthesia as described previously (8).

Electrical stimulation

In acute stimulation, rats were anesthetized with urethane and artificially respirated. Sciatic nerve was stimulated with train pulses (50 Hz, 4.2 V) applied once every 2 s for 60 ms. The duration of the individual pulse was 1 ms. ^{14}C Taurine (25 μCi/kg) was injected into the right cervical vein immediately after the stimulation. In some experiments, d-tubocurarine chloride (3 mg/kg) was administered just before the stimulation.

In chronic study, stimulation was started one day after the implantation. The animals were connected to a stimulator generating square waves via a plug which was secured to the back of animals by sutures and dental cement. Stimulation consisted of trains of pulses at 10 and 100 Hz of 0.2 ms duration for slow-twitch type and fast-twitch type stimulations, respectively. Each train lasted 1 s and was repeated every 3 s. The amplitude of stimulation was adjusted daily to give maximum contractions without causing any discomfort to the animals. Stimulation was performed for 10 h every day for 1 to 4 weeks.

Assays

The content and the transport of taurine in the muscles were determined as described previously (8).

RESULTS

Acute stimulation

Acute electrical stimulation of sciatic nerve on the uptake of radioactive taurine by the muscles was examined. The uptake of taurine by EDL and soleus muscle was significantly enhanced during the electrical stimulation of nerve. The increase by nerve stimulation corresponded to about 20 to 40 % of the control and was almost same magnitude in EDL and soleus muscles. Since the stimulated leg showed a continuous twitch-response during the stimulation, the possible involvement of muscle contraction in this increase of taurine transport should be established. The same experiments were performed with curare-treated rats. In such rats, the nerve stimulation did not evoke any changes in taurine transport by the muscles.

Chronic stimulation

The taurine contents of each muscle did not change after a 1 week period of the slow-twitch type stimulation. However, stimulation for 2 weeks signi-

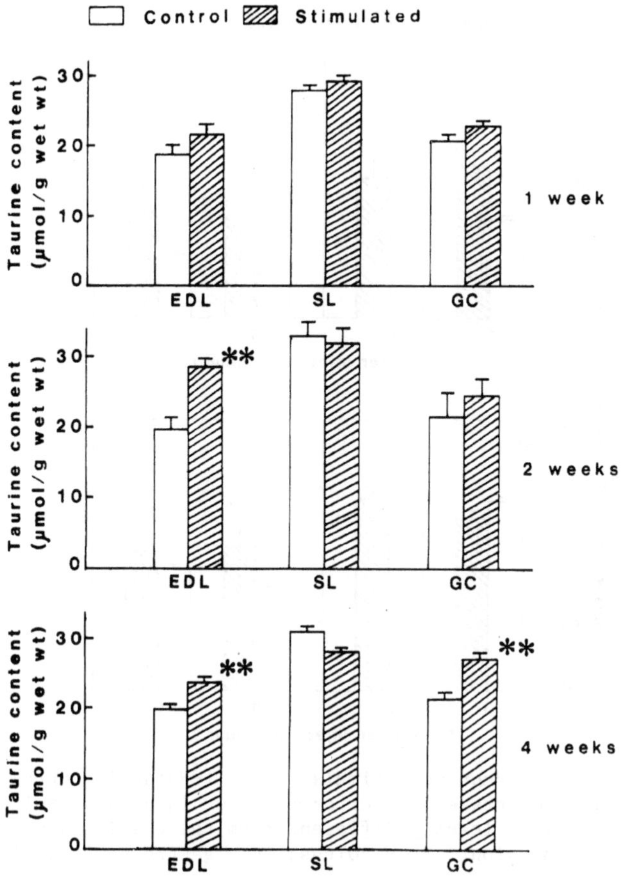

Fig. 1. Effect of low-frequency stimulation on taurine content of rat skeletal muscles. The value represents the mean and standard errors of 4-5 determinations. **$P<0.01$, significantly different from the unstimulated control. From ref. 8 with permission. SL = soleus muscle; GC = gastrocnemius muscle.

ficantly increased the taurine content of EDL. After 4 weeks stimulation the stimulated EDL and gastrocnemius muscles showed significant increases in taurine content. The nerve stimulation did not change the taurine content in soleus muscle at any time during the stimulation (Fig. 1). The effect of slow- twitch type electrical stimulation of the nerve on the transport of taurine was examined (Fig. 2). After 2 weeks stimulation, the transport of radioactive taurine by EDL, and tibialis anterior, another fast-twitch muscle, was increased 3 h after the injection. However, there was no difference in the stimulated soleus muscle. We then examined the effect of fast-twitch type nerve stimulation in the same manner. High frequency stimulation (100 Hz) for 1 week did not evoke any difference of muscle weight between control and stimulated muscles, but stimulated EDL muscle showed a

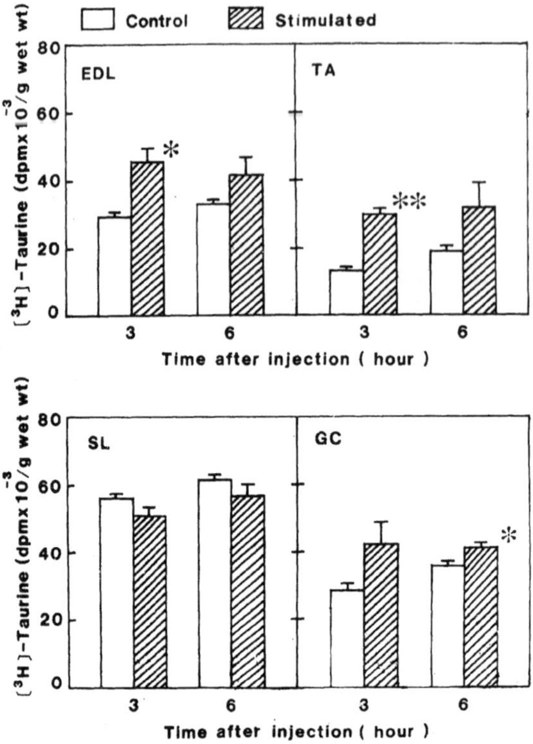

Fig. 2. Effect of low-frequency stimulation on $[^3H]$taurine uptake into rat skeletal muscles. The uptake was performed after 14 days stimulation. $^*P<0.05$, $^{**}P<0.01$, significantly different from the unstimulated control. From ref. 8 with permission. SL = soleus; GC = gastrocnemius; TA = tibialis anterior.

significant increase in the taurine content, while stimulated soleus muscle showed a decrease in the taurine content (Fig. 3). In addition, 100 Hz stimulation for 2 weeks significantly increased the transport of taurine by EDL and tibialis anterior, but not by soleus muscles (Fig. 4).

DISCUSSION

We have previously suggested that transport of taurine into skeletal muscles is differentially regulated by the neural information and the muscular activity and that the regulation is dependent on the muscle phenotypes (7). In an acute experiment, the electrical stimulation of the sciatic nerve accelerated the uptake by EDL and soleus muscles of the noncurarized rats, but not in those of the curare treated rats (7). In chronic experiments, it was found that taurine content and uptake in the EDL muscle but not in the soleus muscles were significantly increased by chronic low frequency stimulation (10 Hz). Increase in the transport of taurine following nerve stimulation was also observed in tibialis anterior muscle, another typical fast-twitch muscle. Thus, chronic nerve stimulation of slow-twitch type modified the transport of taurine of fast-twitch muscle. Mammalian skeletal muscles are physiologically distinguishable into fast-twitch and slow-twitch muscles. Transformation of one muscle type to another is performed by indirect electrical stimulation with a stimulus pattern which resembles the physiological conditions (17). The chronic stimulation at 10 Hz resulted in changes of various parameters of fast-twitch muscle (4,12,15,16). Taking this observation into account, it is reasonable that the transformation of fast- twitch muscle to the slow type by a chronic stimulation of slow-twitch type may account for the increase in the content and the transport of taurine in the EDL muscle. This was also suggested our previous findings in which dedifferentiation of the fast-twitch muscle induced by denervation caused similar changes in the transport of taurine in EDL (7).

In contrast, the reverse experiment, i.e. the transformation of the slow-twitch muscle by stimulation at higher frequency, seems to be unsuccessful,

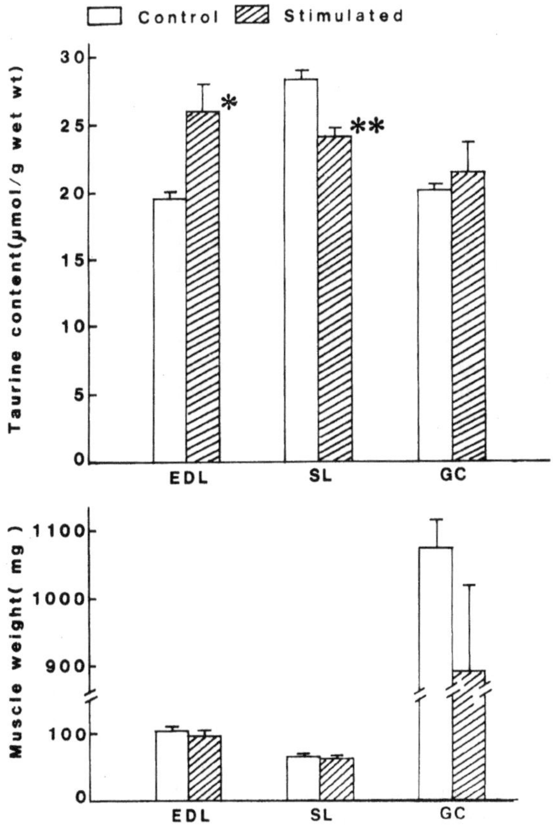

Fig. 3 Effect of chronic-frequency stimulation on taurine content of rat skeletal muscles. The experiments were performed after 14 days-stimulation. *P<0.05, **P<0.01, significantly different from the unstimulated control. From ref. 8 with permission. Abbreviations as for Fig. 2.

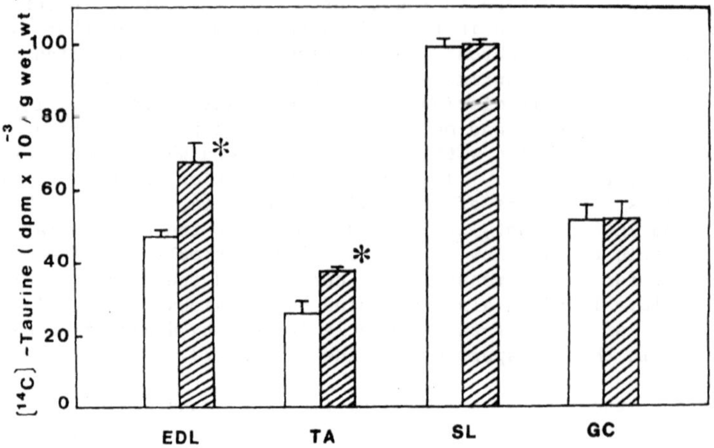

Fig. 4 Effect of chronic high-frequency stimulation on [^{14}C]taurine uptake in rat skeletal muscles. Three hours after injection, the radioactivity in the tissues was counted. *$P<0.05$, significantly different from the unstimulated control. open bars: control; hatched bars: stimulated. From ref. 8 with permission.

because the transport of taurine was also increased in EDL and tibialis anterior muscles but not in the soleus. Some contraindications observed in the present study are not unexpected ones, because equivocal results were also obtained in other cases (13,14,17). Thus, the results obtained with high frequency stimulation can not be explained by a transformation of the muscle phenotypes.

On the basis of the present and the previous communications (6,7,8), it is suggested that there are at least two regulatory systems of taurine transport in skeletal muscles; those are muscular activity and neural information and the latter one is dependent on the muscle phenotypes.

ACKNOWLEDGMENT

This study was supported by grants from the Ministry of Health and Welfare, Japan,and from Taisho Pharmaceutical Co., Tokyo.

REFERENCES

1. Albuquerque, E.X., and Thesleff, A., 1968, A comparative study of membrane properties of innervated and chronically denervated fast and slow skeletal muscle of the rat, Acta Physiol. Scand., 73:471-480.
2. Durelli, L., and Mutani, R., 1979, Myotonia, potassium and taurine. A preliminary report, J. Neurol. Sci., 42:103-109.
3. Gruener, R., and Bryant, H., 1975, Excitability modulation by taurine: action on axon membrane permeabilities, J. Pharmacol. Exp. Ther., 195: 514-521.
4. Heilmann, C., and Pette, D., 1979, Molecular transformation in sarcoplasmic reticulum of fast twitch muscle by electrostimulation, Eur. J. Biochem., 93;437-446.

5. Huxtable, R.J., and Bressler, R., 1973, Effect of taurine on a muscle intracellular membrane, Biochem. Biophys. Acta, 323:573-583.
6. Iwata, H., and Baba, A., Specific increase of taurine in denervated skeletal muscle, in: "Taurine: Biological Actions and Clinical Perspectives", S.S. Oja, L. Ahtee, P. Kontro, and M.L.K. Paasonen, eds., Alan R. Liss, New York (1985), pp. 397-406.
7. Iwata, H., Obara, T., Jim, B.K., and Baba, A., 1986, Regulation of taurine transport in rat skeletal muscle, J. Neurochem., 47:158-163.
8. Kim, B.K., Baba, A., and Iwata, H., 1986, Taurine transport in chronically stimulated fast- and slow-twitch muscles of the rat, Japan. J. Pharmacol., 42:441-446.
9. Lehmann, A., and Hamberger, A., 1983, Inhibition of cholinergic response by taurine in frog isolated skeletal muscle, J. Pharm. Pharmacol., 36:59-61.
10. Margreth, A., Salviati, G., Di Mauro, S., and Turati, G., 1972, Early biochemical consequences of denervation in fast and slow skeletal muscles and their relationship to neural control over muscle differentiation, Biochem. J., 126:1099-1110.
11. Mutani, R., Durelli, L., and Piredda, S., 1982, Normalization of muscle transmembrane ionic fluxes induced by taurine in patients affected with myotonic dystrophy, Acta Neurol. Belg., 82:122-127.
12. Pette, D., Smith, M.E., Staudte, H.W., and Vrbova, G., 1973, Effects of long-term electrical stimulation on some contractile and metabolic characteristics of fast rabbit muscles, Pflugers Arch., 338-257-272.
13. Pette, D., and Tyler, K.R., 1983, Response of succinate dehydrogenase activity in fibers of rabbit tibialis anterior muscle to chronic nerve stimulation, J. Physiol., 338:1-9.
14. Pockett, S., and Pygott, V., 1982, Chronic nerve stimulation increases guantal output in rat muscles, Exp. Neurol., 78:477-480.
15. Salmons, S., and Vrbova, G., 1969, The influence of activity on some contractile characteristics of mammalian fast and slow muscles, J. Physiol., 210:535-549.
16. Sreter, F.A., Gergely, J., Salmons, S., and Romanul, R., 1973, Synthesis by fast muscle of myosin light chains characteristics of slow muscle in response to long-term stimulation, Nature New Biol., 241:17-19.
17. Streter, F.A., Pinter, K., Jolesz, F., and Mabuchi, K., 1982, Fast to slow transformation of fast muscle in response to long-term phasic stimulation, Exp. Neurol., 75:95-102.
18. Wilson, B.W., Peterson, D.W., and Lilyblade, A.L., 1965, Free amino acids of developing skeletal musculature of normal and genetically dystrophic chickens, Proc. Soc. Exp. Biol. Med., 119:104-108.

EFFECT OF TAURINE ON CHLORIDE CONDUCTANCE AND EXCITABILITY OF RAT SKELETAL MUSCLE FIBERS

Diana Conte-Camerino[1], Flavia Franconi[3], Mercede Mambrini[1], Delia Mitolo-Chieppa[2], Federico Bennardini[4], Paola Failli[4], Shirley H. Bryant[5], and Alberto Giotti[4]

[1]Cattedra di Farmacologia e Farmacognosia, Facoltà di Farmacia, Universita di Bari, Italy; [2]Istituto di Farmacologia, Facoltà di Medicina, Università di Bari, Italy; [3]Cattedra di Saggi e Dosaggi Farmacologici, Facoltà di Farmacia, Università di Sassari, Italy; [4]Dipartimento di Farmacologia Preclinica e Clinica "M. Aiazzi Mancini", Università di Firenze, Italy; [5]Department of Pharmacology and Cell Biophysics, College of Medicine, University of Cincinnati, Cincinnati, OH.

Taurine is found in mammalian skeletal muscle fibers and all other excitable tissues (17), but its function is not well understood (16). Abnormal taurine contents whether pathologically or experimentally induced are often associated with changes of excitability in heart (9), nerve (8,21), and retina (4). The taurine content is increased in skeletal muscle from the dystrophic chicken (23).

There is only a small amount of information on the possible role of taurine in the function of mammalian skeletal muscle or, more specifically, on the effects of taurine on ionic channels of skeletal muscle membranes. Signs, however, of myotonia in human patients can be decreased markedly by administration of taurine (11). It was earlier reported that taurine hyperpolarized rat extensor digitorum longus fibers and increased the imput resistance and tendency to fire repetitively (15). These latter results, therefore, do not provide us with an explanation for the anti-myotonic action of taurine which was of interest to us. This discrepancy may be related to the fact these observations were made in taurine-free medium in muscles from rats previously loaded with taurine. Thus, in the experiments which we report in this paper we studied the effects of taurine in the extracellular medium on the electrical parameters and on the component conductances of fibers from rat extensor digitorum longus muscle in vitro. In addition, we compared the effects of taurine to those of gamma-aminobutyric acid (GABA), an agent shown to affect G_{Cl} in other tissues (18).

The experiments were performed "in vitro" on extensor digitorum longus (EDL) muscles removed (10) from male Wistar rats (180-220 g) and set in a muscle bath heated at 30 $\pm$ 1°C. The normal physiological solution had the following mM composition: NaCl 148; KCl 4.5; $CaCl_2$ 2.0; $MgCl_2$ 1.0; $NaHCO_3$ 12.0; NaH_2PO_4 0.44; and glucose 5.55. The chloride-free solution was made by equimolar substitution of methylsulfate salts for NaCl and KCl and nitrate salts for $CaCl_2$ and $MgCl_2$. All solutions were continuously gassed with 95% O_2 and 5% CO_2; the pH was carefully maintained between 7.2-7.3 during each experiment. Taurine, β-alanine and GABA were dissolved in normal or chloride-free medium.

Conventional microelectrode methods were used for intracellular stimulation and recording (5,7). The measurements included membrane resting potentials, intracellular excitability, cable parameters and component conductances. Excitability was determined by observing the intracellular membrane potential response recorded from one microelectrode to a square wave constant current delivered by a second microelectrode inserted within 100 um from the voltage electrode. In each fiber the membrane potential was set by a steady holding current to -80 mV before passing the depolarizing pulses (7). The resting membrane component conductances were calculated (7) from the cable parameters (5) measured in chloride-containing and chloride-free Ringer solution before and after the addition of taurine, β-alanine or GABA. The reciprocal of membrane resistance (R_m) in chloride-containing solution was assumed to be total membrane conductance (G_m) and the same parameter measured in chloride-free medium was considered largely potassium conduct ance, G_K.

The mean chloride conductance (G_{Cl}) of each experimental group of fibers was estimated as the mean G_m minus the mean G_K; sodium and other small conductances were neglected. To test for the influence of osmotic pressure, measurements were made in medium containing sucrose at equimolar concentration to the tested agent.

The intracellular concentrations of taurine were determined using a Perkin-Elmer 53B Liquid Chromatograph according to the method of Krusz et al. (20), after perchloric homogenization of the tissue.

The data are expressed as means ± SEM. Significance of differences between group means was calculated by Student's t test. The estimates for the SEM of G_{Cl} were obtained from the variances of G_m and G_K, assuming no covariance, using standard methods (14).

After 20 minutes incubation with 60 mM taurine the rat EDL fibers hyperpolarized from a control mean of -62 ± 1.0 mV (n = 16) to -72 ± 0.6 mV (n = 20). The hyperpolarization was still present after multiple impalements and it was fully reversible after 20 min of washing. At the 30 mM concentration the increase of resting potential was not significant but there appeared to be a tendency to retard the steady decline with time.

Taurine at 60 mM produced a decrease in excitability characterized by significant increases in current required for double firing (I_{2nd}) and a decrease in latency for a rheobasic stimulus (Table 1). Also the mean of the maximal number of APs that could be produced by raising the intensity of a long pulse was significantly reduced by 60 mM taurine (Fig. 1). On the other hand there were no significant changes in rheobase (I_{th}), firing potential at rheobase, the AP magnitude and the AP overshoot (OS). AP duration, however, was significantly decreased.

Addition of 60 mM taurine to the normal bathing solution produced a statistically significant decrease in R_m (Table 2). There was no effect on R_m at 30 mM nor were there any effects on the capacitance or fiber diameter over this concentration range. The cable parameters were also compared to those measured in a chloride-free medium in the presence and absence of taurine. In this experimental condition, the amino acid had no effects on R_m, which had values of 4321 ± 615 $\Omega \cdot cm^2$ (n = 21) and 3499 ± 424 $\Omega \cdot cm^2$ (n = 27), respectively, before and after the addition of 60 mM taurine. Also, the other cable parameters were unchanged. Converting these parameters to conductances (Table 2), it is clear that the effect of taurine

Table 1. Effect of taurine on excitability characteristics of rat EDL muscle fibers

PREPARATION		I_{th} (nA)	I_{2nd} (nA)	OS (mV)	Lat (msec)	Dur (msec)	N° spikes
Mean of 8 controls	(34)	82±7	110±8	35±1	7.4±0.5	10.1±0.5	7.0±0.3
Mean of 5 taurine treated (60 mM)	(25)	84±4	142±12[a]	30±3	4.2±0.3[b]	7.4±0.3[b]	4.6±0.2[b]
Mean of 2 β-alanine treated (60 mM)	(12)	87±6	130±11	33±2	8.3±2.0	11.9±0.7	6.2±0.6
Mean of 4 sucrose treated (60 mM)	(21)	87±6	118±6	34±2	7.1±0.4	11.2±0.8	6.9±0.8

Values (mean ± SEM for number of observations shown in parenthesis) are as follows: Preparation (drug concentration used, in parenthesis); Ith: threshold current; I2nd: current to elicit the 2nd action potential; OS: overshoot of action potential; Lat: latency at rheobase; Dur: duration of action potential; N°: maximum number of spikes elicited.

Significantly different from mean of control group ([a]$P < 0.5$; [b]$P < 0.001$).

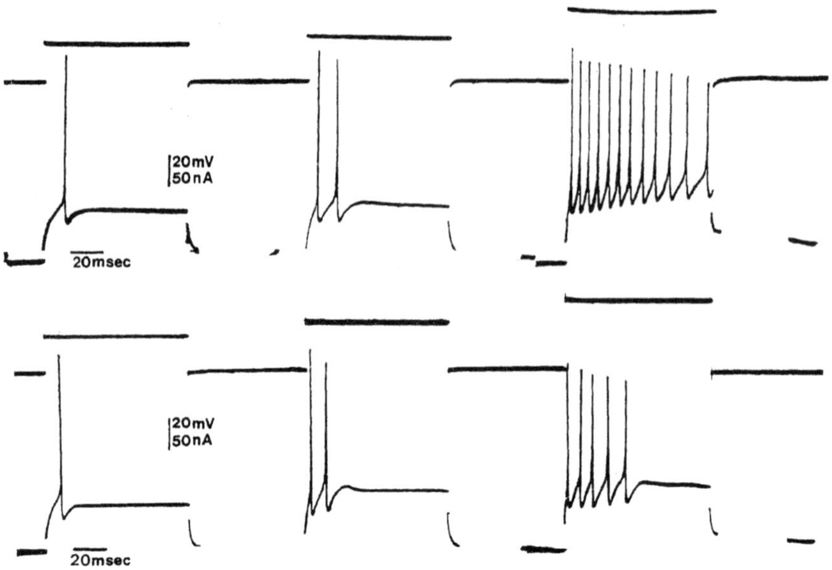

Figure 1. Effect of taurine on intracellular action potentials of rat EDL muscle fibers: upper part, in normal Ringer solution; lower part, 20 min after 60 mM taurine addition to bath. Upper trace, constant current pulse; lower trace, membrane potential. Initial resting potentials: -70 mV (A) and -74 mV (B); membrane potential was set to -80 mV before passing the current pulse. Temperature 30°.

on G_m at 60 mM is entirely due to a significant increase in G_{Cl} with no significant effect on G_K.

In one set of experiments GABA was added to the bath at 30-60 mM in order to compare directly any effects to those of taurine. GABA significantly reduced R_m at both of these concentrations without affecting the capacitance or the diameter of the fiber. Moreover, like taurine this change did not occur in chloride-free medium. The analysis of component conductances given in Table 3 shows that GABA increased G_{Cl} significantly at 60 mM without affecting G_K.

Unlikely taurine, GABA caused a significant depolarization: 5 min after exposure to 30 and 60 mM GABA, the fibers were already depolarized from a control value of -63 ± 2 mV (n = 10) to -58 ± 0.9 (n = 10) and to -56 ± 0.5 (n = 16), respectively. All of the observed effects of GABA were reversible. The rates of onset and offset of the GABA effect on all of the observed parameters were faster than those observed with taurine.

Equimolar concentrations of β-alanine or of sucrose were used as controls for nonspecific amino acid or osmotic pressure effects. None of the actions of either taurine or GABA were mimicked in these controls experiments (Tables 1 and 2).

The effects of taurine on the resting potential, excitability and component conductances were strictly pH dependent. The actions of taurine were fully evident in the pH range 7.2 to 7.3, but disappeared at pH higher than 7.5. This is shown for the case of total membrane conductance (G_m) in Fig. 2. Here the G_m in control solution is nearly constant over the range pH 7.2 to 7.6 whereas G_m in 60 mM taurine is 38% higher at pH 7.2-7.3 but is not significantly different from control in the range pH 7.4-7.6.

Table 2. Effect of taurine on membrane resistance and component resting conductances of rat EDL muscle fibers

PREPARATION	N	R_m ($\Omega \cdot cm^2$)	G_m ($\mu S/cm^2$)	N	G_K ($\mu S/cm^2$)	G_{Cl} ($\mu S/cm^2$)
Mean of 7 controls	26	427 ± 29	2598 ± 368	21	303 ± 88	2295 ± 279
Mean of 2 taurine treated (30 mM)	10	309 ± 29	3260 ± 273	12	291 ± 39	2969 ± 234
Mean of 5 taurine treated (60 mM)	20	267 ± 13^b	3897 ± 174^a	27	359 ± 35	3538 ± 116^b
Mean of 3 sucrose treated (60 mM)	16	477 ± 89	2321 ± 181	11	270 ± 67	2051 ± 145
Mean of 3 β-alanine treated (60 mM)	12	475 ± 102	2355 ± 425	12	267 ± 71	2088 ± 304

Values (mean $\pm$ SEM) are as follows: Preparation (drug concentration used, in parenthesis); N: number of fibers for R_m and G_m, and for G_K respectively. R_m (membrane resistance); G_m (total membrane conductance); G_K (potassium conductance), and G_{Cl} (chloride conductance) are the component conductances.

Significant differences from mean of control group: $^aP < 0.01$; $^bP < 0.001$.

Table 3. Effect of 30 and 60 mM GABA on membrane resistance and component resting conductances of rat EDL muscle fibers

PREPARATION	N	R_m ($\Omega \cdot cm^2$)	G_m ($\mu S/cm^2$)	N	G_K ($\mu S/cm^2$)	G_{Cl} ($\mu S/cm^2$)
Mean of 4 controls	20	402 ± 26	2784 ± 248	10	434 ± 93	2350 ± 209
Mean of 2 GABA-treated (30 mM)	8	282 ± 54^a	3822 ± 738	10	407 ± 75	3415 ± 495
Mean of 3 GABA-treated (60 mM)	12	216 ± 17^b	4745 ± 376^b	11	394 ± 74	4351 ± 276^b

Values (mean $\pm$ SEM) are as follows: Preparation (drug concentration used, in parenthesis); N: number of fibers for R_m and G_m, and for G_K respectively. R_m (membrane resistance); G_m (total membrane conductance); G_k (potassium conductance), and G_{Cl} (chloride conductance) are the component conductances.

Significant difference from mean of control group: $^a p < 0.05$; $^b p < 0.001$.

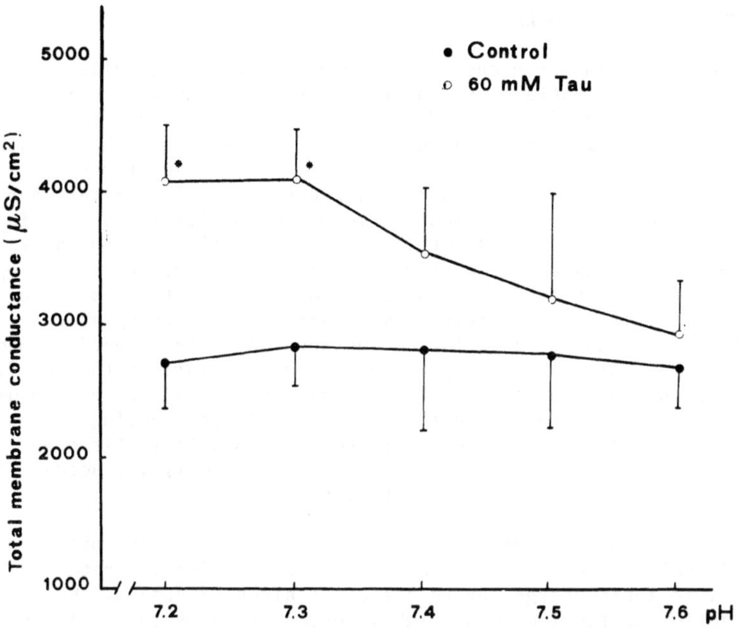

Figure 2. The influence of pH on the effect of 60 mM taurine on total resting membrane conductance (G_m) of rat EDL fibers. Each symbol indicates the mean of 9-12 fibers from 3 muscles. Vertical bars indicate the SEM. *P < 0.01.

Taurine content of the right EDL muscle was measured immediately after dissection and compared with that of the left EDL muscle from the same animal after it had been superfused for one hour in a taurine-free medium. The mean taurine concentrations from 9 rats expressed as µmol/g wet tissue was 24.14 ± 1.67 and 18.40 ± 1.24 for the fresh tissue and the superfused tissue, respectively. The means were significantly different by a paired comparison ($P < 0.02$).

TAURINE INCREASES G_{Cl}

Under the conditions of these experiments, relatively high extracellular concentrations of taurine (30 to 60 mM) decreased R_m in normal medium but did not change R_m in a chloride-free medium. When component resting membrane conductances are calculated from these data it is found that the taurine effect is almost entirely due to a significant increase in G_{Cl}. At 60 mM taurine, G_{Cl} significantly increased by 54% whereas there was a non-significant 18% increase in G_K which is normally only 13% of the total conductance in these fibers.

The high concentrations required to observe the effects of taurine and the other agents made it necessary to test for the influence of osmotic pressure. Addition of equimolar sucrose to our highest concentration of taurine (60 mM) was without effect on any of the parameteres measured. We also found no significant effects of 60 mM β-alanine, a compound which differs from taurine in having a carboxyl rather than a sulfonic acid moiety. Thus, the $-SO_3H$ group may impart this specificity to taurine. On the other hand, GABA, an omega amino acid of about the same molecular weight as β-alanine, produced a larger increase in G_{Cl} than did taurine (85% compared with 54% at 60 mM concentration of each).

TAURINE HYPERPOLARIZES, GABA DEPOLARIZES

Taurine action in nerve and other tissues is often associated with a change in membrane potential. Frequently, but not always, this is a hyperpolarization (15, 22). In our studies, the rat fibers tended to hyperpolarize at the higher 60 mM concentration, or at least decrease the run-down of resting potential at the 30 mM concentration. The reasons for the hyperpolarization is not clear. It has been suggested that the primary action of taurine is to increase G_{Cl} to which the membrane potential responds in a direction depending on the value of the chloride equilibrium potential relative to the resting potential (19). Although this mechanism may operate for transient responses such as occur at synapses or in tissues where chloride permeability plays a role in maintaining the resting state, in vertebrate skeletal muscle chloride conductance may only affect the membrane transiently (10). Also in our experiments, GABA depolarized the rat fibers under conditions where the chloride equilibrium potential may be presumed to be the same as with the taurine experiments. It is interesting to note that G_K decreased with GABA, whereas G_K increased with taurine. Although these G_K changes were not significant, they could conceivably account for the opposite effects on membrane potential (1).

GABA and taurine also differ in their rates of onset and offset of action, GABA being more rapid. This may relate to different rates of penetration, to ionic charge, or to different sites of action. We have also shown the action of taurine to be highly dependent on extracellular pH. The increase in G_m by taurine, presumably mainly G_{Cl}, is greatly reduced by increasing the external pH by only a few tenths of a unit. A similar effect of pH on taurine uptake into cardiac muscle has been reported (13).

TAURINE DECREASES EXCITABILITY

Taurine significantly affected the excitability characteristics of the rat EDL fibers, mainly seen as a shortened latency and decreased repetitive firing capability. In essence, the increased G_{Cl} produced by taurine would act as a shunt conductance shortcircuiting the small steady state sodium currents that exist just below threshold, thus preventing the assistance of these currents in sustaining the depolarization during long pulse stimulation. The result is to force a shorter latency and to decrease the ability to fire repetitively. These currents are postulated on the basis of the "overlap theory" of sodium activation and inactivation (2,3). The hyperpolarization induced by taurine would allow less sodium inactivation with a consequent shorter duration of the AP. Our experimental findings can also account for an antimyotonic action of taurine (11). However, the patients required a lower effective concentration in the plasma. The channels from diseased membrane appear to be more sensitive to the action of taurine.

INCUBATION VERSUS PRE-LOADING

It is necessary to mention the apparent disparities between our results and those reported by Gruener et al. (15). In their studies of rat muscle fibers, the hyperpolarization and decreased duration of the AP did occur, but they observed opposite effects on membrane resistance and excitability. They recorded the hyperpolarization with taurine in the bath and this agrees with our results, but input resistance and excitability were measured in taurine-free medium in fibers from rats pre-loaded with taurine. The indirect method they used to estimate membrane resistance may have produced an erroneous estimate of this variable. Regarding the increased excitability that they observed, it is reminiscent of data on the myotonia of the dystrophic chicken (12) which was suppressed by previously loading the animals with the antimyotonic agent phenytoin. However, when the excitability of these fibers was measured in a phenytoin-free solution, it was found to be greatly increased compared to the nontreated muscles. Possibly the chronic presence of taurine also induces some counter adaptation which can be seen under those conditions.

TAURINE, GABA, AND CHLORIDE CHANNELS

The principal effect of both taurine and GABA in these experiments has been to specifically increase G_{Cl} and this factor alone can account for much of what was observed on muscle fiber excitability. Thus, it is reasonable to propose that the increased G_{Cl} is mediated through specific membrane ionic channels.

The concentrations necessary for these effects are too great to consider seriously the presence of taurine or GABA receptor-activated channels as found in nerve cells or in vertebrate stretch receptors. It seems more likely that taurine or GABA in skeletal muscle have their effects through the same chloride channels that constitute the macroscopic G_{Cl}. We have given preliminary evidence elsewhere that once these channels are blocked by the specific blocker, anthracene-9-carboxylic acid, taurine has no effect (6). We, therefore, propose that taurine and GABA act directly on the mammalian skeletal muscle chloride channel to affect its kinetics or unit conductance or some combination of these parameters. We consider less likely the possibility that taurine or GABA operate indirectly through altering the lipid membrane since channels other than chloride-specific ones were not significantly affected.

ACKNOWLEDGMENT

This research was supported by Italian CNR grant 85426 and MPI 85 grant; and NIH grant NS 03178 (SHB). We thank Mr. M. Persichella for expert assistance.

REFERENCES

1. Adrian, R.H., Chandler, W.K., and Hodgkin, A.L., 1970, Slow changes in potassium permeability in skeletal muscle, J. Physiol. (London), 208:645-668.
2. Adrian, R.H., 1982, Voltage-dependent behavior in excitable membranes, in: "Abnormal Nerve and Muscle as Impulse Generators", J.W. Culp and J. Ochoa, eds., Oxford Press, New York, pp. 54-61.
3. Adrian, R.H., Action currents in transverse tubules, 1982, in: Abnormal Nerve and Muscle as Impulse Generators", J.W. Culp and J. Ochoa, eds., Oxford Press, New York, pp. 595-701.
4. Berson, E.L., Hayes, K.C., Rabin, A.R., Schmidt, S.Y., and Watson, G., 1976, Retinal degeneration in cats fed casein: II Supplementation with methionine, cysteine or taurine, Invest. Ophthalmol., 15:52-58.
5. Bryant, S.H., 1969, Cable properties of external intercostal muscle fibers from myotonic and nonmyotonic goats, J. Physiol. (Lond), 204:539-550.
6. Bryant, S.H., Conte-Camerino, D., Franconi, F., Giotti, A., and Mambrini, M, 1985, Lack of antagonism by taurine on chloride channel block by anthracene-9-carboxylic acid in rat skeletal muscle, Brit. J. Pharmacol., 86:586P.
7. Conte-Camerino, D., Bryant, S.H., and Mitolo-Chieppa, D., 1982, Electrical properties of rat extensor digitorum longus muscle after chronic application of emetine to the motor nerve, Exp. Neurol., 77:1-11.

8. Davison, A.N., and Kaczmarek, L.K., 1971, Taurine. A possible neurotransmitter?, Nature, 234:107-108.
9. Dolara, P., Ledda, F., Mugelli, A., Mantelli, L., Zilletti, L., Franconi, F., and Giotti, A., 1978, Effect of taurine on calcium inotropism and electrical activity of the heart. in: "Taurine and Neurological Disorders, A. Barbeau and R.J. Huxtable, eds., Raven Press, New York, pp. 151-159.
10. Dulhunty, A.F., 1978, The dependence of membrane potential on extracellular chloride concentration in mammalian skeletal muscle fibers, J. Physiol. (Lond), 276:67-82.
11. Durelli, L., Mutani, R., and Fassio, F., 1983, The treatment of myotonia: evaluation of chronic oral taurine therapy. Neurol., 33:559-603.
12. Entrikin, R.K., and Bryant, S.H., 1979, Suppression of myotonia in dystrophic chicken muscle by phenytoin, Am. J. Physiol., 237:C131-C136.
13. Franconi, F., Martini, F., Bennardini, F., Mangiapane, A., and Giotti, A., 1981, Taurine uptake in sarcolemma vesicles isolated from hypertrophic guinea pig hearts, Biochem. Pharmacol., 30:693-695.
14. Green, J.R., and Margerison, D., 1978, Statistical Treatment of Experimental Data, Elseiver, New York, pp. 86-88.
15. Gruener, R., Bryant, H.J., Markovitz, D., Huxtable, R., and Bressler, R., 1976, Ionic actions of taurine on nerve and muscle membranes: electrophysiologic studies. in: "Taurine and Neurological Disorders", A. Barbeau and R.J. Huxtable, eds., Raven Press, New York, pp. 225-246.
16. Huxtable, R.J., 1980, Does taurine have a function?, Fed Proc., 39:2678-2679.
17. Jacobson, J.G., and Smith, L.L.H., 1968, Biochemistry and physiology of taurine and taurine derivatives, Physiol. Rev., 48:424-511.
18. Krnjevic, K., 1974, Vertebrate synaptic transmission, Physiol. Rev., 54:418-540.
19. Krnjevic, K., and Puil, E., 1976, Electrophysiological studies on actions of taurine, in: "Taurine", R.J. Huxtable, and A. Barbeau, eds., Raven Press, New York, pp. 179-189.
20. Krusz, J., Kendrick, Z.V., and Baskin, S.I., 1978, Taurine content in the pineal of the aging monkey, in: Aging in the Nonhuman Primate, Raven Press, New York, pp. 106-111.
21. Okamoto, K., Kimura, H., and Sakai, Y., 1983, Ionic mechanisms of the action of taurine on cerebellar Purkinje cell dendrites in vitro: intradendritic study, Brain Res., 260:261-269.
22. Nistri, A., and Costanti, A., 1976, The action of taurine on the lobster muscle fiber and the frog spinal cord. Neuropharmacol., 15:635-641.
23. Wilson, B.W., Peterson, D.W., and Lilyblade, A.L., 1965, Free amino acids of developing skeletal musculature of normal and genetically dystrophic chickens, Proc. Soc. Biol. Med., 119:104-108.

INSULIN-LIKE ACTIVITY OF TAURINE

Joseph Maturo, III[1] and Elliott C. Kulakowski[2]

[1]Biology Department, C. W. Post College, P.O. Greenvale, New York
[2]Hypertension Endocrine Branch, National Heart, Lung and Blood Institute, National Institutes of Health, Bethesda, Maryland

There is now an increasing body of data that taurine (2-aminoethanesulfonic acid) at pharmacological concentrations alters carbohydrate metabolism. Early reports from Ackerman and Heinsen (1) and Macallum and Sivertz (10) found that taurine was potent a hypoglycemic agent. Several teams of investigators have confirmed these initial results. Donadio and Fromageot (4) confirmed the hypoglycemic properties of taurine and presented evidence that taurine increased glucose utilization in the rat diaphragm. Lampson et al. (9) reported that taurine was capable of enhancing the effects of insulin on cardiac glycolysis, glycogenesis and oxygen consumption. Moreover, when fasting rats were given a bolus injection of glucose, Kulakowski and Maturo (6) observed that taurine alone was capable of reducing serum glucose levels while stimulating 3-O-methylglucose uptake in skeletal muscle and liver, and glycogenesis in the liver. This occurred without a concomitant increase in serum insulin levels.

These latter observations indicated that taurine acted at the level of the cell membrane to stimulate glucose accumulation and metabolism. Dokshina et al. (3) initially proposed that the action of taurine was mediated through cyclic 3', 5'-adenosine monophosphate (cyclic AMP). However, the finding that taurine potentiates the action of insulin in cardiac muscle (9) and the preliminary data by Kulakowski et al. (7) showing that taurine binds to the purified insulin receptor are strong evidence in favor of the hypothesis that the action of taurine is mediated through its binding to the insulin receptor.

In this chapter, we present results from several recent experiments comparing the effects of taurine and insulin at both the molecular and cellular levels to determine if 1) taurine possesses many of the hormonal actions of insulin and 2) its mechanism of action is mediated through an interaction with the insulin receptor. Results are presented from studies in skeletal muscle and adipose cells showing that taurine can stimulate amino acid uptake, glucose uptake, and glycolysis. In addition, we present evidence that taurine binds specifically and reversibly to the purified insulin receptor.

METHODS

CD-1 male rats (200 - 250 g) were obtained from Charles River. The animals were maintained on a twelve hour light - dark cycle and given food and water ad libitum.

Adipocytes were obtained from rat epididymal fat pads and prepared according to the method of Rodbell (16). Taurine and insulin stimulation of the conversion of [U-^{14}C]-glucose to $^{14}CO_2$ in the isolated adipocytes and taurine binding to the isolated adipose cells was performed according to the methods described by Maturo et al. (12).

Intact rat diaphragms were used to measure 3-O-methylglucose and amino acid uptake according to the methods of Rieser and Maturo (15).

The insulin receptor was isolated and purified from fresh human placenta according to Maturo et al. (12). Fresh human placenta was homogenized in 50 mM Tris HCl, pH 7.4 containing 250 mM sucrose, 2 mM EDTA, 2 mM EGTA, 1 mM phenyl-methanesulfonyl fluoride (PMSF), 25mM benzamidine HCl, 10 μM leupeptin, 0.05 trypsin inhibitor units/ ml protinin, 1mM 1, 10 - phenanthrolinic acid, and 1 mM pepstatin A. The membranes were then isolated by differential centrifugation.

The insulin receptor was extracted from the placental membranes by detergent solubilization using Triton X-100. The solubilized receptor was purified by sequential chromatography on DEAE-cellulose, wheat germ agglutin, ricinus communis and an insulin agarose affinity column. The binding of both taurine and insulin to the solubilized receptor was measured according to the polyethyleneglycol method, which was initially described by Cuatrecasas (2).

All radiolabelled compounds were obtained from New England Nuclear. Chemicals were obtained from Sigma Chemical Company or Fisher Scientific.

Assay values are reported as mean $\pm$ SEM. Significance is listed as $p \leq 0.05$ according to the Student's t-test.

RESULTS

We have observed that *in vivo* taurine mimics the action of insulin on glucose uptake and metabolism (6). We sought to determine if we could reproduce the *in vitro* results of Donadio and Fromogeot (4) and demonstrate glucose uptake in the isolated rat diaphragm. These results are shown in Table 1. Insulin produced a six-fold stimulation of 3-O-methylglucose uptake in this preparation while taurine also significantly produced a three-fold increase in 3-O-methylglucose uptake in the isolated rat diaphragm.

Since taurine increased glucose uptake, we explored the possibility that taurine may mimic other activities of insulin such as stimulation of amino acid uptake. Using the isolated rat diaphragm preparation, we measured taurine and insulin stimulated proline uptake (Table 2). Both taurine and insulin were shown to increase proline uptake by nearly two-fold over basal levels. However, taurine and insulin together failed to stimulate a further increase in proline uptake.

TABLE 1

TAURINE STIMULATED UPTAKE
OF 3-O-METHYLGLUCOSE IN RAT DIAPHRAGM

	CONCENTRATION	N	DISTRIBUTION RATIO
Basal	-	3	1.23 ± 0.13
Insulin	1 mg/ml	3	6.01 ± 0.11*
Taurine	1 mg/ml	3	3.27 ± 0.14*

Intact diaphragms from male CD-1 rats (200-225 g) obtained from Charles River were used to determine 3-O-methylglucose uptake. Diaphragms were incubated for 1 h in Krebs Ringer Bicarbonate buffer, pH 7.4 aerated with 95% O_2 and 5% CO_2 in the presence or absence of the various agents and 3-O-methylglucose (5 mg/ml). The distribution ratio was determined as the mg 3-O-methylglucose/ml cell H_2O :mg 3-O-methylglucose/ml medium. An asterisk indicates significant difference from basal levels ($p<0.01$, using the Student's t-test).

We have previously reported that taurine is capable of stimulating glycogen production in the liver (6). We also sought to determine if taurine similar to insulin could stimulate glycolysis in isolated epididymal rat adipocytes. From preliminary data in Table 3, it can be seen that insulin produces 3.25 μmoles CO_2 per ml triglycerides per hour compared to 2.02 μmoles CO_2 produced by taurine. Both values are significantly higher than basal levels.

Having observed that taurine can stimulate glycolysis in isolated adipocytes, we decided to see if taurine could bind to adipocytes. In Table 4, we present a comparison of the data we obtained from isolated adipocytes with data previously reported for taurine binding to isolated cardiac sarcolemma. The data are quite comparable. Taurine binding to the adipocytes is specific, saturable, reversible and bi-phasic. The high affinity dissociation constant of 92 μM for the fat cells is comparable to that observed for the heart membranes (5). In addition, the binding to the low affinity binding site has a dissociation constant in the millimolar range and a Scatchard analysis reveals positive cooperativity. Taurine binding to the low affinity binding sites was inhibited by hypotaurine but not beta-alanine. In addition, taurine binding to adipocytes was inhibited by insulin.

Specific insulin binding to the purified human placental insulin receptor is observed over the range of 0.03 to 14.4 ng/ml [^{125}I]-insulin (Figure 1a). A Scatchard plot of the binding data reveals a typical profile for negative cooperativity and the dissociation constant (K_d) was calculated to be 1.2×10^{-9}M (Data not shown).

TABLE 2

TAURINE STIMULATED UPTAKE OF PROLINE IN RAT DIAPHRAGM

	CONCENTRATION	N	DISTRIBUTION RATIO
Basal	-	3	1.29 ± 0.10
+Insulin	1 mg/ml	3	2.13 ± 0.08*
+Taurine	1 mg/ml	3	2.12 ± 0.10*
+Insulin + Taurine	1 mg/ml +1 mg/ml	3	2.10 ± 0.08*

Intact diaphragms from male CD-1 rats (200-225 g) obtained from Charles River were used to determine proline uptake. Diaphragms were incubated for 1 hr in Krebs Ringer Bicarbonate buffer (pH 7.4) aerated with 95% O_2, 5% CO_2 in the presence or absence of the various agents and ^{3}H-proline (1 mM). The distribution ratio determined as the mg proline/ml cell H_2O :mg proline/ml medium. An asterisk indicates significant difference from basal levels ($p<0.01$, using the Student's t-test).

TABLE 3

TAURINE STIMULATED METABOLISM OF GLUCOSE TO CARBON DIOXIDE IN ISOLATED ADIPOSE CELLS

	CONCENTRATION	N	GLUCOSE METABOLISM ($\frac{\mu\text{moles } CO_2\text{/ml Triglyceride}}{h}$)
Basal	-	3	0.84 ± 0.09
+Insulin	1mg/ml	3	3.25 ± 0.12*
+Taurine	1mg/ml	3	2.02 ± 0.12*

CD-1 male rats (220-225 g), obtained from Charles River, were used to determine CO_2 production. The preparation of isolated fat cells and assay for the conversion of [U-^{14}C]-glucose are described in Methods. An asterisk indicates significant differences from basal levels ($p<0.01$, using Student's t-test).

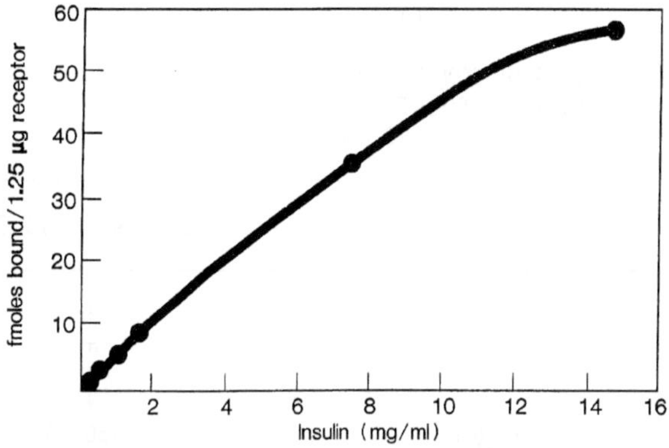

FIGURE 1: Insulin binding to the purified insulin receptor

In contrast to the high affinity binding of insulin to the purified insulin receptor, taurine exhibited lower affinity binding to the receptor (Figure 2a). The dissociation constant was calculated from the Scatchard plot to be 130 x 10^{-9} M and the B_{max} was 1.6 nmole/mg receptor protein (Data not shown).

Table 5 illustrates that taurine binding to the purified insulin receptor was pH dependent. Maximum binding of taurine occurred at pH 7.5 and the binding profile was bell shaped. The pH depbendency appears to be related to changes in the conformation of the insulin receptor.

We examined the effect of various taurine analogues to asses the specificity of taurine binding to the insulin receptor (Table 6). To assess the ability of taurine analogs to displace [^{3}H]-taurine bound to the receptor, we incubated equal molar concentrations of the unlabelled taurine, hypotaurine, and beta-alanine with the receptor-ligand complex. The carboxylic acid derivative of taurine, beta-alanine was found to only minimally displace bound taurine, while hypotaurine, the sulfonic acid derivative reduced taurine binding by 65%. Thus, it appears that the nature of the acid group on the alpha-carbon appears to influence the binding properties of taurine to the insulin receptor.

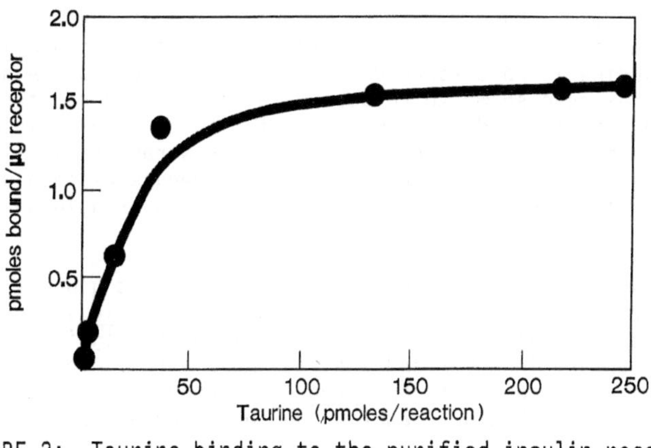

FIGURE 2: Taurine binding to the purified insulin receptor

TABLE 4

COMPARISON OF TAURINE BINDING BETWEEN ISOLATED FAT CELLS AND CARDIAC MUSCLE

	FAT CELLS	CARDIAC MUSCLE*
Binding	+	+
Saturable	+	+
Reversible	+	+
Biphasic	+	+
K_{d1}	92 μM	350 μM
K_{d2}	~58 mM	~3.5 mM
Inhibitors of low affinity site		
Insulin	+	+
Hypotaurine	+	+
Beta-alanine	-	-

*Kulakowski et al.(5)

DISCUSSION

We have presented substantial evidence that taurine binds to the purified human placenta insulin receptor. The binding is specific, reversible, saturable and exhibits positive cooperativity. The dissociation constant was calculated to be 130 x 10^{-9} M with the maximum number of binding sites being 1.6 nmole/mg receptor. These data, together with our previous data showing that taurine binding can be displaced from the insulin receptor with insulin or inhibited by insulin receptor antibodies (7), provides strong evidence that taurine binds to the human insulin receptor. If taurine interacts with the insulin receptor, as we have described, it is only logical that taurine may be an agonist or antagonist to the actions of insulin.

Several laboratories, including our own have demonstrated that taurine can indeed mimic several of the major actions of insulin. These effects are summarized in Table 7. In vivo experiments have provided evidence that taurine is able to reduce blood glucose levels. A single dose of taurine has been shown to reduce normal blood glucose levels in rabbits (10) and rats (3) and in normal rats following a bolus injection of glucose (6) or rats made hyperglycemic due to cold stress (13). In addition, data have shown that taurine is able to reduce blood glucose levels in mice, rats or rabbits made diabetic with either alloxan (17), ditisone (18), or streptozotocin (19).

The decrease in blood glucose levels indicates that glucose was taken up into various tissues. In the only study examining the in vivo distribution of 3-O-methylglucose, taurine was shown to significantly increase glucose uptake in liver and skeletal muscle (6). These observations were consistent with in vitro data reported by Dokshina et al (3) and in this chapter showing that taurine increased glucose uptake in the rat diaphragm.

TABLE 5

pH EFFECTS OF TAURINE BINDING
TO THE PURIFIED INSULIN RECEPTOR

pH	fmoles bound/assay
5.5	4.3
6.0	8.7
6.5	18.6
7.0	55.5
7.5	80.3
8.0	68.2
8.5	50.3

The effect of pH on taurine binding to the purified insulin receptor. The insulin receptor (0.1 μg receptor/assay) was incubated with [^{3}H]-taurine (17.2 pmoles/assay) in 50 mM Tris, 0.03% Trition X-100, 0.5M NaCl at 24°C at the various pH's listed. Samples were analyzed for taurine binding as described in Methods.

TABLE 6

DISPLACEMENT OF TAURINE
BINDING FROM THE INSULIN RECEPTOR

	N	PERCENT OF CONTROL
Control	3	100
Taurine	3	11.0 ± 0.8
Hypotaurine	3	35.5 ± 1.8
Beta-Alanine	3	85.1 ± 2.9

Sepharose-insulin purified insulin receptor was measured for taurine binding [^{3}H]-taurine. After equilibrium, the taurine-receptor complex was incubated in the presence or absence of 50 mg of unlabelled taurine, hypotaurine or Beta-alanine for an additional 10 min. Taurine binding was measured by the polyethylenegylcol method.

TABLE 7

COMPARISON OF THE METABOLIC EFFECTS OF TAURINE AND INSULIN

Metabolic Actions	Insulin	Taurine
Reduction in Blood Glucose Levels	+	+
Glucose uptake	+	+
Glucose metabolism	+	+
Glycogen production	+	+
Amino acid uptake	+	+
Protein synthesis	+	?

Since glucose is taken up and metabolized in cells, it would be expected that taurine, similar to insulin, would increase glucose utilization in those cells. Several investigators have shown that indeed taurine does stimulate glucose utilization. Donadio and Fromageot (4) reported an increase in glucose utilization in rat diaphragms. In this chapter, we presented data showing that both taurine and insulin enhance the conversion of glucose to CO_2 in rat adipocytes. Elegant metabolic studies reported by Lampson et al. (9) showed that in the isolated perfused rat heart, taurine in the presence of insulin increased glucose utilization and lactate production.

However, Laborit and Thuret (8) reported that taurine did not stimulate glucose utilization in the brain. These results are not inconsistent with the present literature. Kulakowski and Maturo (6) reported that taurine did not increase the uptake of taurine in the brain. Furthermore, insulin, which can not cross the blood brain barrier, is not necessary for glucose uptake in the brain. If taurine mediates its action through the insulin receptor, then it would not be expected that taurine would enhance glucose utilization in the brain.

In addition to stimulating glucose utilization, taurine is capable of enhancing glycogen production. In normal rats, taurine was shown to increase glycogenesis in the liver (3,6) and in diabetic rats and rabbits (18). In addition, taurine was shown to enhance the effect of insulin on glycogen synthesis in the rat heart (9). These investigators showed that taurine in the presence of insulin was capable of increasing the activity of glycogen synthetase I, while producing a decrease in phosphorylase a activity.

If, indeed, taurine plays a pharmacological or physiological role similar to insulin in regulating blood glucose levels, then it may be expected that changes in taurine or insulin levels may alter the concentration of the other agent. In rats made diabetic with alloxan, the intracellular concentration of taurine was shown to increase in the heart

(14). Furthermore, Kulakowski and Maturo (6) demonstrated that in the presence of high blood glucose levels, taurine was capable of suppressing the release of insulin from the beta - cells of the Islets of Langerhans in the pancreas.

Taurine not only mimics the action of insulin on carbohydrate metabolism but is capable of stimulating the uptake of amino acids. We have shown that taurine enhances the uptake of proline in the rat diaphragm. While proline may be metabolized for energy, amino acids are primarily incorporated into proteins. The effect of taurine on stimulation of protein synthesis is currently being investigated in our laboratories.

Thus, it appears that taurine may be a potent agonist of insulin. Taurine has been shown to mimic several of the major effects of insulin on carbohydrate metabolism and stimulate amino-acid uptake. In addition, these effects of taurine appear to be directly related to its interaction in the insulin receptor. Research efforts are concentrating on the effects of taurine in diabetic animals.

ACKNOWLEDGEMENT

The authors wish to express our thanks to Kathy Kirschbaum for preparation of the manuscript.

REFERENCES

1. Ackerman, D., and Heinsen, H. A., 1935, Uber die physiologische wirkung des asterubins und anderer, zum tiel neu dargestellter, schwefelhaltiger guanidinderivate, Z. Physiol. Chem., 235:115-121.

2. Cauatrecasas, P., 1972, Isolation of the insulin receptor of liver and fat cell membranes, Proc. Nat. Acad. Sci. U.S.A., 69:318-322.

3. Dokshina, G.A., Silaeva, T.Y., and Yartsev, E.I., 1976, Some insulin-like effects of taurine, Vopr. Med. Khim., 22:503-507.

4. Donadio, G., and Fromageot, P., 1964, Influence exercise par la taurine sur l'utilisation du glucose par le rat, Bull. Soc. Chim. Biol., 46:293-302.

5. Kulakowski, E.C., Maturo, J., and Schaffer, S.W., 1978, Identification of Taurine Receptors from Rat Heart Sarcolemma, Biochem. Biophys. Res. Common., 80:936-941.

6. Kulakowski, E.C. and Maturo, J., 1984, Hypoglycemic properties of taurine: not mediated by enhanced insulin release. Biochem Pharmacol., 33:2835-2838.

7. Kulakowski, E.C., Maturo, J., and Schaffer, S.W., 1985, The low affinity taurine-binding protein may be related to the insulin receptor in Taurine: Biological Actions and Clinical Perspectives, eds. Oja, S.S., Ahtee, L., Kontro, P., and Paasonen, M.K., Allan R. Lis, Inc., New York, 127-136.

8. Laborit, H., and Thuret., F., 1974, Action de la taurine sur certaines activities metaboliques, Agressologie., 15: 183-186.

9. Lampson, W.G., Kramer, J.H., and Schaffer, S.W., 1983, Potentiation of

9. Lampson, W.G., Kramer, J.H., and Schaffer, S.W., 1983, Potentiation of the actions of insulin by taurine, Can. J. Physiol. Pharmcol., 61:457-463.

10. Macallum, A.B., and Sivertz, C., 1942, The potentiation of insulin by sulfones, Can. Chem. Process., 26:669.

11. Maturo, J.M.,III, Siegel, G. and Furino, C. (1975) Studies on the effect of 2-hydroxy-5-nitroenzyl bromide on the mechanism of insulin. Life Science 16:415-422.

12. Maturo, J.M., Hollenberg, M.D., and Aglio, L.S., 1983, Insulin receptor: Insulin-modulated interconversion between distinct molecular forms involving disulfide-sulfhydryl exchange, Biochem., 22:2579-2586.

13. Nakagawa, K., and Kuriyama, K., 1975, Effect of taurine on alteration in adrenal functions induced by stress, Jap. J. Pharmacol., 25:737-746.

14. Reibel, D.K., Shaffer, J.E., Kocsis, J.J., and Neely, J.R., 1979, Changes in taurine content in heart and other organs of diabetic rats, J. Mol. Cell. Cardiol., 11:827-830.

15. Rieser, P., and Maturo, J., 1969, Insulin, substrate transport, and protein synthesis in muscle, Arch. Int. Med., 123:267-271.

16. Rodbell, M., 1964, Metabolism of isolated fat cells, J. Biol. Chem. 239:375-380.

17. Shimuza, T., Shimizu, S., Nagami, S., and Wada, M., 1960, Effect of hypo, taurine and cysteine on alloxan- and dehydroxyascorbic acid-diabetes, Niigata Med. J., 74:931-939.

18. Silaeva, T.Y., Dokshina, G.A., Yartsev, E.I., Yakoleva, V.V. and Arkhangelskaya, T.E., 1976, The effect of taurine on the carbohydrate metabolism of diabetic animals, Prob. Endokrinol., 22:99-103.

19. Tokuwaga, H., Yoneda, Y., and Kuriyama, K., (1979) Protective actions of taurine against streptozotocin-induced hyperglycemia, Biochem. Pharmacol., 28:2807-2811 (1979).

TAURINE INHIBITS WET-DOG SHAKES AND HIPPOCAMPAL SEIZURES INDUCED BY OPIOID PEPTIDES IN RATS

Kanji Izumi[1], Motoaki Yoshida[2], Takeshi Koja[1], Eisuke Munekata[3], Takao Nakanishi[2] and Takeo Fukuda[1]

[1]Department of Pharmacology, Faculty of Medicine, Kagoshima University, Kagoshima 890, Japan. [2]Department of Neurology, Institute of Clinical Medicine, University of Tsukuba, Ibaraki 305, Japan. [3]Institute of Applied Biochemistry, University of Tsukuba, Ibaraki 305, Japan

Wet dog shakes (WDS) are paroxysmal shudders of head, neck and trunk. This behavior has been observed when dipping animals into cold water (41), application of xylene to the fur (39) and tactile stimulation around ears (1). Pharmacological studies revealed that WDS can be produced by administration of thyrotropin-releasing hormone (9,24,29,41,42), 5-Hydroxytryptamine (3,12), or kainic acid (5,25). WDS are also known as a classical sign of morphine abstinence in rats (40,43). In 1976, Bloom et al. first reported that this shaking behavior was elicited by acute administration of β-endorphin into the lateral ventricle in rats. Similar observations were made in rats with enkephalin (15). In addition, it was found that epileptic discharges were provoked by the intraventricular (icv) injection of opioid peptides and WDS always accompanied the onset of seizure discharges (15,16,37). These findings suggest that there may be an association between electrical seizure activities and WDS.

Taurine is a sulfur amino acid known to possess antiepileptic properties against various types of experimentally induced seizures in mammals (2,14,17, 32,38). Taurine also counteracts certain actions of morphine or opioid peptides, as shown by (a) the icv administration of taurine completely blocked the increment in plasma growth hormone induced by morphine in rats (6); (b) taurine given icv reduced the duration time of akinesia and analgesia produced by [D-Ala^2,Met^5]enkephalinamide (DAME) in rats (22); (c) analgesic effects of morphine were antagonized by taurine administered icv (44), or intraperitoneally (ip) (7) in mice; (d) pretreatment with taurine (icv) suppressed development of tolerance to both akinetic and analgesic effects of DAME in rats (21); and (e) taurine induced a dose-related antagonism of morphine tolerance and attenuated abstinence behavior in mice (7). In the present study, we investigated effects of taurine on DAME-induced WDS in rats, with special reference to its effects on epileptic discharges.

BEHAVIORAL METHODS

Male adult Wistar rats (300-400 g) were cannulated into the left lateral ventricle under anesthesia with pentobarbital as described previously (23). DAME and naloxone (HCl form) were dissolved in 0.85% saline solution.

Taurine, GABA and L-leucine were dissolved in 5 mM phosphate buffer in NaCl solution. The amino acid solutions were adjusted to be neutral (pH 7.2-7.4) and iso-osmotic (280-290 mosM). All amino acids and DAME were administered through the cannula in a volume of 10 µl at a rate of 10 µl/30 s. Rats were treated with each amino acid in a dose of 0.95, 0.475 or 0.2375 umol, or 10 ul of 5 mM phosphate buffer in NaCl solution, 10 min prior to the injection of 50 µg of DAME. These doses of taurine were based on the previous studies on the suppressive effects of taurine on experimental epilepsies (18, 19, 20). The dose of DAME is equivalent to the ED_{99} for both akinesia and analgesia. Naloxone (10 ug) was also given icv in a volume of 5 µl, immediately after the administration of 0.95 µmol of DAME. The experiments were carried out from 1:00 to 5:00 pm. The number of shaking movements of the head alone or even the whole body was counted for the first 10 min after the icv injection of DAME.

ELECTROPHYSIOLOGICAL METHODS

Rats were immobilized by an ip injection of decamethonium bromide (20 mg/kg). Under artificial respiration with a Harvard Model 680, animals were placed on a stereotaxic apparatus. Bipolar electrodes made of stainless steel wires (0.1 mm in diameter and 1.5 mm in tip separation) were used for the electroencephalogram (EEG) recordings. EEG was led from both the frontal cortex and the hippocampus. Monitoring was done with a polygraph (Nihon Kohden RM 6000) and a data recorder (TEAC R70). Pulse density variation curves were made with a signal processor (SAN-EI Instrumental Co. 7S06). Rats were pretreated icv with 10 µl of either 5 mM phosphate buffer in NaCl solution, or taurine in a dose of 0.95 µmol (119 µg), then 10 min later they were given 50 µg of DAME, icv in the same volume. Naloxone (10 µg) in a volume of 5 µl was injected during the time when epileptic discharges induced by 50 µg of DAME were persisting.

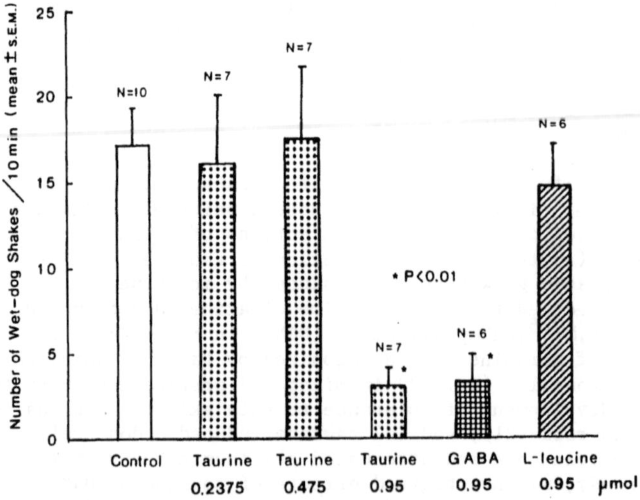

Figure 1. Effects of taurine, GABA or L-leucine on wet dog shakes (WDS) induced by DAME. Rats were pretreated with each amino acid or 5 mM phosphate buffer in NaCl solution and 10 min later injected with 50 µg of DAME in a volume of 10 µl. All substances were administered intraventricularly. WDS were counted for 10 min after the DAME injection. The numbers of WDS in rats receiving 0.95 µmol of taurine or GABA were significantly decreased, compared with those in control animals with 5 mM phosphate buffer in NaCl solution * ($p < 0.01$). N: number of rats.

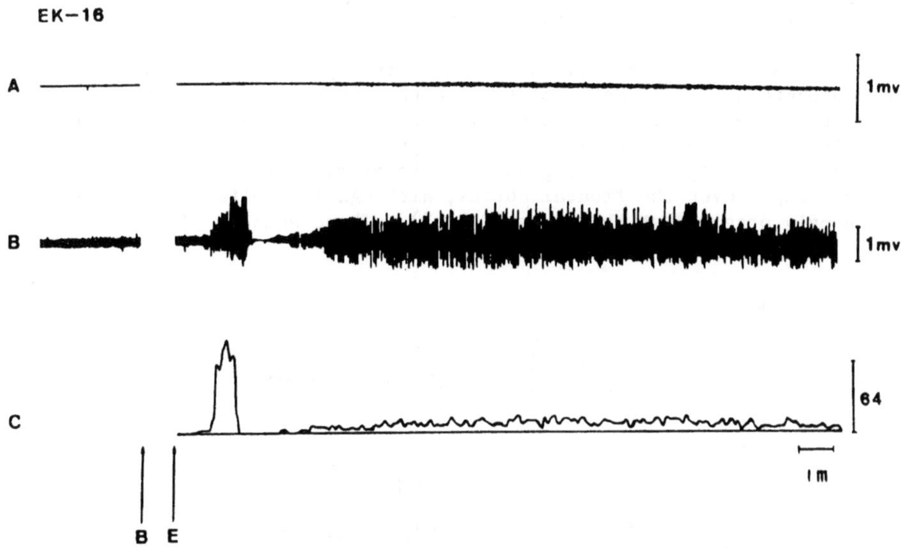

Figure 2. A specimen electroencephalogram taken from a rat given 10 µl of 5 mM phosphate buffer in NaCl solution 10 min prior to the intraventricular injection of 50 µg of DAME in a volume of 10 µl. A: EEG recorded at the frontal cortex, B: EEG recorded at the hippocampus, C: pulse density variation curve in the hippocampus (number of pulses exceeding 300 µV in amplitude for every 6 s). B and E with vertical arrows indicate when 5 mM phosphate buffer in NaCl solution and DAME were administered, respectively. Seizure activities are observed in the hippocampus as shown in B.

EFFECTS OF TAURINE ON WDS

Immediately after the icv administration of 50 µg of DAME, control rats, which had been pretreated with 10 µl of 5 mM phosphate buffer in NaCl solution, frequently exhibited WDS. The shaking movements mostly involved the whole body, but movements of the head alone were also evident. The number of WDS counted for 10 min following the administration of DAME was 17.1 ± 2.2 (mean ± SEM, N=10) (Fig. 1). The WDS were completely inhibited by an ip injection of naloxone, indicating that this behavior is mediated by opioid receptors. These results were consistent with previous findings observed in rats given enkephalin (15,28), β-endorphin (4,16), or DAME (10,36). After about 10 min, all rats (N=10) fell into a marked state of akinesia with muscle rigidity and no more WDS appeared.

The number of WDS was markedly decreased in test rats pretreated with 0.95 µmol of taurine (3.0 ± 1.1, N=7). The difference in the mean number of WDS between the test and control rats was statistically significant ($p < 0.01$). However, small doses of taurine (0.475 and 0.2375 µmol) did not significantly inhibit the WDS, indicating that the dose-dependent inhibitory effect of taurine may be steep. The number of WDS was also markedly decreased in rats given 0.95 µmol of GABA (3.3 ± 1.5, N=6) ($p < 0.01$). On the other hand, the number of WDS was unchanged by pretreatment with L-leucine given in the same dose as was taurine or GABA (Fig. 1). Therefore, the inhibitory effects of taurine and GABA appear to be specific.

EFFECTS OF TAURINE ON HIPPOCAMPAL SEIZURES

As shown in Fig. 2 and 3, in 8 control rats pretreated icv with 10 µl of 5 mM phosphate buffer in NaCl solution, high voltage spikes appeared immediately after the icv injection with 50 µg of DAME and evolved to seizure activities from 1 to 6 min later in the hippocampus. The seizure activities appeared recurrently for as long as 1 h. However, they did not spread from the hippocampus over the frontal cortex, although low voltage spikes were continuously observed at this area (Fig. 2). These seizure discharges

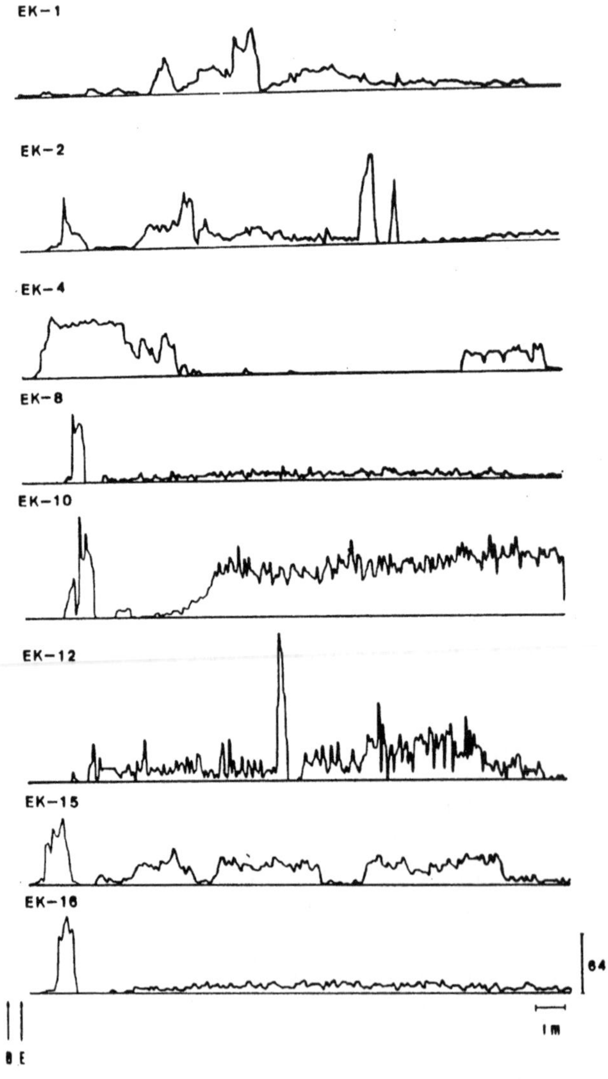

Figure 3. Pulse density variation curves at the hippocampus of 8 rats pretreated with an intraventricular injection of 10 µl of 5 mM phosphate buffer in NaCl solution 10 min prior to an intraventricular administration of 50 ug of DAME. EKs correspond to pulse density variation curves of individual rats. See Fig. 2 for explanation of the symbols B and E. The curves are abnormal in all rats.

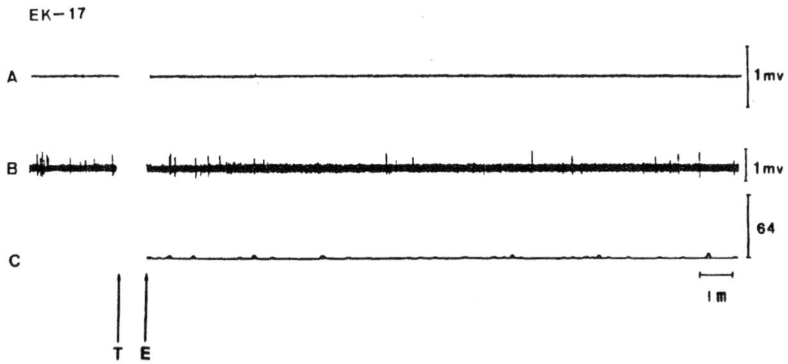

Figure 4. A specimen electroencephalogram taken from a rat given 10 µl of taurine at a dose of 0.95 µmol intraventricularly 10 min prior to an intraventricular injection of 50 µg of DAME in a volume of 10 µl. A, B and C are defined in Fig. 2. T and E with vertical arrows indicate when taurine and DAME were administered, respectively. Seizure activities in the hippocampus are supressed as shown in B.

provoked by DAME were readily inhibited by naloxone, indicating that theseizures are mediated by opioid receptors. These results are consistent with the findings that the icv administration of β-endorphin or DAME elicited epileptic responses characterized by a prolonged sequence of high voltage paroxysmal waves, particularly prominent in the hippocampus and amygdala in rats (16). Similar epileptic discharges have been also recorded from the cortex of rats after injections into the lateral ventricle of a large dose of leucine-enkephalin, methionine-enkephalin (15) and DAME (36).

Fig. 4 and 5 show effects of taurine on hippocampal seizures induced by DAME. In 5 (EK-3, 5, 7, 13, 17) of the 9 rats pretreated icv with taurine (0.95 µmol), there was no seizure activity in the hippocampus after the icv injection of 50 µg of DAME. The inhibitory effects of taurine on seizure activities were statistically significant ($p < 0.05$). In one rat (EK-14), there was evidence of seizure 2 to 6 min following the injection of DAME, but thereafter no seizure activities were seen for the one hour observation period. In the remaining 3 rats, the recurrent seizures were evident as observed in control animals.

DISCUSSION

In the present study, we could not analyze directly the relationship between WDS and hippocampal seizure discharges since the rats were immobilized with decamethonium bromide during the EEG recordings. However, in this respect, we will refer to known data; (a) WDS were observed in association with epileptoid discharges just following the icv administration of DAME (36); (b) WDS appeared at the onset of limbic ictal episodes following the icv administration of β-endorphin (16); (c) WDS were typically observed during the first minute of the seizure following the icv injections of leucine- and methionine-enkephalin, while the median duration of seizure discharges was 6 and 13 min for methionine- and leucine-enkephalin, respectively. Furthermore, WDS have been observed in amygdaloid kindling (35) and septal kindling (26) in rats. On the bases of these descriptions, we proposed that there is an association between WDS and epileptoid discharges, particularly in the limbic regions, as induced by opioid peptides.

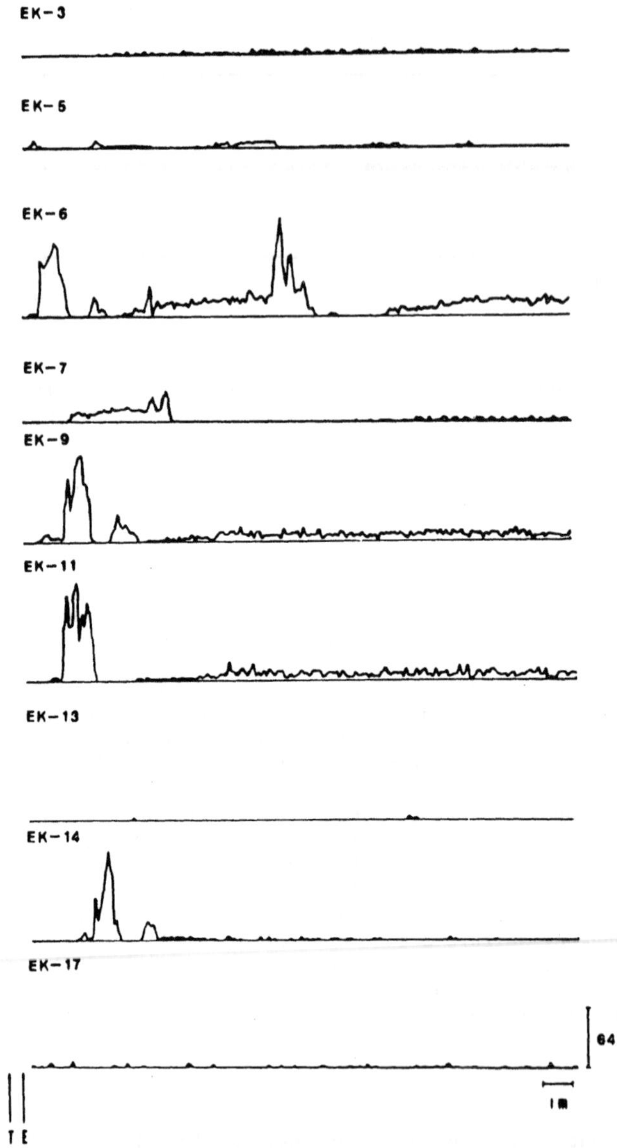

Figure 5. Pulse density variation curves at the hippocampus of 9 rats pretreated with an intraventricular injection of 10 µl of taurine in a dose of 0.95 µmol 10 min prior to an intraventricular injection of 50 ug of DAME. EKs correspond to pulse density variation curves of individual rats. T and E are identical to those in Fig. 4. Pulse density variation curves were normalized in 5 rats (EK-3,-5,-7,-13,-17).

In spite of these considerations, WDS appeared only for several minutes following the appearance of epileptoid discharges, whereas the latter persisted for longer period. This dissociation may suggest that WDS and hippocampal seizure discharges do not depend on a common underlying mechanism. However, it is still possible that the dissociation may be due to akinetic effects which came out uniformly approximately 10 min after the DAME injection, since a high dose of the peptide equivalent to the ED_{99} for akinesia was given. This is supported by an observation (Yoshida and Izumi, in preparation) that WDS reappeared when akinesia was removed by 2 mg/kg of apomorphine, a dopamine receptor stimulant, a dose which does not appear to affect epileptic discharges induced by opioid peptides (16). These findings suggest that WDS might have been masked behind the extrapyramidal symptoms.

Pretreatment with a high dose of taurine (0.95 μmol; 119 μg) markedly inhibited hippocampal seizure activities provoked by DAME (50 μg) in 5 of 9 rats ($p < 0.05$). Taurine has an anticonvulsive action against experimental epilepsy (2,14,17,32,28). If WDS induced by DAME really occur in association with hippocampal seizure discharges, then the inhibition of WDS by taurine may be associated with the suppression of hippocampal seizure activities. Another possibility could be that taurine inhibited the spread of seizure activities to other brain regions responsible for WDS and as a consequence the WDS were suppressed. The last possibility that taurine might have acted separately on WDS and hippocampal seizures cannot be ruled out. We found previously that taurine reduced the duration of both akinesia and analgesia produced by DAME in rats (22). Although the mechanisms of such actions of taurine remain to be elucidated, it is possibly that similar mechanisms may contribute to the inhibitory effects of tauirne on DAME-induced WDS.

GABA is a postsynaptic inhibitory neurotransmitter with putative anticonvulsive actions (18). Hippocampal seizure activities provoked by opioid peptides are presumably due to disinhibition of inhibitory interneurons which contain GABA (11,13,27,30,31,33,45). GABA partially or fully reverses the opioid effects on evoked CA1 field potentials (8). Sodium valproate, which increases synaptosomal GABA, has been reported to abolish the electrical seizure activity elicited by leucine-enkephalin in rats (34). If there is an association between WDS and hippocampal seizure activities both induced by DAME, it may be that GABA suppresses electrical excitation in the hippocampus, thereby resulting in the inhibition of WDS.

REFERENCES

1. Askew, H.R., Librecht, B.C., and Ratner, S.C., 1969, Effects of stimulus duration and repeated sessions on habituation on the head-shake response in the rat, J. Comp. Physiol. Psychol., 67:497-503.
2. Barbeau, A., and Huxtable, R.J., eds., 1978, "Taurine and Neurological Disorders", Raven Press, New York.
3. Bedard, P., and Pycock, C.J., 1977, 'Wet-dog' shake behaviour in the rat: a possible quantitative model of central 5-hydroxytryptamine activity, Neuropharmacology, 16:663-670.
4. Bloom, F., Segal, D., Ling, N., and Giullemin, R., 1976, Endorphins: profound behavioral effects in rats suggest new etiological factors in mental illness, Science, 194:630-632.
5. Collins, R.C., McLean, M., and Olney, J., 1980, Cerebral metabolic response to systemic kainic acid: 14-C-deoxyglucose studies, Life Sci., 27:855-862.
6. Collu, R., Charpente, G., and Clermont, M.J., 1978, Antagonism by taurine of morphine induced growth hormone secretion, Canad. J. Neurol. Sci., 5:139-142.
7. Contreras, E., and Tamayo, L., 1984, Effects of taurine on tolerance to and dependence on morphine in mice, Arch. Int. Pharmacodyn., 267:224-231.
8. Corrigall, W.A., and Linseman, M.A., 1980, A specific effect of morphine on evoked activity in the rat hippocampal slice, Brain Res., 192:227-238.
9. Costall, B., Hui, S.C.G., Metcalf, G., and Naylor, R.J., 1979, A study of the changes in motor behavior caused by TRH on intracerebral injection, Eur. J. Pharmacol., 53:143-150.

10. Cowan, A., and Tortella, F.C., 1982, A quantitative analysis of the shaking behavior induced in rats by β-endorphin and [D-Ala^2,Met^5] enkephalinamide, Life Sci., 30:171-176.
11. Dingledine, R., 1981, Possible mechanisms of enkephalin action of hippocampal CA1 pyramidal neurons, J. Neurosci., 1:1022-1035.
12. Drust, E.G., Sloviter, R.S., and Connor, J.D., 1979, Effect of morphine on 'wet-dog' shakes caused by cerebroventricular injection of serotonin, Pharmacology, 18:299-305.
13. Dunwiddie, T., Mueller, A., Palmer, M., Stewart, J., and Hoffer, B., 1980, Electrophysiological interactions of enkephalins with neuronal circuitry in the rat hippocampus. I. effects on pyramidal cell activity, Brain Res., 184:311-330.
14. Durelli, L., and Mutani, R., 1983, The current status of taurine in epilepsy, Clin. Neuropharmacology, 6:37-48.
15. Frenk, H., Urca, G., and Liebenskind, J.C., 1978, Epileptic properties of leucine- and methionine-enkephalin: comparison with morphine and reversibility by naloxone, Brain Res., 147:327-337.
16. Henriksen, S.J., Bloom, F.E., McCoy, F., Ling, N., and Guillemin, R., 1978, β-Endorphin induces nonconvulsive limbic seizures, Proc. Natl. Acad. Sci. USA, 75:5221-5225.
17. Huxtable, R.J., 1981, Insights on function: metabolism and pharmacology of taurine in the brain, in: "The Role of Peptides and Amino Acids as Neurotransmitters", J.B. Lombardini, and A.D. Kenny, eds., Alan R. Liss, Inc., New York, pp.53-97.
18. Izumi, K., Donaldson, J. Minnich, J.L., and Barbeau, A., 1973, Ouabain-induced seizures in rats: suppressive effects of taurine and γ-aminobutyric acid, Canad. J. Physiol. Pharmacol., 51:885-889.
19. Izumi, K., Igisu, H., and Fukuda, T., 1974, Suppression of seizures by taurine - specific or non-specific ?, Brain Res., 76:171-173.
20. Izumi, K., Kishita, C., Koja, T., Shimizu, T., Fukuda, T., and Huxtable, R.J., 1985, Effects of guanidinoethane sulfonate and taurine on electroshock seizures in mice, in: "Guanidines", A. Mori, B.D. Cohen, and A. Lowenthal, eds., Plenum Publishing Corporation, New York, pp. 227-234.
21. Izumi, K., Munekata, E., Barbeau, A., Nakanshi, T., Yoshida, M., Yamamoto, H., and Fukuda, T., 1982, Effects of taurine on tolerance to [D-Ala^2,Met^5]enkphalinamide in rats, Eur. J. Pharmacol., 82:55-63.
22. Izumi, K., Munekata, E., Yamamoto, H., Nakanishi, T., and Barbeau, A., 1980, Effects of taurine and γ-aminobutyric acid on akinesia and analgesia induced by D-Ala^2-Met-enkephalinamide in rats, Peptides, 1:139-146.
23. Izumi, K., Munekata, E., Yamamoto, H., Nakanishi, T., and Barbeau, A., 1980, Possible involvements of taurine and GABA in morphine-like peptide actions, Int. J. Neurol., 14:253-268.
24. Kalivas, P.W., and Horita, A., 1980, Thyrotropin-releasing hormone: neurogenesis of actions in the pentobarbital narcotized rat, J. Pharmacol. Exp. Ther., 212:203-210.
25. Lanthorn, T., and Isaacson, R.L., 1978, Studies of kainate-induced wet-dog shakes in the rat, Life Sci., 22:171-178.
26. Le Gal La Salle, G., and Cavalheiro, E.A., 1981, Stimulation of septal and amygdaloid nuclei: EEG and behavioral responses during early development of kindling with special reference to wet-dog shakes, Exp. Neurol., 74:717-727.
27. Lee, H.K., Dunwiddie, T., and Hoffer, B., 1980, Electrophysiological interactions of enkephalins with neuronal circuitry in the rat hippocampus. II. effects of interneuronal excitability, Brain Res., 184:331-342.
28. Leybin, L., Pinsky, C., LaBella, F.S., Havlicek, V., and Rezek, M., 1976, Intraventricular Met^5-enkephalin causes unexpected lowering of pain threshold and narcotic withdrawal signs in rats, Nature (Lond.), 264:458-459.

29. Martin, B.R., Dewey, W.L., Chau-Pham, T., and Prange, Jr., A.J., 1977, Interactions of thyrotropin releasing hormone and morphine sulfate in rodents, Life Sci., 20:715-722.
30. Masukawa, L.M., and Prince, D.A., 1982, Enkephalin inhibition of inhibitory input to CA1 and CA3 pyramidal neurons in the hippocampus, Brain Res., 249:271-280.
31. Nicoll, R.A., Alger, B.E., and Jahr, C.E., 1980, Enkephalin blocks inhibitory pathways in the vertebrate CNS, Nature (Lond.), 287:22-25.
32. Oja, S.S., and Kontro, P., 1983, Taurine, in: "Handbook of Neurochemistry", Vol. 3 (2nd Ed.), A. Lajtha, ed., Plenum Publishing Corporation, New York, pp.501-533.
33. Siggins, G.R., and Zieglgansberger, W., 1981, Morphine and opioid peptides reduce inhibitory synaptic protentials in hippocampal pyramidal cells in vitro without alteration of membrane potential, Proc. Natl. Acad. Sci. USA, 78:5235-5239.
34. Snead III, O.C., and Bearden, L.J., 1980, Anticonvulsants specific for peti mal antagonize epileptogenic effect of leucine enkephalin, Science, 210:1031-1033.
35. Squillace, K.M., Post, R.M., and Pert, A., 1980, Development of wet-dog shakes during amygdala kindling in the rat, Exp. Neurol., 70:487-497.
36. Tortella, F.C., Moreton, J.E., and Khazan, N., 1979, Electroencephalographic and behavioral tolerance to and cross-tolerance between D-Ala^2-Methionine-Enkephalinamide and morphine in the rat, J. Pharmacol. Exp. Ther., 210:174-179.
37. Urca, G., Frenk, H., and Liebeskind, J.C., 1977, Morphine and enkephalin: analgesic and epileptic properties, Science, 197:83-86.
38. Van Gelder, N.M., 1972, Antagonism by taurine of cobalt-induced epilepsy in cat and mouse, Brain Res., 47:157-165.
39. Wei, E., 1975, Resemblance of morphine antinociception to the central depressant actions of norepinephrine, Life Sci., 17:17-18.
40. Wei, E., Loh, H.H., and Way, E.L., 1973, Quantitative aspects of precipitated abstinence in morphine-dependent rats, J. Pharmacol. Exp. Ther., 184:398-403.
41. Wei, E., Loh, H.H., and Way, E.L., 1973, Neuroanatomical correlates of wet shake behaviour in the rat, Life Sci., 12:489-496.
42. Wei, E., Sigel, S., Loh, H.H., and Way, E.L., 1975, Thyrotropin-releasing hormone and shaking behaviour in the rat, Nature (Lond.), 253:739-740.
43. Wikler, A., Green, P.D., Smith, H.D., and Pescor, F.T., 1960, Use of a benzimidazole derivative with potent morphine-like properties as a presumptive reinforcer in conditioning of drug seeking behavior in rats, Fed. Proc., 19:22.
44. Yamamoto, H., McCain, H.W., Izumi, K., Misawa, S., and Way, E.L., 1981, Effects of amino acids, especially taurine and γ-aminobutyric acid (GABA), on analgesia and calcium depletion induced by morphine in mice, Eur. J. Pharmacol., 71:177-184.
45. Zieglgansberger, W., French, E.D., Siggins, G.R., and Bloom, F.E., 1979, Opioid peptides may excite hippocampal pyramidal neurons by inhibiting adjacent inhibitory interneurons, Science, 205:415-417.

POSSIBLE INVOLVEMENT OF BRAIN CALCIUM METABOLISM IN THE ACTION OF TAURINE IN MAMMALIAN THERMOREGULATION

M. Palmi, M. Frosini and G.P. Sgaragli

Instituto di Scienze Farmacologiche, Universita di Siena, Via E.S. Piccolomini, 170 - Villa Betlem, 53100 Siena, Italy

Much attention in the last few years has been devoted to the clarification of the possible role of taurine in thermoregulation. One of the most striking and consistent effects of central (icv) administration of taurine in mammals is hypothermia. The primary action of taurine on thermoregulatory effectors is expressed peripherally mainly by the promotion of heat dissipation which takes place through a reduction in peripheral vasomotor tone particularly evident in rabbits at the ear vascular bed (22) and by the inhibition of shivering, as shown in primates (11). The participation of the central serotonin system in mediating, at least in part, the taurine effect on thermoregulation has also been demonstrated (22). This serotonergic component is probably responsible for a decrease in metabolic heat production and for an increase in the evaporative heat loss (9) which also accompany taurine-induced hypothermia in the sheep (1). In addition, structure-action correlation studies (21,23) performed on a number of amino-acid compounds structurally related to taurine suggested the existence of two inhibitory synaptic mechanisms, one taurine- and the other GABA-mediated, centrally acting on thermoregulatory effectors by working either independently or synergistically.

Involvement of endogenous brain taurine in mammalian thermoregulation

The possible involvement of endogenous brain taurine in the central mechanism of thermoregulation has been investigated by studying the consequences on thermal behavior of injecting icv 6-aminomethyl-3-methyl-4h-1,2,4,-benzothiadiazine-1,1,-dioxide (TAG). This compound, synthesized as a potential analogue of GABA, has been recently shown to be a potent and relatively specific antagonist of taurine action on the CNS (7,27). In particular, it has been demonstrated that TAG antagonizes taurine but not GABA inhibitory actions on either cortical or cerebellar neurons (16,27). The aim of this study with TAG was to succeed in blocking the action of endogenous taurine and to ascertain the consequences of this block on thermal behavior. Since the synaptic mimicry of taurinergic pathways performed by exogenous taurine injected icv causes a fall in core temperature, one would therefore expect that the block of endogenous taurine causes a rise in core temperature; this was, in fact, what it has been observed when TAG was injected icv in rabbits (18,23).

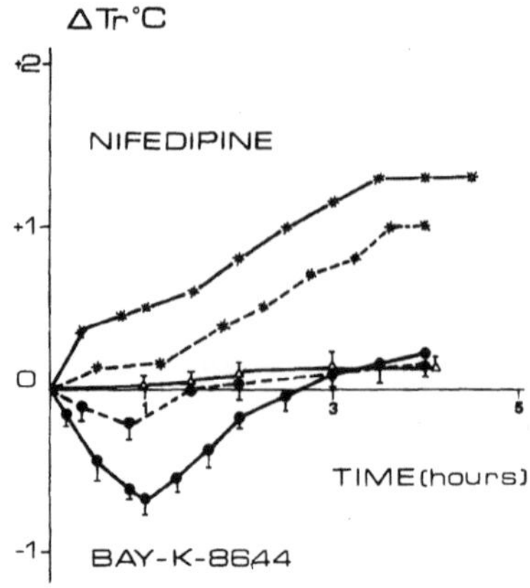

Fig 1. Time course of the effects elicited by nifedipine and BAY-K-8644 on rectal temperature of rabbits. A single animal was icv injected with either 17 nmol/kg (*----*) or 70 nmol/kg (*——*) of nifedipine, while three rabbits were injected icv with either 5.6 nmol/kg (●----●) or 40 nmol/kg (●——●) of BAY-K-8644. Control animals (n=6), were injected with the same volume(15 μl) of vehicle (Δ——Δ) (water: ethanol mixture, 1:1=v:v) used for administering the drugs; bars indicate the SEM. The experiments were performed at an ambient temperature of 20°C.

The role of brain calcium metabolism in mammalian thermoregulation

An important point which needs to be clarified is the biochemical chain of events at the brain level which drives taurinergic effects on thermoregulation. Numerous reports have appeared in recent years suggesting that calcium metabolism is important for taurine to act as a neurotransmitter, and that taurine influences calcium metabolism of excitable tissues.

Any condition, in fact, which leads to a variation in calcium availability in brain cells has been reported to have clear consequences on thermoregulation. The omission of calcium from the solution perfused from the lateral ventricles to the cisterna magna results in a rise of temperature of the unanesthetized cat (4) or monkey (13). Furthermore, the presence of calcium at concentration exceeding the physiological level in the solution perfusing isolated regions of the posterior hypothalamus produces a sharp fall in the body temperature which is accompanied by vasodilation and decrease in the activity of the cat (12). On the contrary, an intense hyperthermia develops if the endogenous sodium is artifically increased in this hypothalamic region (12). Consequently, calcium has been suggested to participate with sodium in an ionic mechanism which regulates the internal temperature set point.

RESULTS AND DISCUSSION

In order to give further support to this hypothesis, two sets of experiments have been performed.

Effects on thermal behavior elicited by some organic calcium antagonists and agonists

By using calcium antagonists one would expect to impair calcium entry into the cells, thus mimicking the experimental conditions where nervous cells were deprived of calcium by using calcium-free perfusing solutions. It has been reported that the icv injection of verapamil caused in rabbits a dose-related hyperthermic response (18,23) which displayed a delay of 1-2 h. We find, on the contrary, as shown in Fig 1, that the hyperthermia elicited by nifedipine was characterized by a prompt onset. Replacement in nifedipine of the nitro group by a $-CF_3$ group is known to convert this blocker to the calcium channel activator known as Bay-K-8644 (19). As illustrated in Fig 1, this compound injected icv in rabbits elicited a prompt hypothermic response whose extent was related to the dose administered. Another organic calcium antagonist, cinnarizine, when given icv at doses similar to those of verapamil and nifedipine elicited hyperthermia in rabbits (data not shown). The possibility that the pharmacological property shared by these compounds - i.e. calcium antagonism - is responsible for this effect is sustained by abundant evidence. Firstly, they have a common effect on thermoregulation in spite of their structural differences. Secondly, in our experiments these compounds were active at low doses. If one admits that once injected icv these compounds are evenly distributed in the brain, their concentrations would then be similar to those reported in the literature for these compounds to be active as organic calcium antagonists on isolated muscle preparations (5,6,24). Finally, the calcium channel activator Bay-K-8644, which is structurally related to nifedipine, when injected icv into rabbits elicited an hypothermic response. It was also noted that the animals appeared sedated. This sedative effect of Bay-K-8644 is in agreement with the observation that rats injected ip with this compound show "inter alia" ataxia and decreased motor activity (2). Interestingly, suppression of behavioral activity, and the development of sleep and a catatonia-like state was observed when calcium concentration in the fluid perfusing the hypothalamus of cats was raised above the normal value (26).

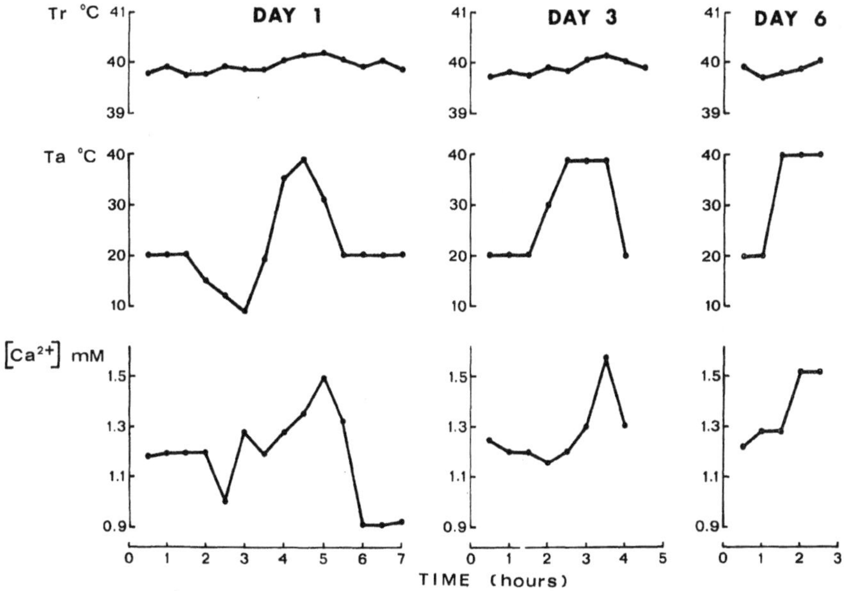

Fig 2. Time course of the effects elicited by the exposure to different ambient temperatures (Ta) on calcium content and cerebrospinal fluid and on rectal temperature (Tr) of a single rabbit. Calcium content was determined colorimetrically by the method of Stern and Lewis (25) on samples collected from the cisterna magna (5 µl. min^{-1}) through a cannula chronically implanted.

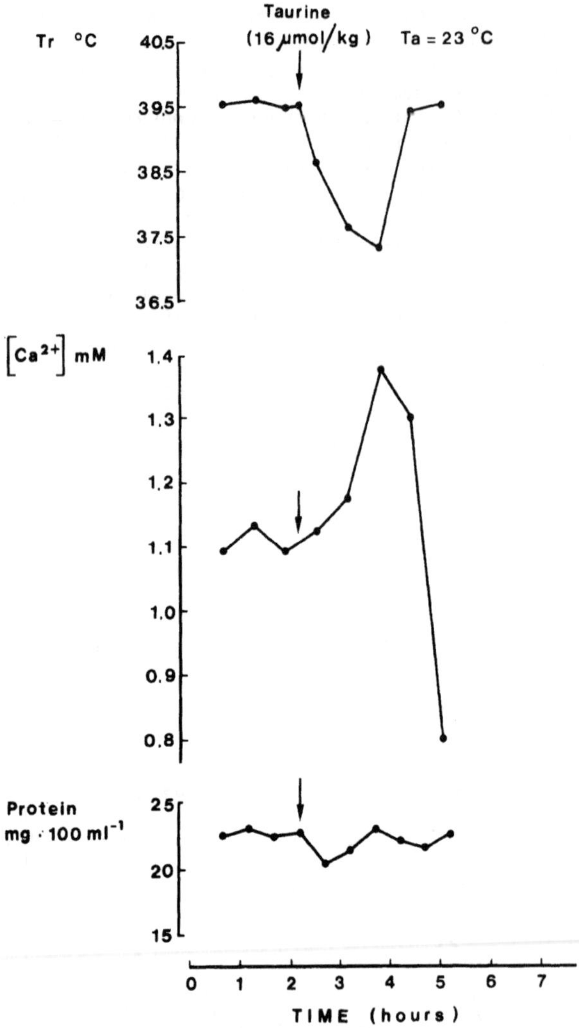

Fig 3. Time course of the effects elicited by ic injection of taurine on rectal temperature (Tr) and calcium and protein content of cerebrospinal fluid, of a single rabbit. Protein content was determined colorimetrically using the coomassie blue binding method as reported by Scopes (20). Taurine was injected through a cannula chronically implanted into the lateral ventricle.

Effects on calcium content of cerebrospinal fluid elicited by the exposure of mammals to different ambient temperatures

If the afore-mentioned hypothesis is correct, calcium concentration in the cerebrospinal fluid might have a relevant role in thermoregulation and it should be finely controlled. Therefore, it should be expected that, when exposing animals to high ambient temperatures, the ensuing stimulation of

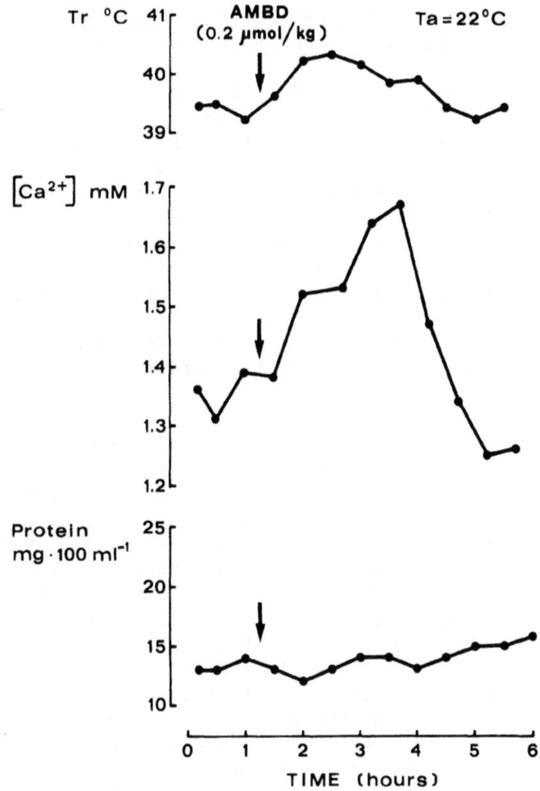

Fig 4. Time course of the effects elicited by icv injection of AMBD(=TAG) on rectal temperature (Tr) and calcium and protein content of cerebrospinal fliud of a single rabbit.

peripheral thermoeffectors towards heat loss is accompanied by an increase in calcium content of cerebrospinal fluid. As shown in Fig. 2, when the ambient temperature was increased from 20°C to 40°C, there was a sudden increase of calcium concentration in cerebrospinal fluid from basal values of 1.15-1.20 mM up to 1.5 mM. Similar results were obtained in the same animal when the experiment was repeated on two different days.

This observation, though preliminary, promisingly outlines the importance in thermoregulation of the exchange of calcium between the blood, brain and cerebrospinal fluid. This matter has long been debated in the attempt to elucidate the biochemical mechanisms of fever (3). Unfortunately, the experiments conducted either to affect body temperature by iv administration of large amounts of Na^+ and Ca^{2+} (8,14) or to relate fever in man to changes in concentrations of Na^+ and Ca^{2+} in cerebrospinal fluid (15) have given poor results. It is clear, however, from previous studies that the maintenance of Ca^{2+} concentration in cerebrospinal fluid is under the control of an active transport mechanism (17). This observation led us to hypothesize that the action of exogenous taurine on thermal behavior in mammals involves also the mechanisms which regulate brain Ca^{2+} metabolism. To test this hypothesis the following experiments have been performed.

Effects on calcium content of cerebrospinal fluid and on thermal behavior elicited by icv injection of taurine and some taurine analogs

As shown in Fig 3, the icv injection of taurine was followed by a fall of body temperature of about 2°C. This was accompanied by an increase of calcium concentration in cerebrospinal fluid from the basal vaule of 1.1 mM up to about 1.4 mM. As the body temperature returned to basal values, there was a sharp decline of calcium concentration down to 0.8 mM. As shown in Fig 4, the icv injection of TAG at a dose of about 2 orders of magnitude lower than that of taurine, caused a taurine-like effect on calcium concentration of cerebrospinal fluid which accompanied the expected increase in body temperature. 2-Aminoethanephosphonic acid given at the same dose of taurine did not modify body temperature; however, it caused a taurine-like effect on calcium concentration (data not shown). Lysine, injected icv as control at the same dose of taurine, did not modify body temperature nor calcium concentration.

The demonstration that either icv taurine injection or the exposure to high ambient temperatures elicit an increase of Ca^{2+} concentration in cerebrospinal fluid while peripheral thermoeffectors are operating strongly support the hypothesis that the mechanism which controls CSF calcium content plays a role in thermoregulation. Future experiments will identify which structure (perhaps the choroid plexus) is responsible for this function. In this context, it is remarkable that taurine prolongs, in a probenecid-like fashion, leukocytic pyrogen-induced fever, probably by occupying a carrier system normally required for termination of the effects of endogenous pyrogens or related central mediators of fever (10). Finally, the fact that some taurine analogs do modify Ca^{2+} content of the cerebrospinal fluid in a taurine-like fashion, in spite of their null (2-aminoethanephosphonic acid) or opposite effect (TAG) on thermal behavior, gives further support to the hypothesis tht at least two target sites for taurine are present in the brain, one dealing with the regulation of Ca^{2+} level in cerebrospinal fluid and the other regulated to the synaptic mechanisms involved in thermoregulation.

ACKNOWLEDGEMENTS

The authors wish to acknowledge Dr. G.G. Yarbrough and Dr. G. Franckowiak for supplying TAG and Bay-K-8644, respectively, used in these experiments. This work was supported by Ministero Pubblica Istruzione (40% + 60%), Roma, Italy.

REFERENCES

1. Bligh, J., Silver, A., Smith, C.A., and Bacon, M.J., 1979, The effects on thermoregulation in the sheep of intracerebroventricular injection of taurine, glycine, GABA and muscimol, J. Therm. Biol., 4:9-14.
2. Bolger, G.T., Weissman, B.A., and Skolnick, P., 1985, The behavioral effects of the calcium agonist Bay-K-8644 in the mouse: Antagonism by the calcium antagonists nifedipine, Naunyn-Schmiedeberg's Arch. Pharmacol., 328:373-377.
3. Cooper, K.E., and Veale, W.L., 1973, Exchange between the blood-brain and cerebrospinal fluid of substances which can induce or modify febrile responses. in: "The pharmacology of thermo- regulation", E. Schonbaum and P. Lomax, eds., S. Karger, Basel, pp. 278-288.
4. Feldberg, W., Myers, R.D., and Veale, W.L., 1970, Perfusion from cerebral ventricle to cisterna magna in the unanaesthetized cat. Effect of calcium on body temperature, J. Physiol. (London), 207: 403-416.

5. Fleckenstein, A., 1968, Experimental heart failure due to disturbances in high-energy phosphate metabolism, Proc. Vth Eur. Cong. Cardiol, Athens, pp. 255-269.
6. Fleckenstein, A., Tritthart, H., Doring, H.J., Byron, Y. K., 1972, Bay-a-1040 ein hochaktiver Ca^{++}-antagonistischer inhibitor der elecktro-mechanischer Koppelungsprozesse im warmbluter-myokard, Arzneim Forsch., 22:22-23.
7. Girard, Y., Atkinson, J.G., Williams, M., Haubrich, D.R., and Yarbrough, G.G., 1982, Aminomethyl-1,2,4-benzothiadiazines as potent analogues of γ-aminobutyric acid. Unexpected discovery of a taurine antagonist, J. Med. Chem., 25:113-116.
8. Greenleaf, J.E., 1978, Thresholds for Na^+ and Ca^{++} effects on thermoregulation, in: "Effectors of thermogenesis", L. Girardier and J. Seydoux, eds., Birkauser Verlag, Basel, pp. 33-43.
9. Lin, M.T., Pang, I.H., Chern, S.I., and Chia, W.Y., 1978, Changes in serotonin contents in brain effect metabolic heat production of rabbit in cold, Am. J. Physiol., 235:R41-R47.
10. Lipton, J.M., and Ticknor, C.B., 1979, Central effect of taurine and its analogues on fever caused by intravenous leukocytic pyrogen in the rabbit, J. Physiol. (London), 287:535-543.
11. Murphy, M.T., Lipton, J.M., Longhran, P., and Giesecke, A.H., 1985, Postanesthetic shivering in primates: Inhibition by peripheral heating and by taurine, Anesthesiology, 63:161-165.
12. Myers, R.D., and Veale, W.L., 1971, The role of sodium and calicum ions in the hypothalamus in the control of body temperature of the unanaesthetized cat, J. Physiol. (London), 212:411-430.
13. Myers, R.D., Veale, W.L., and Yaksh, T.L., 1971, Changes in body temperature of the unanaesthetized monkey produced by sodium and calcium ions perfused through the cerebral ventricles, J. Physiol. (London), 217:381-392.
14. Nielsen, B., 1974, Actions of intravenous Ca^{2+} and Na^+ on body temperature in rabbits, Acta Physiol. Scand., 90:445-450.
15. Nielsen, B., Schwartz, P., and Alhede, J., 1973, Is fever in man reflected in changes in cerebrospinal fluid concentrations of sodium and calcium ions?, Scand. J. Clin. Lab. Invest., 32:309-310.
16. Okamoto, K., Kimura, H., and Sakai, Y., 1983, Antagonistic action of 6-aminomethyl-3-methyl-4H,1,2,4-benzothiadiazine-1,1-dioxide (TAG), and evidence for a transmitter role of taurine in stellate interneurons in the cerebellum. in: "Sulfur aminoacids: Biochemical and clinical aspects", K. Kuriyama, R.J. Huxtable and H. Iwata eds., Alan R. Liss, Inc., New York, pp. 151-160.
17. Oppelt, W.W., Owens, E.S. and Rall, D.P., 1963, Calcium exchange between blood and cerebrospinal fluid, Life Sci., 8:599-605.
18. Palmi, M., and Sgaragli, G.P. 1985, Hyperthermia induced in rabbits by AMBD and organic calcium antagonists: its reversal by taurine, Br. J. Pharmac., 86:693P.
19. Schramm, M., Thomas, G., Towar, R., and Franckowiak, G., 1983, Novel dilydropyridines with positive inotropic action through activation of Ca^{++} channels, Nature (London), 303:535-537.
20. Scopes, R.K., 1982, Solutions for measuring protein concentration. in: "Protein purification. Principles and practice", Springer Verlag, New York, pp. 265-266.
21. Saragli, G.P., Carla, V., Magnani, M., and Giotti, A., 1978, Homotaurine and muscimol mimic taurine and GABA effects on muscle tone and temperature regulation, Naunyn-Schmiedeberg's Arch. Pharmacol., 305:155-158.
22. Sgaragli, G.P., Carla, V. Magnani, M., and Galli, A., 1981, Hypothermia induced in rabbits by intracerebroventricular taurine: Specificity and relationship with central serotonin (5-HT) systems, J. Pharmacol. Exp. Ther., 219:778-785.

23. Sgaragli, G.P., and Palmi, M., 1985, The role and mechanisms of action of taurine in mammalian thermoregulation. in: "Taurine: Biological actions and clinical perspectives", S.S. Oja, L. Ahtee, P. Kontro and M.K. Paasonen, eds., Alan R. Liss, Inc., New York, pp. 343-357.
24. Spedding, M., 1982, Assessment of Ca^{2+} antagonist effects of drugs in K^+-depolarized smooth muscle. Differentiation of antagonist subgroups, Naunyn-Schmiedeberg's Arch. Pharmacol., 318:234-340.
25. Stern, J., and Lewis, W.H.P., 1957, The colorimetric estimations of calcium in serum with O-cresolphthalein complexone, Clin. Chim. Acta, 2:576-582.
26. Veale, W.L., and Myers, R.D., 1971, Emotional behavior, arousal and sleep produced by sodium and calcium ions perfused within the hypothalamus of the cat, Physiol. Behav., 7:601-607.
27. Yarbrough, G.G., Singh, D.K., and Taylor, D.A., 1981, Neuropharmacological characterization of a taurine antagonist, J. Pharmacol. Exp. Ther., 219:604-613.

SECTION IV: NEUROCHEMISTRY

INTRODUCTION: NEUROCHEMISTRY

Ryan J. Huxtable[1] and Flavia Franconi[2]

[1] Department of Pharmacology, College of Medicine, University of Arizona, Tucson, Arizona 85724
[2] Cattedra di Saggi e Dosaggi Farmacologici, Università di Sassari, Sassari, Italy

Methodology and insight go together like lock and key, each useless without the other. The first three chapters in this section present novel methodology, each of which promises to increase our ability to probe the biological functions of taurine. It is a well known story how the high concentration of acetylcholine receptors in the electroplax of *Torpedo californica* allowed their ready isolation, which in turn yielded tremendous progress in the understanding of acetylcholine receptors in mammals, where such receptors are present only in low concentrations. The discovery of taurine receptors on the sensilla of the spiny lobster has the promise of being of equal value in taurine research. Gleeson et al. show how this enriched source of taurine receptors may be readily obtained in a few seconds without decreasing the market value of the lobster. The taurine receptor on the sensilla is exposed to the environment, where it serves as a chemoceptor for detecting taurine in sea water. This system works together with other chemoceptors on the sensilla as a means for finding food sources. The sensitivity of the receptor is extreme, taurine concentrations as low as 10^{-12} M producing a response. Gleeson et al. have done exquisitely nice work recording electrical responses triggered by the receptor in single neurons. An other observation illustrating the potential significance of this receptor for mammalian studies is that the structure activity requirements are the same as those found for mammalian binding sites for taurine. The authors anticipate that this model will yield valuable insight in the understanding of taurine in recognition and response systems. The capability is there. Their hope, however, will only be realized if the appropriate laboratories take the appropriate steps to involve themselves in this problem.

Wu et al., in the following paper, map out potential taurinergic networks in the brain by the relatively novel technique of immunocytochemistry of cysteine sulfinic acid decarboxylase, an enzyme involved in taurine synthesis. Most of the taurine in the brain of the rat derives either from the diet, or from hepatic biosynthesis, that organ having a far, far greater decarboxylase activity than does the CNS. However, the enzyme in the brain may control the production of a specifically neurotransmitter pool of taurine. This supposition is reinforced by the localization of cysteine sulfinic acid decarboxylase to neuronal cells. Another function of the taurine pool controlled by this enzyme is indicated by the interesting findings of these authors that cysteine sulfinic acid decarboxylase-containing neurons are less vulnerable to ischemic damage. The protection of such neurons may be a complementary or alternative function of this pool.

Madsen et al. apply a similar technique to the localization of taurine itself in the brain. As is to be expected with the high concentrations of taurine found in whole brain, compared to the low concentrations of decarboxylase activity, the distribution of taurine is more widespread. Immunocytochemical techniques demonstrate taurine in both glia and neurons, more being found in the glia in some regions, and more in the neurons in others. This is in keeping with our suggestion above that cysteine sulfinic acid decarboxylase is "responsible" for one particular pool of taurine in the brain, which may have specific functions not shared by the general pool of taurine.

The next group of three chapters in this section are concerned with various aspects of taurine transport into the CNS. There has been a long-lasting interest in the regulation and mechanism of taurine transport into the brain, but despite the wealth of literature, there are still many questions to be resolved. Della Corte et al. have carried out autoradiography of the striatum and substantia nigra, following injection of tritiated taurine into these regions. In the striatum, they find labelling in both neurons and glia. However, even close to the injection track, they find that up to 25% of neurons remain unlabelled. Neurons not accumulating taurine were differentiated structurally and by size. Does this finding indicate that some neurons lack a transport system for taurine? This possibility should be investigated. In the substantia nigra, only glial cells accumulated radioactive taurine, and not the cell bodies of neurons. Neurons only showed accumulation into the axons and boutons. Paralleling these studies, Della Corte et al. carried transport and efflux experiments on tissue slices from these brain regions. Transport in the striatum is high affinity, and release from this area is calcium dependent, suggesting a possible neurotransmitter or neuromodulator role.

Del Rio et al. have examined the uptake and release of taurine by the hippocampus using the recently introduced technique of brain dialysis (2). The glutamate transport inhibitor, dihydrokainate increased extracellular concentrations of taurine as well as glutamate. GABA transport inhibitors also increase extracellular concentration of taurine. The authors suppose that these compounds, by some undetermined mechanism, are acting as taurine efflux stimulators. A possible relationship between taurine, glutamate and GABA touches on one of the mysteries and controversies concerning free amino acids in the brain. Van Gelder was the first to propose a relationship between taurine and glutamate concentrations in the brain (12), a proposal that was substantiated by Rassin's work on beagle (11). Huxtable et al. also demonstrated high correlations between taurine and glutamate in brains of individual rats (5). The latter workers interpreted the observations as an indication that glutamate and taurine shared the same cytoarchitectural distribution within the brain. Del Rio et al. suggest the beginnings of a hypothesis to explain how this interdependency works, via some mutual actions of these neuroactive amino acids on their transport systems other than that of being antagonists. They propose that taurine is a "universal coupling substrate" in anti-port systems for amino acids transport. We know that as an osmoregulator, taurine is universally responsive to cellular changes in concentrations of inorganic ions and cytoplasmic osmolarity. The interaction of taurine and the transport of other free amino acids may be associated with this, part of a generalized function for taurine in maintaining cellular homeostasis of free amino acids, inorganic ions, osmolarity, and electrical chemical gradients. It is noteworthy that the insights derived in this exciting and provocative chapter are based on the application of a novel methodology, that of brain dialysis (2,9,13).

One of the authors of the final chapter of this trio, Lombardini, is one of the stalwarts in the area of taurine transport in the brain. Lombardini (10) was the first to systematize and clarify the somewhat

Table 1. Spontaneous release of taurine from synaptosomes

	Taurine: In Pellet (nmol.mg^{-1} Protein)	Taurine: In Supernatant (μM)
I. Synaptosomal P_2B (5.77 mg/ml)	27.8±2.2	
II. Incubated 10 min on ice in Tris (100 mM, pH 7.5) - NaCl (1.4 M), and centrifuged	13.3±1.6	43.0±19.4
III. Pellet from II resuspended in fresh buffer, and procedure repeated	7.0±0.7	76.9±9.0

From Huxtable, unpublished observations.

confusing literature of taurine transport in the brain. In the chapter herein, Hanretta and Lombardini introduce not a new method, but a new theory: a model of hypothalamic taurine compartmentalization. They elegantly integrate a mass of experimental observations, and propose that the high affinity transport site in the hypothalamus is located in neurons, and the low affinity site in glia. Release from the latter site is calcium independent, and they suggest that glia taurine has no neurotransmitter functions.

Although we are in general agreement with the conclusions of these authors, we would make one caveat. Even washed, sonicated synaptosomes release considerable amounts of taurine into a incubation medium. As Table 1 indicates, incubation of as little as 6 mg protein/ml of synaptosomes at 4°C in taurine-free buffer can result in 40-80 μM taurine being released into the medium (Huxtable, unpublished observations). Thus, kinetic parameters obtained from experiments in which taurine levels were not measured absolutely by such means as amino acid analysis must be viewed with suspicion. The kinetic constants, Km and Bmax, calculated from such experiments are numerically too low.

In general, the abundance of new methodology and new hypotheses presented in this section, and, indeed, illustrated throughout this volume, shows the healthy and growing state of taurine research.

The neurochemistry of excitatory and inhibitory amino acids cannot be disengaged. Brain excitability is determined by dynamic equlibrium of the two groups, and the groups are biochemically related, in that decarboxylation of excitatory amino acids yields inhibitory amino acids. In addition, the substances mutually affect each others transport, are electrophysiological antagonists of each other, and modify each others neurochemistry. Thus, the inhibitory substance taurine modulates the excitatory actions of its precursor, cysteine sulfinic acid (8).

The next three chapters in this section consider the so called, "excitotoxins", or excitatory, acidic amino acids, and it is appropriate that the first chapter is by Baba et al. They report a puzzling phenomena; cysteine sulfinate-induced release of GABA. It is difficult to comprehend the functionality of the release of an inhibitory substance by an excitatory substance. However, today's inconsistencies are the foundations of tomorrow's theories. An intriguing new area is open for investigation. Conversely, GABA inhibits the release of cysteine sulfonate and glutamate, pointing to cross-innervation of excitatory and inhibitory circuits in the hippocampus.

In the following chapter, Bennardini et al. also examine taurine-excitotoxin interaction in the hippocampus, this time effects on binding rather than release. Whereas calcium increases glutamate binding, calcium and taurine decrease it, due to a decrease in Bmax. Taurine is without effect by itself, and is not able to reverse the calcium-induced stimulation once it is produced. There are indications that there is a shift in membrane binding of calcium from hydrophobic to hydrophilic regions in the presence of taurine. This suggests that the findings of these authors reflects an effect of taurine on phospholipid membranes, in keeping with the proposal of Huxtable and Sebring (7), expanded on in the last chapter of this book.

The last chapter in this trio is also a report on excitatory amino acids in the hippocampus. Lehmann et al. take advantage of the brain dialysis technique to examine the influence of guanidinoethane sulfonate on extracellular levels of free amino acids. Guanidinoethane sulfonate has been widely used as a taurine depleting agent since its introduction for this purpose by Huxtable et al. (6). Taurine is the only amino acid specifically depleted in whole tissue, an observation confirmed by the present investigators. However, following guanidinoethane sulfonate administration, the intracellular to extracellular ratio for glutamate falls. This indicates that guanidinoethane sulfonate, either directly or indirectly, is modifying the neurochemistry of excitatory amino acids. It is possible that the depolarizing action of this increased extracellular glutamate is responsible for the powerful convulsant activity of guanidinoethane sulfonate given icv.

If, as has often been proposed, taurine is a neuromodulator, one would expect to find neurochemical actions of taurine on other transmitter systems. The final three chapters in this section are a catchpenny collection of studies on adrenergic and dopaminergic systems, and brain dipeptides. There are various reports of interactions of taurine and adrenergic systems in brain (1), pineal (14,15), and heart (3,4). It appears that some of these interactions, at least, may involve modulation of calcium binding to intracellular calmodulin. Segawa et al. find that taurine antagonizes the calcium-induced difference spectrum of calmodulin. In other words, it is blocking the conformational shift produced by calcium binding. The authors suggest that taurine is affecting the positive cooperativity normally exhibited by the calcium binding sites on calmodulin. The authors further suggest that this is one putative mechanism by which taurine regulates beta adrenergic activity, including the release of norephinephrine. The specification of an intracellular action of taurine is in keeping with the accumulating evidence from the heart and brain that the significant actions of taurine are mediated intracellularly, in keeping with its distribution.

Taurine also influences the dopaminergic system, as shown by the studies of Kontro. In fact, here, as in other systems, there is neurochemical crosstalk, there being mutual interactions between taurine and dopamine. For example, the spontaneous efflux of taurine from striatal slices is stimulated by dopamine, but unaffected by norephinephrine and serotonin. The potassium stimulated efflux of taurine, on the other hand, is supressed by dopamine. Conversely, potassium-stimulated the release of dopamine is stimulated by taurine. The physiological significance of this remains to be ascertained, but such findings are the building blocks from which new theories will be built.

In the final chapter of this section, Varga et al. examine the neurochemistry of some of the numerous dipeptides found in the brain. Thus, glutamate binding is stimulated by calcium, as others have found, but inhibited by acidic dipeptides containing glutamyl or aspartyl residues. Certain gamma glutamyl peptides also inhibited GABA binding. However, Kd constants were numerically high, throwing doubt on the possible physiological relevance.

If reading these chapters has taught us anything, it is that no chemical is left to work alone in the brain. All phenomena are expressed in the midst of a crowd of chemicals nudging a response this way or that. The multiple interactions that occur make the neurochemistry of amino acids difficult to study. Tabulation of the interactions that do occur is a first step in deciding which ones are relevant and which ones are physiologically unimportant. Such decisions, in turn, yield the substratum of data necessary for the erection of those structures so beloved by scientists; theories, hypothesis, models, and concepts.

REFERENCES

1. Aldegunde, M., Miquez, I., Martin, I., and Fernandez Otero, M.P., 1983, Changes in brain monoamine metabolism associated with hypothermia induced by intraperitoneally administered taurine in the rat, IRCS Med. Sci. 11:258-259.
2. Hamberger, A., Berthold, C.H., Jacobson, I., Karlsson, B., Lehmann, A., Nystrom, B., and Sandberg, M., 1985, In vivo brain dialysis of extracellular nontransmitter and putative transmitter amino acids, in: "In Vivo Perfusion and Release of Neuroactive Substances", Bayon, A., and Drucker-Colin, R. eds., Academic Press, New York, pp. 119-139.
3. Huxtable, R., and Chubb, J., 1977, Adrenergic stimulation of taurine transport by the heart, Science 198:409-411.
4. Huxtable, R. J., Chubb, J., and Azari, J., 1980, Physiological and experimental regulation of taurine content in the heart, Fed. Proc. 39:2685-2690.
5. Huxtable, R.J., Laird, H., Bonhaus, D., and Thies, A.C., 1982, Correlations between amino acid concentrations in brains of seizure-susceptible and seizure-resistant rats, Neurochem. Int. 4:73-78.
6. Huxtable, R.J., Laird, H.E., and Lippincott, S.E., 1979, The transport of taurine in the heart and the rapid depletion of tissue taurine content by guanidinoethyl sulfonate, J. Pharmacol. Exptl. Therap. 211:465-471.
7. Huxtable, R.J., and Sebring, L.A., 1986, Towards a unifying theory for the action of taurine, TIPS 7:481-485.
8. Iwata, H., and Baba, A., 1983, Neurochemical basis of cysteine sulfinic acid in the central nervous system, in: "Sulfur Amino Acids: Biochemical and Clinical Aspects", Kuriyama, K., Huxtable, R.J., and Iwata, H. eds., Alan R. Liss, Inc., New York, pp. 141-150.
9. Lerma, J., Herranz, A.S., Herreras, O., Abraira, V., and Del Rio, M., 1986, In vivo determination of extracellular concentration of amino acids in the rat hippocampus, A method based on brain dialysis and computerized analysis, Brain Research 384:145-155.
10. Lombardini, J.B., 1978, High-affinity transport of taurine in the mammalian central nervous system, in: "Taurine and Neurological Disorders", Barbeau, A., and Huxtable, R.J. eds., Raven Press, New York pp. 119-136.

11. Rassin, D.K., 1982, Taurine, cysteine sulfinic acid decarboxylase and glutamic acid in brain, in: "Taurine in Nutrition and Neurology", Huxtable, R.J., and Pasantes-Morales, H. eds., Plenum Press, New York, pp. 257-268.
12. Van Gelder, N.M., 1978, Taurine, the compartmentalized metabolism of glutamic acid, and the epilepsies, Can. J. Physiol. Pharmacol. 56:362-374.
13. Vezzani, A., Ungerstedt, U., French, E.D., and Schwarcz, R., 1985, In vivo brain dialysis of amino acids and simultaneous EEG measurements following intrahippocampal quinolinic acid injection: evidence for a dissociation between neurochemical changes and seizures, J. Neurochem. 45:335-344.
14. Wheler, G.H.T., and Klein, D.C., 1979, Cyclic AMP-induced release of ^{14}C taurine from pinealocytes, Biochem. Biophys. Res. Comm. 90:20-27.
15. Wheler, G.H.T., Weller, J.L., and Klein, D.C., 1979, Taurine: stimulation of pineal N-acetyltransferase activity and melatonin production via a beta-adrenergic mechanism, Brain Res. 166:65-74.

A TAURINE RECEPTOR MODEL: TAURINE-SENSITIVE OLFACTORY CELLS IN THE LOBSTER

R. A. Gleeson, H. G. Trapido-Rosenthal and
W. E. S. Carr

C. V. Whitney Laboratory
University of Florida
Route 1, Box 121
St. Augustine, Florida 32086, U.S.A.

Taurine is a widely distributed amino acid which has been associated with a broad range of physiological actions, particularly in excitable tissues. These include, for example, a role in heart function (17), retinal function (24,27,31,32), and neural development (9,38,39). In addition, neurophysiological and biochemical evidence has suggested that taurine acts as an inhibitory neurotransmitter or neuromodulator in the central nervous system (21,26,29,30). In no case, however, has the specific function of taurine been elucidated, and progress towards deciphering its role in these various systems has been considerably retarded by the lack of selective agonists and antagonists. A major factor which has hampered development of such probes has been the absence of a receptor model that is specifically sensitive to taurine.

In this report, we show that both the physiological and biochemical details of taurine reception can be explored with a novel model system: the taurine-sensitive olfactory cells of the spiny lobster, Panulirus argus. As discussed below, chemosensory cells of the lobster have features which confer certain experimental advantages for studying the ligand-receptor interactions of taurine and a number of other neuroactive substances as well [e.g., glutamate, glycine, AMP, and ATP (4)]. Based on the concept that systems for the recognition of small molecules, once evolved, may be conserved in evolution and used for a variety of functions, we expect that the crustacean model can provide valuable insight into the nature of the taurine reception process, and facilitate the identification of pharmacological agents useful in elucidating the function(s) of taurine in vertebrates.

LOBSTER OLFACTORY CELLS AS MODEL POSTSYNAPTIC NEURONS

The olfactory system of the spiny lobster is comprised of a pair of antennules each of which bifurcates to filaments approximately 5-8 cm in length (Fig. 1A). Located on the lateral filament of each antennule is a prominent tuft of about 2000 olfactory sensilla (aesthetascs); the lumen of each sensillum contains the extensively branched dendritic terminals of some 350 bipolar primary sensory neurons (Fig. 1B) (23). The axons of these neurons form the antennular nerve which projects to the brain. As in vertebrates, the terminal branching of the sensory dendrites is in the form

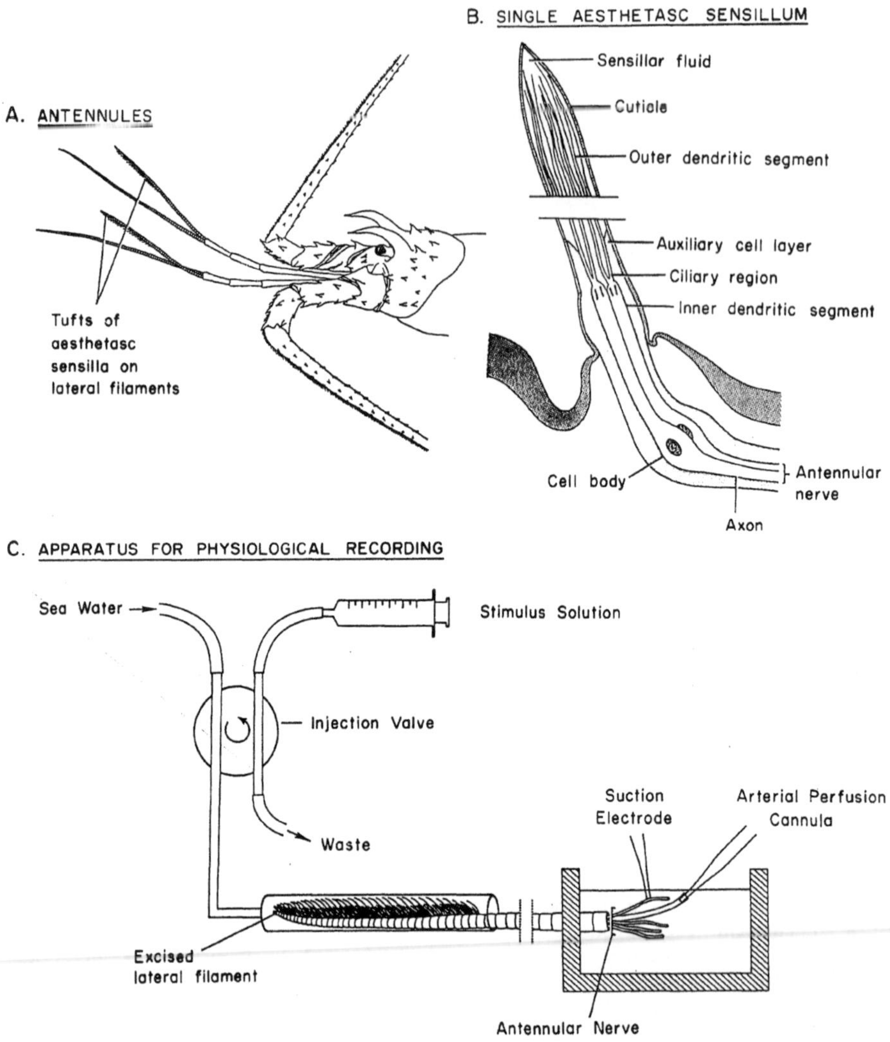

Fig. 1. A. Head of the spiny lobster showing tufts of olfactory sensilla (aesthetascs) on the lateral filaments of the antennules. B. Schematic diagram of an aesthetasc sensillum depicting the branched cilia of the dendrites within the lumen. Two of 350 cells shown. C. Apparatus for stimulus delivery and recording from olfactory cells in the antennule of the lobster. For further details, see Gleeson and Ache (10).

of nonmotile cilia (based on the lack of dynein arms) which, in the lobster, arise in pairs from each dendrite and divide forming 10-11 branches per cilium. The ciliary branches terminate near the tip of the sensillum and are bathed by a sensillar fluid of unknown composition (11, U. Grünert, personal communication). Various lines of evidence from vertebrate studies suggest that the receptors for chemical stimuli reside on the membranes of these cilia (2,28,34).

Chemical stimulation of a lobster olfactory cell gives rise to a graded receptor potential which in turn results in the generation of action potentials in the axon (3). As this process is analogous to neurotransmitter activation of a postsynaptic neuron, the olfactory cell of the lobster represents a potential postsynaptic model system. Indeed, it has been hypothesized that internal receptors (e.g., neurotransmitter receptors) may have evolved from external chemoreceptors of primitive aquatic animals (4). The anatomy of the lobster's olfactory cells, coupled with the finding that many are narrowly "tuned" in their response spectrum to neuroactive substances (4), imparts the following advantages to using them as model postsynaptic neurons. Firstly, using both extra- and intracellular techniques, the electrical responses of single neurons can be recorded as a measure of ligand-receptor interactions. Secondly, since these olfactory cells are primary sensory neurons, their responses are first order (i.e., the chemical stimulus acts directly on the cells) and are therefore not confounded by secondary influences such as inputs from other neurons. Thirdly, in physiological experiments the receptor-containing dendritic processes of these cells are directly and exclusively accessible, thus facilitating precise control of stimulus dose and permitting manipulation of the receptor environment without altering the medium bathing the remainder of the cell. Finally, biochemical analyses of receptor processes, including binding and inactivation/uptake of ligands, can be performed to complement the physiological studies. Moreover, the biochemical analyses are not complicated by the presence of presynaptic neurons as is usually the case with preparations from internal tissues.

ELECTROPHYSIOLOGICAL PROCEDURES

For extracellular recording (Fig. 1C) the lateral filament of the antennule is excised and perfused with oxygenated lobster saline to maintain its viability. In a saline bath, single fascicles containing axons of the sensory neurons are separated from the antennular nerve and a suction electrode is applied _en passant_ for recording evoked action potentials by conventional electrophysiological methods. The cluster of aesthetasc sensilla on the distal portion of the filament is continuously flushed with clean artificial seawater (ASW) into which various amounts of chemical stimuli can be introduced by means of an injection valve.

Fascicles containing axons of cells sensitive to a particular substance in ASW are identified by systematically searching through the olfactory nerve for bundles whose axons alter their frequency of discharge when pulses of the substance are introduced to the ASW flowing over the filament. ASW alone serves as a control. Single cells responding to the substance are isolated either by anatomical separation of fascicles or resolved from multiunit recordings using an amplitude/time window discriminator. In all recordings, verification of single unit status is made by monitoring the waveforms of events that trigger the window discriminator. Interspike intervals are measured on line by a computer and the data stored for immediate or subsequent analysis of the response.

A preparation for intracellularly recording from the sensory neurons of aesthetascs has recently been developed in _P. argus_ (3). Recordings are obtained with patch electrodes using the whole cell configuration. This technique should prove useful in exploring the biophysical basis of the electrical events initiated by chemical stimulation.

PHYSIOLOGY OF TAURINE-SENSITIVE CHEMORECEPTORS

Chemoreceptors exhibiting remarkably selective sensitivity to taurine have been found on the antennules of the spiny lobster, _P. argus_ (7,40), as well as the antennules and walking legs of the American lobster,

Homarus americanus (6,18,19,37). In the spiny lobster where these taurinergic cells have been studied most extensively, semilog plots of the average dose-response relationships based on spike frequency, response duration, or total number of spikes per response, all showed an approximately linear increase over 10^{-9} to 10^{-4} M taurine (7). This range reflects threshold differences among cells as the responses of individual cells generally saturate over a two to three log unit range in taurine concentration. In some cells threshold responses to taurine have been recorded at concentrations as low as 10^{-12} M (40).

Fuzessery et al. (7) examined the structure activity relationships of taurine receptors on olfactory cells of the spiny lobster. In this study all compounds were presented at a concentration of 10^{-5} M and the total number of spikes generated compared with that of taurine as an index of stimulatory capacity. A summary of the results based on tests with the compounds shown in Fig. 2 are as follows (bracketed numbers refer to the compounds in Fig. 2):

1) Compounds with a terminal basic group and a terminal acidic group separated by 2 carbon (C) atoms are the most stimulatory [1,2]. The substitution of a phosphonic acid group for the sulfonic acid group results in an inactive compound [3]. Compounds with the basic and acidic groups separated by greater than 2C atoms are less stimulatory with the potency decreasing as a function of the distance of separation [e.g., 2>15>16>17]. Compounds having a 1C atom separation are ineffective [4,14].

2) Compounds without the terminal amino group [5,6,7,9,10] or with an alpha amino group in addition to the terminal group are virtually inactive [9,11,12]. Substituting a guanidino group for the terminal amine yields an inactive molecule [20,21].

3) Neutral side chains markedly decrease the activity [15>18; 2>19].

4) Compounds with the carboxyl group linked in a peptide bond are ineffective [22,23].

In tests to define the specificity of these taurine-sensitive cells further, it was found that three organic acids (citric, propionic and succinic), twelve α-amino acids, and glycine betaine were all nonstimulatory.

Taken together the data show that these olfactory cells have a rather selective specificity for taurine which has interesting parallels with taurine recognition systems found in vertebrate tissues. For example, β-alanine and hypotaurine, the most stimulatory of the taurine analogs examined in the lobster, are also potent competitors for taurine binding sites in rat brain synaptosomal fractions (14,36) as well as taurine uptake sites in membrane vesicles prepared from heart tissue (35). In addition, uptake of taurine by human blood platelets is competitively inhibited by β-alanine and hypotaurine, but not by 2-aminoethylphosphonic acid, an analog which is also nonstimulatory in the lobster (8).

A second type of taurine-sensitive cell has recently been described for the lobster antennule (10). These cells characteristically exhibit higher thresholds (ca. 10^{-5} M) and have dose-response curves of a lesser slope than the cells described above. Another property of these cells is that their response to taurine is suppressed when taurine is presented in mixture with certain amino acids. An analysis of this suppression using binary mixtures of taurine with either glycine or L-alanine revealed parallel displacement in the dose-response function for taurine in both cases (Fig. 3). These

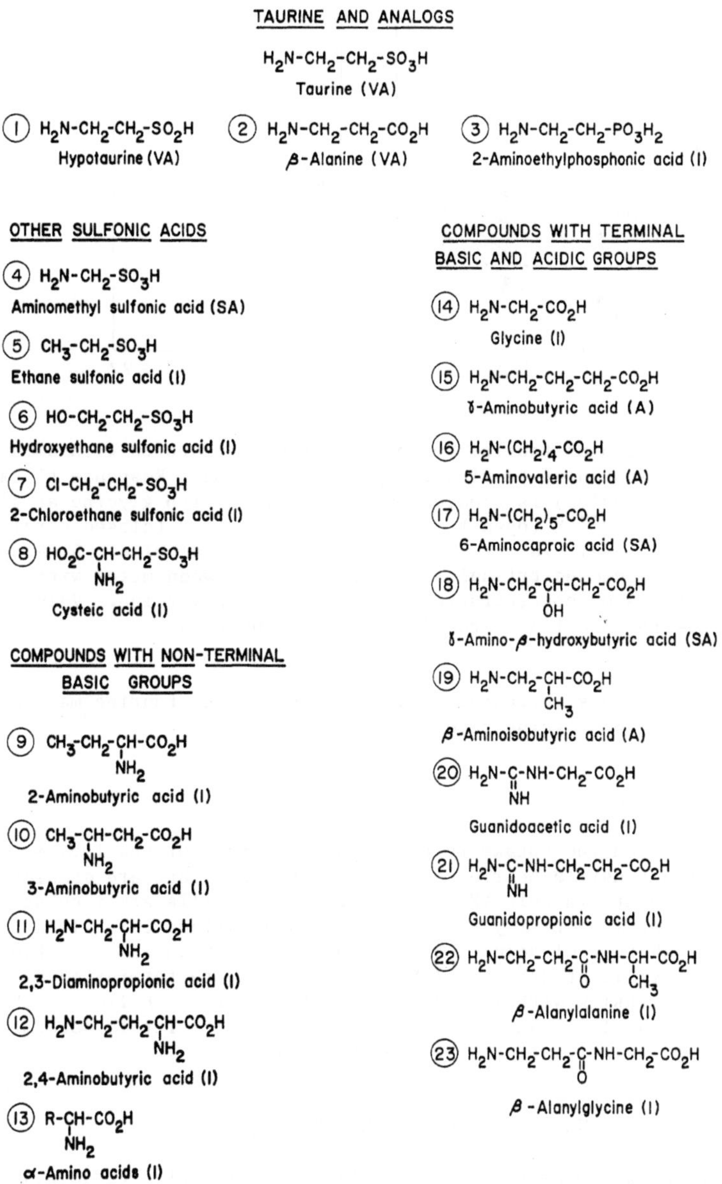

Fig. 2. Structures of compounds tested on taurine-sensitive olfactory receptors of the spiny lobster. Indices of relative activity are as follows: VA = very active; SA = slightly active; I = virtually inactive. Text refers to numbers designating individual compounds. Modified from Fuzessery et al. (7).

results suggest that both glycine and alanine may compete with taurine for common receptor sites. Indeed, evidence is accumulating which suggests competitive inhibition may be an important mechanism underlying the phenomenon of mixture suppression for other types of antennular chemoreceptors as well (1). The fact that glycine is one of the most potent suppressants of these taurine-sensitive cells is of particular interest in

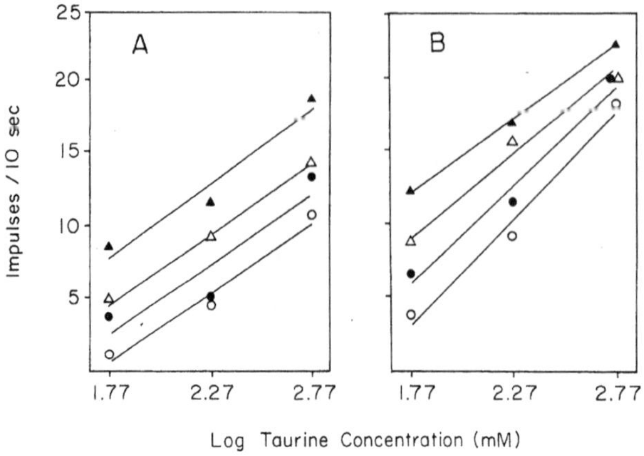

Fig. 3. Suppression of taurine response in binary mixtures with amino acids. A. Suppression of taurine response by glycine. Response to taurine alone (closed triangles) is greater than when mixed with glycine at glycine concentrations of 10^{-4} M (open triangles), 3.6 x 10^{-4} M (closed circles) or 10^{-3} M (open circles). Values are mean responses of 8 cells. B. Response to taurine alone (closed triangles) is greater than when mixed with L-alanine at the same concentrations given above for glycine. Values are mean responses for 6 cells. From Gleeson and Ache (10).

that data from various vertebrate preparations suggest taurine may exert its effects by acting on glycine receptors (5,12,22).

BIOCHEMICAL PROCEDURES

Studies of taurine uptake by the olfactory organ of the lobster are performed using sensilla-bearing sections of cuticle that are dissected from antennular filaments in cold ASW (Fig. 4). The sensilla are incubated with ^{3}H-taurine (specific activity = 20 Ci/mmole) and various unlabelled compounds in 0.5 to 1 ml volumes of ASW, or ASW in which NaCl had been replaced by Tris-Cl, at 22°C, pH 7.8. Incubations are terminated by filtration and rinsing with ice cold incubation medium. Following the digestion of sensilla in NaOH, the amount of radioactivity is determined by means of liquid scintillation spectrophotometry (42).

BIOCHEMISTRY OF TAURINE-SENSITIVE CHEMORECEPTORS

Incubation of the aesthetasc sensilla with ^{3}H-taurine (from 0.05 to 10 μM), in the absence and the presence of an excess of unlabelled taurine, revealed an uptake system that internalizes this compound (41). This uptake system has a Km of 1.57 μM, and a Vmax of 59 fmole/sensillum/h (2.06 nmole/mg protein/h). No specific association of ^{3}H-taurine occurs with sections of sensilla-free antennular cuticle. The uptake of ^{3}H-taurine is quite specific (Fig. 5). Guanidinoethane sulfonic acid (GES) is the most potent competitor for this uptake process, with other effective competitors being, in order, taurine itself, GABA, β-alanine, and D,L-diaminopropionic acid. The amino acids L-alanine and glycine are only moderately effective competitors for the uptake of labelled taurine, and aminoethylphosphonic acid, the phosphonic acid analog of taurine, is a poor competitor. The nucleoside adenosine, a compound for which the sensilla have a separate,

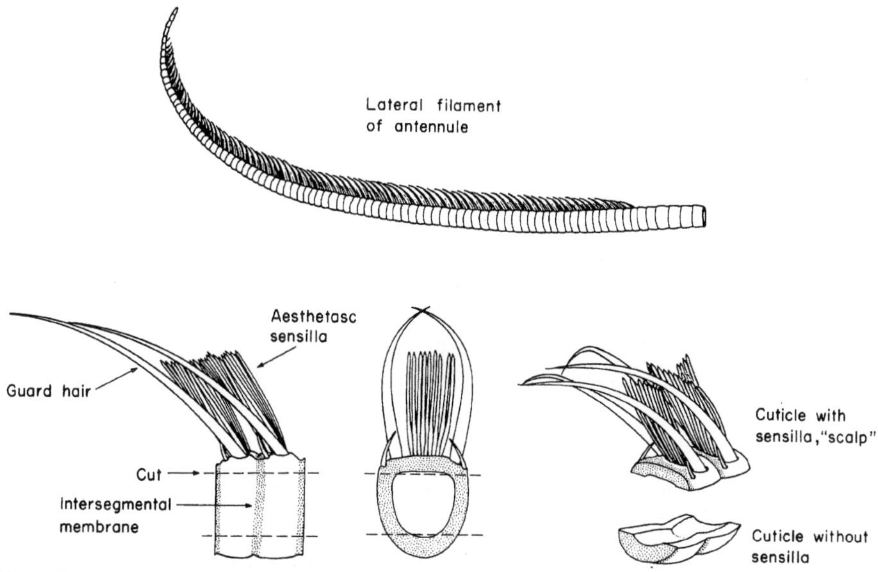

Fig. 4. Preparation of sensilla-bearing sections of cuticle for biochemical studies. Lateral antennular filaments are excised from the head of the lobster and dissected into pairs of segments by cutting through the intersegmental membrane between every second segment. A tangential cut through the cuticle provides sections of cuticle with intact sensilla attached. Dendrites are the only portions of sensory neurons present in sensilla (see Fig. 1B).

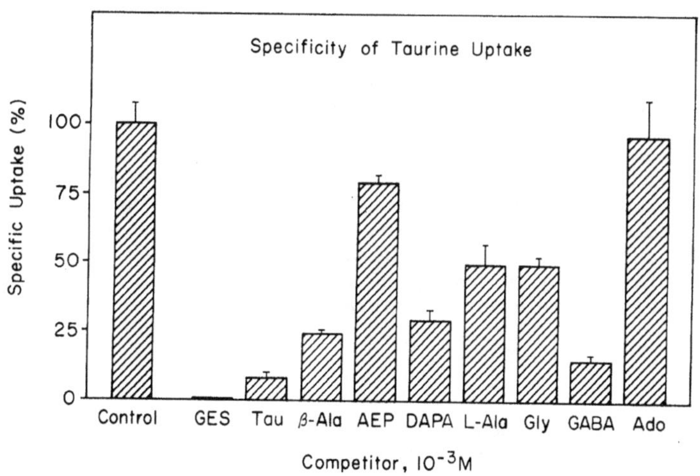

Fig. 5. Competition by the indicated compounds for the specific uptake of ^{3}H-taurine by olfactory sensilla. Sensilla were incubated for 1 h in ASW containing 5×10^{-7} M ^{3}H-taurine with and without competing compounds (GES = guanidinoethane sulfonate; β-ala = β-alanine; AEP = aminoethylphosphonic acid; DAPA = diaminopropionic acid; L-ala = L-alanine; Gly = glycine; GABA = γ-aminobutyric acid; Ado = adenosine). Specific uptake is defined as that competed for by 10^{-3} M GES. Values are means (± SEM) of triplicate determinations.

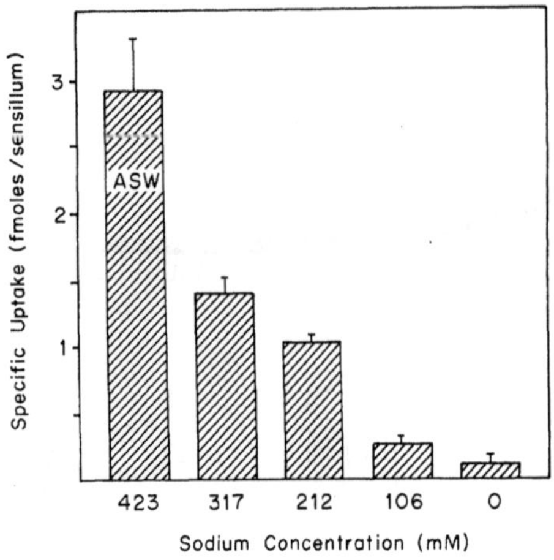

Fig. 6. Sodium dependence of taurine uptake. Olfactory sensilla were incubated for 15 min in 5 x 10^{-7} M ^{3}H-taurine in ASW (concentration of NaCl = 423 mM) and in ASW in which the concentration of NaCl was progressively decreased, and replaced with approximately equimolar Tris-Cl. Taurine uptake decreases in a concentration-dependent manner as the amount of sodium in the ASW is decreased. Values are means (± SEM) of quadruplicate determinations.

highly specific uptake system (42), is completely ineffective in competing for the uptake of taurine. The specificity of this uptake system is quite similar to those that have been characterized from several mammalian sources (16,33,13,25,20).

Uptake of taurine by the sensilla is a sodium-dependent process. In isosmotic incubation media in which NaCl is replaced by Tris-Cl, the specific interaction of labelled taurine with the aesthetasc sensilla is dramatically reduced in a dose-dependent manner (Fig. 6) (41). The sodium-dependency of taurine uptake by the aesthetasc sensilla of the lobster is thus similar to that of synaptosomes from the mammalian CNS (15). Very recent work in our laboratory indicates that the small amount of labelled taurine which is associated with sensilla in the absence of sodium saturates with time of incubation, and that the specificity of this association differs from the specificity of taurine uptake (data not shown). The specific interaction of tritiated taurine with sensilla during sodium-free incubation conditions may represent the reversible binding of taurine to electrophysiologically relevant receptors.

CONCLUSIONS

The reception of exogenous chemical substances by olfactory sensory cells parallels, in many respects, the process of neurotransmitter reception at internal synapses. Hence, olfactory cells may be viewed as functionally analogous to post-synaptic neurons and, as such, represent potentially useful models in this regard. Because of certain anatomical and physiological characteristics, the olfactory cells of the lobster are particularly well suited for this model role. In this report we show that the lobster's taurine-sensitive olfactory cells provide a promising system for exploring taurine reception, transduction, and uptake events in detail. With this system the electrical activity of individual cells can be monitored as an index of ligand-receptor interactions, complemented by biochemical measures of uptake and binding processes. Structure-activity relationships at both the physiological and biochemical levels show interesting parallels with such relationships found in vertebrates. We therefore anticipate that the lobster "taurine receptor" model will yield valuable insight towards understanding the nature of a taurine recognition system. Such insight will likely facilitate the development of agonists and antagonists for use in deciphering the physiological roles of taurine in vertebrates.

ACKNOWLEDGEMENTS

We wish to thank K. Cottrell for technical assistance in the biochemical studies and L. Milstead for technical assistance and preparation of the figures. Thanks to R. Huxtable for supplying the GES. Special thanks is due B. Ache who has been instrumental in developing the electrophysiological procedures for recording from the lobster olfactory cells. This work was supported by NSF Grants BNS-8607513 and BNS-8611256.

REFERENCES

1. Ache, B.W., Gleeson, R.A., and Thompson, H.D., in press, Mechanisms of interaction between odorants at olfactory receptor cells, in: "Proceedings of the Ninth International Symposium on Olfaction and Taste", The New York Academy of Sciences, New York.
2. Adamek, G.D., Gesteland, R.C., Mair, R.G., and Oakley, B., 1984, Transduction physiology of olfactory receptor cilia, Brain Res., 310: 87-97.
3. Anderson, P.A.V., and Ache, B.W., 1985, Voltage- and current-clamp recordings of the receptor potential in olfactory receptor cells in situ, Brain Res., 338: 273-280.
4. Carr, W.E.S., Ache, B.W., and Gleeson, R.A., 1987, Chemoreceptors of crustaceans: Similarities to receptors for neuroactive substances in internal tissues, Environ. Health Perspect., in press.
5. Curtis, D.R., Holsi, L., and Johnston, G.A.R., 1968, A pharmacological study of the depression of spinal neurons by glycine and related amino acids, Exp. Brain Res., 6: 1-18.
6. Derby, C.D., and Atema, J., 1982, Narrow-spectrum chemoreceptor cells in the walking legs of the lobster Homarus americanus: Taste specialists, J. Comp. Physiol., 146: 181-189.
7. Fuzessery, Z.M., Carr, W.E.S., and Ache, B.W., 1978, Antennular chemosensitivity in the spiny lobster, Panulirus argus: Studies of taurine sensitive receptors, Biol. Bull., 154: 226-240.
8. Gant, Z.N., and Nauss, C.B., 1976, Uptake of taurine by human blood platelets: A possible model of the brain, in: "Taurine", R.J. Huxtable and A. Barbeau, eds., Raven Press, New York, pp. 99-121.
9. Gaull, G.E., and Rassin, D.K., 1979, Taurine and brain development: Human and animal correlates, in: "Neural Growth and Differentiation", E. Meisami and M.A.B. Brazier, eds., Raven Press, New York, pp. 461-477.
10. Gleeson, R.A., and Ache, B.W., 1985, Amino acid suppression of taurine-sensitive chemosensory neurons, Brain Res., 335: 99-107.
11. Grünert, U., and Ache, B. W., 1987, Fine structure of the olfactory (aesthetasc) sensilla of the spiny lobster, Chemical Senses, in press.
12. Haas, H.L., and Hosli, L., 1979, The depression of brainstem neurons by taurine and its interaction with strychnine and bicuculline, Brain Res., 52: 399-402.
13. Holopainen, I., and Kontro, P., 1984, Taurine and hypotaurine transport by a single system in cultured neuroblastoma cells, Acta Physiol. Scand., 122: 381-386.

14. Hruska, R.E., Huxtable, R.J., and Yamamura, H.I., 1978, High affinity, temperature-sensitive and sodium-dependent transport of taurine in rat brain, in: "Taurine and Neurological Disorders", A. Barbeau and R.J. Huxtable, eds., Raven Press, New York, pp. 109-117.
15. Hruska, R E., Padjen, A., Bressler, R., and Yamamura, H.I., 1978, Taurine: Sodium-dependent, high-affinity transport into rat brain synaptosomes, Mol. Pharmacol., 14:77-85.
16. Huxtable, R.J., Laird, H.E., and Lippincott, S.E., 1979, The transport of taurine in the heart and the rapid depletion of tissue taurine content by guanidinoethylsulfonate, J. Pharmacol. Exp. Ther., 211:465-471.
17. Huxtable, R.J., and Sebring, L., 1983, Cardiovascular actions of taurine, in: "Sulfur Amino Acids: Biochemical and Clinical Aspects", K. Kuriyama, R. Huxtable, and H. Iwata, eds., Alan R. Liss, New York, pp. 5-37.
18. Johnson, B.R., and Atema, J., 1983, Narrow-spectrum chemoreceptor cells in the antennules of the American lobster, Neurosci. Lett., 45:145-150.
19. Johnson, B.R., Voigt, R., Borroni, P.F., and Atema, J., 1984, Response properties of lobster chemoreceptors: Tuning of primary taste neurons in walking legs, J. Comp. Physiol., 155:593-604.
20. Kontro, P., 1984, Comparison of taurine, hypotaurine and β-alanine uptake in brain synaptosomal preparations from developing and adult mouse, Int. J. Devl. Neuroscience, 2:465-470.
21. Kontro, P., Korpi, E.R., Oja, O.S., and Oja, S.S., 1984, Modulation of noradrenaline uptake and release by taurine in rat cerebral slices, Neuroscience, 13:663-666.
22. Kurachi, M., Yoshihara, K., and Aihara, H., 1983, Effect of taurine on depolarizations induced by L-glutamate and other excitatory amino acids in the isolated spinal cord of the frog, Japan. J. Pharmacol., 33:1247-1254.
23. Laverack, M.S., and Ardill, D.J., 1965, The innervation of the aesthetasc hairs of Panulirus argus, Quart. J. Micr. Sci., 106:45-60.
24. Lin, C.T., Song, G.X., and Wu, J.Y., 1985, Is taurine a neurotransmitter in rabbit retina, Brain Res., 337:393-398.
25. Lombardini, J.B., and Prien, S.D., 1983, Taurine binding by rat retinal membranes, Exp Eye Res., 37:239-250.
26. Mandel, P., and Pasantes-Morales, H., 1978, Taurine in the nervous system, in: "Reviews of Neuroscience", S. Ehrenpreis and I. Kopin, eds., Vol. 3, Raven Press, New York, pp. 157-193.
27. Mandel, P., Pasantes-Morales, H., and Urban, P.F., 1976, Taurine, a putative transmitter in retina, in: "Transmitters in the Visual Process", S.L. Bonting, ed., Pergamon Press, New York, pp. 89-105.
28. Menco, B. Ph. M., 1980, Qualitative and quantitative freeze-fracture studies on olfactory and nasal respiratory epithelial surfaces of frog, ox, rat and dog. II. Cell apices, cilia and microvilli, Cell Tissue Res., 211:5-29.
29. Okamoto, K., Kimura, H., and Sakai, Y., 1983, Evidence for taurine as an inhibitory neurotransmitter in cerebellar stellate interneurons: Selective antagonism by TAG (6-aminomethyl-3-methyl-4H,1,2,4-benzothiadiazine-1,1-dioxide), Brain Res., 265:163-168.
30. Oja, S.S., and Kontro, P., 1983, Taurine, in: "Handbook of Neurochemistry", A. Lajtha, ed., Vol. 3, Plenum Press, New York, pp. 501-533.
31. Pasantes-Morales, H., Ademe, R.M., and Quesada, O., 1981, Protective effect of taurine on the light-induced disruption of isolated frog outer segments, J. Neurosci. Res., 6:337-348.
32. Pasantes-Morales, H., and Cruz, C., 1985, Taurine and hypotaurine inhibit light-induced lipid peroxidation and protect rod outer segment structure, Brain Res., 330:154-157.
33. Quesada, O., Huxtable, R.J., and Pasantes-Morales, H., 1984, Effect of guanidinoethane sulfonate on taurine uptake by rat retina, J. Neurosci. Res., 11:179-186.

34. Rhein, L.D., and Cagan, R.H., 1980, Biochemical studies of olfaction: Isolation, characterization, and odorant binding activity of cilia from rainbow trout olfactory rosettes, Proc. Natl. Acad. Sci. USA, 77:4412-4416.
35. Schaffer, S.W., Kulakowski, E.C., and Kramer, J.H., 1982, Taurine transport by reconstituted membrane vesicles, in: "Taurine in Nutrition and Neurology, Adv. Exp. Med.", R. Huxtable and H. Pasantes-Morales, eds., Vol. 139, Plenum Press, New York, pp. 143-160.
36. Segawa, T., Inoue, A., Ochi, T., Nakata, Y., and Nomura, Y., 1982, Specific binding of taurine in central nervous system, in: "Taurine in Nutrition and Neurology, Adv. Exp. Med.", R. Huxtable and H. Pasantes-Morales, eds., Vol. 139, Plenum Press, New York, pp. 311-324.
37. Shepheard, P., 1974, Chemoreception in the antennule of the lobster, Homarus americanus, Mar. Behav. Physiol., 2:261-273.
38. Sturman, J.A., Moretz, R.C., French, J.H., and Wisniewski, H.M., 1985, Taurine deficiency in the developing cat: Persistence of the cerebellar external granule cell layer, J. Neurosci. Res., 13:405-416.
39. Sturman, J.A., Moretz, R.C., French, J.H., and Wisniewski, H.M., 1985, Postnatal taurine deficiency in the kitten results in a persistence of the cerebellar external granule cell layer: Correction by taurine feeding, J. Neursci. Res., 13:521-528.
40. Thompson, H., and Ache, B.W., 1980, Threshold determination for olfactory receptors of the spiny lobster, Mar. Behav. Physiol., 7:249-260.
41. Trapido-Rosenthal, H.G., Gleeson, R.A., Carr, W.E.S., Lambert, S.M., and Milstead, M.L., 1986, Purinergic and taurinergic chemoreception in the spiny lobster: Physiology and biochemistry, Soc. Neurosci. Abst., 12:1353.
42. Trapido-Rosenthal, H.G., Gleeson, R.A., Carr, W.E.S., Lambert, S.M., and Milstead, M.L., in press, The biochemistry of the olfactory purinergic system, in: Proceedings of the Ninth International Symposium on Olfaction and Taste, The New York Academy of Sciences, New York.

TAURINE SYSTEM IN THE NORMAL AND ISCHEMIC RAT HIPPOCAMPUS

Jang-Yen Wu[1], Flemming F. Johansen[2], Chin-Tarng Lin[1], and Jie-Wen Liu[1]

[1]Department of Physiology, The Pennsylvania State University, Hershey, Pennsylvania, USA and [2]Institute of Neuropathology, University of Copenhagen, Denmark

Taurine (2-aminoethanesulfonic acid) is one of the most abundant amino acids in brain (14,19). There is a growing body of evidence indicating that taurine has significant effects on various aspects of CNS function. Taurine has been suggested as a putative inhibitory transmitter based on the following observations: First of all, taurine has anti-epileptic actions (9,34), depresses specific retinal responses (5), and is inhibitory when applied directly onto neurons in spinal cord (6,7,22), cerebral cortex (4), cerebellum (10,23-27), and hippocampus (33,43). Secondly, high affinity uptake into neuronal terminals in cerebral cortex (20,29,36) and calcium-dependent release in retina have been reported (43). Thirdly, specific binding has been reported in several areas of the CNS including retina, cerebral cortex and cerebellum (21,28). Finally, the taurine synthetic enzyme, cysteine sulfinic acid decarboxylase (CSAD), has been localized by immunocytochemical methods to certain neurons of cerebellum (2,3), retina (17,18), and hippocampus (33,34); in retina, the enzyme has been further localized to synaptic vesicles (18). In addition to a possible transmitter role, taurine has also been suggested to play an important role in the development of mammalian CNS (11,32,35), in calcium metabolism (8,16,21) and in maintaining the structural integrity of the retina, at least in some species (1,13).

In light of its wide range of functions in the mammalian CNS, it is important to assess the changes of taurine neurons under pathological conditions such as ischemia. We have focussed on the ischemic changes in the hippocampus because hippocampus is one of the most vulnerable brain regions to cerebral ischemia, at least in man and experimental animals. In addition, we have recently shown the inhibitory action of taurine in the hippocampal slices (33,43) and also identified taurine neurons immunocytochemically in the hippocampus (33,43). In this communication, we present evidence to support the conclusion that in general, pyramidal cells are more vulnerable to ischemia than taurine neurons.

MATERIALS AND METHODS

Introduction of transient ischemia

Twenty min of transient cerebral ischemia was induced in eight male adult Wistar rats (200-300 g) by means of 4-vessel occlusion. Another four rats were sham-operated but had no ischemia. Briefly, rats were deprived of food for 24 h and then anesthetized with 2% Halothane in 98% oxygen prior to electrocauterization of the vertebral arteries. A Martin Electrotom 30 (setting 4) and 0.5 mm thick cauterization needle was applied through the skin at the site of incision. The animals were then kept food-deprived with free access to water until the next day. They were reanesthetized with Halothane in oxygen, and the common carotid arteries were gently exposed before anesthesia was switched off. Two minutes later the common carotids were clamped with metal clasps for 20 min. The rat body temperature was maintained at 37°C with an incandescent lamp. The rats were allowed to survive for 1 day. All rats were then injected with 0.3 ml of colchicine (0.1% in distilled water) intraperitoneally two days before they were sacrificed.

Production and characterization of antiserum against cysteine sulfinic acid decarboxylase

The methods for purification of CSAD and production and characterization of its antiserum have been described previously (37,41). Briefly, CSAD was purified from bovine brain by a combination of ammonium sulfate fractionation, column chromatographies on Sephadex G-200, hydroxylapatite, DEAE-cellulose and preparative polyacrylamide gel electrophoresis. The purity of CSAD preparations was judged by polyacrylamide gel electrophoresis in which a single protein band with co-migrating CSAD activity was obtained. The purified CSAD preparations were injected into rabbits for the production of antibodies against CSAD. The specificity of anti-CSAD was established by immunodiffusion, immunoelectrophoresis, immunoprecipitation and Western immunoblotting tests.

Immunohistochemistry

The immunohistochemical procedure used in the present study was similar to the method described previously (33,38,40,43). Adult male Wistar rats (200-300 g) were anesthetized with ether and perfused through left cardiac ventricle with 200 ml of fixative (4% paraformaldehyde and 0.01% glutaraldehyde in 0.1 M phosphate buffer, pH 7.4). Following perfusion the brain was removed from the skull and the hippocampus was dissected free. Tissue blocks from the central portion of the hippocampus were immersed overnight in 2% paraformaldehyde in 0.1M phosphate buffer, pH 7.4. The blocks were rinsed in 0.1M phosphate buffer, pH 7.4 and immersed in 0.1M phosphate buffer, pH 7.4 containing 0.025 M hydroxylamine for 30 min to block excess aldehyde groups. Tissue sections (50 µm) were then cut with a vibratome. The sections were incubated with normal goat IgG (0.1 mg/ml) for 10 min, followed by incubation with rabbit antiserum against CSAD for 16 h at 4°C (1:500 dilution). After two buffered washes (10 min each), the sections were further incubated with biotinylated goat anti-rabbit IgG (1:200 dilution) followed by additional wash and incubation in avidin-biotin-peroxidase (ABC) (Vector Laboratories, Burlingame, CA) complex solution for 30 min. Sections were washed again in buffer and refixed in 2.5% glutaraldehyde for 5 min. After extensive wash, sections were incubated with 0.05% 3,3'-diamino-benzidine-4 HCl in 0.2 M Tris HCl buffer (pH 7.2) plus 0.01% H_2O_2 for 5 min. Sections were then washed, counterstained in 0.01% OsO_4 for 1 min, rinsed in distilled water and mounted on glass slides with 50% glycerin in buffer. Control sections were incubated with either normal rabbit serum or antiserum against CSAD which had been pretreated with excess CSAD.

Several recent reviews on the production and characterization of antibodies against amino acid transmitters synthesizing enzymes and their application for immunohistochemical and immunocytochemical localization of these enzymes should be consulted for the detailed discussion on the pitfalls and specificity of the technique (38-41).

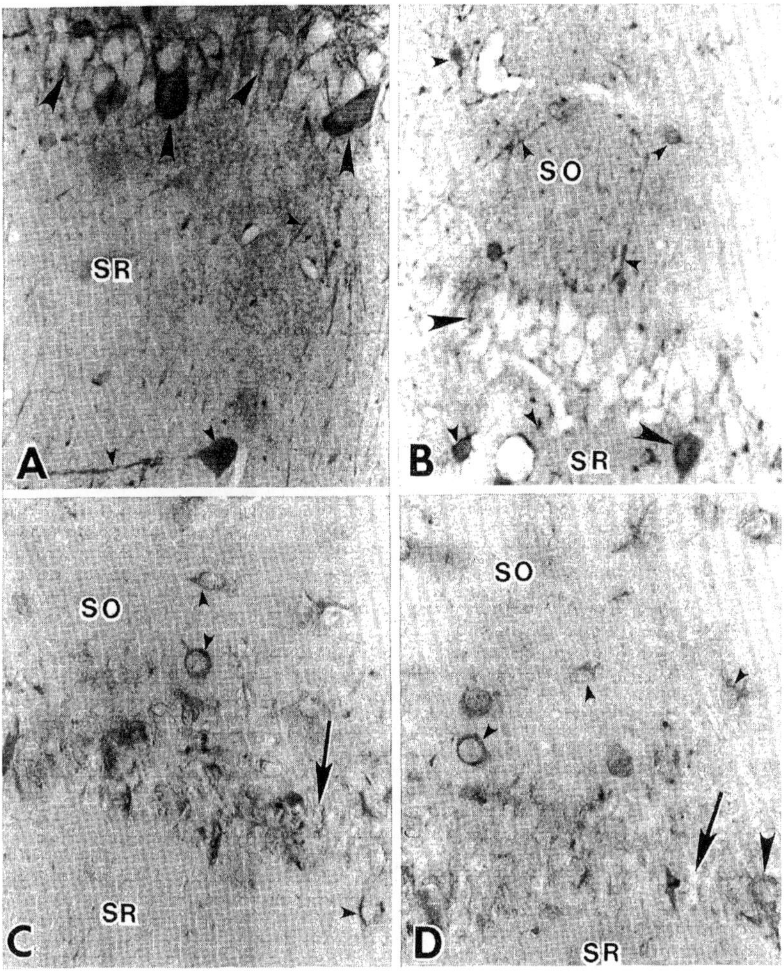

Fig. 1 Localization of CSAD in area CA1. A and B: Sham-operated. In the pyramidal cell layer, CSAD-IR is seen mainly in the somata of some pyramidal basket neurons and their neuronal processes surrounding the pyramidal neurons (large arrowheads). In the stratum oriens (SO) and stratum radiatum (SR), some somata of interneurons and their processes also contain reaction product (small arrowheads). A (X 1,100), B (X 1,100). C and D: Ischemic. CSAD-IR is found in some somata of interneurons in stratum oriens and stratum radiatum (small arrowheads) and of pyramidal basket neurons (large arrowheads) and their neuronal processes. However, the intensity of reaction product is markedly reduced. Most of the pyramidal neurons show severe degeneration (arrows) C (X 1,100), D (X 1,100).

RESULTS

In the CA1 region

In the sham-operated animals, CSAD-immunoreactivity (CSAD-IR) was found in all strata of CA1 including stratum oriens, stratum pyramidal and stratum

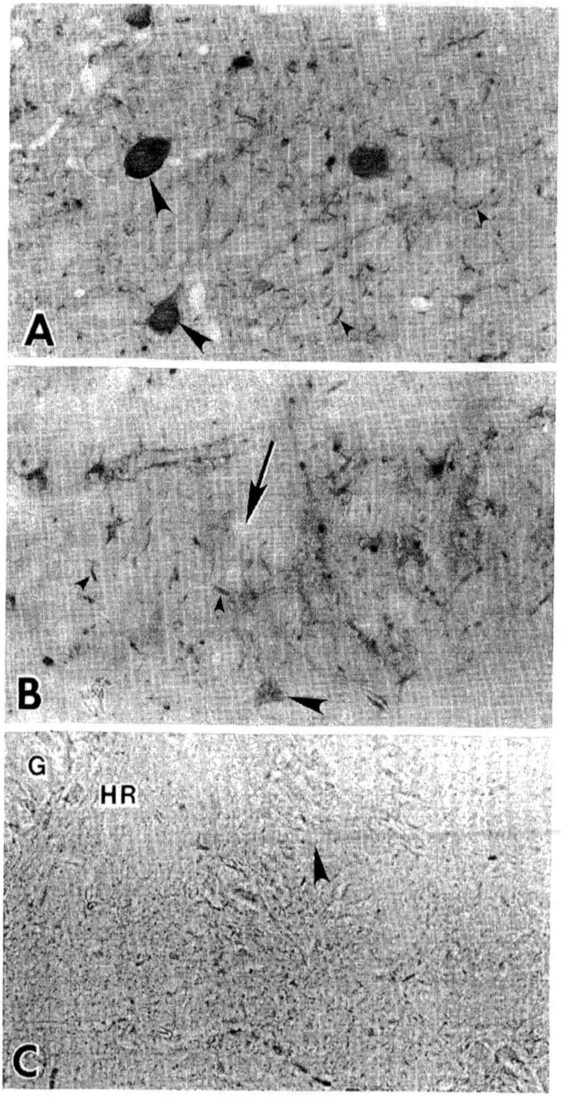

Fig. 2. Immunostaining of CSAD in area CA4. A: Sham-operated. Some pyramidal basket neurons (large arrowheads) and neuronal processes surrounding pyramidal neurons (small arrowheads) are stained (X 1,100). B: Ischemic. Some neuronal processes containing CSAD-IR are seen around pyramidal neurons (small arrowheads). One pyramidal basket neuron containing weak reaction product is also shown (large arrowheads). The outlines of pyramidal neurons (arrows) are not clear (X 1,000). C: A control section from ischemic rat, stained with preimmune serum, shows no specific staining. Arrowhead: ischemic pyramidal neuron; HR: hilar region; G: granule cell (X 1,700).

radiatum (Figs. 1A and B). Certain larger stained perikarya could be the pyramidal basket neurons (Fig. 1A) (large arrowheads). Many smaller, non-pyramidal cellular profiles were stained in all strata (small arrowheads). In some cases, reaction product was localized in stratum pyramidale as puncta which often appeared to surround the somata of pyramidal neurons. In the ischemic animals, a massive degeneration of neurons was found in all layers of CA1 region. No clear neuronal profile could be seen in most of the neurons in the stratum pyramidale (arrow). Some of CSAD-positive neurons in the stratum pyramidale (large arrowheads). stratum oriens and stratum radiatum (small arrowheads) appeared to be relatively intact (Figs. 1C and D). Numerous CSAD-positive neuronal processes could still be seen in the stratum radiatum. A control section treated with preimmune serum showed no specific staining but a marked degeneration of neurons could be easily identified (data not shown).

In the CA4 region

Similar results were obtained in the CA4 region (Figs. 2A-B). Although in the sham-operated animals, CSAD-positive neurons in the CA4 region are not as numerous as those in the CA1 region, several stained pyramidal basket neurons (large arrowheads) and numerous stained neuronal processes surrounding the pyramidal cells could also be easily discerned (Fig. 2A). Again ischemia caused a massive destruction of neurons in CA4 region (arrows). Some neuronal processes containing CSAD-IR around pyramidal neurons (small arrowheads) and one pyramidal basket neuron containing weak reaction product (large arrowheads) were found. However, the staining of CSAD in the perikarya as well as the processes was also greatly reduced (Fig. 2B). Sections treated with preimmune serum as control showed no specific staining (Fig. 2C).

In the dentate gyrus

In the sham-operated dentate gyrus CSAD-IR was seen in the somata and neuronal processes of some interneurons in the molecular layer (arrow), in the hilar region (small arrowheads) and in the granule cell layer (large arrowheads) (Figs. 3A-B). In the stained neurons, reaction product was found mostly in perikaryal region. In the granule cell layer some CSAD-positive neurons appeared to be morphologically indistinguishable from CSAD-negative neurons. Some stained neuronal processes derived from the interneurons in the hilar region were seen passing by the granule cell bodies and extending to the molecular layer (arrow) (Fig. 3B). No clear punctate reaction product surrounding the granule cell was shown. Similarly, in the molecular layer (small arrowheads) and granule cell layer (large arrowheads), CSAD-IR was seen most often in the interneurons and processes (Fig. 3B). It was rather rare to find CSAD-IR as puncta suggesting the lack of synaptic interaction between CSAD-positive neuronal processes and the target neurons. No specific staining was found in the sections treated with preimmune serum. In the ischemic animals, most of CSAD-positive neurons in the granule cell layer, molecular layer and the hilar region were spared (small arrowheads) while CSAD-negative neurons showed severe degeneration (Figs. 3C and D). A few stained granule cell-like neurons (large arrowheads) in the granule cell layer were also discernible. Even those spared CSAD-containing neurons also showed mild shrinkage. Unlike the perikaryal staining, CSAD-positive neuronal processes in the hilar region, granule cell layer and molecular layer appeared to be markedly reduced. Sections treated with preimmune serum as control showed no specific staining. However, severe degeneration of granule cells were observed.

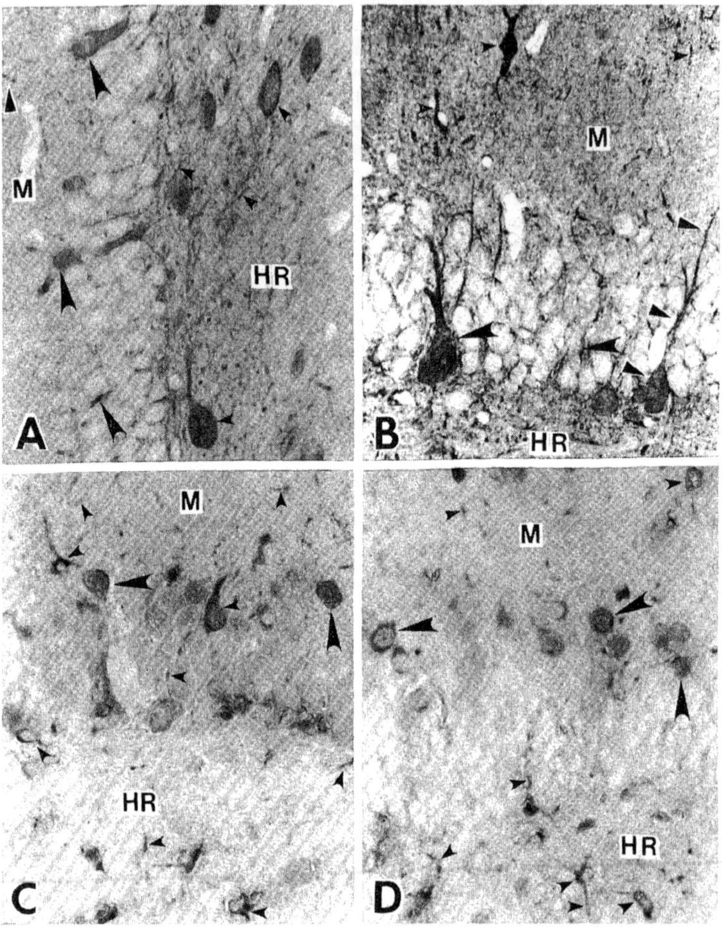

Fig. 3. Localization of CSAD in dentate gyrus. A and B: Sham-operated. A: Several interneurons in the hilar region (HR) and some neuronal processes (small arrowheads) are CSAD-positive. In the granule cell layer, some interneurons and their processes are also stained (large arrowheads). In the molecular layer (M), a few stained neuronal processes are noted (black-and-white arrowhead) (X 1,100). B: In the molecular layer (M), some interneurons and neuronal processes are stained (small arrowheads). Similarly, some interneurons and neuronal processes (large arrowheads) in the granule cell layer also contain CSAD-IR. The neuronal process from one stained interneuron in the hilar region close to granule cell layer is seen to pass through the whole granule cell layer is seen to pass through the whole granule cell layer toward molecular layer (black-and-white arrowheads) (X 1,100). C and D: Ischemic. Stained interneurons and neuronal processes are still visible in the molecular layer, granule cell layer and in the hilar region (small arrowheads). A few stained granule cell-like neurons (large arrowheads) in the granule cell layer are also discernible. However, the staining intensity is weak (C: X 1,100); (D: X 1,100).

DISCUSSION

There are several important observations from the present study. First of all, a twenty minute transient ischemia does cause a massive degeneration of neurons in all areas examined including CA1, CA4 and dentate gyrus. Secondly, the CSAD-positive neurons and their processes in general are less vulnerable to ischemic damage than other large neurons such as pyramidal cells. The mechanism underlining this protective function of taurine is not clear. It could be due to its role as an inhibitory transmitter. However, we have shown recently that in the rat hippocampus, CSAD is concentrated in the cell body and dendritic processes and only a few axonal terminals contain CSAD (33) suggesting that only a small number of CSAD-positive neurons may use taurine as neurotransmitter and the majority of CSAD-positive neurons may use taurine for other functions such as regulation of calcium (8,16,21) and in maintaining the structural integrity of the cell membrane (1,13). The latter property of taurine as a membrane stabilizer is thought to be responsible for the integrity of photoreceptor cells and the lack of it in the taurine-deprived kitten is believed to be the main cause of photoreceptor degeneration (1,13). Recently, we have shown that, under the same conditions, there is no appreciable loss of GABAergic interneurons in the rat hippocampal formation (42). It appears that GABAergic neurons are even more resistant to ischemic change than taurine neurons. This protective function of GABA may be related to its possible neurotrophic action as recently reported (12). Thirdly, CSAD-positive neurons in the dentate gyrus seem to be less affected by ischemia than those in the CA1 and CA4 regions. This differential effect of ischemia on taurine neurons in different parts of hippocampus is not clear. One possible explanation is that taurine neurons in different hippocampal regions may contain different neuroactive substances. For instance, some may also contain other inhibitory neurotransmitters such as GABA, others may contain excitatory transmitters such as glutamate or neuropeptides. This notion is supported by the observation that most CSAD-positive neurons and GAD-positive neurons are interneurons in the hippocampus and dentate gyrus (15,30,31). Previously, we have shown that CSAD is also present in the known inhibitory GABAergic neurons such as Purkinje, basket and stellate cells in the cerebellum (2,3) and amacrine cells in the retina (17,18) as well as in non-GABAergic excitatory neurons such as granule cells in the cerebellum (2, 3) and photoreceptor cells in the retina (17,18). It is conceivable that CSAD-positive neurons which also contain GAD are the ones suffering the least from ischemic insult while those do not contain GAD are more vulnerable to ischemia.

ACKNOWLEDGEMENT

This work was supported in part by grants NS 20978, NS 20922 and EY 05385 from National Institutes of Health. Skillful technical help from Diane Evans and excellent secretarial service from Pat Gering is greatly appreciated.

REFERENCES

1. Berson, E.L., Hayes, K.C., Rabin, A.R., Schmidt, S.Y., and Watson,G., 1976, Retinal degeneration in cats fed casein. II. Supplementaion with methionine, cysteine, or taurine, Invest. Ophthalmol., 15:52-58.
2. Chan-Palay, V., Lin, C.-T. Palay, S., Yamamoto, M.,and Wu, J.-Y., 1982, Taurine in the mammalian cerebellum: Demonstration by autoradiography with [^{3}H]taurine and immunocytochemistry with antibodies against the taurine-synthesizing enzyme, cysteine sulfinic acid decarboxylase, Proc. Natl. Acad. Sci., USA, 79:2695-2699.

3. Chan-Palay, V., Palay, S.L., and Wu, J.-Y., 1982, Sagittal cerebellar microbands of taurine neurons: Immunocytochemical demonstration by using antibodies against the taurine-synthesizing enzyme, cysteine sulfinic acid decarboxylase, Proc. Natl. Acad. Sci., USA, 79:4221-4225.
4. Crawford, J.M., and Curtis, D.R., 1964, The excitation and depression of mammalian cortical neurons by amino acids, Br. J. Pharmacol., 23:313-329.
5. Cunningham, R., and Miller. R.F., Taurine: Its selective action on neuronal pathways in the rabbit retina, Brain Res., 117:341-345.
6. Curtis, D.R., Hosli, L., and Johnston, G.A.R., 1968, A pharmacological study of the depression of spinal neurons by glycine and related amino acids, Exp. Brain Res., 6:1-8.
7. Curtis, D.R., Phillis, J.W., and Watkins, J.C., 1961, Actions of amino acids on the isolated hemisected spinal cord of the toad, Br. J. Pharmacol., 16:262-283.
8. Dolara, P., Agresti, A., Giotti, A., and Pasquini, G., 1973, Effect of taurine on calcium kinetics of guinea-pig heart, Eur. J.. Pharmacol., 24:352-358.
9. Durelli, L., and Mutani, R., 1983, The current status of taurine in epilepsy, Clin. Neuropharmacol., 6:37-48.
10. Frederickson, R., Neuss, M., Morzorati, S., and McBride, W., 1978, A comparison of the inhibitory effects of taurine and GABA on identified Purkinje cells and other neurons in the cerebellar cortex of the rat, Brain Res., 145:117-126.
11. Gaull, G.E., 1982, Taurine nutrition in man, Adv. Exp. Biol. Med., 139:89-95.
12. Hansen, G.H., Meier, E., and Schousboe, A., 1985, High affinity GABA receptors mediate the neurotrophic action of GABA, J. Neurochem., 44 (Suppl.):S80A.
13. Hayes, K.C., Carey, R.E., and Schmidt, S.Y., 1975, Retinal degeneration associated with taurine deficiency in the cat. Science, 188:949-951.
14. Jacobson, J.G., and Smith, L.H., Jr., 1968, Biochemistry and physiology of taurine and taurine derivatives, Physiol. Rev., 48:424-511.
15. Kosaka, T., Kosaka, K., Tateishi, K., Hamaoka, Y., Yanaihara, N., Wu, J.-Y., and Hama, K., 1985, GABAergic neurons containing CCK-8-like and or VIP-like immunoreactivities in the rat hippocampus and dentate gyrus, J. Comp. Neurol., 239:420-430.
16. Kuriyama, K., Muramatsu, M., Nakagawa, K., and Kakita, K., 1978, Modulating role of taurine on release of neurotransmitters and calcium transport in excitable tissue, in: "Taurine and Neurological Disorders", A. Barbeau and R.J. Huxtable, eds., Raven Press, New York, pp. 201-216.
17. Lin, T.-T., Li, H.-Z., and Wu, J.-Y., 1983, Immunocytochemical localization of L-glutamate decarboxylase, gamma-aminobutyric acid transaminase, cysteine sulfinic acid decarboxylase, aspartate aminotransferase and somatostatin in rat retina, Brain Res., 270:273-283.
18. Lin, C.-T., Song, G.-X., and Wu, J.-Y. 1985, Ultrastructural demonstration of L-glutamate decarboxylase and cysteine sulfinic acid decarboxylase in rat retina by immunocytochemistry, Brain Res., 331:71-80.
19. Lombardini, J.B., 1976, Regional and subcellular studies on taurine in the rat central nervous system, in: "Taurine", R.J. Huxtable and A. Barbeau, eds., Raven Press, New York, pp. 311-326.
20. Lombardini, J.B., 1978, High affinity transport of taurine in mammalian central nervous system, in: "Taurine in Neurological Disorders", A. Barbeau and R.J. Huxtable, eds., Raven Press, New York, pp. 119-135.
21. Lopez-Colome, A.M., 1982, Taurine receptors in CNS membranes: Binding studies, Adv. Exp. Med. Biol., 139:293-310.
22. Nicoll, R.A., Padjen, A., and Barker, J.L., 1976, Analysis of amino acid response on frog motoneurons, Neuropharmacology, 15:45-53.
23. Okamoto, K., Kimura, H., and Sakai, Y., 1983, Taurine-induced increase of the Cl-conductance of cerebellar Purkinje cell dendrites in vitro, Brain Res., 259:319-323.

24. Okamoto, K., Kimura, H., and Sakai, Y., 1983, Effects of taurine and GABA on Ca spikes and Na spikes in cerebellar Purkinje cells in vitro: Intrasomatic study, Brain Res., 260:249-259.
25. Okamoto, K., Kimura, R., and Sakai, Y., 1983, Ionic mechanism of the action of taurine on cerebellar Purkinje cell dendrites in vitro: Intradendritic study, Brain Res., 260:261-269.
26. Okamoto, K., Kimura, H., and Sakai, Y., 1983, Evidence for taurine as an inhibitory neurotransmitter in cerebellar stellate interneurons: Selective antagonism by TAG (6-aminomethyl-3-methyl-4H, 1,2,4-benzothiadiazine-1,1-dioxide), Brain Res., 265:163-168.
27. Okamoto, K., and Sakai, Y., 1981, Inhibitory actions of taurocyamine, hypotaurine, homotaurine, taurine and GABA on spike discharges of Purkinje cells, and localization of sensitive sites, in guinea pig cerebellar slices, Brain Res., 206:371-386.
28. Sagawa, T., Inoue, A., Ochi, T., Nakata, Y., and Nomura, Y., 1981, Specific binding of taurine in central nervous system, Adv. Exp. Med. Biol., 139:311-324.
29. Schmid, R., Sieghart, W., and Karobath, M., 1975, Taurine uptake in synaptosomal fractions of rat cerebral cortex, J. Neurochem., 25:5-9.
30. Somogyi, P., Hodgson, A.J., Smith, A.D., Nunzi, M.G., Gorio, A., and Wu, J.-Y., 1984, Different populations of GABAergic neurons in the visual cortex and hippocampus of cat contain somatostatin- or cholecystokinin-immunoreactive material, J. Neuroscience, 4:2590-2603.
31. Somogyi, P., Smith, A.D., Nunzi, M.G., Gorio, A., Takagi, H., and Wu, J.-Y., 1983, Glutamate decarboxylase immunoreactive neurons and distribution of their synaptic terminals on pyramidal neurons in the hippocampus of the cat with special references to the axon initial segment, J. Neuroscience, 3:1450-1468.
32. Sturman, J., 1981, Is taurine essential for development? Adv. Exp. Med. biol., 139:487-512.
33. Taber, K.H., Lin, C.-T., Liu, J.-W., Thalman, R., and Wu, J.-Y., Taurine in the hippocampus: Localization and postsynaptic action, Brain Res., in press.
34. Toth, E., Lajtha, A., Sarahan, S., and Seiler, N., 1983, Anti-convulsant effects of some inhibitory neurotransmitter amino acids, Neurochem. Res., 8:291-302.
35. Van Gelder, N.M., and Parent, M., 1982, Protein and taurine of maternal diets during the mouse neonatal period: Permanent effects on cerebellar-brainstem amino acid levels in mature offspring, Neurochem. Res., 7:987-998.
36. Wheler, G.H.T., Bradford, H.F., Davison, A.N., and Thompson, E.J., 1981, Uptake and stimulated release of taurine by preparations of cerebral cortex, in: "The Effects of Taurine on Excitable Tissues", S.W. Schaffer, S.I. Baskin, and J.J. Kocsis, eds., Spectrum Publications, Inc., Jamaica, New York, pp. 93-106.
37. Wu, J.-Y., 1982, Purification and characterization of cysteic acid and cysteine sulfinic acid decarboxylase and L-glutamate decarboxylase from bovine brain, Proc. Nat. Acad. Sci., USA, 79:4270-4274.
38. Wu, J.-Y., 1983, Preparation of glutamic acid decarboxylase as immunogen for immunocytochemistry, in: "Neuroimmunocytochemistry", (IBRO Handbook Series: Methods in Neurosciences), A.C. Cuello, Ed., John Wiley and Sons, Ltd., Sussex, pp. 159-191.
39. Wu, J.-Y., Denner, L.A., Wei, S.C., Lin, C.-T., Song, G.-X., Xu, Y.F., Liu, J.W., and Lin, H.S., 1986, Production and characterization of polyclonal and monoclonal antibodies to rat brain L-glutamate decarboxylase, Brain Res., 373:1-14.
40. Wu, J.-Y., and Lin, C.-T., 1985, Immunocytochemical techniques, in: "Neuromethods, Vol. 3, Amino Acids", A.A. Boulton, G.B. Baker, and J.D. Wood, eds., The Humana Press, pp. 155-178.
41. Wu, J.-Y., Lin, C.-T., Hwang, B., Wei, S., and Lin, H.S., 1986,

Antibodies against enzymes synthesizing amino acid neurotransmitters, in: "Histochemistry Modern Methods and Applications, Neurology and Neurobiology, Vol. 16", P. Panula, H. Paivarinta, and S. Soinila, eds., Alan R. Liss, Inc., New York, pp. 21-47.

42. Wu, J.-Y., Lin, C.-T., Liu, J.W., and Johansen, F.F., 1986, GABA and taurine systems in controls and ischemic rat hippocampal formation. Soc. Neurosci, Abstr. Vol. 12, Part 2, p. 1532.

43. Wu, J.-Y., Lin, C.-T., Thalman , R., Taber, E., Song, G.-X., and Su, Y.Y. T., 1985, Immunocytochemical and physiological identification of taurine neurons in mammalian CNS. in: "Symposium: Taurine: Biological Actions and Clinical Perspectives", Alan R. Liss, Inc., Helsinki, Finland, pp. 261-270.

IMMUNOCYTOCHEMICAL DEMONSTRATION OF TAURINE

S. Madsen, O.P. Ottersen and J. Storm-Mathisen

Anatomical Institute, University of Oslo, Karl Johansgate 47
N-0162 Oslo 1, Norway

The exact localization of a compound in normal tissue and its redistribution in neurological diseases and disease models can give important clues as to the functional role of the compound in question. In the case of taurine (2-aminoethanesulfonic acid), so far only biochemical methods have been available for its localization. Such methods will necessarily have a limited anatomical resolution and may easily fail to detect important pathological alterations in the cellular and subcellular distribution of taurine. A much better anatomical resolution is provided by the immunocytochemical technique, which has recently been adapted to amino acids (24). Antisera against conjugated amino acids have been employed to map the light- and electronmicroscopic localization of GABA, glutamate, glutamine, asparatate, and glycine in normal nerve tissue (1,2,6,9,10,17,18,19,22,24), and have also been used to study the redistribution of transmitter amino acids in experimental models of epilepsy (14).

We report here that the immunocytochemical technique can be extended to taurine to provide new information about the distribution of this amino acid in the CNS.

METHODS

Preparation of antiserum

Taurine was conjugated to bovine serum albumin (BSA) by glutaraldehyde (13,19,24). The product was dialyzed until free amino acids were no longer detectable, mixed with adjuvant and injected in rabbits. Crude sera were obtained and purified through a sequence of columns containing Sepharose beads bearing glutaraldehyde-treated BSA and glutamate fixed to BSA by glutaraldehyde (13,24). The serum obtained in this manner will be termed standard antiserum. Aliquots of the standard antiserum were further purified by applying it to a column containing BSA-glutaraldehyde-taurine followed by elution of the specific antibodies at low pH, or by absorption of serum with BSA-glutaraldehyde-hypotaurine complexes. Since our original reports were published (13,16), Campistron et al. reported on the preparation and application to the CNS of an antiserum against fixed taurine (3). The latter authors used an approach that was slightly different from that outlined above.

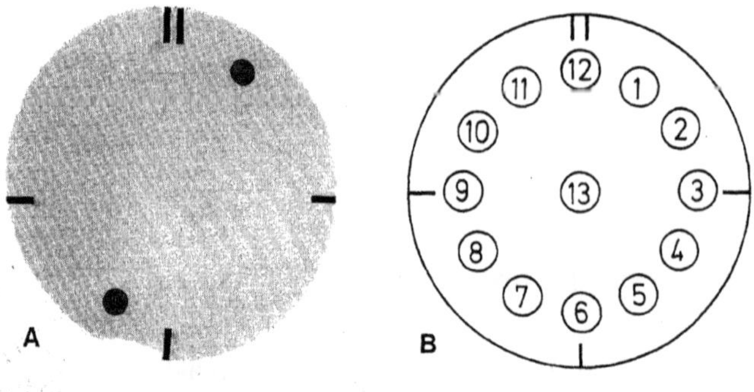

Fig. 1. Specificity of standard antiserum. A: photomicrograph of standard test filter treated with taurine antiserum diluted 1:1000. B: Key to spots in A. 1 and 7, taurine; 2 and 8, GABA; 3 and 9, glutamate; 4 and 10, aspartate; 5 and 11, glutamine; 6 and 12, methionine-enkephalin.

Preparation of tissue for immunocytochemistry

Release, uptake and metabolism of amino acids may lead to rapid changes in their tissue localization. Fixation of tissue must therefore be accomplished as quickly as possible. We use the following perfusion procedure: after anesthesia, a cannula, connected to a peristaltic pump, is placed in the left ventricle. Perfusion is started with a flush of dextran to remove blood and continued with 5% glutaraldehyde in buffer. The neck of the animal usually becomes stiff within 90 s after opening of the thorax (13). Vibratome or frozen sections of the brain are cut at 20 µm.

Immunocytochemical procedure

The immunocytochemical procedure is based on the peroxidase-antiperoxidase technique of Sternberger (23). Tissue sections and test filters (see below) are processed together for optimum specificity control (14,18). Filters and sections were: (1) treated with ethanolamine and methanol/H_2O_2 to remove free aldehyde groups and endogenous peroxidase activity; (2) preincubated with normal sheep serum; (3) incubated for 1-4

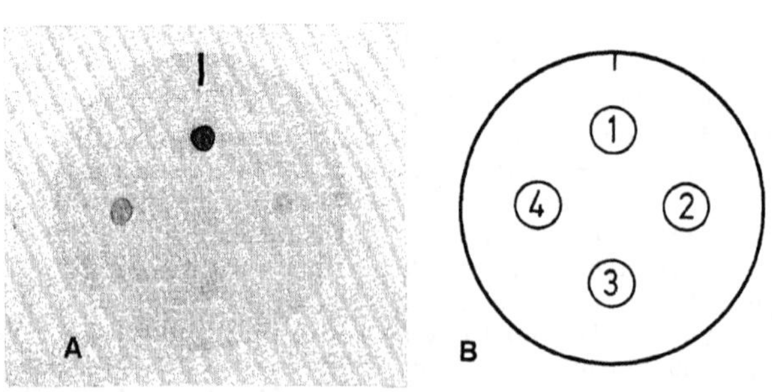

Fig. 2. Specificity of standard antiserum. A: Photomicrograph of standard test filter treated with taurine antiserum diluted 1:1000. B: Key to A. 1, taurine; 2, cysteinesulfinic acid; 3, cysteic acid; 4, hypotaurine. The crossreactivity against hypotaurine could be strongly reduced by absorption with BSA-glutaraldehyde-hypotaurine complexes (see text).

days with the specific taurine antiserum diluted 1:500 to 1:3000; (4) treated with sheep anti-rabbit immunoglobulins; (5) treated with rabbit peroxidase-antiperoxidase complex; and (6) immersed in diaminobenzidine and H_2O_2. The staining process was stopped by addition of H_2SO_4 to the wells. In experiments designed to evaluate quantitative aspects of the method (see below), all incubations were accurately timed in order to make the results reproducible. Sections were then rinsed, soaked in glycerol and mounted in glycerolgelatin, or dried, cleared in xylene and mounted in Eukit.

SPECIFICITY

Testing against amino acid conjugates

To test the specificity of the taurine antiserum and other antisera a technique allowing identical conditions for testing and immunocytochemistry has been developed (13,18).

The small molecules to be tested were conjugated to rat brain proteins by glutaraldehyde. The conjugates were applied to cellulose ester filters (standard filters, diameter 13 mm), which after drying could be processed immunocytochemically together with the sections. Thus, we were able to run a specificity control in all experiments. The degree of crossreactivity could be evaluated by visual inspection of the test filters. For quantitative studies (see below) we have used both standard filters carrying spots of test conjugates or smaller filters (diameter 2.3 mm) that had been immersed in the soluble fraction of the test conjugates (20). To permit optical densitometry, the filters had to be cleared in xylene before mounting.

The compounds that have been tested for crossreactivity by the filter method include all amino acids that occur at a higher average concentration than 1 mmol/kg in the mammalian brain, many structural analogues of taurine and several peptides (Table 1). Only four of the compounds tested (hypotaurine, gamma-glutamyltaurine, cysteic acid and cysteine sulfinic acid) show staining appreciably above background with the standard antiserum (Figs. 1 and 2). The hypotaurine conjugates show the strongest interaction: still the concentration of hypotaurine will have to be 15 times higher than that of taurine to give comparable staining. It should be emphasized in this context that the concentration of taurine in the rat brain is about 80 times that of hypotaurine (4.6 vs. 0.06 μmol/g wet weight) (21).

Absorption of the standard antiserum with BSA-glutaraldehyde-hypotaurine complexes almost abolished crossreactivity against conjugated hypotaurine, but did not lead to any changes in the staining pattern in the cerebellum. However, the titer against fixed taurine was moderately reduced. Affinity purification with BSA-glutaraldehyde-taurine reduced the reactivity against hypotaurine and the other three compounds mentioned above. Again, the immunostaining pattern was indistinguishable from that obtained wih the standard antiserum. The standard antiserum thus appears to show a sufficient specificity for immunocytochemical purposes.

Inhibition of staining

In a first series of experiments we tried to block tissue staining by preincubating standard antiserum with free taurine. Inhibition of staining is observed at taurine concentrations of 50 to 800 mM. Similar effects are not observed with glutamate, GABA, asparate and glycine, thus indicating that free taurine interacts selectively, although at low affinity, with the specific antibodies. Subsequently we tested whether inhibition of staining could be more effectively obtained by preincubating the standard antiserum

Table 1. Compounds that have been tested for crossreactivity

Alanine	Histidine
2-Amino-7-phosphonoheptanoic acid	Homocarnosine
Aspartate	Hypotaurine
Beta-Alanine	Leucine
Carnosine	Leucine-enkephalin
Cysteic acid	Methionine
Cysteine	Methionine-enkephalin
Cysteinesulfinic acid	N-Methyl-DL-aspartate
Cystine	Norepinephrine
D-Aspartate	Oxidized glutathione
Epinephrine	Phosphoethanolamine
GABA	Reduced glutathione
Gamma-Glutamyltaurine	Serine
Glutamate	Serotonin
Glutamine	Threonine
Glycine	Tyrosine

All compounds are L-form if not otherwise stated.

with taurine-glutaraldehyde complexes rather than free taurine. Indeed, such complexes produce a strong inhibition of staining at concentrations above 50 µM, measured as the amount of taurine added. At about 300 uM staining is almost completely abolished. Glutaraldehyde treated glutamate, GABA, glycine and aspartate have no effect on staining intensity. This finding suggests that glutaraldehyde participates in the formation of the specific three-dimensional configuration of the epitope (7,8), and that high affinity binding can be obtained in the absence of protein carriers.

QUANTITATIVE ASPECTS

What is the relationship between tissue levels of taurine and immunostaining intensities? To address this question we prepared a series of test conjugates with different taurine concentrations in the reaction mixture, but

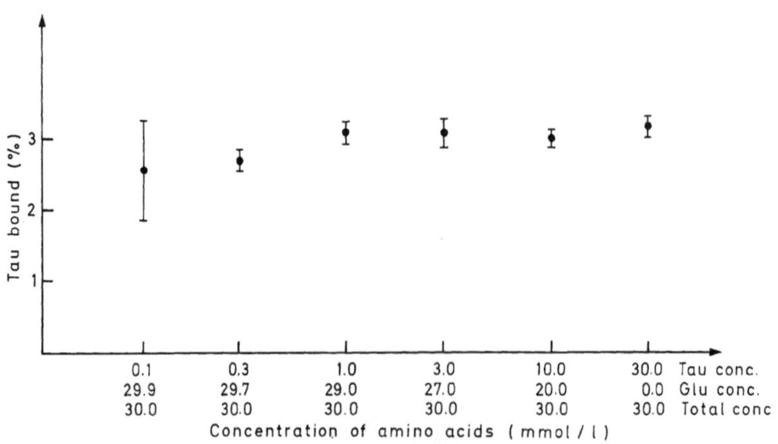

Fig. 3. Binding of radiolabelled taurine to rat brain proteins by glutaraldehyde as a function of amino acid concentration in the reaction mixture. Binding was measured as percent radioactivity remaining after dialysis of the reaction mixture. Bars represent $\pm$ 1 SD (n=4).

kept the total concentration of amino acids constant by adding glutamate. By the use of radiolabelled taurine we found that the amount of taurine that was bound was proportional to the concentration of taurine in the reaction mixture (Fig. 3). The conjugates were applied to test filters that were processed exactly as for immunocytochemistry. In a logarithmic system there seemed to be a linear relationship between taurine concentration (throughout the biologically relevant range) and staining intensity (as assessed by densitometric recordings of the immunostained filters) (Fig. 4). These results suggest that our immunocytochemical approach should be useful for a semiquantitative assessment of experimentally induced changes in tissue levels of taurine (but see below for a discussion on the possiblilty of immunological interference). It should be borne in mind, however, that it is impossible in model systems to reproduce exactly the conditions of the immunocytochemical experiments.

It has been suggested on theoretical and experimental grounds that the antigen combining site is large enough to accomodate 3-8 amino acid residues (11). By extrapolation it can be assumed that a good fit with the specific taurine antibodies requires several conjugated taurine molecules in a row on the polymerized glutaraldehyde backbone (7,8). The results of the inhibition experiments could be interpreted to support this assumption. With a low concentration of taurine and a high concentration of other amino acids in the test tube or in the tissue, the formation of an optimal antigenic site would be difficult, as other amino acids could intervene between the taurine molecules. Hence, in such a situation, the immunostaining would be artifactually reduced. Indeed, we have shown that the addition of an excess of glutamate during the preparation of protein-glutaraldehyde-taurine complexes leads to a reduction in immunostaining that is larger than expected on the basis of reduced taurine binding. This immunological interference is not likely to reduce the strongly staining of taurine-containing neurons, but

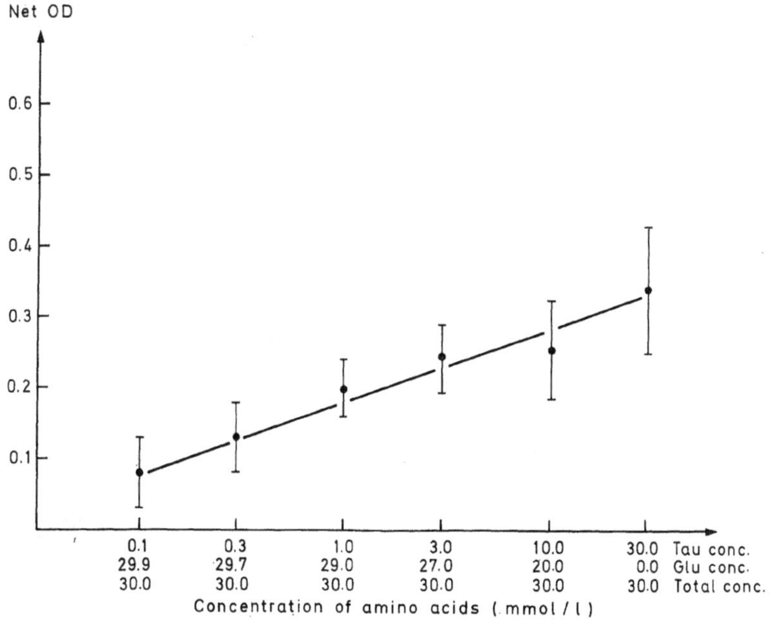

Fig. 4. Relationship between immunostaining intensity and taurine concentration. Cellulose ester filters were immersed in the soluble fraction of conjugates made with different taurine concentrations, and processed with the taurine antiserum (1:1000). Bars represent ± 1 SD (n=4).

should be taken into consideration when interpreting results obtained with amino acid antisera (20).

IMMUNOCYTOCHEMICAL LOCALIZATION OF TAURINE IN THE RAT CNS

An accurate study of the distribution of taurine-like immunoreactivity (Tau-LI) in normal material will form a necessary platform for the study of possible changes under experimental conditions. Pilot studies have attested to the usefulness of the immunocytochemical technique for studying experimentally induced taurine redistribution (12).

There are striking regional differences in the localization of Tau-LI in the rat brain. Interestingly, we have found a predominant glial localization of Tau-LI in the thalamus, hypothalamus and brain stem, and a predominant neuronal localization in the cerebral (Fig. 5) and cerebellar cortices, and in the caudatoputamen.

Hippocampus

In the hippocampus Tau-LI is found mainly in neurons. The pattern observed in the rat is much less variegated than in the baboon (16). In the rat, most pyramidal cells show a medium staining intensity of perikarya and dendrites. The staining is similar in all fields. Intermediate levels of Tau-LI were found in a few cells resembling basket cells at the hilar aspect of the granule cell layer, and in neurons scattered in the deep hilar region. The granule cells vary in their staining intensity, but are generally less strongly stained than the pyramidal cells.

Basal ganglia

Taurine is found in a majority of neurons in the caudatoputamen, including both large, medium and small neurons. Very few glial cells are stained in this region.

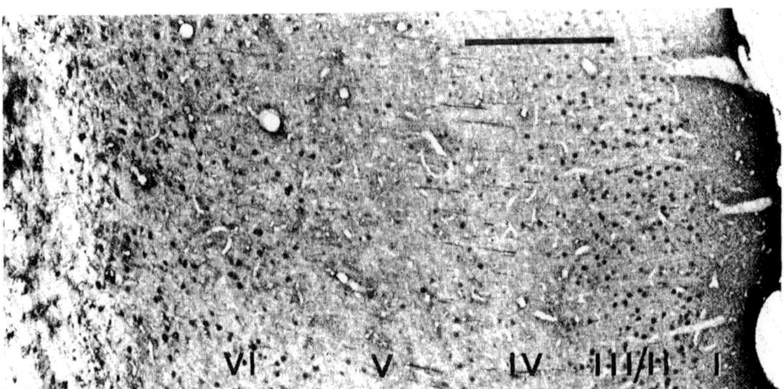

Fig. 5. Tau-LI in the rat cerebral cortex (area 41). Roman numerals indicate cortical layers. Bar: 200 μm.

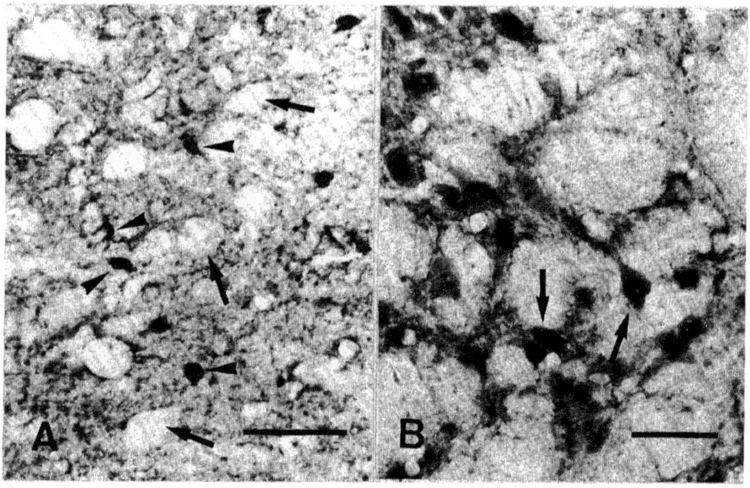

Fig. 6. Tau-LI in the thalamus of the rat. A: Glial staining in the nucleus paracentralis. Arrowheads, glial cells; arrows, unstained neurons. Bar: 50 µm. B: Labelled neurons in nucleus reticularis (arrows). Bar: 50 µm.

Thalamus

Glial cell staining is predominant in the thalamus (Fig. 6A), except in the nucleus reticularis where a subpopulation of the neurons is stained (Fig. 6B).

Cerebellum

Cerebellum is of considerable interest in regard to taurine localization. First, it has been suggested that taurine is an inhibitory transmitter of the stellate cells of the molecular layer (15). Second, the cerebellum has been examined with an antiserum against cysteine sulfinic acid decarboxylase (CSAD), an enzyme involved in taurine synthesis (4,5).

In the rat cerebellum, Tau-LI is found in most or all Purkinje cells (Fig. 7). There is intense staining of dendritic tree, perikaryon and axon (13). With antisera against CSAD, several unstained Purkinje cells were found, the unstained cells forming sagittal microbands. No banding pattern was observed in our material. This suggests that ther must be a differential localization of taurine and CSAD, possibly reflecting existence of taurine synthesizing pathways not involving CSAD. CSAD-like immunoreactivity was also found in a substantial number of stellate cells and basket cells, whereas less than 2% of these cells showed Tau-LI. On the other hand, GABA-LI is strong in most stellate and basket cells (13,18). The granular cells show low to medium staining for Tau-LI, but a number of Golgi cells are strongly stained. Bundles of stained axons can be followed from the Purkinje cell layer to the cerebellar nuclei which contain abundant nerve terminal-like structures surrounding negative cell bodies. Bergmann's glia are unstained, but some stained glial cells, possibly representing oligodendroglia, are found in the white matter. Our results are in good accordance with other data on Tau-LI in the cerebellum (3).

Brain stem

The brain stem contains very few stained neurons and axons. However, Tau-LI occurs in a large number of astroglia and oligodendroglia throughout the brain stem.

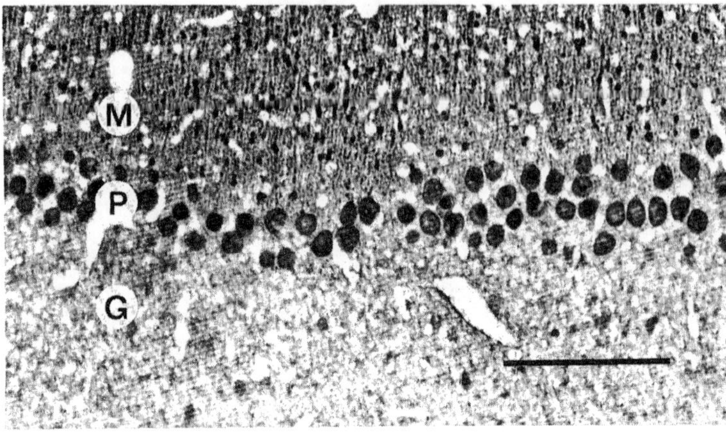

Fig. 7. Transverse section of a folium of the rat cerebellum showing labelled Purkinje cell somata. Profiles of cut dendrites are seen in the molecular layer. M, molecular layer; P, Purkinje cell layer; G, granùlar layer. Taurine antiserum diluted 1:1000. Bar: 150 μm.

SUMMARY

Amino acid immunocytochemistry represents a new and powerful tool in neuroscience. Antisera are now available to a large number of amino acids, including those that are thought to serve a transmitter role. The antiserum described here against conjugated taurine seems to be of good specificity and can be used to demonstrate taurine in the CNS as well as in other organs; in normal tissue and in experimental and pathological conditions.

ACKNOWLEDGEMENTS

Supported by the Norwegian Council for Science and the Humanities, the Norwegian Council on Cardiovascular Disease, and the Norwegian Society for Fighting Cancer.

REFERENCES

1. Campistron, G., Buijs, R.M., and Geffard, M., 1986, Specific antibodies against aspartate and their immunocytochemical application in the rat brain, Brain Res., 365:179-184.
2. Campistron, G., Buijs, R.M., and Geffard, M., 1986, Glycine neurons in the brain and spinal cord. Antibody production and immunocytochemical localization, Brain Res., 376:400-405.
3. Campistron, G., Geffard, M., and Buijs, R.M., 1986, Immunological approach to the detection of taurine and immunocytochemical results, J. Neurochem., 46:862-868.
4. Chan-Palay, V., Lin, C.-T., Palay, S., Yamamoto, M., and Wu, J.-Y., 1982, Taurine in the mammalian cerebellum: demonstration by autoradiography with [^{3}H]taurine and immunocytochemistry with antibodies against the taurine-synthesizing enzyme cysteinesulfinic acid decarboxylase, Proc. Natl. Acad. Sci. USA, 79:2695-2699.
5. Chan-Palay, V., Palay, S.L., and Wu, J.-Y., 1982, Saggital cerebellar

microbands of taurine neurons: immunocytochemical demonstration by using antibodies against the taurine-synthesizing enzyme cysteinesulfinic acid decarboxylase, Proc. Natl. Acad. Sci. USA, 79:4221-4225.

6. Dale, N., Ottersen, O.P., Roberts, A., and Storm-Mathisen, J., 1986, Inhibitory neurones of a motor pattern generator in Xenopus revealed by antibodies to glycine, Nature (London), 324:255-257.
7. Hardy, P.M., Nicholls, A.C., and Rydon, H.N., 1976, The nature of the cross-linking of proteins by glutaraldehyde. I. Interactions of glutaraldehyde with the amino-groups of 6-aminohexanoic acid and of alpha-N-acetyl-lysine, J. Chem. Soc. Perkin Trans., 9:958-962.
8. Hardy, P.M., Hughes, G.J., and Rydon, H.N., 1979, The nature of the cross-linking of proteins by glutaraldehyde. Part 2. The formation of quarternary pyridinium compounds by the action of glutaraldehyde on proteins and the identification of a 3-(2-piperidyl)-pyridinium derivative, anabilysine, as a crosslinking entity, J. Chem Soc. Perkin Trans., 1:2282-2288.
9. Hodgson, A.J., Penke, B., Erdei, A., Chubb, I.W., and Somogyi, P., 1985, Antisera to gamma-aminobutyric acid. I. Production and characterization using a new model system, J. Histochem. Cytochem., 33:229-239.
10. Laake, J.H., Gundersen, V., Nordbo, G., Ottersen, O.P., and Storm-Mathisen, J., 1986, An antiserum against glutamine, in: "Excitatory amino acids," P.J. Roberts, J. Storm-Mathisen and H.F. Bradford, eds., Macmillan, London pp .448-450.
11. Larsson, L.-I., 1983, Methods for immunocytochemistry of neurohormonal peptides, in: "Handbook of chemical neuroanatomy. Volume 1: Methods in chemical neuroanatomy," A. Bjorklund and T. Hokfelt, eds., Elsevier, Amsterdam, pp. 147-209.
12. Madsen, S., Lehmann, A., Ottersen, O.P., and Storm-Mathisen, J., N-methyl-DL-aspartate (NMDA) induced redistribution of amino acids in the hippocampal formation visualized by immunocytochemistry, in: "Excitatory amino acids," P.J. Roberts, J. Storm-Mathisen and H.F. Bradford, eds., Macmillan, London pp. 444-445.
13. Madsen, S., Ottersen, O.P., and Storm-Mathisen, J., 1985, Immunocytochemical visualization of taurine: neuronal localizaton in the rat cerebellum, Neurosci. Lett., 60:255-260.
14. Meldrum, B.S., Swan, J.H., Ottersen, O.P., and Storm-Mathisen, J., 1986, Redistribution of transmitter amino acids in rat hippocampus and cerebellum during seizures induced by L-allylglycine and bicuculline: an immunocytochemical study with antisera against conjugated GABA, glutamate, and aspartate, Neuroscience, (in press).
15. Okamoto, K., Kimura H., and Sakai, Y., 1982, Evidence for taurine as an inhibitory neurotransmittor in cerebellar stellate interneurones: selective antagonism by TAG (6-aminomethyl-3-methyl-4H,1,2,4-benzothiadiazine-1,1-dioxide), Brain Res., 265:163-168.
16. Ottersen, O.P., Madsen, S., Meldrum, B.S., and Storm-Mathisen, J., 1985, Taurine localization in the hippocampal formation of the senegalese baboon Papio Papio: an immunocytochemical study with an antiserum against conjugated taurine, Exp. Brain res., 59:457-462.
17. Ottersen, O.P., and Storm-Mathisen, J., 1984, GABA-containing neurones in the thalamus and pretectum of the rodent, Anat. Embryol., 170:197-207.
18. Ottersen, O.P., and Storm-Mathisen, J., 1984, Glutamate- and GABA-containing neurones in the mouse and rat brain as demonstrated with a new immunocytochemical technique, J. Comp. Neurol., 229:374-392.
19. Ottersen, O.P., and Storm-Mathisen, J., 1985, Different neuronal localization of aspartate-like and glutamate-like immunoreactivities in the hippocampus of rat, guinea-pig and Sengalese baboon (Papio papio), with a note on the distribution of gamma-aminobutyrate, Neuroscience, 16:589-606.
20. Ottersen, O.P., Storm-Mathisen, J., Madsen, S., Skumlien, S., and Stromhaug, J., 1986, Evaluation of the immunocytochemical method for amino acids, Med. biol., 64:147-158.

21. Perry, T.L., Kish, S.J., and Hansen, S.H., 1979, Gamma-vinyl GABA: effects of chronic administration on the metabolism of GABA and other amino compounds, J. Neurochem., 32:1641-1645.
22. Seguela, P., Geffard, M., Buijs, R.M., and Le Moal, M., 1984, Antibodies against gamma-aminobutyric acid: Specificity studies and immunocytochemical results, Proc. Natl. Acad. Sci. USA, 81:3888-3892.
23. Sternberger, L.A., 1979, Immunocytochemistry, 2nd ed., John Wiley, New York, 354 pp.
24. Storm-Mathisen, J., Leknes, A.K., Bore, A.T., Vaaland, J.L., Edminson, P., Haug, F.-M.S., and Ottersen, O.P., 1983, First visualizaion of glutamate and GABA in neurones by immunocytochemistry, Nature (London), 301:517-520.

UPTAKE, LOCALIZATION AND RELEASE OF TAURINE IN THE RAT BASAL GANGLIA

L. Della Corte, D.J. Clarke, J.P. Bolam and A.D. Smith.

University Department of Pharmacology, South Parks Road, Oxford, U.K.

L.D.C. Permanent address: Dip.to di Farmacologia Preclinica e Clinica, Universita degli Studi di Firenze, Viale G.B. Morgagni, 65, 50134 Firenze

Although taurine (2-aminoethanesulfonic acid) has long been known to be present in high concentrations in the mammalian CNS, the suggestion that it may act as an inhibitory neurotransmitter or neuromodulator in the brain is relatively recent (1,6,23). Among other areas in the brain, especially high concentrations of taurine and its synthetic enzyme, cysteine sulfinic acid decarboxylase have been found in the neostriatum (ST) and substantia nigra (SN) (7,8,17,30,31). In spite of this, very little is known about the function of taurine in the basal ganglia. The inhibitory action exerted on nigral neurons by iontophoretic application of taurine (5) and the contralateral circling behavior induced by taurine injections into the SN (12, 13,20) suggest some role for this amino acid in nigral functions.

In the ST endogenous levels of taurine are decreased following kainic acid lesions (15,22,25) suggesting that at least part of the pool of this amino acid is within neurons. Measurements of the levels of taurine in the SN following striatal lesions have led to conflicting results. Decreased levels of taurine in the SN of rats with striatal lesions produced by kainic acid led to the suggestion that a significant portion of this amino acid may be present in nerve terminals of the inhibitory striatonigral pathway (8,16). In addition to its presence in the ST a high affinity taurine uptake system has been described (4,18,19) and furthermore it has been shown to be released from the rat ST in vitro from slices (14) and cultured neurons (24) and in vivo (10). Both the uptake and release in the ST were found to be markedly reduced by prior administration of kainic acid (21,25).

The objectives of the present experiment were 1) to characterize the type of neuron in the striatum that takes up exogeneously administered radiolabelled taurine, 2) to characterize morphologically the structures that take up radiolabelled taurine in the substantia nigra, 3) to analyze the uptake mechanism and the conditions of release of exogenous taurine from slices of the SN.

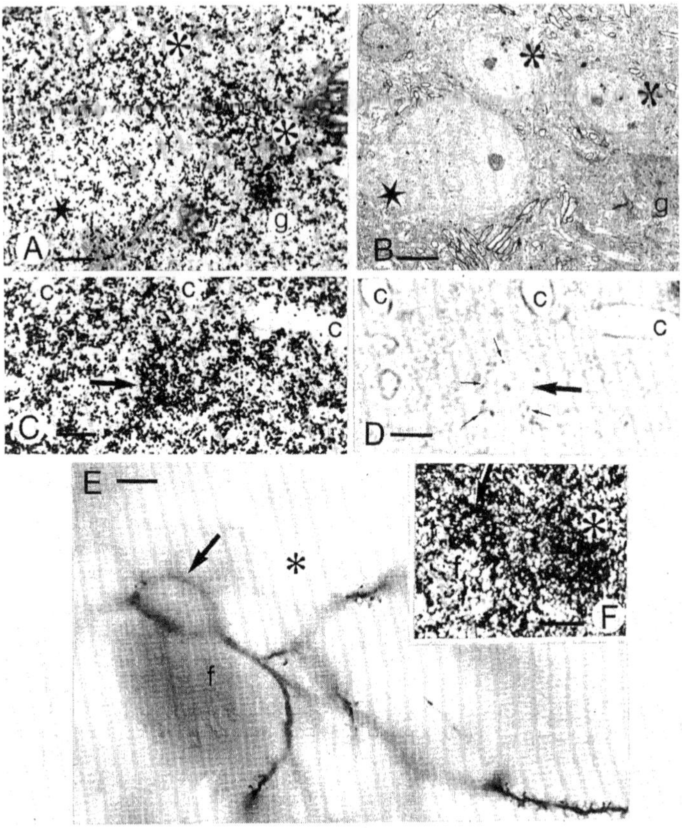

Fig. 1A: Light microscopic autoradiograph of a 1 µm section of the ST close to the site of injection of $[^3H]$taurine. Three structures have a high density of silver grains overlaying them. These structures were subsequently identified as two neurons (asterisks) and one glial cell (g) (see 1B). The star indicates a large neuron that has not accumulated the radiolabel, it is overlayed by the same density of silver grains as the surrounding neurophil. B: An electron micrograph of a section that was adjacent to the one in 1A. The structures that accumulated the radiolabel are identified as two neurons (asterisks) and a glial cell (g). The large neuron that did not accumulate the radiolabel is also present (star) and is of a different morphological class to the radiolabelled neurons. C,D: Autoradiograph of a 1 µm section of rat ST close to the injection of $[^3H]$ taurine. Only one structure shows an accumulation of the radiolabel (arrow). This section was taken from a rat that had been injected with horseradish peroxidase into the SN and the neuron that has accumulated the radiolabel is shown in micrograph 1D before the autoradiography was carried out. The neuron contains granules of the HRP reaction product and is thus identified as a striatonigral neuron. Three capillaries (C) are labelled for correlation between the two micrographs. E: Light micrograph of the perikaryon and part of the dendritic tree of a Golgi-impregnated neuron in the ST. The perikaryon is indicated by an arrow. Several darkly stained spiny dendrites are seen on the right of the micrograph. These structures are part of the same neuron but the connecting segments of dendrites are out of the plane of focus. The inset F is an autoradiograph of a 1 µm section through the perikaryon of this neuron. The Golgi-impregnated neuron (arrow) shows a selective accumulation of the radiolabel as does an unstained neuron that appears in both micrographs (asterisk). A fiber bundle (f) is labelled in both micrographs for correlation. Scale bars: A, 3 µm; B-E, 7 µm; F, 10 µm.

METHODS

Autoradiographic localization of [^{3}H]taurine in the neostriatum and substantia nigra

Rats (160-180g; female Wistar or male Sprague Dawley) were injected with 0.2 µl of a solution of [^{3}H]taurine in distilled water (40 µCi/µl, 2.85 mM) into the ST or SN. Two of the rats that received striatal injections had received injections of horseradish peroxidase conjugated with wheatgerm agglutinin (HRP/WGA) into the SN 24 h previously. Twenty to thirty minutes following the taurine injection the rats were anaesthetized and then fixed by perfusion and the brains were removed, dissected and sectioned at 70 um on a vibrating microtome. Those sections that contained injected or transported HRP were incubated with 3'3'-diaminobenzidine, the striatal sections were Golgi-impregnated and gold toned (9) and all sections were prepared for light and electron microscopy.

As controls for the selectivity of the uptake of [^{3}H] taurine, rats were injected in the ST with [^{3}H]glutamic acid (35 µCi/ul, 0.78 mM) or [^{3}H]leucine (35 µCi/µl, 0.76 mM) in the same manner as the [^{3}H]taurine.

The regions around the injection track and particularly regions in the ST that contained retrogradely-labelled and/or Golgi-impregnated neurons close to the injection track (which were previously drawn and photographed) were re-embedded for autoradiography and electron microscopy. Light microscopic autoradiography was then carried out on the 1 µm sections of the injection track and, in the case of the ST, 1 µm sections of the perikarya of the previously identified neurons (i.e. HRP-labelled or Golgi-impregnated).

To examine in the electron microscope structures that were identified in the autoradiographs of 1 µm sections, ultrathin sections were cut from the surface of the block from which the 1 µm sections had been cut. These sections were collected on Formvar-coated single slot grids, stained with lead citrate and examined in a Philips 201 electron microscope. Electron microscopic audoradiography was carried out on ultrathin sections obtained from the SN.

Preparation of brain slices for uptake and release studies

Male Sprague Dawley rats (350-550g) were killed by decapitation and the brain quickly removed. Coronal sections (0.3 mm thick) were obtained using a vibrating microtome and various brain regions were dissected out bilaterally on a chilled glass surface. The slices were weighed and then transferred to precooled vessels containing the incubation medium. The medium was a Krebs Ringer solution (normal Krebs) containing: 118 mM NaCl, 1.2 mM KH_2PO_4, 4.7 mM KCl, 1.1 mM $MgSO_4$, 2.5 mM $CaCl_2$, 25 mM $NaHCO_3$ and 11.1 mM glucose, gassed with a mixture of 5% carbon dioxide-95% oxygen. For the experiments involving perfusion with Ca^{2+}-free media, the $CaCl_2$ was omitted and $MgSO_4$ increased to 3.6 mM. The medium used in experiments involving high Mg^{2+} perfusion contained 20 mM $MgSO_4$ and 99 mM NaCl. High potassium stimulation was obtained with a medium containing 47 mM KCl and 71 mM NaCl. Within each experiment the slices obtained from one rat were randomly allocated to the designed control and test groups.

Uptake studies

Slices (0.3 mm) obtained from SN, hippocampus, cortex and ST were pre-incubated for 15 min at 37°C in 5 ml of normal Krebs with continuous

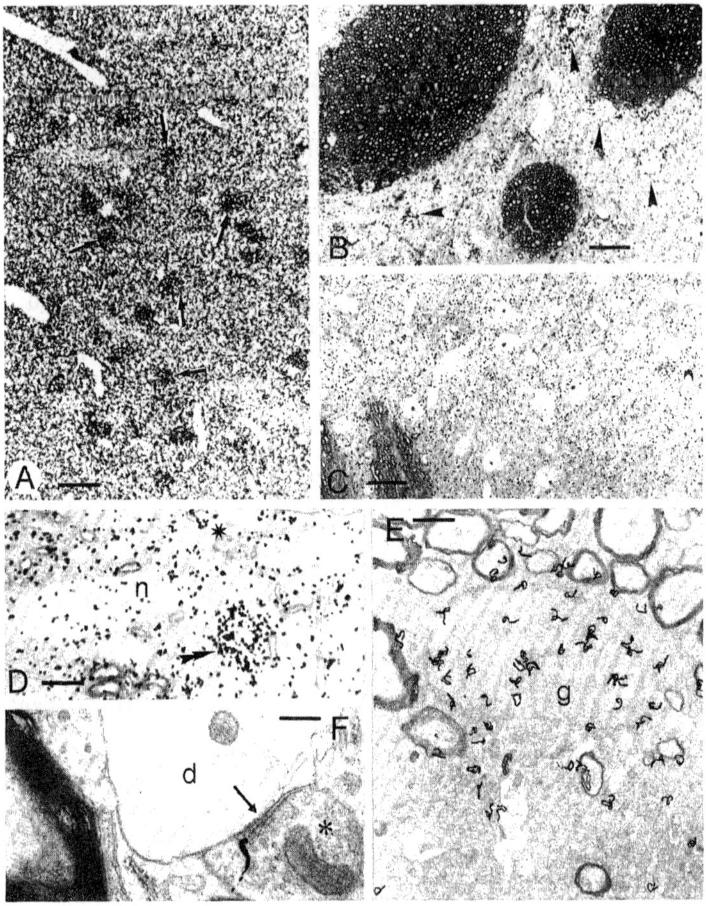

Fig. 2. A: Autoradiograph of a 1 μm section of rat ST close to the injection of [^{3}H]taurine. At this magnification the large number of neurons that accumulate the radiolabel can be seen, some of which are indicated by arrows. B: Autoradiograph of a similar area but from a rat that has been injected with [^{3}H]leucine. In this experiment all neurons (some of which are indicated by arrowheads) accumulated the radiolabel. The large dark regions are myelinated fiber bundles. C: Autoradiograph of a similar region but from a rat that had been injected with [^{3}H]glutamate. In this case there was no selective accumulation of the radiolabel. D: High power light micrograph of an autoradiograph of a 1 μm section of rat SN close to the site of injection of[^{3}H]taurine. One glial cell has accumulated the radiolabel as indicated by the high density of silver grains overlying it whereas a second glial cell (star) and a neuronal perikaryon (n) do not show any accumulation. E: Electron microscopic autoradiograph of SN close to the site of injection of [^{3}H]taurine. A glial cell (g) (probably an oligodendrocyte) has a high density of silver grains overlying it indicating a selective accumulation of the radiolabel. F: Electron microscopic autoradiograph of a region similar to that in E. The micrograph shows an axonal bouton (asterisk) in synaptic contact (arrow) with a dendritic shaft (d). Overlying the bouton is a silver grain which suggests that it may have taken up [^{3}H]taurine. Scale bars: A-C 2.5 μm; D, 6 μm; E, 0.8 μm; F, 0.3 μm.

shaking and gassing with O_2/CO_2. Preliminary experiments had shown that when tissue slice preparations were incubated in the presence of [^{3}H]taurine (2.7×10^{-7}M) the accumulation of radioactivity was linear up to 30 min. Thus, after preincubation the tissue slices were further incubated for 15 or 30 min following the addition of different concentrations of [^{3}H]taurine (3-6 μCi) and [^{14}C]leucine (1-3 μCi). The slices were then recovered and radioactivity measured. The results were either expressed as tissue:medium ratios or, after subtraction of the radioactivity accumulated in the tissue slice at 0°C, expressed as nmole.g^{-1}.h^{-1}. Values for Km and Vmax were then estimated by linear regression analysis of double reciprocal plots.

Superfusion release studies

One to three slices of SN were pre-incubated in 2 ml normal Krebs for 20 min at 37°C with continuous shaking and gassing with 5% CO_2 in O_2. [^{3}H]taurine (10 μCi; 3.5×10^{-7}M) and [^{14}C]leucine (5 μCi; 3.5×10^{-7} M) were then added and the incubation was continued for 30 min to allow the uptake of the labelled amino acids. At the end of the incubation the slices were rinsed with warm medium, transferred to superfusion chambers and superfused at a rate of 0.3 ml.min^{-1} at 37°C. After a washing period of 40 min, 2 min fractions were collected. When 4-6 fractions had been collected the superfusion medium was replaced by stimulating solutions for a period of 4 min.

The radioactivity released in each fraction and that remaining in the slices was counted. The results were expressed as a percentage of the prestimulation release (26). Fractional release figures were obtained for each collection period, this being the amount of radioactivity released during that collecion period divided by the total radioactivity remaining in the slice at the start of that period. For each chamber the mean of the initial 4-6 fractional releases was calculated to obtain a pre-stimulation mean baseline release. Each of the subsequent calculated fractional releases was then divided by this prestimulation mean and multiplied by 100 to give the fractional release as a percentage of basal outflow.

RESULTS

Autoradiographic localization of [^{3}H]taurine in the striatum

Autoradiographs of the injection site of [^{3}H]taurine contained a high concentration of silver grains around the capillary track (Fig. 2A). Up to about 500 μm from the injection site there were numerous neurons that showed a selective accumulation of the radioactivity interspersed with a smaller proportion that did not show any accumulation (Fig. 1A, 1B). 'Close to the injection site, the labelled neurons comprised up to 75% of the total population. Within this same area glial elements were also radiolabelled. The perikarya of 77 radiolabelled neurons were examined in the electron microscope. The neurons had round or oval unindented nuclei surrounded by a thin rim of cytoplasm. These features are characteristic of medium-sized densely-spiny neurons (27-29) (Fig. 1B). Eleven neurons that did not accumulate the [^{3}H]taurine were also examined in the electron microscope. These neurons varied from medium-size (i.e. the same size as those neurons showing accumulation of the radioactivity) to large neurons (Fig. 1A, 1B) and all possessed ultrastructural features that were markedly different from those of the taurine-accumulating cells. The most noticeable difference was that the nucleus of each nonaccumulating neuron was indented.

Autoradiographic localization of [^{3}H]taurine in identified striatonigral neurons or Golgi-impregnated neurons

Eight neurons that were retrogradely-labelled following injection of HRP/WGA into the SN were also processed for autoradiography. Of these eight neurons, five showed an accumulation of [^{3}H]taurine (Figs. 1C and 1D), the remaining three were beyond the region of neuronal labelling. Seven Golgi-impregnated medium-sized spiny neurons were drawn using a camera lucida and photographed prior to re-embedding. Autoradiographs of 1 μm sections of the perikarya of those neurons were prepared and five of them were found to selectively accumulate the radiolabel (Fig. 1E and 1F).

Control experiments

The experiments with [^{3}H]glutamic acid and [^{3}H]leucine gave distinctly different patterns of perikaryal labelling. There was no selective accumulation of [^{3}H]glutamic acid into neurons (Fig. 2C) whereas the [^{3}H]leucine was accumulated in a nonspecific manner into all neurons (Fig. 2B).

Autoradiographic localization of [^{3}H]taurine in the substantia nigra

As in the ST, autoradiographs of the site of injection of [^{3}H]taurine in the SN contained high concentrations of silver grains extending to a distance of 300-400 μm. Selective accumulations of the silver grains were clearly visible. However, in contrast to the ST, neuronal perikarya were not labelled, only glial cells showing accumulations of the radiolabel (Fig. 2D). Nonaccumulating glial cells were also seen (Fig. 2D).

In confirmation of the light microscope observations (Fig. 2D), electron microscopic autoradiographs revealed that the highest grain densities were over glial cells. Most examples were astroglia or pericytes, although several oligodendrocytes (Fig. 2E) were also radiolabelled. Nonlabelled astroglia and oligodendrocytes were also observed. In addition to the glial cells, axons and axon terminals were also radiolabelled. The [^{3}H]taurine-accumulating boutons (Fig. 2F) were large (approximately 1 μm diameter) with many large round or pleomorphic synaptic vesicles. Some of them formed symmetrical synaptic specializations with dendrites (Fig. 2F). Very few grains were found over dendrites whereas neuronal cell bodies were not labelled at all. Preliminary results from the quantitative analysis of electron microscope autoradiographs according to the "hypothetical grain" analysis method of Blackett and Parry (2) indicate that in the neuropil 31% of the radioactivity could be attributed to glial processes, 58% to axonal elements, 11% to dendrites and none to neuronal cell bodies. Glial processes showed the highest specific activity and dendrites the lowest.

Uptake of [^{3}H]taurine in tissue slices from various regions of the rat brain

The accumulation of [^{3}H]taurine was measured simultaneously with the accumulation of [^{14}C]leucine which is presumed not to be a neurotransmitter. As shown in Fig. 3, slices from the four areas of the rat CNS accumulated both radiolabels following incubation in a medium containing the two amino acids. In each area studied [^{3}H]taurine accumulation was consistently greater than that of [^{14}C]leucine although there were considerable regional differences in the uptake of radioactive taurine. Slices of SN and ST showed a much higher capacity to accumulate taurine than did slices of cortex and hippocampus.

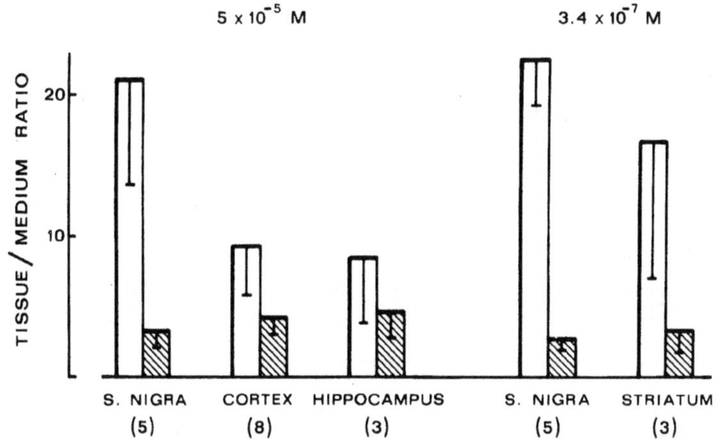

Fig. 3. Accumulation of [^{3}H]taurine (white columns) and [^{14}C]leucine (dashed columns) in rat brain slices following 30 min incubation at 37°C in a medium containing the two radiolabelled amino acids at the indicated concentrations. Tissue/medium ratio represents dpm.min^{-1}.g^{-1} of slice wet weight divided by dpm.min $^{-1}$.ml^{-1}. Bars represent 95% confidence limits of the ratio calculated by the Fieller's theorem (11). Numbers in brackets represent number of experiments.

Preliminary experiments on the dependence of the [^{3}H]taurine rate of uptake on substrate concentration were performed. The "initial rate" of accumulation of this amino acid into slices was measured at 3-5 different substrate concentrations in the range 0.066-5.0 µM. An estimate of the Kinetic parameters Km and Vmax was obtained from linear regression analysis of the Lineweaver-Burk plots. Apparent Km values were below 0.1 mM in all four brain areas studied. Maximal velocity values were found to be in the range 50-100 nmol.g^{1-}.h^{-1} in cortex and hippocampal slices; however, as anticipated fom the differences in tissue:medium ratios shown in Fig. 3, Vmax values 10 times higher were observed when measured in ST and SN nigra slices.

Release of preloaded [^{3}H]taurine from substantia nigra slices

Release of [^{3}H]taurine previously accumulated by SN slices was studied simultaneously with the release of preloaded [^{14}C]leucine to test its specificity. As it has been observed in other superfused CNS tissues, there was a rapid initial efflux of labelled taurine and leucine. A stable basal release was reached approximately 30-40 min after initiation of superfusion. Samples were collected therefore after a 40 min wash out period. At this time the amount of radioactivity remaining in the tissue was 88% and 30%, respectively, of the radioactivity taken up during the incubation period with [^{3}H]taurine and [^{14}C]leucine; thereafter basal [^{14}C]leucine efflux remained significantly higher than taurine efflux (see Fig. 4). Superfusion for 4 min with 50 µM veratridine in the presence of Ca^{2+} produced a 4-13 fold increase in the rate of efflux of preloaded [^{3}H]taurine. The time following stimulation when [^{3}H]taurine efflux reached a peak varied from as early as 4 min when the stimulus was still being applied, up to 40 min later.

Mean increases in fractional efflux of radiolabelled taurine and leucine over prestimulation values with time are shown in Fig. 4. The veratridine-stimulated release did not show any sign of return to prestimulaton values

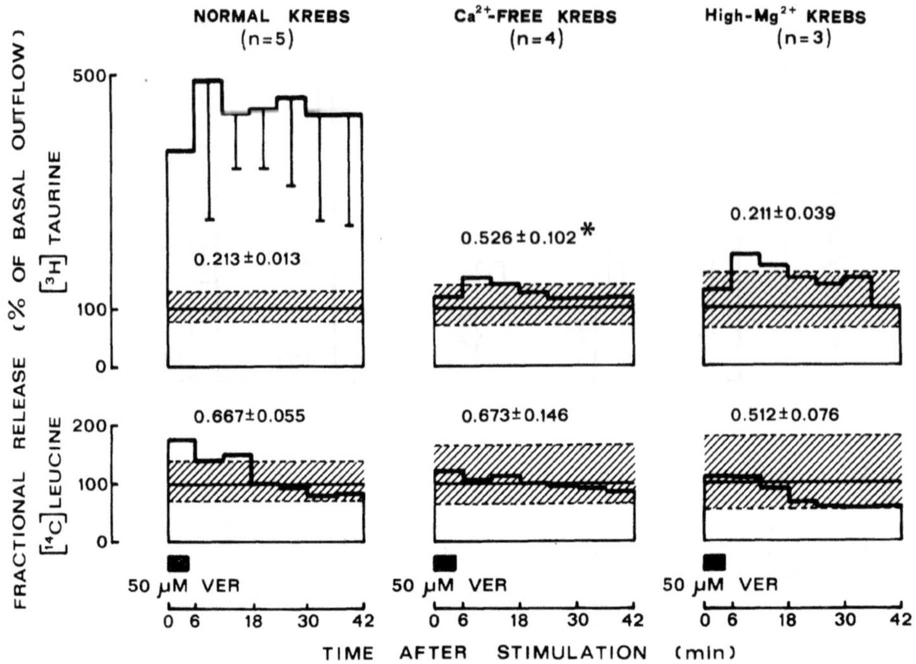

Fig. 4. Time profiles of mean increases in fractional efflux of preloaded [^{3}H]taurine and [^{14}C]leucine from SN slices induced by veratridine under different experimental conditions. Each segment represents the mean value obtained from 3 consecutive 2 min fractions. Basal prestimulation values are indicated by the 100% line, the hatched areas or bars representing 95% confidence intervals or limits of the ratio. Figures (± SEM) represent absolute values of prestimulation fractional release x 100. * Significantly ($P<0.05$) different from the value obtained under normal Krebs superfusion.

for at least 40 min after removal of the stimulus, thus indicating a very slow recovery from veratridine depolarization. The release of [^{14}C]leucine was not significantly increased. The veratridine-stimulated [^{3}H]taurine efflux was abolished in the absence of Ca^{2+} and in the presence of high Mg^{2+}. The basal efflux of labelled taurine was significantly increased when the slices were superfused with calcium-free medium (Fig. 4). Furthermore the veratridine-stimulated [^{3}H]taurine release was abolished by tetrodotoxin; high potassium also enhanced [^{3}H]taurine release in a Ca^{2+}-dependent manner (data not shown).

DISCUSSION

Further to our previous work (3), the combination of Golgi impregnation with the local uptake of [^{3}H]taurine suggests that in the ST the type of neurons that selectively accumulate radioactive taurine are medium-sized densely spiny striatonigral neurons. In the SN, the observation that, apart from glial cells and processes, axonal elements were preferentially radiolabelled rather than dendrites or neuronal cell bodies, suggests that afferent terminals rather than local terminals become labelled, possibly arising from stratonigral neurons.

Results obtained from experiments with brain slices are in keeping with the suggestion that taurine is present in striatonigral neurons. A high affinity system for the accumulation of exogenous taurine was confirmed to be present in ST slices, and the presence of a system characterized by similar kinetic parameteres was demonstrated in SN slices. Furthermore, preloaded [^{3}H]taurine was shown to be released from SN slices by veratridine in a Ca^{2+}-dependent and tetrodotoxin-sensitive manner.

Since intranigral injections of taurine elicit contralateral rotation in rats, and in electrophysiological experiments in an inhibitory effect of taurine on nigral neurons has been observed, it is possible that taurine plays a neurotransmitter or neuromodulator role in this striatonigral pathway.

ACKNOWLEDGMENTS

L.D.C. was variously supported by a PGT research fellowship (1985) from the European Science Foundation, a NATO Senior Fellowship, the Consiglio Nazionale delle Ricerche (Art. 20/2-7. 26/1/67) and MPI (60%), Roma. Drs. D. Parry, and T. Hattori are acknowledged for their kind help with the analysis of the electron microscopic autoradiographs, and M. Appleyard for assistance in the release studies.

REFERENCES

1. Barbeau, A., Inoue, N., Tsukada, A., and Butterworth, R.F., 1975, The neuropharmacology of taurine, Life Sci., 17:669-678.
2. Blackett, N.M., and Parry, D.M., 1977, A simplified method of "hypothetical grain" analysis of electron microscope autoradiographs, J. Histochem. Cytochem., 25:206-214.
3. Clarke, D.J., Smith, A.D., and Bolam, J.P., 1983, Uptake of [^{3}H]taurine into medium-size neurons and into identified striatonigral neurons in the rat neostriatum, Brain Res., 289:342-348.
4. Collins, G.G.S., 1974, The rates of synthesis, uptake and disappearance of [^{14}C]-taurine in eight areas of the rat central nervous system, Brain Res., 76:447-459.
5. Davies, J., and Dray A., 1976, Changes in amino acid sensitivity during polypeptide 'desensitisation', Nature, 262:606-607.
6. Davison, A.N., and Kaczmarek, L.K., 1971, Taurine - a possible neurotransmitter?, Nature, 234:107-109.
7. Dray, A., and Straughan, D.W., 1976, Synaptic mechanisms in the substantia nigra, J. Pharm. Pharmac., 28:400-405.
8. Emson, P.C., 1978, "Biochemical and metabolic changes in epilepsy" in: "Taurine in Neurological Disorders", R. J. Huxtable, and A. Barbeau eds., Raven Press, New York, pp. 319-338.
9. Freund, T.F., and Somogyi, P., 1983, The section-Golgi impregnation procedure. 1. Description of the method and its combination with histochemistry after intracellular iontophoresis or retrograde transport of horseradish peroxidase, Neuroscience, 9:463-474.
10. Girault, J.A., Barbeito, L. Spampinato, U., Gozlan, H., Glowinski, J., and Besson, M-J., 1986, In vivo release of endogenous amino acids from the rat striatum: further evidence for a role of glutamate and asparate in corticostriatal neurotransmission, J. Neurochem., 47:98-106.
11. Goldstein, A., Biostatics, an Introductory Text, Macmillan, New York (1964).
12. Kaakkola, S., and Kaariainen, I., 1980, Contralateral circling behavior induced by intranigral injection of taurine in rats, Acta Pharmacol. Toxicol., 46:293-298.

13. Kamata, K. Kameyana, T., Okuyama, S., Hashimoto, S., and Aihara, H., 1985, Contralateral circling behavior induced by intranigral microinjections of taurine and GABA in rats, Brain Res., 343:275-282.
14. Katz, R.I., Chase, T.N., and Kopin, I.J., 1969, Effect of ions on stimulus induced release of amino acids from mammalian brain slices, J. Neurochem., 16:961-967.
15. Korf, T., and Venema, K., 1983, Amino acids in the substantia nigra of rats with striatal lesions produced by kainic acid, J. Neurochem., 40:1171-1173.
16. Lindgren, S., and Anden, N.E., 1985, Effect of the normal nerve impulse flow in the synthesis and utilization of GABA in the rat substantia nigra, J. Neurol. Transmission, 61:21-34.
17. Lombardini, J.B., 1976, Regional and subcellular studies on taurine in the rat central nervous system. in: "Taurine", R.J. Huxtable, and A. Barbeau, eds., Raven Press, New York, pp. 311-326.
18. Lombardini, J.B., 1977, High affinity uptake systems for taurine in tissue slices and synaptosomal fractions prepared from various regions of the rat central nervous system. Correction of transport data by different experimental procedures, J. Neurochem., 29:305-312.
19. Mandel, P., and Pasantes-Morales, H., 1978, Taurine in the nervous system, Rev. Neurosci., 3:157-193.
20. Martin, G.E., Benedesky, R.J., and Williams, M., 1981, Further evidence for selective antagonism of taurine by 6-aminomethyl-3-methyl-4H-1,2,4-benzothiadiazine-1,1-dioxide, Brain Res., 229:530-535.
21. Meiners, B.A., Speth, R.C., Bresolin, N., Huxtable, R.J., and Yamamura, H.I., 1980, Sodium-dependent, high affinity taurine transport into rat brain synaptosomes, Federation Proc., 39:2695-2700.
22. Nicklas, W.J., Duvoisin, R.C., and Berl, S., 1979, Amino acids in rat neostriatum: alterations by kainic acid lesions, Brain Res., 167:107-117.
23. Oja, S.S., and Kontro, P., 1978, Neurotransmitter actions of taurine in the central nervous system. in: "Taurine in Neurological Disorders", A. Barbeau and R.J., Huxtable eds., Raven Press, New York, pp. 181-200.
24. Pin, J.P., Weiss, S., Sebben, M., Kemp, D.E., and Bockaert, J., 1986, Release of endogenous amino acids from striatal neurons in primary culture, J. Neurochem., 47:594-603.
25. Placheta, P., Singer, E., Schonbeck, G., Heckl, K., and Karobath, M., 1979, Reduction of endogenous level, uptake and release of taurine after intrastriatal kainic acid injection, Neuropharmacology, 18:339-402.
26. Pycock, C.J., and Smith, L.F.P., 1983, Interactions of dopamine and the release of [^{3}H]-taurine and [^{3}H]-glycine from the isolated retina of the rat, Br. J. Pharmacol., 78:395-404.
27. Somogyi, P., Bolam, J.P., and Smith, A.D., 1981, Monosynaptic cortical input and local axon collaterals of identified striatonigral neurons. A light and electron microscopic study using the Golgi-peroxidase transport-degeneration procedure, J. Comp. Neurol., 195:567-584.
28. Somogyi, P., Hodgson, A.J., and Smith, A.D., 1979, An approach to tracing neuron networks in the cerebral cortex and basal ganglia. Combination of Golgi staining, retrograde degeneration of synaptic boutons in the same material, Neuroscience, 4:1805-1852.
29. Somogyi, P., and Smith, A.D., 1979, Projection of neostriatal spiny neurons to the substantia nigra. Application of a combined Golgi-staining and horseradish peroxidase transport procedure at both light and electron microscopic levels, Brain Res., 178:3-15.
30. Spears, R.M., and Martin, D.L., 1982, Resolution and brain regional distribution of cysteine sulphinate decarboxylase isoenzyme from hog brain, J. Neurochem., 38:985-991.
31. Wu, T.Y., Bird, E.D., Chen, M.S., and Huang, W.M., 1979, Abnormalities of neurotransmitter enzymes in Huntington's chorea, Neurochem. Res., 4:575-586.

BASAL CONCENTRATION AND EVOKED CHANGES OF EXTRACELLULAR TAURINE IN THE RAT HIPPOCAMPUS IN VIVO

R.M. del Rio, A.S. Herranz, J.M. Solis, O. Herreras and J. Lerma

Depto. Investigacion, Hosp. "Ramón y Cajal", Crta. Colmenar Km 9, 28034-Madrid, Spain

The brain extracellular fluid constitutes the microenvironment on which specific nervous tissue functions are highly dependent (27). Thus its composition should be strictly maintained. Cellular processes of secretion and uptake as well as the homeostatic mechanisms of blood and cerebrospinal fluid (CSF) probably contribute to buffering changes in the extracellular fluid composition evoked by cellular activities. The extracellular space is also the main medium of communication between nerve cells through which their chemical signals are transmitted. For these reasons, many studies have been performed to determine the "in vivo" extracellular fluid concentrations of neuroactive substances such as catecholamines, their metabolites (5) and ions (9), during basal and altered cellular activities, since such information will improve significantly our understanding of CNS physiology.

The release of endogenous neuroactive amino acids into extracellular fluid after chemical or electrical stimulation has been described using cortical cup (25) or push-pull (30,37,39) methods, but no suitable procedure has been available to measure actual extracellular fluid concentrations of these compounds under physiological conditions. Recently, based on the dialytrode technique (3,4), a method for brain dialysis has been developed which permits not only detection of different amino acids present in the extracellular fluid but also determination of their precise concentration (12,18).

The amino acid taurine has been proposed as an inhibitory transmitter in the CNS (28), and is believed to be the transmitter released from stellate cells of the cerebellum (24). The high concentration of this amino acid in the hippocampus (e.g. 22) indicates its possible role in inhibitory phenomena in this structure. In addition, among other neuroactive amino acids, taurine has been reported to be present in the extracellular fluid of several structures in the CNS (7,19,31,36) at moderate concentrations compared with other physiological fluids (blood, CSF) (18). The present study describes basal extracellular fluid taurine concentration in the "in vivo" rat hippocampus as well as changes evoked by several cell microenvironment modifications. Results are discussed on the basis of a functional model.

IN VIVO DETERMINATION OF EXTRACELLULAR FLUID BASAL CONCENTRATION OF AMINO ACIDS

The precise extracellular concentration of several amino acids was determined by means of computerized nonlinear regression analysis of brain dialysis data, using a simple diffusion model (18). The dialysis device consisted of a Cuprophan hollow fiber, 200 µm i.d., bent into a "U" shape loop. A steel wire, 140 um thick, was inserted into the fiber to facilitate bending. The fiber, isolated with cyanocrilate, was introduced into a 22G stainless steel cannula and sealed with epoxy, forming a "U" loop, 2.05-2.4 mm long, as the active surface with an internal volume of 0.033-0.038 µl. The dialysis device was implanted in the dorsal dentate gyrus of the hippocampal formation in urethane-anesthetized rats and perfused with a Krebs-Ringer bicarbonate solution by means of a peristaltic pump operating at different flow rates. The time that a specific volume remained in the loop during perfusion was considered dialysis time (t_d) and varied inversely proportional to the flow rate. In order to increase t_d, the perfusion liquid was in some cases circulated for long periods in a closed system; i.e. after connecting the outlet and inlet of the perfusion line.

Assuming that compounds cross the dialysis membrane by simple diffusion and that their extracellular fluid concentration remains constant around the dialysis probe, so that the system is in a steady state, concentration variation through time of any substance in the perfusate can be expressed as:

$$dc/dt=K(A-c)$$

where A is the basal extracellular concentration of any compound, c is its perfusate concentration at t time and K represents an apparent diffusion constant through the dialysis membrane. The resulting integrated form of the above equation is:

$$c=A(1-e^{-Kt})$$

where t corresponds to the dialysis time (t_d). As can be seen in this model, at the beginning of the perfusion ($t_d=0$), c is 0. c will attain the A value after prolonged t_d (i.e. $t_d \to \infty$). A then corresponds to the asymptote of the function.

To determine the parameters of this equation we used the Gauss-Newton method, which consists of an optimization procedure for determining the parameter values of a nonlinear model from experimental data. This method assigns arbitrary initial values to the function parameters and iteratively minimizes the mean squares from a first order Taylor's series approximation to the considered function (see ref. 33). We then used a computer program which by this method arrives at the best estimates of unknown parameter values, calculates the corresponding standard errors and concomitantly performs its variance analysis (33).

Perfusate samples were analysed for amino acid content using reversed-phase HPLC with the O-phthalaldehyde (OPA) precolumn derivatization procedure, as in previous studies (19,20,21), except that 3-mercaptopropionic acid was used instead of 2-mercaptoethanol in the derivatization reaction. This change improved the analysis sensitivity as much as two-fold (10). Tissue amino acids were determined by ion-exchange chromatography using an automatic amino acid analyzer, after their extraction with 8 vols of 5% sulfosalicylic acid solution.

Nonlinear regression analyses of experimental values obtained in the "in vivo" basal situation were performed for the following amino acids: aspartate, glutamate, serine, glutamine, arginine, taurine, alanine and GABA.

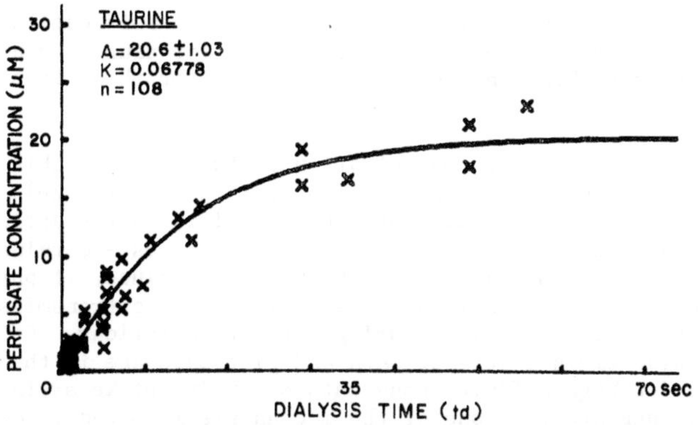

Figure 1. Computer nonlinear regression analysis of "in vivo" brain dialysis taurine data. The estimated figures for equation parameters (A± SEE in μM: K in s^{-1}) are indicated at the top; n is the number of samples analyzed from pooled data of 21 animals.

Fig. 1 shows the progressive increase of perfusate taurine concentration with dialysis time. Single experimental values of c are indicated by "x" while the solid line corresponds to the theoretical curve which best fits the experimental data ($p < 0.0001$, ANOVA of regression test). The values found for A and K should therefore be considered the most likely for both the taurine extracellular fluid concentration and its apparent diffusion constant through the dialysis membrane, respectively. Table I shows the extracellular fluid concentration (A) of several amino acids, as estimated by computer analysis from the rat dentate gyrus data, as well as the hippocampal tissue content (Ts). The amino acid profile of the extracellular fluid compartment was quantitatively and qualitatively different from that of tissue. The hippocampal amino acid content can be considered to represent the intracellular amino acid pool (see ref. 8). Taking tissue density as 1.0, the amino acid tissue content expressed as μmol/g of fresh tissue may be defined as concentration units (μM), from which the intra/extracellular amino acid concentration ratios can be obtained (Table I, Ts/A). Ratio values in the dentate gyrus were extremely high for those amino acids considered putative neurotransmitters (>2000), while nonneuroactive amino acids presented moderate ratios (<100) and the concentration ratio of taurine was intermediate. Considering both high intra/extracellular "in vivo" ratios and

Table I. Amino acid concentrations in the dentate gyrus extracellular fluid (A) and hippocampal tissue (Ts) and the corresponding concentration ratios (Ts/A)

	A (μM)	Ts (μmol/g)	Ts/A
Aspartate	1.7±0.22	3.59±0.13	2111
Glutamate	2.9±0.38	11.80±0.38	4069
Serine	24.9±1.42	1.05±0.03	42
Glutamine	193.4±10.9	5.08±0.19	26
Arginine	6.9±0.52	0.09±0.01	13
Taurine	20.6±1.03	8.43±0.57	409
Alanine	7.7±0.51	0.71±0.03	92
GABA	0.8±0.15	2.08±0.15	2600

Values are expressed as the means ± SEM of 5 animals for tissue data. A corresponds to the computer estimated values obtained by nonlinear regression analysis of data pooled from 21 animals.

low extracellular level as additional criteria for classification of any substance as a neurotransmitter (18), the figures for taurine would indicate a neuromodulator rather than a neurotransmitter role for this amino acid in the rat dentate gyrus.

Basal extracellular fluid concentration of the most probable amino acid neurotransmitters in the dentate gyrus (i.e. GABA and glutamate) are much higher than the reported K_d values for their specific postsynaptic receptors (23,38). Therefore, it appears that K_d values - obtained in vitro - are meaningless for the in vivo situation, because, if physiologically valid, they would imply a blockade in synaptic transmission which did not occur (see refs. 19-21,34) during our determinations. On the other hand, extracellular fluid levels of GABA and glutamate are in the range of reported values for high affinity constants of their uptake systems (1,11). This similarity suggests that one of the mechanisms involved in extracellular fluid homeostasis is the transport of these amino acids through specific carriers into cellular structures, indicating that high affinity uptake systems are determining the extracellular fluid levels of amino acid neurotransmitters. Therefore, it could be hypothetized that low extracellular fluid levels of amino acid neurotransmitters should be increased by inhibiting their respectives high affinity uptake systems. A concomitant change in the population neuronal excitability should also be expected.

EVOKED CHANGES OF EXTRACELLULAR TAURINE LEVELS IN THE IN VIVO RAT HIPPOCAMPUS

In order to test the above hypothesis, the effects of perfusion with different solutions, including nipecotic and dihydrokainic acids, GABA or moderately elevated [K^+], on the extracellular fluid levels of amino acids were evaluated. In this set of experiments, field potentials, evoked by perforant pathway stimuli and recorded through a 60 um electrode incorporated to the dialysis probe, served to monitor simultaneously the response of dentate gyrus cells to monosynaptic activation.

Perfusion with several nipecotic acid concentrations greatly enhanced the extracellular levels of taurine and GABA in a dose-dependent manner, simultaneously decreasing the dentate gyrus response to excitatory afferent volleys (Fig. 2). However, the extracellular fluid taurine concentration

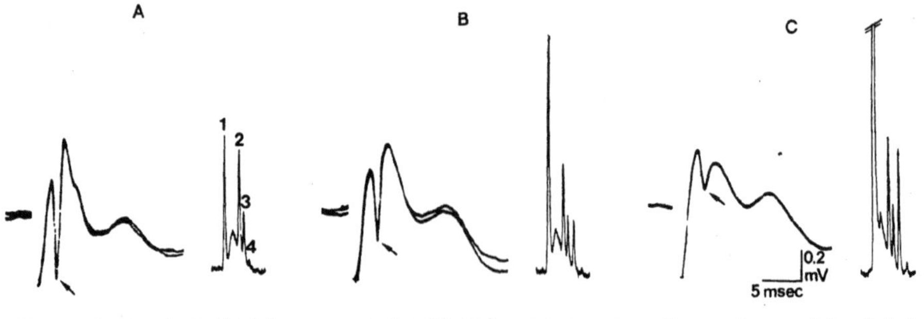

Figure 2. Evoked field potentials (left) and spectra for amino acids (right) during perfusion with (A) Krebs-Ringer bicarbonate, (B) 5 mM and (C) 20 mM nipecotic acid solutions. The population spike is indicated by an arrow. The levels of taurine (1) and GABA (4) increased in the presence of nipecotic acid, while β-alanine (2) and tyrosine (3) did not change. From ref. 20.

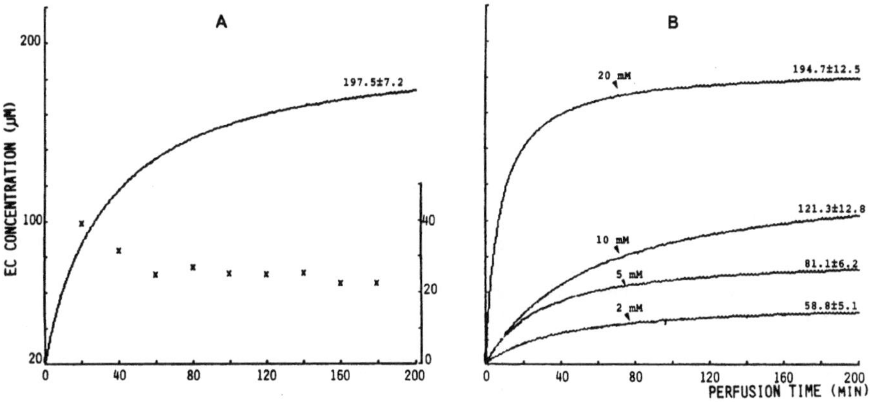

Figure 3. Nonlinear regression analysis of extracellular taurine concentrations obtained during long perfusion periods of (A) 20 mM nipecotate or (B) several concentrations of GABA. The concomitant enhancement of extracellular GABA levels induced by nipecotate is showed in A by "x" as the mean of two animals. Figures on the curves refer to the asymptotic values of the function.

increased with time reaching a plateau, whereas extracellular GABA concentration rose to a maximum and then decreased to a lower steady level. Figure 3A shows the nonlinear regression curve of data obtained from long perfusion periods of 20 mM nipecotate solution. The extracellular fluid concentrations of taurine attained were 53.0±3.3, 85.7±7.1 and 197.5±7.2 uM for 2, 5 and 20 mM nipecotate respectively.

Since GABA enhancement by nipecotic acid apparently occurs prior to that of taurine (7), we tested the possibility that the taurine increase was caused by elevated GABA in the extracellular fluid. Several concentrations of GABA were perfused while measuring extracellular taurine concentration and as seen in Fig. 3B, the extracellular concentrations of taurine were 58.8±5.1, 81.1±6.2, 121.3±12.8 and 194.7±12.5 μM for 2, 5, 10 and 20 mM GABA respectively. These taurine figures did not differ from those obtained when nipecotic acid was perfused and, in both cases, the dose-response effect was rather linear.

While studying the effect of inhibited uptake of glutamic acid, considered the neurotransmitter of the main pathways in the hippocampal formation (35), on the extracellular amino acid levels and neuronal excitability in the hippocampus, we found that dihydrokainic acid, a potent inhibitor of high affinity glutamate uptake (13), increased the extracellular levels not only of glutamate but also of taurine. Fig. 4 shows the time course changes in the dialyzate concentration of taurine, glutamate and glutamine evoked by perfusion of 5 mM dihydrokainate through dialysis tubes implanted either in the dentate gyrus (top) or CA1 field (bottom). In the dentate gyrus, dihydrokainate enhanced the extracellular concentration of glutamate 2.5-fold and taurine two-fold, while glutamine levels remained unaffected. Increased levels returned to control values after dihydrokainate withdrawal, this remission being slower for taurine. In CA1, the glutamate increase was similar, whereas taurine concentration rose almost 400% above the control figure.

The exposure of the CNS to high concentrations of K^+ (40-100 mM) evokes the efflux of putative amino acid neurotransmitters, including glutamate, GABA and taurine. It is well known, however, that extracellular K^+ concentration ($[K^+]_o$) increases up to 12 mM as a consequence of neuronal activity (9,27) and that these elevated concentrations (6-12 mM) enhance the amplitude of hippocampal evoked field potentials (6). We therefore studied the possible influence of physiological $[K^+]_o$ on extracellular amino acid levels, particularly that of taurine, in the in vivo dentate gyrus of the rat. After four 15 min perfusion samples were collected, perfusate KCl concentration was increased (with equimolar reduction of NaCl) up to 100 mM. Figure 5 shows the significant and specific increase in extracellular taurine concentration evoked by perfusion with successive KCl concentrations, introduced in 3 mM steps up to 12 mM. In another group of similarly treated animals, perfusate $[K^+]$ was elevated up to 100 mM. Taurine levels increased with $[K^+]$ reaching a plateau at 25 mM (Fig. 5). Subsequently perfused higher $[K^+]$ (i.e. 50 and 100 mM) evoked increased in both glutamate and GABA levels (not shown) but did not further increase the taurine levels. Population spikes (Fig. 5, bottom) rose in parallel with increased $[K^+]$ in the perfusate, with the maximum amplitude at 25 mM. Higher $[K^+]$ markedly reduced the evoked potential amplitude, preventing cell discharge.

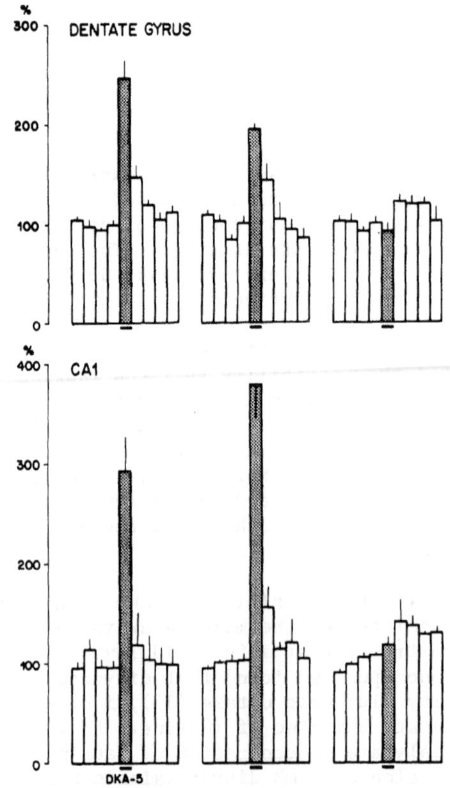

Figure 4. Time course changes of the dialyzate concentration of amino acids evoked by dihydrokainate. Means $\pm$ SEM are expressed as percentages, taking the average concentration value of controls as 100%. After 1 h of control perfusion, dihydrokainate (5mM) was introduced for 15 min (shaded bar) and fresh Krebs-Ringer bicarbonate buffer was then substituted. After a washing period of 15 min, four dialysate samples were taken. From ref. 26.

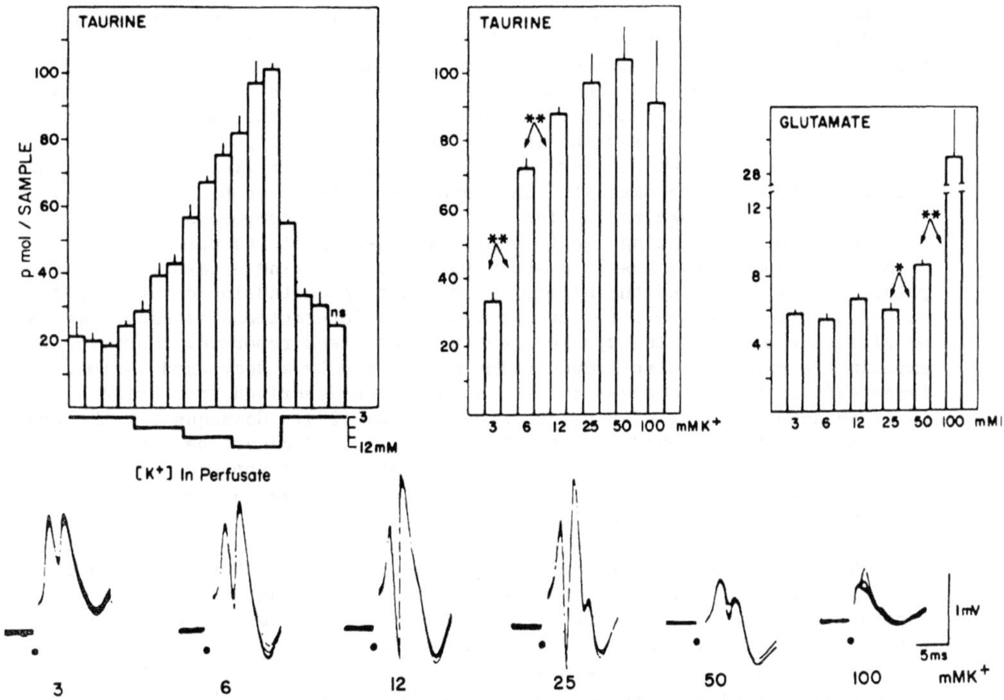

Figure 5. Effects of different perfusate $[K^+]$ on the extracellular levels of taurine and glutamate (top), and dentate gyrus evoked field potentials (bottom). Perfusate $[K^+]$ was sequentially varied as indicated at the bottom of histograms. Bars of the left histogram represent the Mean $\pm$ SEM in 15 min samples (n=4). In intermediate and right histograms, each bar corresponds to the Mean $\pm$ SEM of the dialyzate levels in the last sample of every set at indicated $[K^+]$. Field potential records correspond to six superimposed traces on the storage screen of an oscilloscope at each $[K^+]$ studied. Student t-test, *$p<0.05$; **$p<0.01$. From ref. 34.

DISCUSSION

The present results demonstrate that the hippocampal extracellular fluid compartment has a characteristic amino acid composition, different from that of tissue, in which putative amino acid neurotransmitters have very low concentrations and present very high intra/extracellular concentration ratios. As recently suggested (18), these facts could be considered as additional criteria in supporting their neurotransmitter role. Therefore, brain dialysis permits evaluation of steady state concentrations in the extracellular fluid of endogenous compounds by nonlinear regression analysis as well as determination of the extent of their evoked changes.

Since amino acid neurotransmitters specifically increased in the extracellular fluid when inhibitors of their high affinity uptake systems were perfused (i.e. nipecotic acid for GABA; dihydrokainate for glutamate), respectively decreasing (Fig. 2) or augmenting (26) the excitability, we can conclude that amino acid neurotransmitter inactivation occurs through uptake processes which, by maintaining extracellular concentrations at certain levels, modulate neuronal excitability.

While the increase of GABA and glutamate content caused by their uptake inhibitors was an expected result, the concomitant rise in extracellular taurine concentration constitutes a finding which demands further explanation, inasmuch as the endogenous compounds GABA (Fig. 3B; see also ref. 19) and glutamate (unpublished results) themselves also evoke the enhancement of extracellular levels of taurine in the "in vivo" rat hippocampus. The augmented concentration of taurine in the extracellular fluid after perfusion with nipecotate, dihydrokainate, GABA and glutamate is probably caused by an increased efflux/uptake ratio, indicating that these compounds may be either taurine-efflux-stimulating agents and/or inhibitors of its uptake. For example, although GABA has been shown to inhibit taurine uptake (17), but rather weakly (15), and the nipecotic acid action on taurine can be explained as mediated by GABA (20), to our knowledge, no interactions of dihydrokainate or glutamate with taurine uptake processes have been reported. Moreover to explain the extracellular accumulation of taurine by a decreased inflow, a large efflux would be necessary and, as has been described from brain slices, the spontaneous efflux of taurine is quite slow (29). Consequently the great increase of taurine found in the extracellular fluid cannot be explained only on the basis of an impairment of taurine uptake and it appears more likely that all these compounds stimulate the release "in vivo" of taurine by means of a still undetermined mechanism.

The alterations of extracellular basal concentration of taurine evoked by nipecotic acid and/or GABA can be explained by a counter-transport process (anti-port) mediated by a carrier of GABA, which, as has been proposed, is mobile (14), bidirectional (32) and able to transport both GABA and taurine (16). Hypothetically, GABA would bind to the carrier loci faced to the extracellular space, while the portion of carrier pointing toward the intracellular compartment would incorporate cytoplasmic taurine. Then, GABA would be internalized at the same time that taurine is transported to the extracellular fluid compartment. Intracellularly, where the $[K^+]$ concentration is high compared with that of Na^+, recognition of taurine by this common carrier should be predominant over GABA binding. The model explains the fact that enhancement of extracellular taurine levels follows that of GABA when nipecotic acid is perfused.

Perfusion with the glutamate uptake inhibitor dihydrokainate also elevates the extracellular levels of taurine. This phenomenon is probably mediated by glutamate because dihydrokainate also enhanced the extracellular concentration of this amino acid and glutamic acid alone was able to elicit similar increments in extracellular fluid taurine. Since this fact resembles the effect of nipecotic acid/GABA action on taurine release, we think the above proposed model could also be applied to the glutamate uptake system. A role may thus be indicated for taurine as a universal coupling substrate which may function in uptake anti-port systems of amino acid neurotransmitters. In this sense, it should be noted that the high ($\sim$400) intra/extracellular concentration ratio of taurine, an outward directed gradient, may add driving force to the inward transport of the amino acids when the electrochemical sodium gradient is decreased (e.g. during depolarization).

According to this model, moderate membrane depolarization should evidence the postulated role of taurine as an exchange substrate. Tentatively, such a membrane depolarization would augment release of amino acid neurotransmitters which, under physiological conditions, should be buffered by a concomitant increase of their uptake rate in order to avoid neuronal functional blockade. In this situation a specific increase of extracellular taurine would be expected. In fact, perfusion with physiologically occurring $[K^+]_o$ produces an increase in neuronal excitability, as may be predicted from a moderate depolarization (see ref. 2), and a specific increase in extracellular taurine, while amino acid neurotransmitters (i.e. glutamate and GABA) remain extracellularly unaltered. Larger depolarizations would create a situation in which the neuronal function would become seriously impaired. As a consequence, there would be both a massive and nonphysiological release of neurotransmitters and a lack of their uptake. All these phenomena seemed to occur when 50 and 100 mM K^+ were perfused (Fig. 5). However, extracellular taurine levels were not increased further, supporting the assumption that the proposed model only operated under physiological situations. It is therefore suggested that during uptake of synaptically released amino acid neurotransmitters, an increase in extracellular taurine concentration is produced which may also represent an autoregulative system of the excitability, taking into account that taurine exerts modulatory actions in several parts of the CNS.

ACKNOWLEDGMENTS

We thank Ms. C. Stoddard Delgado for editorial help, V. Abraira for computer programs design, M.D. Munoz for collaborating in some experiments and A. Latorre for technical assistance. This work was partially supported by the Fondo de Investigaciones Sanitarias (FISSS 85/1070 and 83/0623).

REFERENCES

1. Baudry, M, and Lynch, G., 1979, Two glutamate binding sites in rat hippocampal membranes, Eur. J. Pharmacol., 57:283-285.
2. Baylor, D.A. and Nicholls, J.G., 1969, Aftereffects of nerve impulse on signalling in the central nervous system of the leech, J. Physiol. (Lond.), 203:571-589.
3. Delgado, J.M.R., DeFeudis, F.V., Roth, R.H., Ryugo, D.K., and Mitruka, B.M., 1972, Dialytrode for long-term intracerebral perfusion in the awake monkeys, Arch. Int. Pharmacodyn. Therm., 198:9-21.
4. Delgado, J.M.R., Lerma, J., Martín del Río, R., and Solís, J.M., 1984, Dialytrode technology and local profile of amino acids in the awake cat brain, J. Neurochem., 42:1218-1228.
5. Gonon, F., Buda, M., and Pujol, J.F., 1984, Treated carbon fibre electrodes for measuring catechols and ascorbic acid., in: "Measurements of Neurotransmitter Release In Vivo", C.A. Marsden ed., Wiley, New York, pp. 153-172.
6. Hablitz, J.J., and Lundervold, A., 1981, Hippocampal excitability and changes in extracellular potassium, Exp. Neurol., 71:410-420.
7. Hamberger, A., Berthold, C.H., Jacobson, I., Karlsson, B., Lehmann, A., Nystrom, B., and Sandberg, N., 1985, In vivo brain dialysis of extracellular non transmitter and putative transmitter amino acids, in: "In Vivo Perfusion and Release of Neuroactive Substances", A. Bayón and R. Drucher-Colin eds., Academic Press, Orlando, Florida, pp. 119-139.
8. Hamberger, A., and Nystrom, B., 1984, Extra- and intracellular amino acids in the hippocampus during development of hepatic encephalopathy, Neurochem. Res., 9:1181-1192.
9. Heinemann, U., Lux, H.D., and Gutnich, M.J., 1978, Changes in extracellular free calcium and potassium activity in the somatosensory cortex of cats, in: "Abnormal Neuronal Discharges", N. Chalazonitis and M. Boisson eds., Raven Press, New York, pp. 329-345.
10. Herranz, A.S., Cristín, J.L.R., Lerma, J., and Martín del Río, R., 1985, Incremento de sensibilidad en los analisis por CLAE de los OPA-aminoácidos usando como reactivo de derivación el ácido 3-mercaptopropiónico, Resúmenes de la Reunión Científica Anual del Grupo de Cromatografia y Tecnicas Afines, R.S.E.Q., Sevilla, pp. 58-59.
11. Hertz, L., and Schousboe, A., 1980, Interactions between neurons and astrocytes in the turnover of GABA and glutamate, Brain Res. Bull., 5, Suppl. 2:389-395.

12. Jacobson, I., Sandberg, M., and Hamberger, A., 1985, Mass transfer in brain dialysis devices -a new method for the estimation of the extracellular amino acids concentration, J. Neurosci. Meth., 15:262-268.
13. Johnston, G.A.R., Kennedy, S.M.E., and Twitchin, B., 1979, Action of the neurotoxin kainic acid on high affinity uptake of L-glutamic acid in rat brain slices, J. Neurochem., 32:121-127.
14. Johnston, G.A.R., Krogsgaard-Larsen, P., Stephanson, A.L., and Twitchin B., 1976, Inhibition of the uptake of GABA and related amino acids in the rat brain slices by the optical isomers of nipecotic acid, J. Neurochem., 26:1029-1032.
15. Kontro, P., and Oja, S.S., 1981, Hypotaurine transport in brain slices: comparison with taurine and GABA, Neurochem. Res., 6:1179-1191.
16. Kontro, P., and Oja, S.S., 1983, Mutual interaction in the transport of taurine, hypotaurine and GABA in brain slices, Neurochem. Res., 8:1377-1387.
17. Lähdesmäki, P., and Oja, S.S., 1973, On the mechanism of taurine transport at brain cell membranes, J. Neurochem., 20:1411-1417.
18. Lerma, J., Herranz, A.S., Herreras, O., Abraira, V., and Martín del Río, R., 1986, In vivo determination of extracellular concentration of amino acids in the rat hippocampus. A method based on brain dialysis and computerized analysis, Brain Research, In Press.
19. Lerma, J., Herranz, A.S., Herreras, O., Munoz, D., Solis, J.M., del Rio, R.M., and Delgado, J.M.R., 1985, γ-aminobutyric acid greatly increases the in vivo extracellular taurine in the rat hippocampus, J. Neurochem., 44:983-986.
20. Lerma, J., Herreras, O., Herranz, A.S., Munoz, D., and del Rio, R.M., 1984, In vivo effects of nipecotic acid on levels of extracellular GABA and taurine, and hippocampal excitability, Neuropharmacology, 23:595-598.
21. Lerma, J., Herreras, O., and Martín del Río, R., 1985, Electrophysiological evidence that nipecotic acid can be used in vivo as a false transmitter, Brain Research, 335:377-380.
22. Lombardini, J.B., 1976, Regional and subcellular studies in taurine in the central nervous system, in: "Taurine", R. Huxtable and A. Barbeau eds., Raven Press, New York, pp. 311-326.
23. Madtes, P., Jr., 1984, Chloride ions preferentially mask high-affinity GABA binding sites, J. Neurochem., 43:1434-1437.
24. McBride, W.J., and Frederickson, R.C.A., 1978, Neurochemical and neurophysiological evidence for a role of taurine as an inhibitor neurotransmitter in the cerebellum of the rat, in: "Taurine and Neurological Disorders", A. Barbeau and R.J. Huxtable, eds., Raven Press, New York, pp. 415-527.
25. Moroni, F. Mulas, A., Moneti, G., and Pepeu, G., 1982, In vivo changes in GABA output from the cerebral cortex induced by inhibitors of GABA uptake and metabolism, J. Neurochem., 39:582-585.
26. Muñoz, M.D., Herreras, O., Herranz, A.S., Solís, J.M., Martín del Río, R., and Lerma, J., 1986, Effects of dihydrokainic acid on extracellular amino acids and neuronal excitability in the in vivo rat hippocampus, Neuropharmacology, In Press.
27. Nicholson, C., 1980, Dynamics of the Brain Cells Microenvironment Neurosci. Res. Prog. Bull., MIT Press, Cambridge, Massachusettes, 18:322.
28. Oja, S., and Kontro, P., 1978, Neurotransmitter actions of taurine in the central nervous system, in: "Taurine and Neurological Disorders", A. Barbeau and R.J. Huxtable eds., Raven Press, New York, pp. 181-200.
29. Oja, S.S., Korpi, E.R., Holopainen, I., and Kontro, P., 1985, Mechanisms of stimulated taurine release from nervous tissue, in: "Taurine, Biological Actions and Clinical Perspectives", S.S. Oja, L. Ahtee, P. Kontro, and M.K. Paasonen eds, Alan R. Liss, New York, pp. 237-247.
30. Philippu, A., 1985, The use of push-pull cannula for superfusing various hypothalamic areas in anaesthetized and conscious, freely moving animals, in: "In Vivo Perfusion and Release of Neuroactives Substances", A. Bayón

and R. Drucker-Colín eds., Academic Press, Orlando, Florida, pp. 221-232.

31. Sandberg, M., and Lindstrom, S., 1983, Amino acids in the dorsal lateral geniculate nucleus of the cat - collection in vivo, J. Neurosci, Meth., 9:65-74.
32. Sellstrom, A., and Hamberger, A., 1977, Potassium-stimulated γ-aminobutyric acid release from neurons and glia, Brain Research, 119:189-198.
33. Snedecor, G.W., and Cochran, W.G. 1980, Metodos Estadisticos, CECSA, Mexico, pp. 703.
34. Solís, J. M., Herranz, A.S., Herreras, O., Muñoz, M.D., Martín del Río, R., and Lerma, J., 1986, Variation of potassium ion concentrations in the rat hippocampus specifically affects extracellular taurine levels, Neurosci. Lett., 66:263-268.
35. Storm-Mathisen, J., 1978, Localization of putative transmitter in the hippocampal formation, in: "CIBA Foundation Symposium, Functions of the Septo-Hippocampal System", Elsevier, Amsterdam, pp. 49-79.
36. Tossman, U., Eriksson, S., Delin, A., Hagenfeldt, L., Law, D., and Ungerstedt, U., 1983, Brain amino acids measured by intracerebral dialysis in portacaval shunted rats, J. Neurochem., 41:1046-1051.
37. van der Heyden, J.A.M., Venema, K., and Korf, J., 1985, Push-pull perfusion studies on the in vivo release of endogenous transmitter and non-transmitter amino acids in the rat brain, in: "In vivo Perfusion and Release of Neuroactive Substances", A. Bayón and R. Drucker-Colín eds., Academic Press, Orlando, Florida, pp. 51-68.
38. Werling, L.L., and Nadler, J.V., 1982, Complex binding of L-[^{3}H] glutamate to hippocampal synaptic membranes in the absence of sodium, J. Neurochem., 38:1050-1062.
39. Wolfensberger, M., 1984, Gas-chromatographic and mass-fragmentographic measurement of amino acid released into brain perfusate collected in vivo by push-pull cannula techniques, in: "Measurement of Neurotransmitter Release In Vivo", C.A. Marsden ed., Wiley, New York, pp. 39-62.

A MODEL OF THE COMPARTMENTALIZATION OF TAURINE IN RAT HYPOTHALAMIC NEURONAL AND GLIAL CELL PARTICLES

Allan T. Hanretta and J. Barry Lombardini

Department of Pharmacology
Texas Tech University Health Sciences Center
Lubbock, Texas 79430

In the central nervous system, taurine has been postulated to be an inhibitory neurotransmitter and/or a neuromodulator of nerve activity (4,15). Recent studies on the effects of taurine on hypothalamic function indicate that taurine may influence temperature regulation either producing a decrease in body temperature if injected intraventricularly (23), or producing a biphasic temperature effect depending upon the dose when injected directly into the preoptic region of the anterior hypothalamus of the rat (9). Low doses of taurine produce hyperthermia while high doses result in hypothermia (9). In addition, exogenous taurine has been shown to alter hypothalamic hormonal release mechanisms of prolactin (21), growth hormone (2), and luteinizing hormone (20). Taurine inhibits the in vitro release of somatostatin from median eminences, and thus may be acting as a hypothalamic neurotransmitter to inhibit anterior pituitary release of growth hormone (1). There are also a number of reports that taurine may modulate hypothalamic neuronal activity by interacting with dopaminergic (5), GABAergic (16), glycinergic (6), and serotonergic systems (23).

The possibility that these effects are mediated by a neurotransmitter/neuromodulatory role for taurine prompted a number of studies in our laboratory over the past few years to characterize the specific binding, uptake and release of taurine in hypothalamic membrane and crude P_2 synaptosomal preparations. These studies are summarized herein and we propose a model based on the results.

PROPOSED MODEL

A model (Fig. 1) is presented for the binding, uptake and release of taurine in the rat hypothalamus. Experimental evidence will be presented which suggests that high affinity taurine binding and uptake are located in neuronal tissue while the low affinity system is located in glial cells which are contained in the P_2 synaptosomal preparations used in these experiments. In the following sections data are presented which support this model.

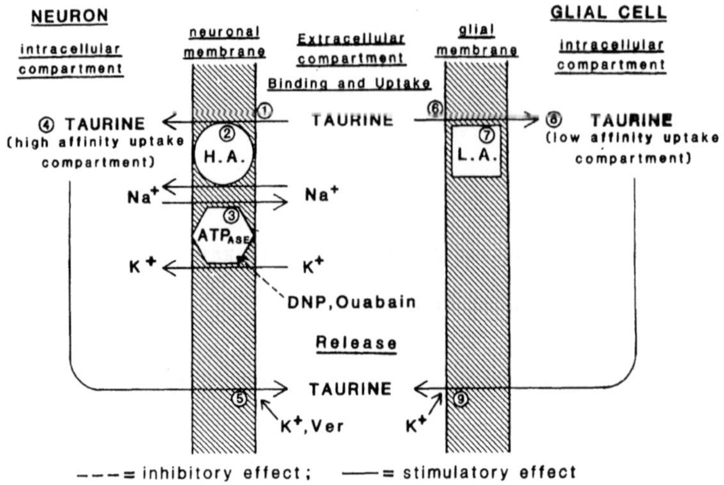

Fig. 1. Model for the binding, uptake and release of taurine in the hypothalamus based on experimental observations. The numbered sites shown in the diagram correspond to the numbered sites discussed in the text. (H.A. = high affinity uptake carrier, L.A. = low affinity uptake carrier).

HIGH AFFINITY TAURINE BINDING SITE (site 1)

Hypothalamic membranes were prepared as follows: rat hypothalami were pooled and homogenized in 40 volumes of ice-cold 50 mM Tris-HCl, pH 7.7, using a Polytron homogenizer on setting 6 for 12 seconds. The homogenate was centrifuged at 17,000 g for 20 min and the supernate was discarded; the pellet was washed in fresh buffer and then recentrifuged. The final pellet was homogenized in 5 volumes of Krebs-Tris buffer, pH 7.4. A Millipore vacuum filtration system using Whatman GF/B filters was utilized for the binding (and uptake) assay (7).

The dissociation constant of the high affinity taurine binding site ranges from 27-98 μM with a maximum binding constant of 57-120 nmoles of taurine/g protein. The high affinity taurine binding has an essential requirement for sodium ions but no dependence on calcium or potassium ions. Taurine binding follows Michaelis-Menton kinetics which presupposes that only one taurine molecule binds to one carrier site and cooperativity is not involved.

HIGH AFFINITY TAURINE UPTAKE CARRIER SYSTEM (site 2)

Hypothalamic crude P_2 synaptosomal pellets were prepared as follows: rat hypothalami were placed in 9 volumes of ice-cold 0.32 M sucrose buffered with 2 mM Tris-HCl (pH 7.0). The hypothalami were homogenized in a Potter Elvehjem glass homogenizer and centrifuged at 1000 g for 10 min to remove nuclei and cellular debris. Synaptosomes were obtained by centrifuging the supernate at 17,500 g for 20 min. The resulting crude P_2 synaptosomal pellet was washed in the sucrose-Tris buffer and recentrifuged. The synaptosomal pellet was suspended in Krebs-Ringer medium.

Taurine uptake as a function of the taurine concentration is shown in Fig. 2A. Taurine uptake into rat synaptosomes is composed of two saturable transport systems, and a nonsaturable process due to diffusion (7). Calculation of a diffusion constant at high medium concentrations of taurine allows the nonsaturable component to be subtracted from the total influx leaving the two saturable transport processes (observed in a Lineweaver-Burk plot of the data - Fig. 2B) which have been designated high affinity and low affinity. Each of these transport systems contribute to the other so that the apparent kinetic constants of either system represent contributions from both. Thus a curve fitting method for a two parameter uptake system was employed to determine the true kinetic constants. The high affinity uptake system had a Km of 240 ± 118 μM with a Vmax of 400 ± 135 nmole/g protein/min while the low affinity uptake system (site 7) had a Km of 5290 ± 1460 μM with a Vmax of 1640 ± 91 nmole/g protein/min (kinetic parameter ± the asymptotic standard error).

The effects of temperature on both the high (site 2) and low (site 7) affinity taurine uptake systems were also studied. In general two transition temperatures, 24.5°C and 33°C, were observed for the high affinity uptake system while only one transition temperature, 31.5°C, was observed for the low affinity uptake system. The high affinity system was observed to have large activation energy (Ea = 23.7 Kcal/mole) and temperature quotient (Q_{10} = 4.3) values obtained at temperatures below the 24.5°C transition temperature, while the Ea and Q_{10} values in the low affinity uptake system were considerably lower (8.5 Kcal/mole and 1.6, respectively).

Hypo-osmotic treatment of the crude P_2 synaptosomal pellet decreases (85.3% reduction) high affinity taurine uptake to a greater extent than it decreases (29% reduction) low affinity taurine uptake (Table 1). This experiment brings up a rather interesting point in that during the osmotic shock treatment endogenous taurine is potentially released which will mix with and thus dilute the radioactive exogenous taurine which is added to the incubation system. The crude P_2 synaptosomal preparation used in these experiments (0.3 mg) contains 1.9 nmole of taurine which is equivalent to 9.5 μM in the incubation system (200 μl). Thus in the extreme condition the potential total taurine concentration (exogenous plus endogenous) could be equal to 19.5 μM or 2-times the planned concentration. In this situation the quantity of taurine taken up by the high affinity system would be 2.58 nmole/mg protein/min or 29.4% of the control. Even under these circumstances the osmotic shock treatment has a significantly greater effect on the high affinity taurine uptake than on the low affinity uptake system.

SODIUM-POTASSIUM ATPase (site 3)

Taurine uptake is dependent upon sodium ions and the driving force for transport is supplied by the Na^+ concentration gradient across the membrane which, in turn, is dependent on a functioning Na^+-K^+ ATPase. Agents such as monensin, a sodium ionophore, which interfere with the sodium gradient, or dinitrophenol and ouabain, metabolic inhibitors which diminish the Na^+ gradient by affecting Na^+-K^+ ATPase activity, inhibit the uptake of taurine (Fig. 3).

HIGH AFFINITY UPTAKE COMPARTMENT (site 4)

Once inside the neuron, taurine is stored in the high affinity uptake compartment.

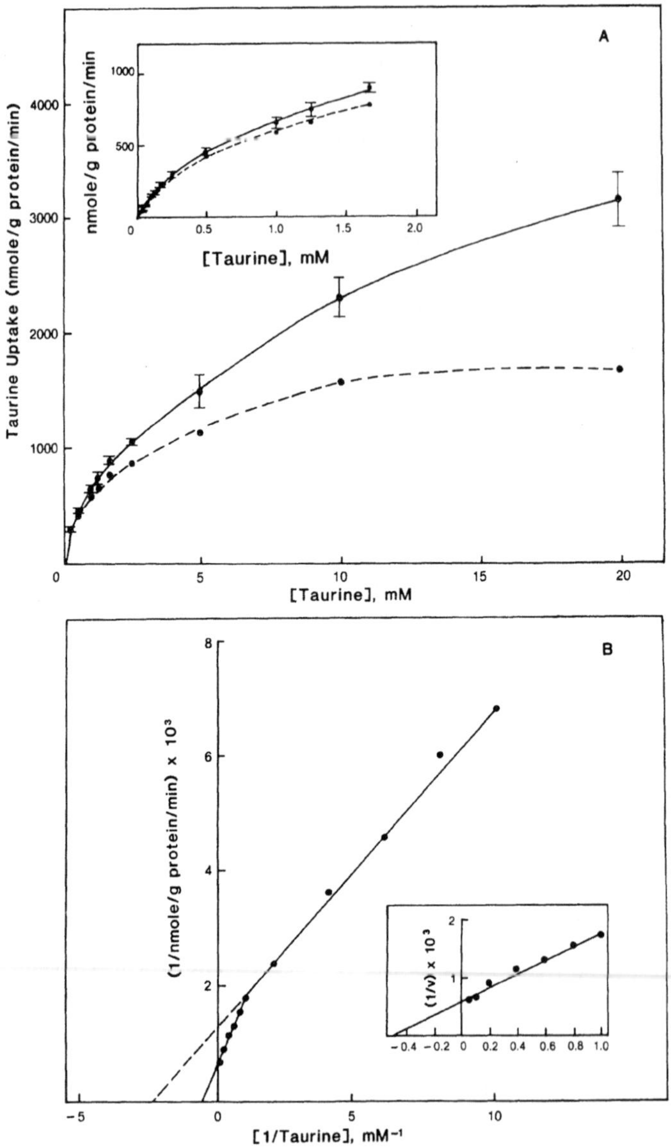

Fig. 2. A: Taurine uptake as function of the taurine concentration. Velocity versus substrate plots uncorrected (solid line) and corrected (dashed line) for diffusion. Na^+ concentration was 128 mM. Assay conditions and calculations of the kinetic constants are reported in reference 7. B: Lineweaver-Burk plot of taurine uptake demonstrating two uptake systems. Data are corrected for diffusion. [Reproduced from Brain Research 378, 205-215 (1986)].

RELEASE OF TAURINE FROM THE HIGH AFFINITY UPTAKE COMPARTMENT (site 5)

In superfusion experiments both potassium ions (56 mM) and veratridine (100 μM) evoke the release of taurine from hypothalamic P_2 prepara-

Table 1. High and low affinity taurine uptake in hypo-osmotically treated crude P_2 synaptosomal preparations.

Treatment	High affinity	Low affinity
	nmole/mg/min	nmole/mg/min
Control	8.79 ± 1.00	386 ± 28
Osmotic shock	1.29 ± 0.19	274 ± 43
% Control	14.7 ± 2.1	71 ± 11[a]

High affinity taurine concentration = 10 μM; low affinity taurine concentration = 2 mM. Means ± SEM of uptake velocities are calculated from 12-14 data points over a temperature range from 15 to 37°C. The means were normalized by converting to percentages and then the significant difference of the hypo-osmotic treatment between the high and low affinity uptake was determined by the Student's t-test ([a]$P < 0.01$).

tions which were preloaded with 1.5 μM [^{3}H]taurine (Fig. 4). This release from the high affinity uptake compartment was unaffected by the omission of calcium ions from the incubation medium (performed in the presence of 1 mM EGTA) indicative of cytoplasmic release as opposed to vesicular release (Fig. 5). Evoked release by potassium ions or veratridine was not affected by the continuous oxygenation of the incubation medium (95% O_2; 5% CO_2) throughout the superfusion period. Additionally, substitution of lithium ions for potassium ions did not produce an evoked release of [^{3}H]taurine from the superfused tissue. When sodium ions were omitted from the superfusion medium there was no effect on the spontaneous release of [^{3}H]taurine thus suggesting that the release of taurine is not caused by simple reversal of the direction of the high affinity uptake carrier system (site 2).

LOW AFFINITY TAURINE BINDING SITE (site 6)

The dissociation constant ranges from 33-40 mM with a maximum binding constant of 25 μmole/g protein. Low affinity binding does not display cooperativity thus suggesting that only one taurine molecule binds to one carrier site. Sodium ions are not an essential requirement for low affinity binding as was found in the high affinity system.

LOW AFFINITY TAURINE UPTAKE CARRIER SYSTEM (site 7)

The low affinity uptake kinetic constants were calculated from the data presented in Figure 2: taurine uptake as a function of the taurine concentration (Km = 5.29 mM, Vmax 1640 nmole/g protein/min).

LOW AFFINITY UPTAKE COMPARTMENT (site 8)

Taurine is stored in a low affinity uptake compartment in the glial cells.

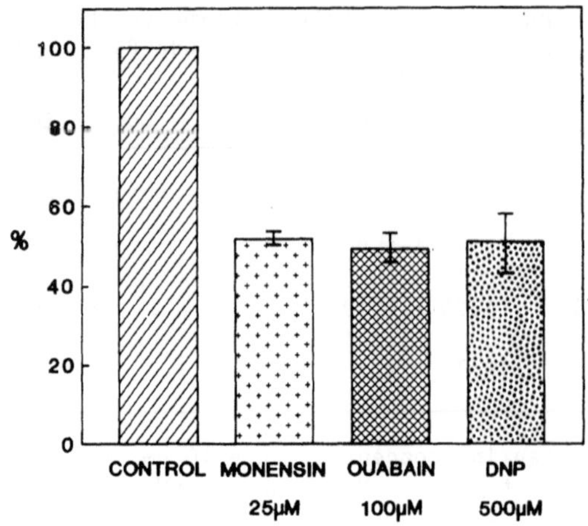

Fig. 3. Effects of fixed concentrations of monensin, ouabain and dinitrophenol (DNP) on high affinity taurine uptake. Means ± SEM of 4 experiments.

RELEASE OF TAURINE FROM THE LOW AFFINITY UPTAKE COMPARTMENT (site 9)

When the two depolarizing agents, veratridine and potassium ions, were utilized to evoke the release of taurine from hypothalamic crude P_2 synaptosomal preparations preloaded with [^{3}H]taurine in the low affinity uptake range only potassium ions had an effect (Fig. 6). The potassium evoked release was Ca^{2+}-independent.

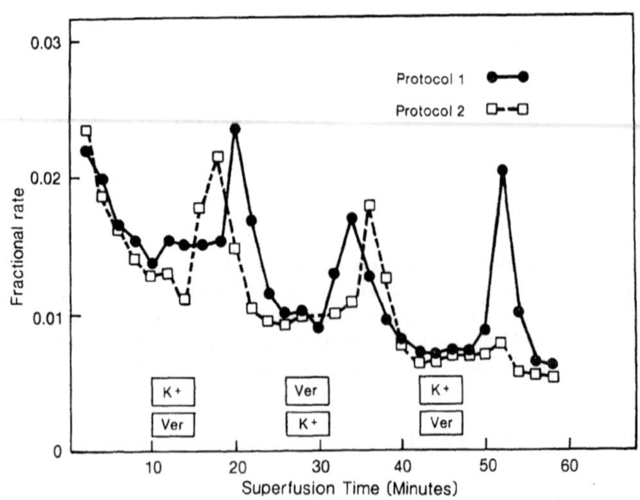

Fig. 4. High affinity potassium- and veratridine-evoked release of [^{3}H]taurine in the absence of calcium. The order of exposure to veratridine (Ver) or potassium (K^+) were: for protocol 1 (solid line), K^+-Ver-K^+; for protocol 2 (dashed line), Ver-K^+-Ver. [Reproduced from Brain Research 378, 205-215 (1986)].

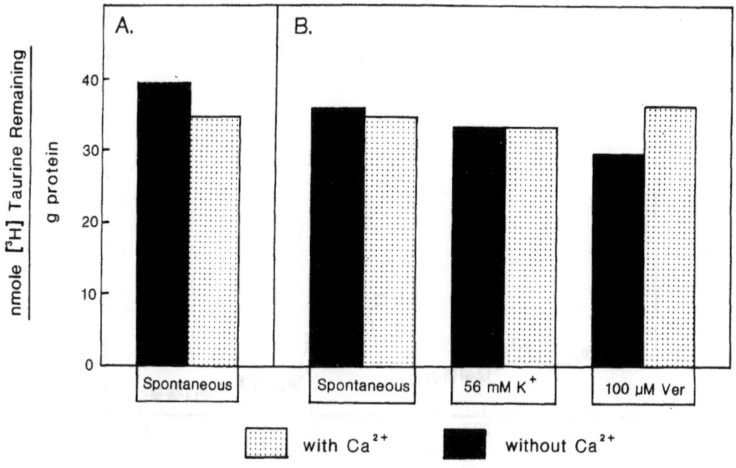

Fig. 5. Effect of calcium on uptake and spontaneous and evoked release of $[^3H]$taurine. A. Synaptosomes were preincubated with 10 μM $[^3H]$taurine for 30 min at 37°C *with* 2.7 mM Ca^{2+}. Uptake was terminated by centrifugation at 17,000 g for 10 min. The final pellets containing $[^3H]$taurine-loaded synaptosomes were resuspended in medium with (dotted bars) or without (+ 1 mM EGTA, solid bars) Ca^{2+}. After a 10 minute incubation period, spontaneous release was terminated utilizing a filtration assay. The bars represent the amount of taurine remaining in the synaptosomes and are the average values of two experiments performed in duplicate. B. Synaptosomes were preincubated with 10 μM $[^3H]$taurine for 30 min at 37°C *without* Ca^{2+}. Uptake was terminated by centrifugation at 17,000 g for 10 min. The final pellets containing $[^3H]$taurine-loaded synaptosomes were resuspended in medium with (dotted bars) or without (+ 1 mM EGTA, solid bars) Ca^{2+} and contained either no additions (spontaneous release or 56 mM potassium chloride or 100 μM veratridine (evoked release). Release was terminated as described in A.

LOCALIZATION OF THE HIGH AND LOW AFFINITY UPTAKE SYSTEMS

It is our hypothesis that the high affinity taurine binding site is located in the neuronal membrane while the low affinity taurine binding site is located in the glial cell membrane. Verifying this hypothesis is difficult since in the experiments presented herein we utilized a crude P_2 synaptosomal preparation, which is known to be contaminated by glial cell particles (3,8,26). However, the following observations support this hypothesis:

1) Hypo-osmotic treatment of the membrane preparations has a greater effect on the high affinity uptake system than on the low affinity uptake system. This is consistent with literature reports that synaptosomes are lysed under hypo-osmotic conditions in contrast to glial cells which are relatively resistant to such treatment (22,24,25).

2) Temperature-dependency studies (Arrhenius plots) indicate that the high affinity uptake system for taurine is highly temperature and energy dependent as indicated by the large activation energy (Ea) and temperature quotient (Q_{10}) obtained for uptake at temperatures below the 24.5°C transition temperature. In contrast we observed low Ea and Q_{10} values for the low affinity uptake below the 31.5°C transition temperature

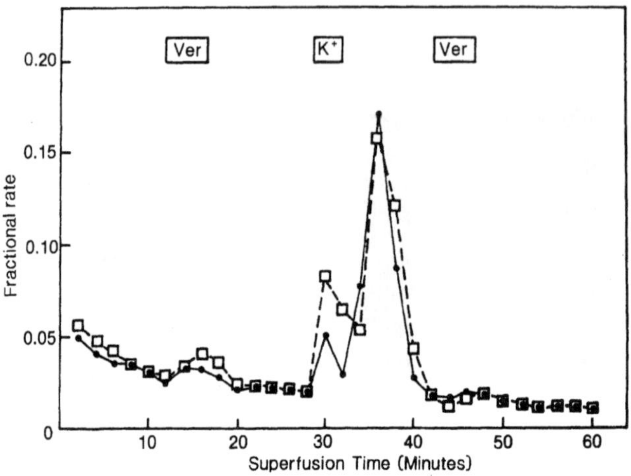

Fig. 6. Low affinity potassium- and veratridine-evoked release of [^{3}H]taurine from the crude synaptosomal preparation which had been prelabeled with [^{3}H]-taurine concentration in the low affinity range (2 mM). Order of exposure was Ver-K^+-Ver. In protocol 1 (solid line), 2.7 mM Ca^{2+} was included in the superfusion medium. In protocol 2 (dashed line), Ca^{2+} was omitted and 1 mM EGTA was present throughout the superfusion period. [Reproduced from Brain Research 378, 205-215 (1980)].

from the Arrhenius plot. These results are in agreement with the studies of Sieghart and Karobath (25) who observed that the temperature-dependency of taurine uptake into a P_2 fraction of C-6 glioma cells was less than in rat cerebral cortex synaptosomes. These results are thus in agreement with our hypothesis that the 24.5°C transition temperature represents synaptosomal membrane (nerve cells) fluidity changes in the high affinity taurine uptake system while the 31.5-33°C transition temperature represents changes in glial particle membrane (glial cells) fluidity in the low affinity taurine uptake system.

3) The differential ability of veratridine to stimulate taurine release from the crude P_2 synaptosomal preparation preloaded with taurine in the high affinity uptake concentration range, but not from the low affinity taurine uptake compartment, indicates that the two compartments are located in different cell types: i.e, high affinity uptake system on neuronal membranes and low affinity uptake system on glial membranes. The selectivity of veratridine for distinguishing between neuronal and glial release of neurotransmitters has been used by other investigators (12).

4) Release of taurine by either veratridine or high potassium ion concentration from preloaded membrane preparations is Ca^{2+}-independent in both the high and low affinity uptake systems. This observation strongly suggests that taurine is released from intracellular cytoplasmic pools of free taurine rather than vesicular pools.

ANATOMICAL SEPARATION OF THE HIGH AND LOW AFFINITY UPTAKE SYSTEMS: IS THERE ANY FUNCTIONAL SIGNIFICANCE?

It is proposed that the high affinity uptake system rapidly removes extracellular synaptic taurine, which has been released by taurinergic neuronal activity, for the purpose of limiting the duration of the action of taurine on postsynaptic receptor sites. The low affinity taurine uptake system located in glial cells would decrease the high concentrations of taurine in the synaptic area (caused by release from both glial and neuronal cells) to a low enough intra-synaptic concentration of taurine that then could be removed by the neuronally located high affinity uptake system. Thus low affinity uptake would function as a backup system in order to preserve the integrity of information which is normally relayed through the taurinergic synapse.

Alternatively, the cellular separation of the high and low affinity uptake systems may be associated with a neuromodulatory role for taurine and its inhibition of neuronal activity. Three proposals are presented:

1. Taurine may act to enhance the synaptic concentration of GABA and thus the action of GABA. This neuromodulatory role for taurine is suggested by the observations of Kontro and Oja (10) that GABA and taurine have a common carrier system and thus compete for reuptake. Also, it is known that neuronal activity produces a rise in the extracellular K^+ concentration which can cause passive depolarization of neighboring glial cells (27) and the release of taurine. Thus, intense GABAergic neuronal activity may cause the neuronal release of both GABA and taurine plus the release of taurine from glial cells, due to their depolarization induced by increased K^+ concentration. The competition of taurine with presynaptic GABA reuptake via the common carrier system allows for a prolongation of the action of GABA at its postsynaptic receptor sites.

2. Numerous investigators have observed that taurine affects calcium ion fluxes in the retina (11,13,14,18) and in the brain (17,19). Thus it is suggested that neuronal activity produces a release of glial taurine which in turn suppresses the Ca^{2+}-dependent vesicular release of neurotransmitters by inhibiting the synaptosomal uptake of calcium.

3. Taurine acts as a feedback inhibitor which prevents neuronal hyperexcitation by a membrane stabilizing action. In this model intense neuronal activity depolarizes neighboring glial cells, taurine is released, and the increased taurine levels dampen the excessive neuronal activity.

SUMMARY

We conclude that the high affinity taurine uptake system is located on neuronal membranes in the hypothalamus and that the low affinity taurine uptake system is located on glial membranes. The evoked release of taurine by either K^+ and/or veratridine is Ca^{2+}-independent and thus taurine, according to classical criteria, does not function as a neurotransmitter in the hypothalamus. This model for the differential cellular transport and compartmentalization of taurine into neuronal and glial cells and for a cytoplasmic location has important implications concerning its possible role in the central nervous system, specifically as a neuromodulator of neuronal activity.

ACKNOWLEDGMENT

The studies presented here were supported in part by the Tarbox Parkinson's Disease Institute at Texas Tech University Health Sciences Center.

REFERENCES

1. Aguila, M.C., and McCann, S.M., 1985, Stimulation of somatostatin release from median eminence tissue incubated in vitro by taurine and related amino acids, Endocrinology, 116:1158-1162.
2. Collu, R., Charpenet, G., and Clermont, M.J., 1978, Antagonism by taurine of morphine induced growth hormone secretion, Can. J. Neurol. Sci., 5:139-142.
3. Cotman, C., Herschman, H., and Taylor, D., 1971, Subcellular fractionation of cultured glial cells, J. Neurobiol., 2:169-180.
4. Davison, A.N., and Kaczmarek, L.K., 1971, Taurine a possible neurotransmitter, Nature, 234:107-108.
5. Garcia de Yebenes Prous, J., Carlsson, A., and Mena Gomez, M.A., 1978, The effect of taurine on motor behaviour, body temperature and monoamine metabolism in rat brain, Naunyn-Schmiedeberg's Arch. Pharmacol., 304:95-99.
6. Glyn, J.R., and Lipton, J.M., 1981, Effects of central administration of alanine on body temperature of the rabbit: comparisons with the effects of serine, glycine and taurine, Brain Res. Bull., 6:467-472.
7. Hanretta, A.T., and Lombardini, J.B., 1986, Properties of spontaneous and evoked release of taurine from hypothalamic crude P_2 synaptosomal preparations, Brain Res., 378:205-215.
8. Henn, F.A., Anderson, D.J., and Rustad, D.G., 1976, Glial contamination of synaptosomal fractions, Brain Res., 101:341-344.
9. Kerwin, R.W., and Pycock, C.J., 1979, Role of taurine as a possible transmitter in the thermoregulatory pathways of the rat, J. Pharm. Pharmacol., 31:466-470.
10. Kontro, P., and Oja, S.S., 1983, Mutual interactions in the transport of taurine, hypotaurine, and GABA in brain slices, Neurochemical Research, 8:1377-1387.
11. Kuo, C.-H., and Miki, N., 1980, Stimulatory effect of taurine on Ca-uptake by disc membranes from photoreceptor cell outer segments, Biochem. Biophys. Res. Commun., 94:646-651.
12. Levi, G., Gallo, V., and Raiteri, M., 1980, A reevaluation of veratridine as a tool for studying the depolarization-induced release of neurotransmitters from nerve endings, Neurochem. Res., 5:281-295.
13. Lombardini, J.B., Effects of taurine on ATP-dependent calcium uptake in rat retinal membrane preparations, in: "Sulfur Amino Acids: Biochemical and Clinical Aspects", K. Kuriyama, R.J. Huxtable, and H. Iwata, eds., Alan R. Liss, New York (1983), pp. 251-262.
14. López-Colomé, A.M., and Pasantes-Morales, H., 1981, Effect of taurine on ^{45}Ca transport in frog retinal rod outer segments, Exp. Eye Res., 32:771-780.
15. Mandel, P., Pasantes-Morales, H., and Urban, P.F., Taurine a putative transmitter in retina, in: "Transmitters in the Visual Process", S.L. Bonting, ed., Pergamon Press, Oxford, New York, (1976) pp. 89-105.
16. Mayer, M.L., 1981, Inhibitory actions of amino acids in the preoptic-anterior hypothalamus of the rat, Exp. Brain Res., 43:154-158.
17. Namima, M., Okamoto, K., and Sakai, Y., 1983, Modulatory action of taurine on the release of GABA in cerebellar slices of the guinea pig, J. Neurochem., 40:1-9.
18. Pasantes-Morales, H., Ademe, R.M., and López-Colomé, A.M., 1979, Taurine effects on $^{45}Ca^{2+}$ transport in retinal subcellular fractions, Brain Res., 172:131-138.
19. Pasantes-Morales, H., Arzate, N.E., and Cruz, C., The role of taurine in nervous tissue: its effects on ionic fluxes, in: "Taurine in Nutrition and Neurology", R.J. Huxtable and H. Pasantes-Morales, eds., Plenum Press, New York, (1982), pp. 273-292.

20. Price, M.T., Olney, J.W., Mitchell, M.V., Fuller, T., and Cicero, T.J., 1978, Luteinizing hormone releasing action of N-methyl aspartate is blocked by GABA or taurine but not by dopamine antagonists, Brain Res., 158:461-465.
21. Scheibel, J., Elsasser, T., and Ondo, J.G., 1980, Stimulation of prolactin secretion by taurine, a neurally depressant amino acid, Neuroendocrinology, 30:350-354.
22. Schmid, R., Sieghart, W., and Karobath, M., 1975, Taurine uptake in synaptosomal fractions of rat cerebral cortex, J. Neurochem., 25:5-9.
23. Sgaragli, G., Carla, V., Magnani, M., and Galli, A., 1981, Hypothermia induced in rabbits by intracerebroventricular taurine: specificity and relationships with central serotonin (5-HT) systems, J. Pharm. Exp. Ther., 219:778-785.
24. Sieghart, W., and Karobath, M., 1974, Evidence for specific synaptosomal localization of exogenous accumulated taurine, J. Neurochem., 23:911-915.
25. Sieghart, W., and Karobath, M., 1976, Uptake of taurine into subcellular fractions of C-6 glioma cells, J. Neurochem., 26:981-986.
26. Sieghart, W., Sellström, A., and Henn, F., 1978, Sedimentation characteristics of subcellular vesicles derived from three glial systems, J. Neurochem., 30:1587-1589.
27. Varon, S.S., and Somjen, G.G., 1979, Neuron-glia interactions, Neurosci. Res. Prog. Bull., 17:7-239.

NEUROCHEMICAL CHARACTERIZATION OF EXCITATORY AMINO ACID RECEPTORS IN HIPPOCAMPUS

Akemichi Baba, Yutaka Koyama, Hiroshi Morimoto and Heitaroh Iwata

Department of Pharmacology, Faculty of Pharmaceutical Sciences
Osaka University, 1-6 Yamada-oka, Suita, Osaka 565 Japan

Glutamate and aspartate may be excitatory neurotransmitters within central nervous system (CNS) (6,9). The recent development of specific antagonists has revealed at least three pharmacologically distinct classes of receptors for excitatory amino acids in the vertebrate CNS (19). Cysteine sulfinic acid (CSA) is an intermediate in taurine biosynthesis (Fig. 1), in which the final product, taurine, has inhibitory effects in the CNS. CSA has a strong excitatory effect in the central neurons when applied iontophoretically, as do glutamate and aspartate, and thus it was classified as an excitatory amino acid (7). Among the metabolites appearing in Fig 1, on intraventricular injection, only CSA caused severe behavioral and EEG seizures, which were markedly reduced by a simultaneous injection of taurine (14,15). With respect to the neurochemical aspects of CSA, several studies were reported during the last five years. In brief, CSA is unevenly distributed in rat brain but its content is much lower than that of glutamate (5,8,12). Like many other substances, CSA is accumulated in synaptosomes by a Na^+-dependent high affinity uptake system. The Km value is 12 µM (17). Glutamate, aspartate and cysteic acid are competitive inhibitors for the uptake of CSA, indicating that CSA may be transported into synaptosomes by the high affinity carrier system for acidic amino acids (17).

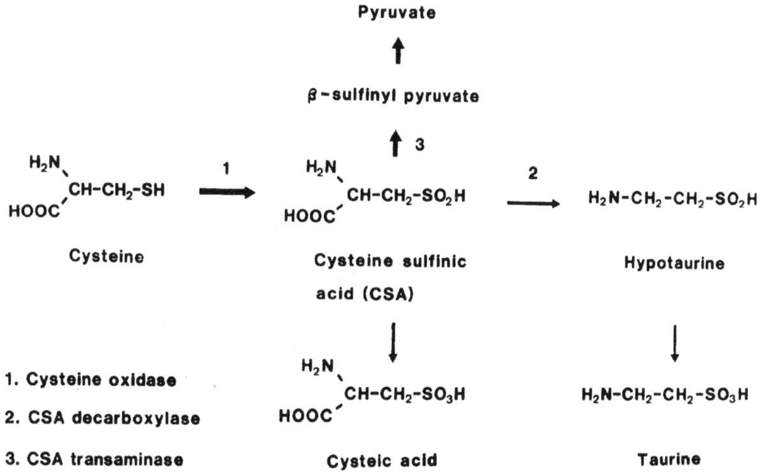

Fig 1. Metabolic pathway of cysteine sulfinic acid

Like many other putative transmitter substances, CSA is released in a Ca^{++}-dependent manner by depolarization (8,17). Synaptic membranes have a Na^{+}-independent specific binding site for CSA which may be distinct from glutamate binding sites (16,22,23). Neurochemical characteristics of CSA have been described in detail elsewhere (13). The neurochemical and pharmacological evidence strongly suggests that CSA is an excitatory neurotransmitter candidate in the CNS. However, to assess CSA as a physiologically active candidate distinct from glutamate, it is required to find some specificities in the pharmacological actions of CSA. Previously we have reported that CSA-induced GABA release (2) and CSA-stimulated formation of cAMP (1,4) in hippocampal slices may be related to specific receptors. In the present communication, we describe further characterization of CSA-induced release of [^{3}H]GABA from preloaded brain slices.

MATERIALS AND METHODS

[^{3}H]GABA Release

Male Sprague-Dawley rats, 4-5 weeks old, were used throughout, The slices of hippocampal formation, consisting primarily of the hippocampus and associated dentate gyrus, were prepared as described previously (17). The slices (0.25 mm thick) were cut nearly perpendicularly to the longitudinal axis of the hippocampus. Release of [^{3}H]GABA from the preloaded slices was examined by a superfusion method as described elsewhere (2). Values for the release activity represent the ratio of the percentage of the radioactivity released during the first and second stimulation (S2/S1) (2).

RESULTS AND DISCUSSION

Glutamate and similar substances are likely to be the neurotransmitters of several afferent and efferent pathways, and GABA seems to be a neurotransmitter for basket cell inhibition of pyramidal and granule cells in the hippocampal formation (25). To differentiate CSA neurons, if any, from

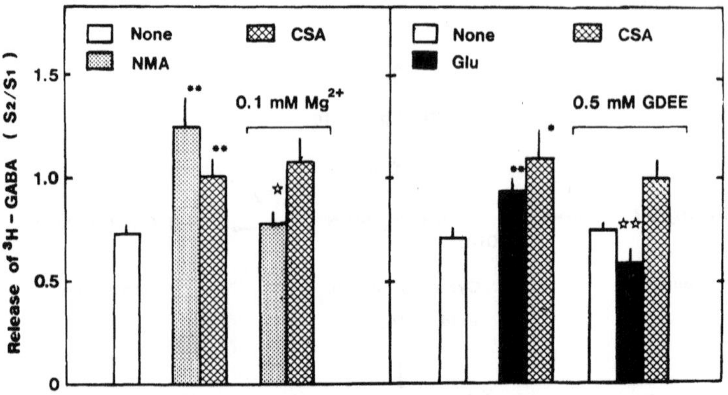

Fig 2. Effects of N-methyl-DL-aspartate and glutamate diethylester on the stimulation by excitatory amino acids of [^{3}H]GABA release from preloaded hippocampal slices. NMA: N-Methyl-DL-aspartate, GDEE: Glutamate diethylester. *$P < 0.05$ vs None, **$P < 0.01$ vs None, one open star:$P < 0.05$ vs NMA, two open stars:$P < 0.01$, vs Glu.

glutamate neurons, we characterized the stimulation of GABA release induced by excitatory amino acids, especially CSA, in rat hippocampal slices (2). Excitatory amino acid agonists, such as CSA, glutamate, kainate and quisqualate, significantly potentiated the high K^+ (25 mM)-induced release of [^{3}H]GABA from preloaded hippocampal slices at concentrations above 0.1 mM. Among these agonists, CSA was the most potent (2). Whereas N-methyl-DL-aspartate, another agonist, did not enhance the release of [^{3}H]GABA in the presence of Mg^{++}, it markedly enhanced the release in the absence of Mg^{++} (Fig. 2, left). In contrast, the effect of CSA was not affected by Mg^{++}. In order to examine whether the stimulatory effect of CSA is restricted to GABA release, the effect of CSA was examined on the release of ACh, one of the neurotransmitters in the hippocampus. Results showed that CSA had no significant effect on the high K^+-induced release of ACh. In addition, the stimulatory effect of CSA on the release of GABA was negated by tetrodotoxin and was not observed in the crude synaptosomal fractions of hippocampus. These results suggest the neuronal circuit being absolutely required for the stimulatory effect of CSA. In a separate experiment, we found that GABA and benzodiazepines at relatively low concentrations markedly reduced the depolarization-induced release of CSA and glutamate through GABA receptors in rat hippocampal slices (3). Thus, these neurochemical evidences also suggest that the GABA and the excitatory amino acid neurons innervate each other in the hippocampal formation.

In view of the electrophysiological studies, N-methyl-aspartate, kainate and quisqualate are selective agonists to three types of receptors for excitatory amino acids (26), which exhibit differential sensitivities to blockage by various antagonists. The effect of glutamate diethylester, which is a relatively nonspecific antagonist for quisqualate receptors, on CSA- and glutamate-induced release of [^{3}H]GABA was examined (Fig. 2, right). Glutamate diethylester (0.5 mM) completely antagonized the stimulatory effect of glutamate but not that of CSA. 2-Amino-4-phosphonobutyrate is an antagonist for excitatory amino acid receptors and selectively blocks the Cl^--dependent binding site for glutamate (20). 2-Amino-4-phosphonobutyrate at concentrations of 1 and 2.5 mM significantly reduced the stimulation of GABA release induced by CSA in a dose dependent manner without affecting that induced by glutamate (Fig 3). Recently it was reported that insulin receptors are present in the CNS (10). In electrophysiological studies, insulin selectively blocked mossy fiber transmission in the hippocampus (21).

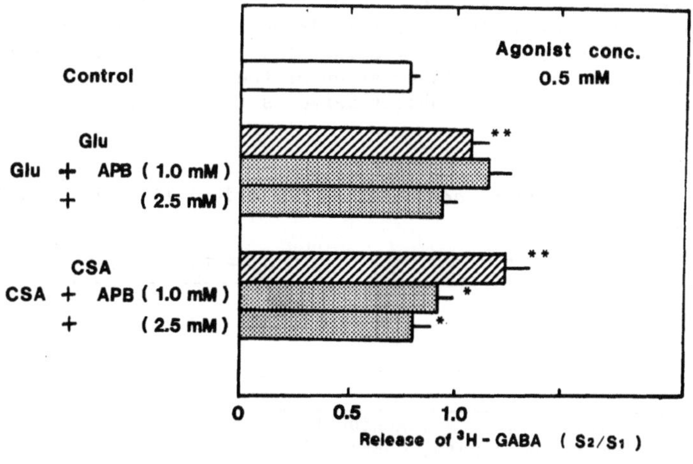

Fig 3. Effects of 2-amino-4-phosphonobutyrate on the stimulation of [^{3}H]GABA release. $^*P < 0.05$, vs CSA, $^{**}P < 0.01$, vs Control.

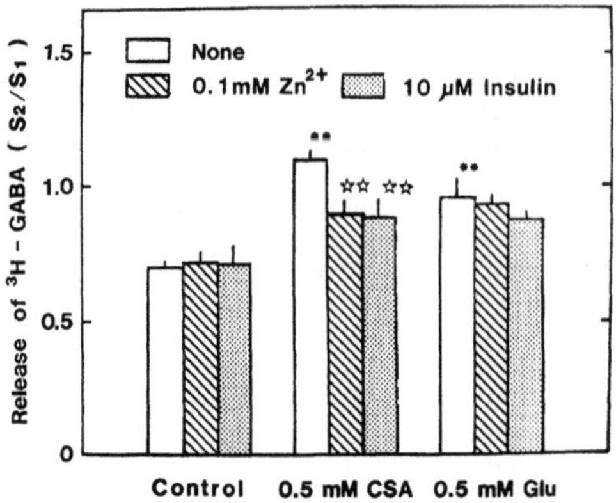

Fig 4. Effects of zinc and insulin on the stimulation of [^{3}H]GABA release. **P < 0.01, vs Control, open stars:P < 0.01, vs None.

In addition, the mossy fiber terminals in hippocampus are known to contain a high amount of zinc. Although the physiological and pharmacological significance of zinc in the hippocampus is not known, zinc has been recently reported to be released by depolarization (11). We have next examined the effects of insulin and zinc on the enhancement of GABA release induced by CSA and glutamate. These compounds had no effect on the basal high K^+-induced efflux of [^{3}H]GABA. As shown in Fig. 4, the stimulation induced by CSA was significantly negated by 0.1 mM zinc and 10 µM insulin. In contrast, these substances had no effect on the stimulation induced by glutamate. Table 1 summarizes the effects of various excitatory amino acid antagonists on CSA- and glutamate-induced release of GABA from preloaded slices (2). It is clear that the stimulatory effects of CSA and glutamate have different sensitivities to the antagonists. Thus, CSA receptors involved in the enhancement of GABA release seem to be distinct from glutamate receptors. However, since these antagonists are relatively nonspecific, it is required to examine with more specific antagonists for a further evaluation of the receptors for CSA.

Excitatory amino acids evoke a release of ACh from the preloaded striatal slices but not from the hippocampal slices through the N-Methyl-DL-aspartate receptor (18,24). In the present communication, excitatory amino acids failed to enhance the release of ACh from the hippocampal slices. Thus, the direction of the effect of excitatory amino acids on transmitter release seems to depend on the neuronal innervation of the regions. As described above, glutamate and aspartate probably serve as transmitters of the

Table 1. Effects of various compounds on the stimulation of [^{3}H]GABA release by CSA and glutamate

Compound	DAA	GDEE	KA	SB	Mg^{++}	APB	Zn^{++}	Insulin
CSA	-	-	+	+	-	+	+	+
Glutamate	-	+	+	-	ND	-	-	-

Antagonism (+), No antagonism (-), Not determined (ND)
DAA: D- -Aminoadipate, GDEE: Glutamate diethylester, KA: Kynurenic acid, SB: Secobarbital, APB: 2-Amino-4-phosphonobutyric acid.

perforant path and the commissural afferent in hippocampus, respectively. In contrast, the transmitter involved in the transmission of mossy fibers is not well identified. Mossy fiber terminals make synaptic connections with basket cells, in which GABA serves as a neurotransmitter (25). On the basis of these observations, taking into account the results in which the stimulatory effect of CSA on GABA release but not that of glutamate was inhibited by insulin and zinc, it is likely that CSA has its own receptor on the basket cell.

ACKNOWLEDGMENT

This study was supported by a grant from Taisho Pharmaceutical Co., Tokyo.

REFERENCES

1. Baba, A., Lee, E., Tatsuno, T., and Iwata, A., 1982, Cysteine sulfinic acid in the central nervous system: Antagonistic effect of taurine on cysteine sulfinic acid-stimulated formation of cyclic AMP in guinea pig hippocampal slices, J. Neurochem., 38:1280-1285.
2. Baba, A., Morimoto, H., and Iwata, H., 1985, Neurochemical relation between excitatory and inhibitory amino acids in hippocampus, in: "Taurine: Biological Actions and Clinical Perspectives", S.S. Oja, L. Ahtee, P. Kontro, and M.L.K. Paasonen, eds., Alan R. Liss, New York, pp.397-406.
3. Baba, A., Okumura, S., Mizuo, H. and Iwata, H., 1983, Inhibition by diazepam and γ-aminobutyric acid of depolarization-induced release of [^{14}C]cysteine sulfinic acid and [^{3}H]glutamate in rat hippocampal slices, J. Neurochem., 40:280-284.
4. Baba, A., Tatsuno, T., Kumagae, Y., and Iwata, H., 1983, Inhibitory effect of taurine on the formation of cyclic AMP in guinea pig hippocampus, in: "Sulfur Amino Acid: Biochemical and Clinical Aspects", K. Kuriyama, R.J. Huxtable, and H. Iwata, eds., Alan R. Liss, New York, pp.161-168.
5. Baba, A., Yamagami, S., Mizuo, S., and Iwata, H., 1980, Microassay of cysteine sulfinic acid by enzymatic cycling method, Anal. Biochem., 101:288-293.
6. Curtis, D.R., and Johnston, G.A.R., 1974, Amino acid transmitters in mammalian central nervous system, Ergebn. Physiol., 69:94-188.
7. Curtis, D.R., and Watkins, J.C., 1960, The excitation and depression of spinal neurons by structurally related amino acids, J. Neurochem., 6:117-141.
8. Do, K.Q., Mattenberger, M., Streit, P., and Cuenod, M., 1986, In vitro release of endogenous excitatory sulfur-containing amino acids from various rat brain regions, J. Neurochem., 46:779-786.
9. Fonnum, F., 1984, Glutamate: A neurotransmitter in mammalian brain, J. Neurochem., 42:1-11.
10. Havrankova, J., Schmechel, D., Roth, J., and Brownstein, M., 1978, Identification of insulin in rat brain, Proc. Natl. Acad. Sci., (USA), 75:5737-5741.
11. Howell, G.A., Welch, M.G., and Frederickson, C.J., 1984, Stimulation-induced uptake and release of zinc in hippocampal slices, Nature, 308:736-738.
12. Ida, S., and Kuriyama, K., 1983, Simultaneous determination of cysteine sulfinic acid and cysteic acid in rat brain by high-performance liquid chromatography, Anal. Biochem., 130:95-101.
13. Iwata, H., and Baba, A., 1983, Neurochemical basis of cysteine sulfinic acid in central nervous system, in: "Sulfur Amino Acids: Biochemical and Clinical Aspects", K. Kuriyama, R.J. Huxtable, and H. Iwata, eds., Alan R. Liss, New York, pp. 141-150.

14. Iwata, H., and Baba, A., 1982, Interaction of taurine with its precursor, cysteine sulfinic acid in the central nervous system, in: "Taurine in Nutrition and Neurology", R.J. Huxtable, and H. Pasantes-Morales eds., Plenum Pub., New York, pp. 211-219.
15. Iwata, H., Baba, A., and Yamagami, S., 1980, Possible function of cysteine metabolism in rat brain, in: "Modulaton of neurochemical transmission", E.S. Vizi ed., Pergamon Press, New York, pp. 417-426.
16. Iwata, H., Yamagami, S., and Baba, A., 1982, Cysteine sulfinic acid in central nervious system: Specific binding of [^{35}S] cysteic acid to cortical synaptic membranes- An investigation of possible binding sites for cysteine sulfinic acid, J. Neurochem. 38:1275-1279.
17. Iwata, H., Yamagami, S., Mizuo, H., and Baba, A., 1982, Cysteine sulfinic acid in the central nervous system: Uptake and release of cysteine sulfinic acid by rat preparation, J. Neurochem., 38:1268-1274.
18. Lehmann, J., and Scatton, B., 1982, Characterization of the excitatory amino acid receptor-mediated release of [^{3}H]acetylcholine from rat striatal slices, Brain Res., 252:77-89.
19. Mclennan, H., 1983, Receptor for the excitatory amino acids in mammalian central nervous system, Prog. Neurobiol. 20:251-271.
20. Monaghan, D.T., McMills, M.C., Chamberlin, A.R., and Cotman, C.W., 1983, Synthesis of [^{3}H]2-amino-4-phosphonobutyric acid and characterization of its binding to rat brain membranes: a selective ligand for the chloride/calcium-dependent class of L-glutamate binding sites, Brain Res., 278:137-144.
21. Palovick, R.A., Phillips, I., Kappy, M.S., and Raizada, M.K., 1984, Insulin inhibits pyramidal neurons in hippocampal slices, Brain Res., 309:187-191.
22. Recasens, M., Saadoun, F., Baudry, M., Maitre, M., and Lynch, G., 1984, A comparative study of L-[^{3}H]glutamate and L-[^{3}H]cysteine sulfinate binding sites in subcellular fractions of rat brain, J. Neurosci. Res., 11:157-169.
23. Recasens, M., Varga, V., Nenopoulos, D., Saadoun, F., Vinvendon, G., and Benavides, J., 1982, Evidence for cysteine sulfinate as a neurotransmitter, Brain Res., 293:158-173.
24. Scatton, B., and Lehmann, J., 1982, N-methyl-D-aspartate type receptors mediate striatal [^{3}H]acetylcholine release evoked by excitatory amino acids, Nature, 297:422-424.
25. Storm-Mathisen, J., 1977, Localization of transmitter candidates in the brain: The hippocampal formation as a model, Prog. Neurobiol., 8:119-181.
26. Watkins, J.C., Davies, J., Evans, R.H., Francis, A.A., and Jones, A.W., 1981, Pharmacology of receptor for excitatory amino acids, in: "Glutamate as a Neurotransmitter", G. DiChiara, and G.L. Gessa, eds., Raven Press, New York, pp. 263-274.

THE EFFECT OF TAURINE ON EXCITATORY RECEPTOR BINDING IN RAT HIPPOCAMPAL MEMBRANES

F. Bennardini, P. Failli, R. Matucci, A. Bartolini, D. Morandini, L. Polenzani, F. Franconi,[1] and A. Giotti

Dipartimento di Farmacologia Preclinica e Clinica ,"Mario Aiazzi Mancini", Universita degli Studi di Firenze, Firenze, Italy; [1]Cattedra di Saggi e Dosaggi Farmacologici, Università di Sassari, Sassari, Italy

Recent reports have shown that L-^{3}H glutamate binds with high affinity to hippocampal membranes (12). Moreover, this binding is increased by $CaCl_2$ (1), although it is still under discussion whether the calcium increase is mediated by chloride ions (11,16). Many authors suggest that Ca^{2+} can activate neutral protease (2) and that the activity of these proteases could be responsible for learning and memory capability (10). Taurine is a molecule that is able to reduce L-glutamate-induced depolarization in isolated frog spinal cord (7) and to abolish the after-hyperpolarization induced by L-glutamate (8). Moreover, it reduces acquisition of avoidance conditioned reflex (14) and decreases acquisition and, to a lesser extent, retention of a step-down passive avoidance task (13). In the nervous system, taurine interacts with calcium. In fact, 25 mM taurine increases the high affinity Ca^{2+} binding to rat hippocampal membranes but has no effect on low affinity Ca^{2+} uptake to rat brain synaptosomes (6).

METHODS

Membranes were prepared according to the procedure of Enna and Snyder (5). Sprague-Dawley rats weighing 150-250 g were decapitated and the hippocampus was rapidly dissected on ice and homogenized in 20 volumes of ice-cold 320 mM sucrose solution using a glass teflon homogenizer. A pellet was obtained from the 1,000 g supernatant by centrifugation at 14,000 g for 20 min. The pellet was resuspended in 4 ml of cold distilled water containing 1 mM K_2EDTA with vigorous stirring. After 40 min the membranes were pelleted at 7,000 g for 20 min. The supernatant was discharged and the pellet gently washed with cold distilled water to collect only the upper layer. This washing was repeated twice. The suspension was centrifuged at 40,000 g for 30 min, the pellet was resuspended by sonication in cold distilled water and then centrifuged at 50,000 g for 30 min. The pellet was finally resuspended in 50 mM TRIS-HCl buffer, pH 7.4. Each assay contained 50-60 µg protein measured according to Bradford (4), using bovine serum albumin as a standard, and was done in triplicate. Binding studies were performed according to Baudry and Lynch (4). Nonspecific levels of binding were measured by including 100 µM nonradioactive L-glutamate. Aliquots of membrane preparation were preincubated at 30°C for 30 min. At the beginning of preincubation $CaCl_2$ and the drug under examination were added

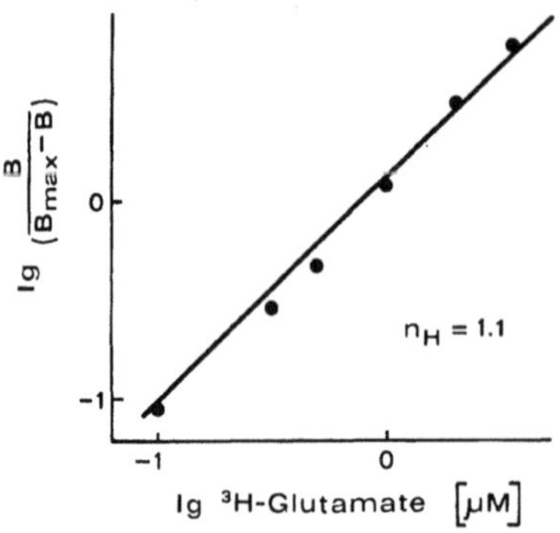

Figure 1. Hill plot of $CaCl_2$-activated L-^3H glutamate binding to rat hippocampal membranes. The B_{max} was calculated by Scatchard analysis.

to the medium. Preliminary experiments indicated that the addition of $CaCl_2$ did not delay the attainment of maximum specific binding. In all our experimental conditions, L-^3H glutamate (46 Ci/mmol, New England Nuclear) was added and incubation continued for 15 min. Incubation was concluded by the addition of the 3 ml ice-cold TRIS-HCl buffer and samples were filtered under vacuum on Millipore cellulose filters. The tubes were rinsed with 3 ml of ice-cold buffer and filters washed with 4 ml of buffer. The radioactivity in the filter was extracted with 3.5 ml of scintillation cocktail (ACS, Amersham) and counted in a scintillation counter (Packard Tricarb 460 C) with an efficiency of 30%.

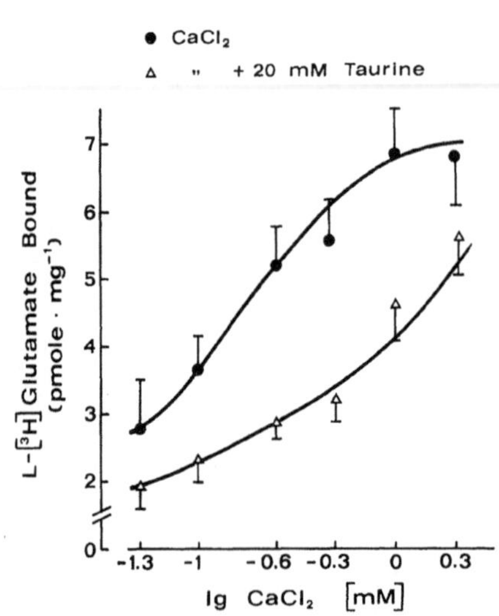

Figure 2. Effect of different $CaCl_2$ concentrations (0.05-2 mM) on L-^3H glutamate binding (0.5 μM) to rat hippocampal membranes in the presence or absence of 20 mM taurine. $CaCl_2$ and taurine were added to the membranes at the beginning of the preincubation period. Results are means ± SEM of at least 7 experiments carried out in triplicate.

Table 1. Effect of taurine on L-^{3}H glutamate binding performed in the presence of $CaCl_2$

	L-^{3}H glutamate bound (pmol.mg^{-1})
1 mM $CaCl_2$	10.43±0.69
20 mM taurine + 1 mM $CaCl_2$ for 30 min	8.24±0.66*
1 mM $CaCl_2$ for 30 min + 20 mM taurine for 30 min	11.28±1.38

Values are means ± SEM of at least 10 experiments, each one performed in triplicate. * $0.05>P>0.01$.

RESULTS AND DISCUSSION

L-^{3}H Glutamate binding to rat hippocampal membranes

The saturation curve of L-^{3}H glutamate binding (0.1-3.5 μM) to rat hippocampal membranes revealed that this binding was saturable. The Scatchard analysis showed a K_D of 0.65 μM and B_{max} of 39.3 pmol.mg^{-1}. The Hill plot gave a slope of 1.1, indicating the absence of cooperativity (Fig. 1). This binding was increased by $CaCl_2$, being augmented 2-3 fold with 1 mM $CaCl_2$. The effect of $CaCl_2$ was studied over the range of 0.05-2 mM. $CaCl_2$ dose-dependently increased the binding reaching a plateau at 1 mM (Fig. 2). This increase was reduced by leupeptin, a specific inhibitor of calcium-dependent neutral protease activity (15). In fact, 40 μM leupeptin reduced calcium-stimulated binding by about 50%, while the two serine protease inhibitors, diisopropylfluorophosphate (DFP) (0.16 mM), and aprotinin (10 trypsin international unit ml^{-1}), were practically ineffective (Fig. 3). The incubation of membrane with 20 mM taurine in the absence of $CaCl_2$ did not modify the L-^{3}H glutamate binding to rat hippocampal membranes, basal and 20 mM taurine values being 4.3±0.4 and 4.2±0.5 pmol.mg^{-1} protein respectively. The incubation of the membranes with 20 mM taurine and $CaCl_2$

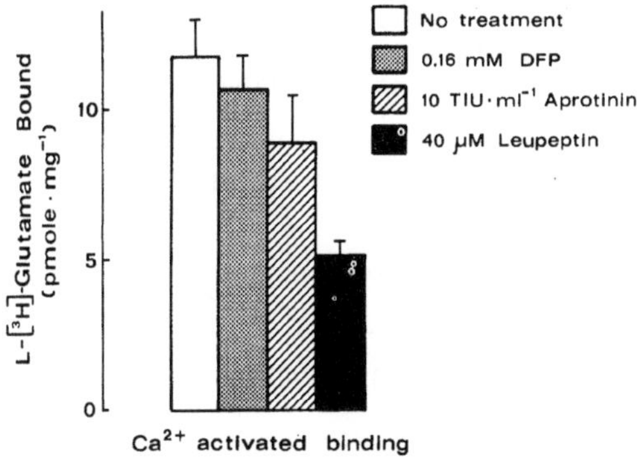

Figure 3. Effect of various protease inhibitors on $CaCl_2$-activated L-^{3}H glutamate binding (0.5 μM) to rat hippocampal membranes. $CaCl_2$ (1 mM), DFP, aprotinin [10 trypsin international units (TIU) per ml] and leupeptin were added to membranes at the beginning of the preincubation period. Values are means ± SEM of 3 different experiments, each one performed in triplicate.

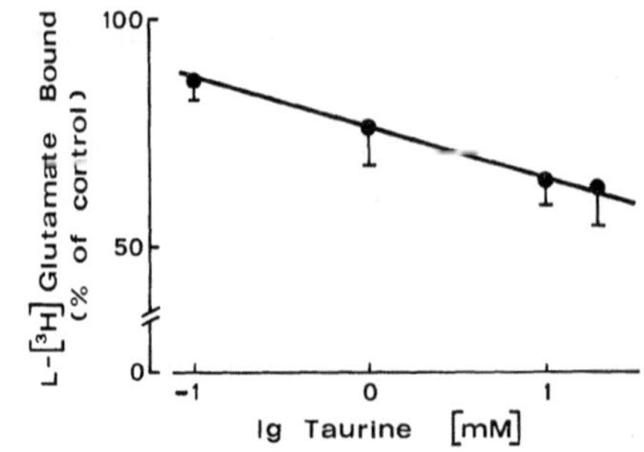

Figure 4. Effect of different concentrations of taurine (0.1-20 mM) on L-^{3}H glutamate binding (0.5 μM) to rat hippocampal membranes performed in the presence of 1 mM $CaCl_2$. Values are means ± SEM of at least 10 experiments, each one performed in triplicate.

reduced the increase of L-^{3}H glutamate binding due to $CaCl_2$. In fact, 20 mM taurine shifted the dose-response curve of $CaCl_2$ to the right without a maximum being reached. The Scatchard analysis of the saturation curve showed that taurine addition did not modify the affinity of L-^{3}H glutamate for its sites, but decreased significantly the maximum number of binding sites (Table 1).

In a second set of experiments performed at 1 mM $CaCl_2$ it has been shown that the inhibition of $CaCl_2$-activated L-^{3}H glutamate binding by taurine was dose-dependent (Fig. 4). Moreover at 1 mM $CaCl_2$ the effect of taurine was present only when the amino acid was incubated together with $CaCl_2$. Taurine was not able to reduce calcium activation, measured when $CaCl_2$ was preincubated with membrane for 30 min before taurine addition (Table 2). The taurine analog, aminomethanesulfonic acid, did not reduce the calcium-activated binding to rat hippocampal membranes (Table 3).

As previously described, 1 mM $CaCl_2$ produces a 2-3 fold increase in L-^{3}H glutamate binding (1,16). This increase is reduced by taurine. In fact the whole dose-response curve of $CaCl_2$ is shifted to the right without a maximum being reached, suggesting a noncompetitive mechanism. Interestingly, taurine does not modify basal binding, suggesting that taurine interacts with a step which is activated by $CaCl_2$. This hypothesis is confirmed by the fact that taurine prevents but does not reverse $CaCl_2$ activation of L-^{3}H glutamate binding. The reduction of binding induced by taurine is ascribed to a reduction of B_{max} while the K_D is not significantly changed. These

Table 2. Effect of taurine on Ca^{2+}-activated L-^{3}H glutamate binding to rat hippocampal membranes

	K_D (μM)	B_{max} (pmol.mg^{-1})
1 mM $CaCl_2$	0.65±0.03	39.27±6.13
1 mM $CaCl_2$ + 20 mM taurine	0.62±0.11	33.07±6.01*

Values are means ± SEM of at least 4 experiments, each one performed in triplicate. * 0.01>P>0.001.

Table 3. Effect of taurine (20 mM) and aminomethanesulfonic acid (20 mM) on L-^{3}H glutamate binding (0.5 uM) to rat hippocampal membranes.

	pmole.mg^{-1} protein			
	Basal binding		$CaCl_2$-activated binding	
Control	(4)	4.19±0.39	(4)	11.83±0.42
Taurine	(10)	4.16±0.33	(10)	8.33±0.66*
Aminomethane-sulfonic acid	(3)	4.47±2.20	(3)	10.20±2.50

Values are means ± SEM with the number of experiments being given in parenthesis, (performed in triplicate). *0.05 P 0.01.

taurine effects could explain the impact of taurine on learning although preliminary experiments have shown that taurine added in the test tube does not modify calcium-activated neutral protease activity. Lazarewicz et al. (13) showed that taurine produces an inhibitory effect on chlortetracycline-calcium fluorescence in rabbit hippocampal membranes, indicating a shift of bound calcium from hydrophobic to hydrophilic regions of the membrane. This different compartmentalization of calcium could explain our results.

ACKNOWLEDGEMENTS

This study was supported by a CNR special project grant "Medicina Preventiva e Riabilitativa" SP 8 (grant #85.00609.56). We would like to thank Paolo Ceccatelli and Mauro Beni for their skillful techincal assistance.

REFERENCES

1. Baudry, M., and Lynch, G., 1979, Regulation of glutamate receptors by cations, Nature, 282:748-750.
2. Baudry, M., and Lynch, G., 1980, Regluation of the hippocampal glutamate receptors: Evidence for the involvement of a calcium-activated protease, Proc. Natl. Acad. Sci., (USA), 77:2298-2302.
3. Baudry, M., and Lynch, G., 1981, Characterization of two ^{3}H-glutamate binding sites in rat hippocampal membranes, J. Neurochem., 36:811-820.
4. Bradford, M.M., 1976, A rapid and sensitive method for the quantification of microgram quantities of protein utilizing the principle of protein-dye binding, Anal. Biochem., 72:248-254.
5. Enna, S.J., and Snyder, S.H., 1975, Properties of γ-aminobutyric acid (GABA) receptor binding in rat synaptic membrane fractions, Brain Res., 100:81-97.
6. Izumi, K., Butterworth, R.F., and Barbeau, A., 1977, Effect of taurine on calcium binding to microsomes isolated from rat cerebral cortex, Life Sci., 20:943-950.
7. Kurachi, M., Yoshihara, K., and Ailara, H., 1983, Effect of taurine on depolarizations induced by L-glutamate and other excitatory amino acids in isolated spinal cord of the frog, Jap. J. Pharmacol., 33:1247-1254.
8. Kurachi, M., and Aihara, H., 1985, The action of taurine on the response to glutamate in the motoneuron of the isolated frog spinal cord, Neuropharmacology, 24:775-781.
9. Lazarewicz, J.W., Noremberg, K., Lehmann, A., and Hamberger, A., 1985, Effects of taurine on calcium binding and accumulation in rabbit

hippocampal and cortical synaptosomes, Neurochem. Int., 7:421-427.
10. Lynch, G., and Baudry, M., 1984, The biochemistry of memory: a new and specific hypothesis, Science, 224:1057-1063.
11. Mena, E.E., Fogg, G.E., and Cotman, C.W., 1982, Chloride ions enhanced L-glutamate binding to rat brain synaptic membranes, Brain Res., 243:378-381.
12. Roberts, P.J., 1974, Glutamate receptors in rat central nervous system, Nature, 252, 399-401.
13. Sanberg, P.R., and Fibiger, H.C., 1979, Impaired acquisition and retention of a passive avoidance response after chronic ingestion of taurine, Psychopharmacology, 62:97-99.
14. Sgaragli, G.P., 1978, Effect of taurine on animal behavior, in: "Taurine and Neurological Disorders", A. Barbeau and R.J. Huxtable eds., Raven Press, New York, pp. 307-318.
15. Toyo-oka, T., and Masaki, T., 1979, Calcium-activated neutral proteases from bovine ventricular muscle: isolation and some of its properties, J. Mol. Cell. Cardiol., 11:769-786.
16. Whittemore, S.R., Mena, E.E., Monaghan, D.T., and Cotman, C.W., 1983, Regional distribution and ionic requirement of Cl^-/Ca^{2+}-activated and Cl^-/Ca^{2+}-independent glutamate receptors in rat brain, Brain Res., 277:99-107.

TAURINE DEFICIENCY IN THE RAT AND CAT: EFFECTS ON NEUROTOXIC AND BIOCHEMICAL ACTIONS OF KAINATE

Anders Lehmann, Ryan J. Huxtable* and Anders Hamberger

Institute of Neurobiology, University of Göteborg
Göteborg, Sweden and *Department of Pharmacology
University of Arizona, Tucson, AZ, U.S.A.

Taurine occurs at high concentration in the brain and has been suspected to be a major inhibitory transmitter. However, only a compound that fulfils all the criteria for transmitter identification can be considered to play this role. One important criterion is that the pharmacological actions of the proposed transmitter be mimicked by the endogenous substance. This, in fact, has been demonstrated for taurine only in the cerebellar cortex (16). Thus, the present state of knowledge does not support a general transmitter role for taurine. Since taurine affects so many different processes in the brain, a "neuromodulator" function has been proposed. The fact that an exogenously added agent influences a process is not necessarily relevant for its normal function. A better understanding of the physiological role for a substance may come from studies where the endogenous level of that compound has been lowered, and physiological perturbations are observed. One of the chief reasons why so little is known about taurine function is that this experimental approach is difficult. There are presently two ways to deplete endogenous taurine selectively. One is to administer the taurine uptake blocker 2-guanidinoethane sulfonate (GES) to rodents (5), and the other is to feed cats a taurine-deficient diet (21,23). The first approach is inexpensive and comparatively rapid, but involves introduction of a neuroactive agent (15) which accumulates in the brain (5,11, 12) and may have side-effects. The second approach leads to a more severe taurine depletion, but it is expensive and time-consuming. In both cases the treatment has to be carried out chronically. Adaptive changes may be induced and might indeed be anticipated if taurine has an important function. One such possibly homeostatic alteration is the claimed enhancement of Na^+-independent binding of taurine after GES treatment (11).

We have reported that excitatory amino acids are potent stimulators of taurine release in vivo (8,9). Inasmuch as exogenously applied taurine protects nerve cells against the toxic action of the excitatory amino acids in certain (3,4,19) situations, we have proposed that endogenously released taurine acts as an anti-excitotoxin (8).

Here we report on intra- and extracellular amino acids and on the toxicity of kainate in the brain of taurine-depleted rats.

EXPERIMENTAL PROCEDURES

GES Administration to Rats

Male Sprague-Dawley rats (250-350 g) were given water with or without 1% GES. Nine days later, the animals were anesthetized with urethane (1.25 g/kg) and placed in a stereotaxic apparatus. A dialysis probe (20) was inserted into the hippocampus and perfused at 2.5 µl/min with Krebs-Ringer bicarbonate buffer (9). After an equilibration period, perfusates were collected and analysed for amino acids with automatic o-phthaldialdehyde derivatization and HPLC separation (10). The effect on extracellular amino acids of locally perfused 1 mM kainate was followed during a 50 min period. The rats were then decapitated and the hippocampus was dissected and immersed in liquid N_2. The tissue was homogenized in 0.6M $HClO_4$. After centrifugation at 10,000 g for 10 min, the supernatant was neutralized and analyzed for amino acids.

In a separate series of experiments, rats (control or GES-treated) were injected under pentobarbital anesthesia with 0.5 nmol kainate in the dorsal hippocampus. One week later, the animals were perfusion-fixed with 4% paraformaldehyde and the hippocampus was embedded in Historesin after appropriate dehydration and infiltration. Sections (2 µm) were stained with hematoxylin-eosin and studied with light microscopy.

Taurine Deficiency in Kittens

Ten kittens were divided into two groups with 4 males and one female in each group. Directly after weaning, they were given a synthetic diet (ref. 22; Astra-Ewos, Södertälje, Sweden) which was taurine deficient or supplemented with 0.4% taurine. Dialysis experiments were performed after 8-12 weeks: the kittens were anesthetized with Saffan (alphaxalone 9 mg/ml and alphadolone acetate 3 mg/ml). 1.5 ml/kg was injected

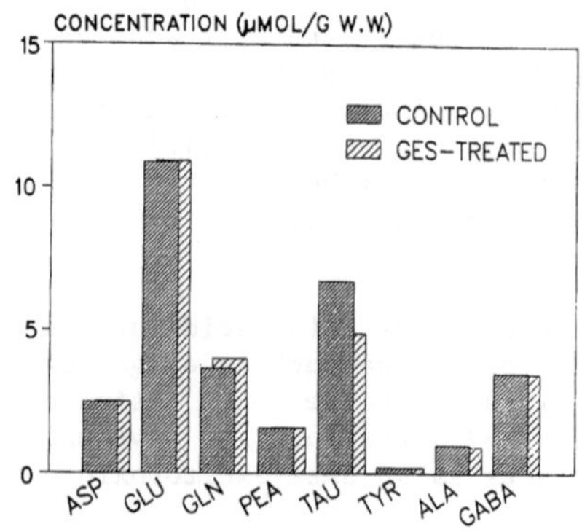

Fig. 1. Amino acid levels in the hippocampus of control and GES treated rats. Only taurine was significantly changed ($p<0.01$) after treatment with 1% GES for 9 days.

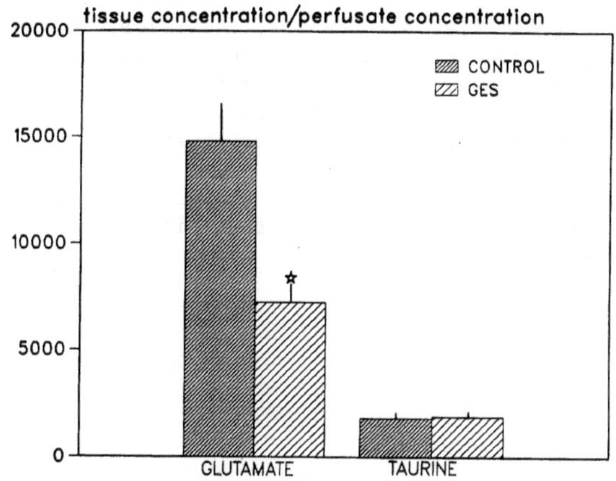

Fig. 2. Intracellular/extracellular concentration ratio for glutamate and taurine in the hippocampus of control and GES treated rats. The experimental conditions are described in the text. Star denotes $p<0.05$.

initially and supportive doses (0.2 mg/kg) were given every 15 min. The kittens were placed in a stereotaxic frame and a dialysis tubing (20) was implanted into the parietal cortex 5 mm posterior to the bregma and 5 mm laterally. The brain dialysis was performed as described above. At the end of the experiment, blood was collected in a heparinized test tube. The blood was centrifuged directly (10,000 g; 5 min) to remove cells and plasma proteins were precipitated with 0.6M $HClO_4$. The neutralized supernatant was analyzed for its amino acid content. The kittens were allowed to recover for 3 days after the dialysis experiment. They were then lightly anesthetized with diethyl ether and placed in a restraining device. Thirty min later, a cannula connected to an infusion pump was inserted intraperitoneally and pentylene tetrazol (PTZ; 20 mg/ml) was infused at 0.2 ml/min. The infusion was stopped when clonic seizures appeared and the dose of PTZ was calculated.

RESULTS

GES Administration to rats

Nine days of oral administration of GES resulted in a moderate and selective fall in the total hippocampal content of taurine (Fig. 1).

In the extracellular space there was a depression of taurine and, additionally, an increase of glutamate. A comparison of the intra-/extracellular gradients for these amino acids revealed that GES did not affect the ratio for taurine while the glutamate ratio was halved (Fig. 2). Kainate (1 mM) evoked a release of predominantly taurine and phosphoethanolamine (PEA; Fig. 3). However, the release of many other amino acids was also increased by kainate. One notable exception was glutamine which was depressed by kainate administration. Extracellular glutamate was higher in the GES rats than in the control rats during kainate perfusion, while taurine levels were lower.

Injection of 0.5 nmol kainate into the hippocampus induced a selective loss of pyramidal cells of the CA3 a-b subfield. This effect was not overtly altered by GES treatment (Fig. 4).

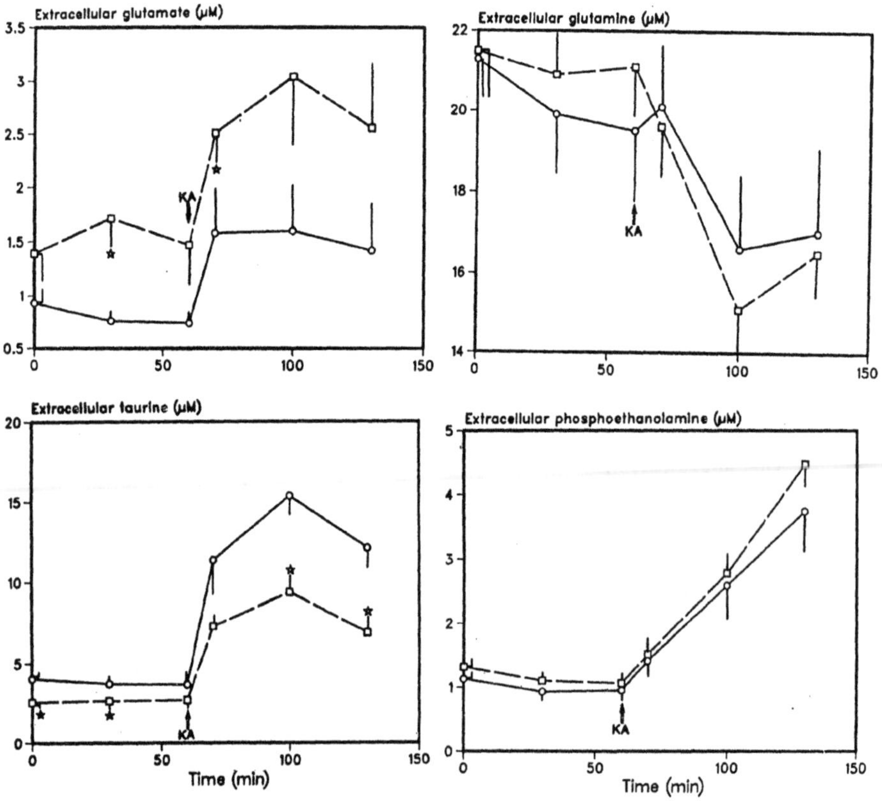

Fig. 3. Basal and kainate-evoked changes in glutamate, glutamine, taurine and phosphoethanolamine in control (solid line) and GES (interrupted line) treated rats. The dialysis-perfusion procedure is described in the text. Kainate was included in the medium at 1 mM. Star denotes p <0.05.

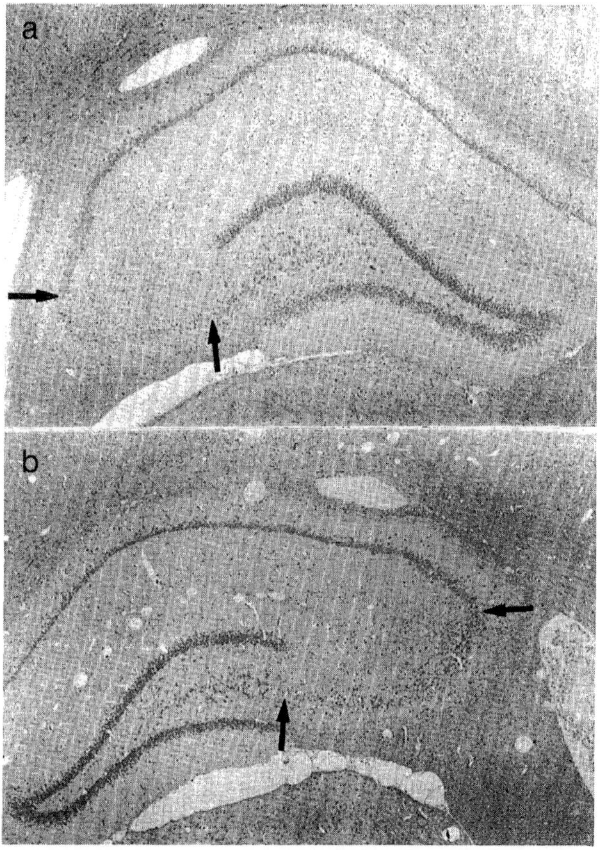

Fig. 4. Effect of locally injected kainate (0.5 nmol) in the rat hippocampus in (a) control and (b) GES treated rats. The kainate lesion encompasses most of the CA3a - b pyramidal cells (area between arrows).

Taurine Deficiency in Kittens

There was no difference in growth rate between the two groups: at the time of sacrifice, taurine-supplemented kittens weighed 1.87±0.08 kg and deficient kittens weighed 2.01±0.10 kg. The kittens raised on a taurine-free regimen developed a marked taurine deficiency: plasma taurine was 123.4 μM in the taurine-supplemented group, and 1.8 μM in the deficient group (Table 1). In addition, the plasma concentration of asparagine was significantly elevated in taurine-deficient kittens (Table 1).

Table 1. Plasma amino acids in taurine supplemented and taurine deficient kittens.

Amino acid	Supplemented (μM)	Deficient (μM)
Asp	14+1	15+2
Asn	42+3	58+6*
Glu	158+18	182+18
Ser	123+22	190+20
PEA	8+1	13+2
Tau	123+40	2+0*
Tyr	70+3	59+9
Ala	292+36	294+45
Met	30+2	43+9
Val	201+21	242+42
Phe	100+7	86+8
Ile	76+11	92+18
Leu	125+21	166+47

Kittens were fed a taurine supplemented (n=5) or taurine deficient (n=5) diet directly after weaning. The animals were used for dialysis experiments 8-12 weeks later, at which time blood samples were taken and analysed as described in the text. Asterisk denotes p <0.05.

Analysis of the extracellular amino acid concentrations of the cerebral cortex showed a selective decrease of taurine in taurine-deficient animals. Taurine decreased to 3.8% of control in the extracellular space of the cortex (Table 2), compared to 1.4% in plasma. Kainate (1 mM) evoked a similar response in terms of amino acid changes in the feline cortex as in the rat hippocampus (Table 2). One exception was a lack of PEA response in the kitten. Extracellular taurine was doubled after kainate perfusion in control kittens, but was virtually unaffected in

Table 2. Effects of kainic acid on extracellular amino acids in the parietal cortex of taurine supplemented and deficient kittens.

Time (min)	Extracellular taurine (μM) Supplemented	Extracellular taurine (μM) Deficient	Extracellular phosphoethanolamine (μM) Supplemented	Extracellular phosphoethanolamine (μM) Deficient
-20	2.13+0.30	0.08+0.01*	0.85+0.12	1.15+0.35
-10	2.01+0.33	0.08+0.01*	0.62+0.09	0.94+0.25
0	2.10+0.25	0.08+0.01*	0.58+0.08	0.83+0.25
10	4.14+0.61	0.11+0.02*	0.64+0.04	0.94+0.21
30	3.78+0.46	0.08+0.01*	0.41+0.02	0.67+0.15
50	3.27+0.22	0.08+0.01*	0.49+0.05	0.81+0.14
Time (min)	**Extracellular glutamine (μM) Supplemented**	**Extracellular glutamine (μM) Deficient**	**Extracellular glutamate (μM) Supplemented**	**Extracellular glutamate (μM) Deficient**
-20	36.3+8.7	43.1+10.0	0.32+0.12	0.46+0.17
-10	35.3+5.9	44.4+11.9	0.30+0.12	0.41+0.18
0	40.3+8.2	43.5+10.9	0.36+0.16	0.35+0.11
10	36.4+7.4	44.6+11.5	0.72+0.23	0.66+0.10
30	30.4+8.1	36.1+13.0	0.52+0.13	0.49+0.08
50	28.6+9.7	38.7+14.0	0.34+0.04	0.48+0.09

Brain dialysis was performed as outlined in "Experimental Procedures". Kainate was included at 1 mM in the perfusion medium at Time =0. Asterisk denotes significant difference between taurine supplemented and deficient kittens (p<0.05).

taurine-deficient animals (Table 2).The threshold for PTZ seizures did not differ between taurine-supplemented and deficient kittens: the threshold was 36.3±3.2 mg/kg in the former group and 37.3±4.0 in the latter.

DISCUSSION

The present results demonstrate some important differences between the two experimental models for taurine deficiency. In rats treated with GES, there was a selective reduction of intracellular taurine. In the extracellular space there was a decrease of taurine. In addition, extracellular glutamate was doubled in GES-treated animals. The effect on interstitial glutamate might be a direct effect of GES, or a secondary response to taurine depletion. Provided this effect is not species- or region-dependent, the results from the kitten experiments (i.e. unaltered extracellular glutamate) strongly suggest that the elevation of glutamate is caused directly by GES. This observation is of importance for the interpretation of results obtained after GES treatment, as glutamate has opposite neurophysiological effects as taurine. However, the matter is further confused by the inhibitory actions of GES per se (15).

Both supportive (3,4,19) and conflicting (17,18) evidence exist for exogenous taurine as an anti-excitotoxin. In the present study we were unable to find a GES-induced neuronal supersensitivity to kainate toxicity. It appears, therefore, that endogenous taurine is no simple modulator of kainate neurotoxicity.

The acute challenge of the hippocampus with kainate elicited an increase of extracellular taurine which was unchanged in relative terms after GES treatment. However, in taurine-deficient kittens the cortex was virtually unresponsive to kainate with respect to taurine release. This may be explained by a better ability of the feline cortical parenchyma to retain taurine in the deficient state, as the reduction of intracellular taurine is far less than the depression of extracellular taurine (cf. ref. 23 and the present work).

Taurine has been widely documented as an effective anticonvulsant when given by a route that by-passes the blood-brain barrier. The interest for the antiepileptic action of taurine has been fueled by the assumptiom that endogenous taurine has a similar effect. This issue has been addressed using the GES model for taurine deficiency. However, while central injections of taurine protect mice against PTZ seizures (7), reduction of brain taurine with GES has no effect on the PTZ threshold in the adult rat (14) and even raises the threshold in the neonatal rat (1). On the other hand, GES treatment leads to a diminished threshold for cysteine sulfinic acid-induced epilepsy (6). Our observation in the kitten that PTZ threshold was unchanged when extracellular taurine is diminished by 96% indicates that this amino acid is not an important endogenous anticonvulsant, at least not in the PTZ model. However, little is known about adaptive changes in the taurine-deficient kitten, and therefore no definite conclusions can be drawn as yet. Another evidence against a taurine mediated control of epileptogenesis is that a taurine antagonist (TAG) evokes electrographic spikes only at very high concentrations (50 mM≤; ref. 2). In comparison, any manipulation which reduces GABAergic inhibition will inevitably evoke seizures (13).

In summary, the role of taurine in excitotoxicity and epilepsy is still uncertain and will remain so until novel appropriate experimental approaches have been developed.

ACKNOWLEDGEMENTS

This work was partially supported by the Swedish Medical Research Council (grant No. 00164). The technical assistance of Ms. B. Karlsson and the secretarial aid of Ms. G. Grönstedt is gratefully acknowledged.

REFERENCES

1. Bonhaus, D.W., Pasantes-Morales, H., and Huxtable, R.J., 1985, Actions of guanidinoethane sulfonate on taurine concentration, retinal morphology and seizure threshold in the neonatal rat, Neurochem. Int. 7:263- 279.
2. Fariello, R.G., Golden, R.T., and McNeal, R.B., Taurine and related amino acids in seizure disorders - current controversies, In "Taurine: Biological Actions and Clinical Perspectives" (Eds. Oja, S.S., Ahtee, L., Kontro, P. and Paasonen, M.K.), Alan R. Liss, New York (1985), pp. 413-424.
3. Fariello, R.G., Golden, R.T., and Pisa, M., 1982, Homotaurine (3-aminopropane-sulfonic acid: 3-APS) protects from the convulsant and cytotoxic effects of systemically administered kainic acid, Neurology 32:241-245.
4. French, E.D., Vezzani, A., Whetsell, W.O., Jr., and Schwarcz, R., 1985, Anti-excitotoxic effects of taurine in the rat hippocampus in vivo and in vitro, Epilepsia 26:506-507.
5. Huxtable, R.J., Laird, H.E., and Lippincott, S.E., 1979, The transport of taurine in the heart and the rapid depletion of tissue taurine content by guanidinoethyl sulfonate, J. Pharmacol. Exp. Ther. 211:465- 471.
6. Iwata, H., Yamamoti, T., Kumagae, Y., and Baba, A., Further study on cysteine sulfinate-induced EEG seizures in rat. In "Sulfur Amino Acids: Biochemical and Clinical Aspects" (Eds. Kuriyama, K., Huxtable, R.J. and Iwata, H.), Alan R. Liss, New York (1983), p. 224.
7. Izumi, K., Igisu, H., and Fukuda, T., 1975, Effect of edetate on seizure suppressing actions of taurine and GABA, Brain Research 88:576-579.
8. Lehmann, A., Hagberg, H., and Hamberger, A., 1984, A role for taurine in the maintenance of homeostasis in the central nervous system during hyperexcitation? Neurosci. Lett. 52:341-346.
9. Lehmann, A., Isacsson, H., and Hamberger, A., 1983, Effects of in vivo administration of kainic acid on the extracellular amino acid pool in the rabbit hippocampus, J. Neurochem. 40:1314-1320.
10. Lindroth, P., Sandberg, M., and Hamberger, A., Liquid chromatographic determination of amino acids after pre-column fluorescence derivatization. In "Neuromethods", Vol. 3 "Amino Acids" (Eds. Boulton, A.A., Baker, G.B. and Wood, J.D.) Humana Press, Clifton, N.J. (1985), pp. 97-116.
11. Marnela, K.-M., and Kontro, P., 1984, Free amino acids and the uptake and binding of taurine in the central nervous system of rats treated with guanidinoethanesulfonate, Neuroscience 12:323-328.
12. Marnela, K.-M., Kontro, P., and Oja, S.S., 1984, Effects of prolonged guanidinoethanesulfonate administration on taurine and other amino acids in rat tissues, Med. Biol. 62:3239-3244.
13. Meldrum, B.S., 1975, Epilepsy and GABA-mediated inhibition, Int. Rev. Neurobiol. 17:1-36.

14. Nakagawa, K., and Huxtable, R.J., 1985, The anticonvulsive actions of two lipophilic taurine derivatives, Neurochem. Int. 7:819-824.
15. Okamoto, K., and Sakai, Y., 1981, Inhibitory actions of taurocyamine, hypotaurine, homotaurine, taurine and GABA on spike discharges of Purkinje cells, and localization of sensitive sites, in guinea pig cerebellar slices, Brain Research 206:371-386.
16. Okamoto, K., Kimura, H., and Sakai, Y., 1983, Evidence for taurine as an inhibitory neurotransmitter in cerebellar stellate interneurons: selective antagonism by TAG (6-aminomethyl-3-methyl-4H, 1,2,4-benzothiadiazinel,1-dioxide), Brain Research 265:163-168.
17. Olney, J.W., and Price, M.T., 1980, Neuroendocrine interactions of excitatory and inhibitory amino acids, Brain Res. Bull. 5 (Suppl. 2):361-368.
18. Olney, J.W., Price, M.T., Fuller, T.A., Labruyère, J., Sanson, L., Carpenter, M., and Mahan, K., 1986, The anti-excitotoxic effects of certain anesthetics, analgesics and sedative-hypnotics, Neurosci. Lett. 68:29-34.
19. Sanberg, P.R., Staines, W., and McGeer, E.G., 1979, Chronic taurine effects on various neurochemical indices in control and kainic acid-lesioned neostriatum, Brain Research 161:367-370.
20. Sanberg, M., Butcher, S.P., and Hagberg, H., 1986, Extracellular overflow of neuroactive amino acids during severe insulin-induced hypoglycemia - in vivo dialysis of the rat hippocampus, J. Neurochem. 47:178- 184.
21. Schmidt, S.Y., Berson, E.L., and Hayes, K.C., 1976, Retinal degeneration in cats fed casein. I. Taurine deficiency, Invest. Ophthal. 15:47-52.
22. Sturman, J.A., Gargano, A.D., Messing, J.M., and Imaki, H., 1986, Feline maternal taurine deficiency. Effect on mother and offspring, J. Nutr. (in press).
23. Sturman, J.A., Rassin, D.K., Hayes, K.C., and Gaull, G.E., 1978, Taurine deficiency in the kitten: exchange and turnover of (^{35}S) taurine in brain, retina and other tissues, J. Nutr. 108:1462-1476.

FURTHER OBSERVATION ON THE INTERACTION OF TAURINE AND CALMODULIN ON THE CENTRAL ADRENERGIC NEURON

Tomio Segawa, Yasuyuki Nomura and Ichiro Shimazaki

Department of Pharmacology, Institute of Pharmaceutical Sciences, Hiroshima University School of Medicine, Hiroshima 734, Japan

Taurine (2-aminoethanesulfonic acid) is present in the CNS in high concentration. Several findings support the hypothesis that taurine might act as an inhibitory transmitter or a modulator in CNS (1,3,9). Taurine decreases the depolarization-induced release of labelled norepinephrine and acetylcholine from rat cerebral cortex, possibly by increasing retention of Ca^{2+} within mitochondria (5). Similarly, taurine reduced the release of [^{3}H]GABA from guinea pig cerebellar slices by reducing the availability of Ca^{2+} (8).

It has become evident that calmodulin (CaM) can mediate various Ca^{2+}-dependent processes in synaptic transmission and can play a role in the complex interactions between neurotransmitters, Ca^{2+} and cyclic AMP (10). However, possible involvement of CaM in a modulatory role of taurine in synaptic transmission has not been considered so far.

We previously presented the evidence suggesting that taurine could function as a neuromodulator which stabilizes and/or reduces β-adrenoreceptor sensitivity by interaction with Ca^{2+} in the CNS (7) and that possibly CaM is involved in this modulatory role of taurine (11). In order to investigate the mechanism of an inhibitory action of taurine with CaM on adrenergic neurons in brain, the following effects were examined: 1) the influence of the CaM inhibitor, N-(6-aminohexyl)-5-chloro-1-naphthalene sulfonamide hydrochloride (W-7) on taurine uptake into nerve terminals; 2) the influence of W-7 and taurine on high potassium-induced uptake of Ca^{2+} into nerve terminals; 3) the influence of taurine on the Ca^{2+}-induced difference spectra of CaM.

METHODS

Male Wistar rats weighing approximately 200 g were used throughout.

Influence of W-7 on taurine uptake into synaptosomal P_2 fractions

After decapitation, cerebral cortex was rapidly dissected. Crude mitochondrial P_2 fractions were prepared by the method of Simon et al. (13). Briefly, the tissue was homogenized in 20 vol of ice-cold 0.32 M sucrose with a Teflon pestle and the homogenate was centrifuged for 10 min at 1,000 g. The pellet (crude nuclear fraction) was discarded and the resulting supernatant was centrifuged for 20 min at 17,000 g to obtain crude mitochondrial P_2 pellets.

P_2 fractions suspended in 2ml Krebs-Ringer solution were centrifuged at 17,000 g for 10 min, after which 2 ml high potassium Krebs-Ringer solution (30 mM KCl) were added to the precipitate, incubated at 37°C for 10 min. Following centrifugation at 17,000 g for 10 min, the precipitate was washed twice with 2 ml cold Krebs-Ringer solution after which [^{14}C]taurine (113 mCi/mmol, Amersham) and W-7 (final concentration 20 μM) in 2 ml Krebs-Ringer solution were added, incubated at 37°C for 15 min. Following centrifugation at 17,000 g for 10 min, the precipitate was washed twice with 2 ml cold Krebs-Ringer solution. Following centrifugation at 17,000 g for 10 min, radioactivity in the pellet was counted in Bray solution.

Influence of taurine and W-7 on high potassium-induced uptake of $^{45}Ca^{2+}$ into synaptosomal P_2 fractions

The method of Mendelson et al. (6) was used. Endogenous substances were removed from P_2 fractions as described above, after which 10 mM taurine or 20 μM W-7 in Krebs-Ringer solution was added to the fractions and the fractions were incubated at 37°C for 15 min. Following centrifugation at 17,000 g for 10 min, the precipitate was washed twice with Krebs-Ringer solution, and suspended in Krebs-Ringer solution. To the suspension 1 ml of high potassium Krebs-Ringer solution containing 10 μCi/assay $^{45}Ca^{2+}$ (18.4 Ci/mg, New England Nuclear), 10 μM imipramine and 10 μM nialamide were added and the mixture was incubated for 10 min at 37°C. The uptake of $^{45}Ca^{2+}$ was terminated by rapid filtration through Whatman GF/C filter and the amount of radioactivity on the filter was counted. The amount of high potassium induced $^{45}Ca^{2+}$ uptake was expressed as the difference between the amount of radioactivities in the presence and the absence of high potassium (30 mM).

Measurement of difference UV absorption spectra of CaM

Difference UV absorption spectra of CaM were measured using a Hitachi Model 200-10 spectrophotometer by the method of Yazawa et al. (15). CaM (100 μg) was dissolved in 10 mM Tris-HCl buffer solution (pH 8.0) containing 0.1 M NaCl. In the sample and reference solution, 9.15 mM $CaCl_2$ and 0.1 mM EGTA was added, respectively. The influence of taurine on Ca^{2+}-induced difference spectra of CaM was investigated by addition of taurine to the sample solution.

RESULTS

Influence of W-7 on taurine uptake into synaptosomal P_2 fractions

As shown in Fig. 1, W-7 (20 μM) did not influence the uptake of taurine (1-20 mM) into synaptosomal P_2 fractions.

Influence of taurine and W-7 on high potassium-induced uptake of $^{45}Ca^{2+}$ into synaptosomal P_2 fractions

One minute after incubation high potassium-induced $^{45}Ca^{2+}$ uptake into synaptosomal P_2 fractions reached the maximum level. Thereafter it decreased gradually over the next 10 min (Fig. 2). There were no significant differences between the amounts of $^{45}Ca^{2+}$ uptake into nontreated fractions and taurine- or W-7-treated fractions.

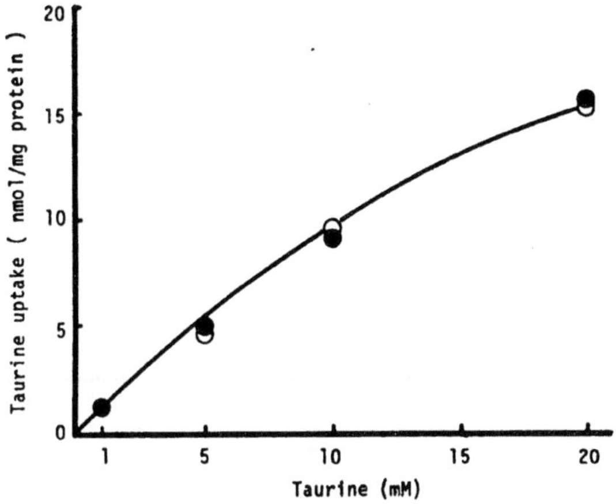

Fig. 1. Influence of W-7 on uptake of taurine into synaptosomal P_2 fractions of rat cerebral cortex in the presence (○) or absence (●) of 20 μM W-7.

Influence of taurine on difference UV absorption spectra of CaM

Fig. 3 shows Ca^{2+}-induced difference UV absorption spectra of CaM. A negative UV difference was observed at between 270-280 nm and 280-290 nm. When 10 mM taurine was added to the medium the negative difference disappeared. Leucine (10 mM) also changed the spectrum to some extent but a negative difference was still observed (Fig. 4).

DISCUSSION

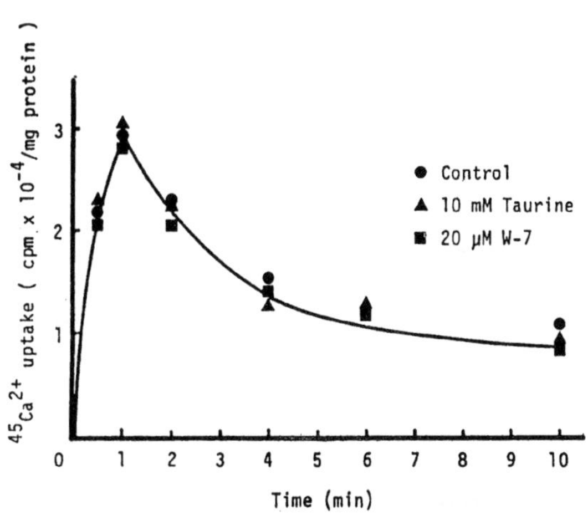

Fig. 2. Influence of taurine and W-7 on high potassium-induced $^{45}Ca^{2+}$ uptake into synaptosomal P_2 fractions of rat cerebral cortex.

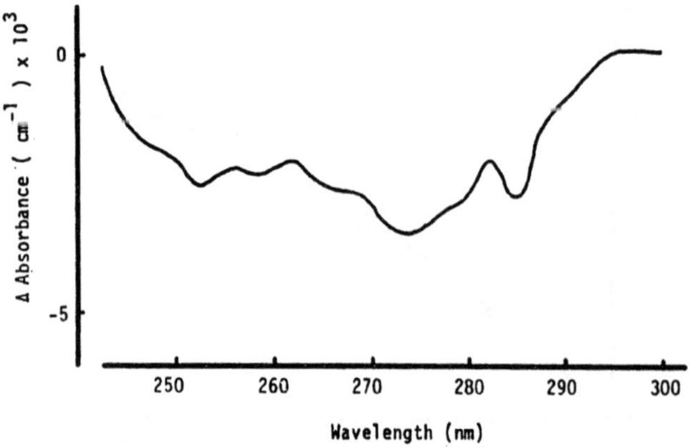

Fig. 3. Ca^{2+}-induced difference UV absorption spectra of CaM.

Much evidence has been presented which suggests that taurine may act as a neuromodulator by a modification in the movement of Ca^{2+} (5,8). We previously reported that taurine stabilizes and/or reduces β-adrenoceptor sensitivity by interaction with Ca^{2+} in CNS (7) and that the Ca^{2+}-CaM system is involved in this mechanism (11). In addition taurine was found to inhibit norepinephrine release from nerve terminal. In the presence of W-7, the inhibitory activity of taurine on norepinephrine release disappeared, suggesting that the Ca^{2+}-CaM system is involved (11). From an equilibrium binding study we obtained evidence suggesting that taurine diminishes affinity of Ca^{2+} binding to CaM without affecting the maximum nunber of binding sites (12). From these results we concluded that taurine regulates β-adrenergic neuron activity by reducing binding affinity of Ca^{2+} to CaM both in presynaptic and postsynaptic areas.

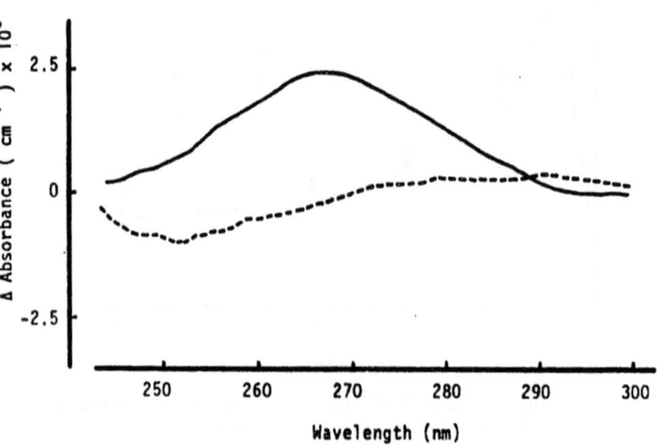

Fig. 4. Influence of taurine or leucine on Ca^{2+}-induced difference UV absorption spectra of CaM. (——); taurine (10 mM), (······); leucine (10 mM)

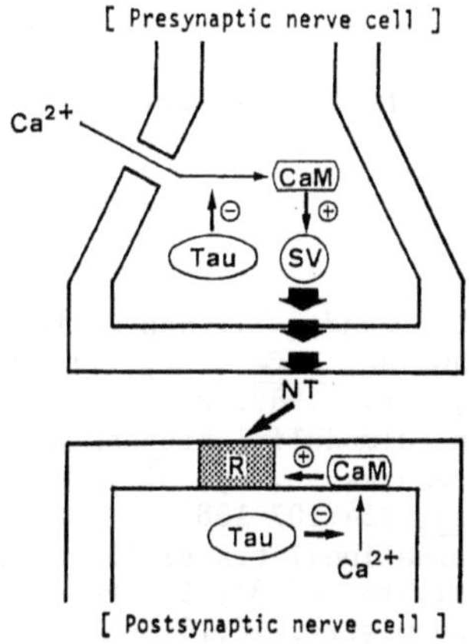

Fig. 5. Schematic diagram for possible interaction of taurine and calmodulin in adrenergic neuron. Tau; taurine, CaM; calmodulin, SV; synaptic vesicle, NT; neurotransmitter, R; receptor.

However, before we reach the conclusion mentioned above, the following two possibilities must be excluded. 1) Pretreatment with W-7 might inhibit taurine uptake into nerve endings; 2) W-7 and/or taurine might inhibit Ca^{2+} uptake into nerve endings. From the results obtained in present experiments these two possibilities were excluded and it became obvious that both W-7 and taurine influence norepinephrine release by interaction with CaM intracellularly.

In order to obtain the precise mechanism for the inhibitory effect of taurine on Ca^{2+}-CaM binding, Ca^{2+}-induced difference UV absorption spectra of CaM were measured. When one or two moles Ca^{2+} bind to one mole CaM, exposure of tyrosine residues of CaM is increased as part of a large conformational change of the CaM molecule. As a result a small increase in α-helix content occurs. The difference UV absorption spectrum of CaM in a solvent containing Ca^{2+}, a negative UV difference being observed at 279 and 287 nm, is usually interpreted as due to an increased exposure of tyrosine residues to the solvent (2,4,14,15). After the addition of taurine (10 mM), however, a positive difference spectrum was observed while the addition of leucine (10 mM) had no influence on the spectrum. It is postulated that there is a positive cooperativity between the two first sites to be occupied by Ca^{2+} (2). When the first Ca^{2+} binding site of CaM is occupied by Ca^{2+}, then a comformational change occurs and as a result, the 2nd, 3rd, and 4th Ca^{2+} easily bind to CaM. Taurine might inhibit such a conformational change by an unknown mechanism.

The possibility that taurine antagonizes Ca^{2+} binding to CaM is ruled out because effective concentration of taurine is mM order which is too high to consider such a mechanism. Chelate formation of taurine with Ca^{2+} is also unlikely because β-alanine, a more potent chelator than taurine, had no effect of Ca^{2+}-CaM system.

From these results, together with our previous observations, it is suggested that in central adrenergic neurons taurine does not interfere with the influx of Ca^{2+} but it does interfere with Ca^{2+}-CaM system, thereby decrease norepinephrine release from nerve terminals. Postsynaptically, taurine may also decrease receptor activity by a similar mechanism (Fig. 5). Details of such a mechanism remains to be determined.

ACKNOWLEDGEMENT

The authors wish to thank Ms. Nariko Irisawa for preparation of the manuscript.

REFERENCES

1. Barbeau, A., Inoue, N., Tsukada, Y., and Butterworth, R.F., 1975, The neuropharmacology of taurine, Life Sci., 17:669-678.
2. Crouch, T.H., and Klee, C.B., 1980, Positive cooperative binding of calcium to bovine brain calmodulin, Biochemistry, 19:3692-3698.
3. Davidson, A.N., and Kaczmarek, L.K., 1971, Taurine, a possible neurotransmitter, Nature, 234:107-108.
4. Klee, C.B., 1977, Conformational transition accompanying the binding of Ca^{2+} to the protein activator of 3', 5'-cyclic adenosine monophosphate phosphodiesterase, Biochemistry, 16:1017-1024.
5. Kuriyama, K., Muramatsu, M., Nakazawa, K., and Kakita, K., 1978, Modulating role of taurine on release of neurotransmitter and calcium transport in excitable tissues, in: "Taurine and Neurological Disorders", A. Barbeau and R. J. Huxtable eds., Raven Press, New York, pp. 201-216.
6. Mendelson, W.B., Skolnich, P., Martion, J.J., Lun, M.D., Wagner, R., and Paul, S.M., 1984, Diazepam-stimulated increases in the synaptosomal uptake of $^{45}Ca^{2+}$: Reversal by dihydropyridine calcium channel antagonists, Eur. J. Pharmacol., 104:181-183.
7. Miura, Y., Nomura, Y., and Segawa, T., 1982, Interaction of taurine with γ-adrenergic receptors in rat central nervous system, Sulfur Amino Acids, 5:77-86.
8. Namina, M., Okamoto, K., and Sakai, Y., 1983, Modulatory action of taurine on the release of GABA in cerebellar slices of the guinea-pig, J. Neurochem., 40:1-9.
9. Okamoto, K., and Sakai, Y., 1980, Localization of sensitive sites to taurine, γ-aminobutyric acid, glycine and β-alanine in the molecular layer of guinea-pig cerebellar slices, Brit. J. Pharmac., 69:407-413.
10. Schaffer, S.W., Kramer, J.H., Lampson, W.G., Kulakowski, E., and Sakane, Y., 1983, Effect of taurine on myocardial metabolism: Role of calmodulin, in: "Sulfur Amino Acids: Biochemical and Clinical Aspects", K. Kuriyama, R.J. Huxtable and H. Iwata eds., Alan R. Liss, Inc., New York, pp. 39-50.
11. Segawa, T., Nomura, Y., and Shimazaki, I., 1985, Possible involvement of calmodulin in modulatory role of taurine in rat cerebral β-adrenergic neurons, in: "Taurine: Biological Actions and Clinical Perspectives", S.S. Oja, L. Ahtee, P. Kontro and M.K. Paasonen eds., Alan R. Liss, Inc., New York, pp. 321-330.
12. Shimazaki, I., Nomura, Y., and Segawa, T., 1984, Interaction of taurine and calmodulin on adrenergic neuron in rat brain, Sulfur Amino Acids, 7:25-31.
13. Simon, J.R., Contrera, J.F., and M.J. Kuhar, 1976, Binding of [^{3}H]kainic acid, an analogue of L-glutamate, to brain membranes, J. Neurochem., 36:141-147.
14. Yazawa, M., Kuwayama, H., and Yagi, K., 1978, Molecular protein as a Ca^{2+}-dependent activator of rabbit skeletal myosine light-chain kinase, J. Biochem., 84:1253-1258.
15. Yazawa, M., Sakuma, M., and Yagi, K., 1980, Calmodulins from muscles of marine invertebrates, scallop and sea anemone, J. Biochem., 87:1313-1320.

INTERACTIONS OF TAURINE AND DOPAMINE IN THE STRIATUM

Pirjo Kontro

Department of Biomedical Sciences
University of Tampere
Box 607, SF-33101 Tampere, Finland

The concentration of taurine in the striatum is high (23, 29). High-affinity taurine uptake has been demonstrated in striatal synaptosomes (5, 24), and selective uptake into a subpopulation of striatal neurons (4). Both the content and K^+-stimulated release of taurine are reduced after an intrastriatal injection of kainate, indicating a neuronal localization of the amino acid (25, 31). Although electrophysiological data on taurine in the striatum are scant, the above evidence suggests that taurine is involved in striatal functions. The striatum is particularly rich in dopaminergic neurons, rendering it possible that taurine could modulate dopaminergic transmission. Indeed, intraventricularly injected taurine increases the synthesis of dopamine in the rat brain and inhibits the firing of dopamine neurons (7). Intraventricular taurine also stimulates prolactin secretion, possibly via interaction with dopamine or serotonin transmitter systems (34), and unilateral intranigral injection of taurine induces a dose-dependent contralateral circling behavior which is partially blocked by haloperidol (12). I have now further elucidated the modulatory role of taurine in the function of striatal dopaminergic systems by studying mutual interactions of taurine and dopamine in release and uptake processes in the rat striatum. The effects of taurine on the binding of tritiated spiperone, a ligand for dopamine receptors, to striatal synaptic membranes was also investigated.

MATERIALS AND METHODS

Materials

[1,2-^{3}H]Taurine (specific activity 1.07 PBq/mol), [7,8-^{3}H]dopamine (1.81 PBq/mol) and [phenyl-4-^{3}H]spiperone (spiroperidol, 0.666 PBq/mol) were obtained from Amersham International, U.K. Taltrimide (2-phthalimidoethane-sulphon-N-isopropylamide, MY-117) and its dealkylated biologically active metabolite (2-phthalimidoethanesulphonamide, MY-103) were synthesized as described in Andersen et al. (1). Other reagents and drugs were from common commercial sources. Adult Sprague-Dawley rats of both sexes were used throughout.

Release of taurine and dopamine

Striatal slices 0.4 mm thick weighing 15-20 mg prepared with a Stadie-Riggs tissue slicer were first preloaded for 30 min at 37°C with 10 μM labelled taurine or 0.1 μM labelled dopamine and then superfused for 50 min in Krebs-Ringer-Hepes-glucose medium at a flow rate of 0.25 ml/min as previously described (22). The superfusion medium was pooled during the first 20 min and thereafter 2-min fractions (0.5 ml) were collected directly into scintillation vials with a fraction collector (Gilson TDC 220). At 30 min the superfusion medium was changed to another medium containing 50 mM K^+ ions (stimulated release) or left unchanged (spontaneous release). The effectors were also added to the medium at 30 min. In dopamine release experiments, the medium always contained 0.1 mM ascorbic acid and 0.02 mM nialamide. After superfusion the slices were weighed, homogenized in 5 % (w/v) ice-cold trichloracetic acid solution, centrifuged and the resulting clear supernatants measured for radioactivity. Any major metabolic breakdown of labelled taurine and dopamine was excluded (22). The efflux rate constants for the time intervals of 20-30 min (k_1), 34-50 min (k_2) for taurine and 32-40 min (k_2) for dopamine were computed from linear regression of the logarithmic contents of labelled taurine or dopamine in slices on superfusion time (21). The constants are given for the experimental phases of 34-50 min or 32-40 min as percentages of the constants of the baseline efflux from 20-30 min.

Uptake of taurine and dopamine

The uptake of taurine and dopamine by rat striatal homogenates was measured as described in Kontro (14) and Kontro and Oja (16). The homogenates (0.8 mg protein/sample) were first preincubated in plain Krebs-Ringer-Hepes-glucose medium for 10 min and then for 10 min with labelled taurine (10 μM) or for 5 min with labelled dopamine (0.1 μM). In all incubations with dopamine the medium contained 0.1 mM ascorbic acid and 0.02 mM nialamide. The effectors were present only during the final incubations. The incubations were terminated by rapid filtration through Whatman glass-fiber filters (CF/B) with subsequent rinsings with cold medium. The filters and samples from the medium were assayed for radioactivity as described in Oja and Kontro (27). The amount of taurine or dopamine taken up by the homogenates was calculated as in Kontro and Oja (15).

Spiperone binding

Striatal membranes were prepared and binding experiments with $[^3H]$-spiperone carried out as described by Kontro and Oja (18). Specific binding was defined as the difference in bound radioactivity between samples incubated without and with 10 μM (+)butaclamol. $[^3H]$Spiperone binding to serotonergic receptors was displaced with 0.1 μM mianserin. The binding parameters, KD and Bmax, and their standard errors (SEM) were computed by the method of least squares from the Scatchard plots. The IC_{50} values and their 95 % confidence intervals were calculated by means of a computer program based on inverse prediction using a fitted single linear regression model.

Statistical comparisons

Significance of differences between the control and test values was assessed by two-tailed Student's t-test.

RESULTS

Taurine efflux from striatal slices was fairly slow, about 75 % of the radioactivity remaining in the slices after the 50 min superfusion. Potassium stimulation (50 mM K^+) enhanced the release more than 2-fold, the response being slow at the onset and only slowly returning to the basal level if the slices were again exposed to low-potassium superfusion medium from 40 min onwards (Fig. 1). Spontaneous taurine efflux was increased by dopamine and apomorphine, but not by norepinephrine and serotonin (Table 1). On the other hand, the potassium-stimulated release was supressed by

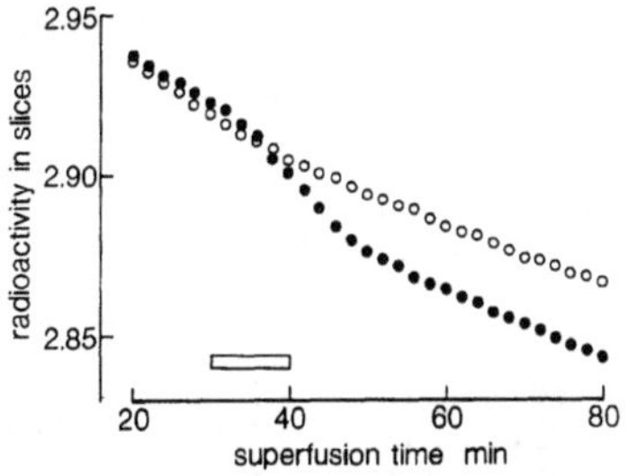

Fig. 1. Spontaneous (-o-) and K^+-stimulated (-•-) release of taurine from rat striatal slices. The logarithm of permillage of radioactivity remaining in the slices is plotted against superfusion time. The period of superfusion with 50 mM K^+ is indicated by the horizontal bar. The negative slopes of the straight regression lines fitted to the experimental points for selected intervals represent efflux rate constants. A single representative experiment is shown.

dopamine and apomorphine as well as by the two other monoamines and the taurine derivatives, taltrimide and MY-103 (Table 1). The decreased release by dopamine was significantly ($p<0.01$) enhanced when both dopamine and haloperidol were present. Haloperidol alone had no significant effect on the evoked taurine release. Furthermore, the maximal taurine release occurred earlier in the presence of dopamine than in unsupplemented superfusion medium (Fig. 2).

The basal spontaneous outflow of dopamine from striatal slices was unaffected by apomorphine, haloperidol as well as taurine and its derivatives. Dopamine release was enhanced over 30-fold by 50 mM K^+, this response being potentiated by apomorphine, haloperidol and isoproterenol (Table 2). Taurine (1.0 mM) was also stimulatory, whereas taltrimide and MY-103 had no effect. Stimulation with 25 mM K^+ caused an 11-fold increase in dopamine release, which effect was also augmented by taurine, whereas apomorphine was now inhibitory (Table 2). Also the anticonvulsive taurine derivatives now decreased dopamine release.

Spiperone binding to striatal membranes was saturable in the studied concentration range (0.01-7.5 nM), consisting of two types of binding sites, high- and low-affinity, as revealed by the constructed Scatchard plots. The estimated binding constants (KD, means ±SEM) were 77 ± 14 pM and 4.8 ± 0.3 nM, and the maximal binding capacities (Bmax) 69 ± 9 nmol/kg and 6.8 ± 0.3 μmol/kg membrane protein, respectively. The inhibitor effects were studied using 7 different inhibitor concentrations at a fixed spiperone concentration of 0.3 nM. The IC_{50} values were calculated from the data obtained (Table 3). The binding was effectively displaced by (+)butaclamol and dopamine. Taurine, taltrimide and MY-103 also inhibited spiperone binding, whereas GABA, β-alanine and homotaurine were ineffective (IC_{50} >10 mM). The effect of taurine on spiperone binding was further characterized by assessing the binding parameters in the presence of 1 μM taurine. The binding still consisted of the two components, but the Bmax (3.8 ± 0.4 μmol/kg) and KD (3.4 ± 0.5 nM) of the low-affinity component were both significantly decreased by taurine, while Bmax was unaltered (63 ± 4 nmol/kg).

Table 1. Taurine release from rat striatal slices

Substance (mM) present in medium 30 - 50 min		Efflux rate constants (%) ±SEM Spontaneous k_2 (34-50 min)		Stimulated k_2 (34-50 min)	
None (control)		88.7 ± 3.1	(9)	201.3 ± 8.8	(9)
Dopamine	1.0	102.2 ± 4.9*	(10)	152.2 ± 3.5**	(8)
	0.1	95.6 ± 1.8	(4)	163.9 ± 10.0*	(5)
	0.01	91.7 ± 3.5	(6)	150.5 ± 8.6**	(5)
Norepinephrine	1.0	91.7 ± 8.2	(5)	156.3 ± 16.9*	(4)
Serotonin	1.0	95.9 ± 4.6	(6)	165.9 ± 8.4*	(5)
Apomorphine	0.1	110.5 ± 8.5*	(5)	159.9 ± 6.1**	(4)
	0.01	110.1 ± 4.8**	(5)	176.3 ± 11.1	(5)
Haloperidol	0.005	82.1 ± 2.0	(4)	228.8 ± 10.9	(7)
Dopamine +haloperidol	0.01 0.005	-		188.1 ± 6.6	(5)
Taltrimide	0.1	-		166.6 ± 6.8*	(5)
MY-103	0.1	-		153.4 ± 6.9**	(5)

The slices were preloaded for 30 min in Krebs-Ringer-Hepes-glucose medium, pH 7.4, with 10 μM [^{3}H]taurine and then superfused for 50 min, from 30 to 50 min with the above effectors (spontaneous release) or with 50 mM K^+ together with the above effectors (stimulated release). The results are calculated as percentages from the average basal rate constant (k_1, 20-30 min) (1.59 ± 0.10) x 10^{-3} min^{-1} (mean ±SEM, n=135). Number of independent experiments in parentheses. Significance of differences compared to the corresponding controls: *$p<0.05$, **$p<0.01$.

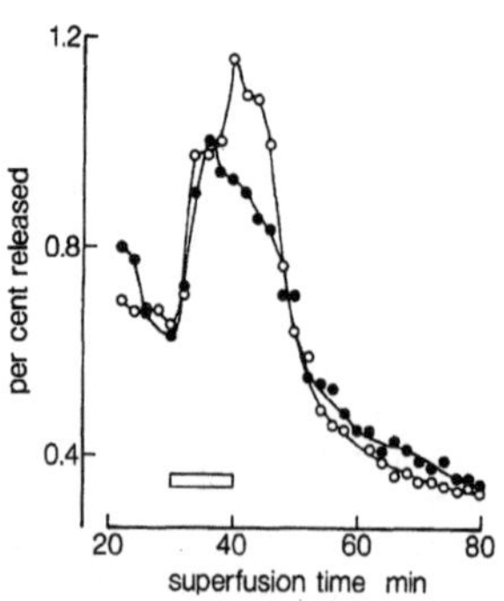

Fig. 2. K^+-stimulated taurine release from rat striatal slices in standard medium (-o-) and in the presence of 10 μM dopamine (-•-). The period of stimulation by 50 mM K^+ is indicated by the horizontal bar. A single representative experiment is shown.

Table 2. Dopamine release from rat striatal slices

Substance (mM) present in medium 30 - 50 min		Efflux rate constants (%) ±SEM k_2 (32-40 min)
None		84 ± 6 (7)
50 mM K^+ (control)		2546 ± 103 (9)
Taurine	1.0	2850 ± 69* (6)
Apomorphine	0.1	2850 ± 47* (5)
Haloperidol	0.005	2920 ± 86* (5)
Isoproterenol	0.01	2881 ± 100* (5)
Clonidine	0.1	2893 ± 115 (3)
Taltrimide	0.1	2595 ± 194 (5)
MY-103	0.1	2611 ± 83 (7)
25 mM K^+ (control)		917 ± 70 (9)
Taurine	1.0	1236 ± 102* (5)
Apomorphine	0.1	570 ± 98* (5)
Haloperidol	0.005	1242 ± 63** (5)
Taltrimide	0.1	632 ± 40** (3)
MY-103	0.1	571 ± 105* (4)

The slices were preloaded for 30 min in Krebs-Ringer-Hepes-glucose medium, pH 7.4, with 0.1 μM [^{3}H]dopamine and then superfused for 50 min, from 30 to 50 min with 50 mM K^+ or 25 mM K^+ together with the above effectors. The results are calculated as percentages from the basal rate constant (k_1, 20-30 min) (1.88 ± 0.02) x 10^{-3} min^{-1} (mean ±SEM, n=121). Number of independent experiments in parentheses. Statistical significance of differences from the corresponding controls: *$p<0.05$, **$p<0.01$.

Table 3. Inhibition of spiperone binding to rat striatal membranes

Inhibitor	IC_{50}	
(+)Butaclamol	2.6 nM	(2.2 - 3.1)
Dopamine	3.0 nM	(2.6 - 3.8)
MY-103	493 nM	(415 - 639)
Taltrimide	11.8 μM	(10.1 - 14.6)
Taurine	34.4 μM	(31.7 - 37.9)

Data are from 2 separate quadruplicate experiments at 7 concentrations of the inhibitors. The spiperone concentration was 0.3 nM. The results are means with 95 % confidence intervals given in paretheses. Data from Kontro and Oja (18).

Taurine uptake by striatal homogenates was only slightly inhibited by dopamine and not at all by taltrimide or MY-103 (Table 4). Dopamine uptake was not affected by taurine and its analogs, whereas the other monoamines were strongly inhibitory.

Table 4. Uptake of taurine and dopamine by rat striatal homogenates

Effector (mM)		Uptake (% of control) Taurine		Dopamine	
None (control)		100.0± 3.7	(19)	100.0± 3.4	(18)
Taurine	1.0	-		95.2± 4.0	(6)
Dopamine	1.0	84.0± 3.9*	(10)	-	
Norepinephrine	1.0	94.1± 4.0	(6)	3.5± 0.5**	(4)
Serotonin	1.0	34.3± 3.0**	(9)	8.9± 0.7**	(4)
Taltrimide	0.1	87.6± 0.4	(4)	107.9± 2.1	(4)
MY-103	0.1	90.8± 2.5	(4)	95.7± 5.4	(4)

Striatal homogenates were incubated in Krebs-Ringer-Hepes-glucose medium, pH 7.4, with [^{3}H]taurine (10 μM) or [^{3}H]dopamine (0.1 μM) for 10 and 5 min, respectively. The results are means ±SEM with number of experiments in parentheses. Taurine uptake without effectors was 4.81 ± 0.34 nmol/(s x kg) (mean ±SEM, n=19) and that of dopamine 5.39 ± 0.18 nmol/(s x kg) (n=18). Significance of differences from the corresponding controls: *p<0.05, **p<0.01.

DISCUSSION

The salient features of stimulated taurine release appeared to be basically similar in the striatum and in other brain regions. The response to K^+-stimulation was characteristically slow in striatal slices, as also in slices from the cerebral cortex, cerebellum and brain stem (19). The same phenomenon has been found in the striatum in vivo (8) and in striatal neurons in primary culture (30). The increase in taurine outflow from the retina is particularly delayed, becoming manifest only after the cessation of K^+-stimulation (32, 37). Stimulated taurine release from the retina is reduced by dopamine (0.8 mM or 4 mM) (32), but there has been no earlier report on similar dopamine effects on striatal taurine release. Dopamine at concentrations prevailing in the rat striatum in vivo (about 100 μmol/kg, 36) now clearly inhibited the evoked taurine release and altered its time course. The dopaminergic systems are thus potentially able to modify stimulated taurine release. In keeping with this, apomorphine mimicked here the suppressive action of dopamine and the effect was also partially antagonized by haloperidol. In contrast to stimulated taurine release, spontaneous taurine efflux was increased dose-dependently by dopamine. The effects of dopamine and apomorphine were again the same. This depressant action of dopamine on stimulated taurine release was shared by the other monoamines, norepinephrine and serotonin, suggesting a partial lack of specificity of the effect. Also haloperidol failed to potentiate significantly evoked taurine release, contrary to observations in the mouse cerebral cortex (20).

Dopamine release from dopaminergic terminals could be modulated by presynaptic autoreceptors. The existence of these is still uncertain, however, the contradictory results possibly resulting from different experimental conditions and tissue preparations used (see 3, 33). In the present study dopamine release from striatal slices evoked by 25 mM K^+ could have been modified by presynaptic dopamine receptors, apomorphine inhibiting and haloperidol potentiating the release. On the other hand, both apomorphine and haloperidol were stimulatory when the release was elicited by 50 mM K^+. In keeping with this, a strong presynaptic inhibition has been found in norepinephrine release induced by 13-26 mM K^+, while the release caused by

56 mM K^+ has been hardly at all affected by α-receptor activation (6). Taurine now facilitated dopamine release stimulated by both K^+ concentrations (25 and 50 mM). This action may not be directly mediated by dopamine autoreceptors. The same statement does not apply to the anticonvulsive taurine derivatives, which both decreased the 25 mM K^+-evoked dopamine release sensitive to autoreceptors.

GABA potentiates dopamine release from rat striatal slices, but it is not clear whether or not the effect involves GABAergic receptors (see 3). Stimulation through glycinergic receptors is among the mechanisms suggested for GABA action (13). In principle, taurine could also exert its effect through GABA and/or glycine receptors (see 17). In addition, a GABA-evoked dopamine release from striatal slices blocked by both strychnine and picrotoxin has been described (13). The present results are not at variance with the proposition of Kerwin and Pycock (13) that GABA may elicit this dopamine release via taurine (or β-alanine) "receptors".

Despite the heterogeneous nature of [^{3}H]spiperone binding sites the ligand is predominantly bound in the striatum to sites displaying characteristics of the postsynaptic D_2 dopamine receptors responsible for the behavioral and clinical effects of neuroleptics (35). The present KD value of the high-affinity spiperone binding tallies with those reported by other investigators (9, 11, 26). This high-affinity site obviously represents mainly dopamine D_2 receptors. The low affinity site probably corresponds to the spiperone binding site (KD >1 nM) of unknown pharmacological character (2, 10, 26).

Taurine was now found to interfere with both spiperone binding sites at concentrations prevailing in the rat striatum in vivo (28). Its mode of action appeared to be rather complex, the binding affinity being more markedly affected than the maximal binding capacity. KD of high-affinity binding was reduced by taurine, suggesting sensitization of the neuroleptic receptors. Bmax of the low-affinity site decreased together with KD, characteristic of an uncompetitive type of inhibition. This could signify that taurine is not capable of interacting directly with the low-affinity binding site. The site may become available for taurine only after spiperone has been bound to the receptor. Taltrimide and MY-103 inhibited spiperone binding even more strongly than taurine. Moreover, oral taltrimide has been shown to affect dopamine-mediated behavior in mice and rats, causing a transient decrease in amphetamine-induced hyperactivity and potentiation of amphetamine-induced stereotypy (38). Taltrimide also potentiates haloperidol-induced catalepsy dose-dependently. The above behavioral effects of this anticonvulsant may also be mediated by dopamine D_2 receptors.

SUMMARY

Both spontaneous and K^+-stimulated taurine release from rat striatal slices were affected by dopamine and apomorphine, suggesting that dopaminergic systems are able to modulate taurine release. K^+-stimulated dopamine release was potentiated by taurine, which effect may not involve dopamine autoreceptors. Taurine was able to inhibit spiperone binding to striatal membranes in a uncompetitive manner and thus interfere with the function of dopaminergic receptors.

ACKNOWLEDGEMENTS

The author thanks Miss Riitta Mero for skilful technical assistance and the Research Council for Medicine of the Academy of Finland and the Scientific Fund of the University of Tampere for financial support.

REFERENCES

1. Andersen, L., Sundman, L.-O., Lindén, I.-B., Kontro, P., and Oja, S.S., 1984, Synthesis and anticonvulsant properties of some 2-aminoethanesulphonic acid (taurine) derivatives, J. Pharm. Sci., 73: 106-108.
2. Banay-Schwartz, M., Wajda, I.J., Manigault, I., DeGuzman, T., and Lajtha, A., 1982, Lithium: effect on [^{3}H]spiperone binding, ionic content, and amino acid levels in the brain of rats, Neurochem. Res., 7: 179-189.
3. Chesselet, M.-F., 1984, Presynaptic regulation of neurotransmitter release in the brain: facts and hypotheses, Neuroscience, 12: 347-375.
4. Clarke, D.J., Smith, A.D., and Bolam, J.P., 1983, Uptake of [^{3}H]taurine into medium-size neurons and into identified striatonigral neurons in the rat neostriatum, Brain Res., 289: 342-348.
5. Collins, G.G.S., 1974, The rates of synthesis, uptake and disappearance of [^{14}C]taurine in eight areas of the rat central nervous system, Brain Res., 76: 447-459.
6. Dismukes, R.K., De Boer, A.A., and Mulder, A.H., 1977, The mechanism of alpha-receptor mediated modulation of [^{3}H]noradrenaline release from slices of rat neocortex, Naunyn-Schmiedebergs Arch. Pharmac., 299: 115-122.
7. Garcia de Yebenes Prous, J., Carlsson, A., and Mena Gomez, M.A., 1978, The effect of taurine on motor behaviour, body temperature and monoamine metabolism in rat brain, Naunyn-Schmiedebergs Arch. Pharmac., 304: 95-99.
8. Girault, J.A., Barbeito, L., Spampinato, U., Gozlan, H., Glowinski, J., and Besson, M.-J., 1986, In vivo release of endogenous amino acids from the rat striatum: further evidence for a role of glutamate and aspartate in corticostriatal neurotransmission, J. Neurochem., 47: 98-106.
9. Hancock, A.A., and Marsh, C.L., 1984, Distinctions between ligand-binding sites for [^{3}H]dopamine and D_2 dopaminergic receptors characterized with [^{3}H]spiroperidol, Mol. Pharmac., 26: 439-451.
10. Howlett, D.R., and Nahorski, S.R., 1980, Quantitative assessment of heterogeneous [^{3}H]spiperone binding to rat neostriatum and frontal cortex, Life Sci., 26: 511-517.
11. Huff, R.A.M., and Molinoff, P.B., 1982, Quantitative determination of dopamine receptor subtypes not linked to activation of adenylate cyclase in rat striatum, Proc. Natl Acad. Sci. USA, 79: 7562-7565.
12. Kaakkola, S., and Kääriäinen, I., 1980, Contralateral circling behavior induced by intranigral injection of taurine in rats. Acta Pharmac. Toxic. (Kbh), 46: 293-298.
13. Kerwin, R., and Pycock, C., 1979, Effects of ω-amino acids on tritiated dopamine release from rat striatum: evidence for a possible glycinergic mechanism, Biochem. Pharmac., 28: 2193-2197.
14. Kontro, P., 1984, Comparison of taurine, hypotaurine and β-alanine uptake in brain synaptosomal preparations from developing and adult mouse, Int. J. Dev. Neurosci., 2: 465-470.
15. Kontro, P., and Oja, S.S., 1978, Taurine uptake by rat brain synaptosomes, J. Neurochem., 30: 1297-1304.
16. Kontro, P., and Oja, S.S., 1984, Binding and uptake of taurine and GABA in developing and adult mouse brain, Acta Univ. Tamper. [B], 21: 60-67.
17. Kontro, P., and Oja, S.S., 1985, Properties of sodium-independent taurine binding to brain synaptic membranes, in:" Taurine: Biological Actions and Clinical Perspectives", S.S. Oja, L. Ahtee, P. Kontro and M.K. Paasonen, eds., Alan R. Liss, New York, pp. 249-260.
18. Kontro, P., and Oja, S.S., 1986, Taurine interferes with spiperone binding in the striatum, Neuroscience, 19: 1007-1010.
19. Kontro, P., and Oja, S.S., 1987, Taurine efflux from brain slices: potassium-evoked release is greater from immature than mature brain tissue, in this volume.

20. Kontro, P., and Oja, S.S., 1987, Taurine and GABA release from mouse cerebral cortex slices: effects of structural analogues and drugs, Neurochem. Res., 12: 475-482.
21. Korpi, E.R., and Oja, S.S., 1979, Efflux of phenylalanine from rat cerebral cortex slices as influenced by extra- and intracellular amino acids, J. Neurochem., 32: 789-796.
22. Korpi, E.R., and Oja, S.S., 1984, Comparison of two superfusion systems for study of neurotransmitter release from rat cerebral cortex slices, J. Neurochem., 43: 236-242.
23. Lombardini, J.B., 1976, Regional and subcellular studies on taurine in the rat central nervous system, in: "Taurine", R. Huxtable and A. Barbeau, eds., Raven Press, New York, pp. 311-326.
24. Lombardini, J.B., 1978, High-affinity transport of taurine in the mammalian central nervous system, in: "Taurine and Neurological Disorders", A. Barbeau and R.J. Huxtable, eds., Raven Press, New York, pp. 119-135.
25. Nicklas, W.J., Duvoisin, R.C., and Berl, S., 1979, Amino acids in the rat neostriatum: alteration by kainic acid lesion, Brain Res., 167: 107-117.
26. Nomura, Y., Oki, K., and Sewaga, T., 1982, Ontogenic development of the striatal [^{3}H]spiperone binding: regulation by sodium and guanine nucleotide in rats, J. Neurochem., 38: 902-908.
27. Oja, S.S., and Kontro, P., 1980, Hypotaurine uptake by brain slices from adult and 8-day-old mice, J. Neurochem., 35: 1303-1308.
28. Oja, S.S., Kontro, P., and Lähdesmäki, P., 1977, Amino acids as inhibitory neurotransmitters, Prog. Pharmac., 1 (3): 1-119.
29. Perry, T.L., Sanders, H.D., Hansen, S., Lesk, D., Kloster, M., and Gravlin, L., 1972, Free amino acids and related compounds in five regions of biopsied cat brain, J. Neurochem., 19: 2651-2656.
30. Pin, J.-P., Weiss, S., Sebben, M., Kemp, D.E., and Bockaert, J., 1986, Release of endogenous amino acids from striatal neurons in primary culture, J. Neurochem., 47: 594-603.
31. Placheta, P., Singer, E., Schönbeck, G., Heckl, K., and Karobath, M., 1979, Reduction of endogenous level, uptake and release of taurine after intrastriatal kainic acid injection, Neuropharmacology, 18: 399-402.
32. Pycock, C.J., and Smith, L.F.P., 1983, Interactions of dopamine and the release of [^{3}H]-taurine and [^{3}H]-glycine from the isolated retina of the rat, Br. J. Pharmac., 78: 395-404.
33. Raiteri, M., Marchi, M., and Maura, G., 1984, Release of catecholamines, serotonin, and acetylcholine from isolated brain tissue, in: "Handbook of Neurochemistry, 2nd edn, vol. 6", A. Lajtha, ed., Plenum Press, New York, pp. 431-462.
34. Scheibel, J., Elsasser, T., and Ondo, J.G., 1980, Stimulation of prolactin secretion by taurine, a neurally depressant amino acid, Neuroendocrinology, 30: 350-354.
35. Seeman, P., 1980, Brain dopamine receptors, Pharmac. Rev., 32: 229-313.
36. Sharp, T., Zetterström, T., and Ungerstedt, U., 1986, An in vivo study of dopamine release and metabolism in rat brain regions using intracerebral dialysis, J. Neurochem., 47: 113-122.
37. Smith, L.F.P., and Pycock, C.J., 1982, Potassium-stimulated release of radiolabelled taurine and glycine from the isolated rat retina, J. Neurochem., 39: 653-658.
38. Toivonen, M.-L., Lindén, I.-B., and Vapaatalo, H., 1984, Effect of taltrimide on dopamine mediated behaviour. Abstracts of the Symposium Taurine: Biological Actions and Clinical Perspectives, p. 49, Hanasaari, Espoo.

EFFECTS OF ACIDIC DIPEPTIDES ON AMINOACIDERGIC NEUROTRANSMISSION IN THE BRAIN

V. Varga[1,2], Kirsi-Marja Marnela[1], Pirjo Kontro[1], J. Gulyás[3], Z. Vadász[2], P. Lähdesmäki[4] and S.S. Oja[1]

[1]Department of Biomedical Sciences, University of Tampere, Box 607, SF-33101 Tampere, Finland,
[2]Institute of Biochemistry, Department I, Semmelweis University Medical School, Box 260, H-1444 Budapest, Hungary,
[3]Institute of Organic Chemistry, Eötvös Lóránd University, Múzeum krt. 6-8, H-1088, Budapest, Hungary,
[4]Department of Biochemistry, University of Oulu, SF-90570 Oulu, Finland

Among the numerous oligopeptides found in the brain the small acidic peptides have received little attention, although their existence has long been known (5, 6, 36, 38). These substances are often acetylated and contain aspartate, glutamate, serine, glycine, alanine and taurine in various combinations (9, 30, 37, 40). Of the taurine-containing peptides the most plentiful and best known is L-glutamyl-taurine (31), which exists in two forms, alpha and gamma, in approximately equal amounts in calf brain synaptic vesicles (32). Their origin is still obscure, although the biosynthesis of γ-glutamyl-taurine has been shown to be catalyzed by γ-glutamyltransferase (42). Various effects have been attributed to γ-glutamyltaurine in different organisms (10). However, the physiological role of all the above mentioned peptides has so far remained entirely unknown in the case of the brain. We have therefore studied the effects of glutamyltaurine and some synthetic analogs on the actions of putative neurotransmitter amino acids.

MATERIALS AND METHODS

Materials

The peptides used in this study were synthesized by us. Their homogeneity was checked with thin-layer chromatography, paper electrophoresis, mass spectrometry and automated amino acid analysis. The peptides and amino acids were in L-form unless otherwise indicated. L-[G-^{3}H]Glutamic acid (spec. act. 1.33 PBq/mol), [2,3-^{3}H]GABA (2.11 PBq/mol), [1,2-^{3}H]taurine (1.07 PBq/mol), [G-^{3}H]kainic acid (0.185 PBq/mol) and [G-^{3}H]diazepam (3.15 PBq/mol) were from Amersham International, U.K. Adult Wistar rats and NMRI mice of both sexes were used.

Tissue preparations

The brains were rapidly excised from decapitated animals. Cerebral cortex slices 0.4 mm thick were prepared at 4°C with a Stadie-Riggs tissue slicer for studies of amino acid efflux. Synaptosomal preparations (P_2 fractions) were isolated by the method of Whittaker and Barker (43) for studies of amino acid uptake. For studies of GABA binding, frozen, thawed, and thoroughly washed membranes were prepared from the crude synaptosomal membrane fraction as described by Zukin et al. (45). For studies of glutamate, kainate and diazepam binding, synaptosomes were prepared from the cerebral cortex by the differential centrifugation method of Hajós (12). The synaptosomal fraction obtained was osmotically shocked in cold distilled water and the membranes pelleted by centrifugation at 50,000 g for 20 min. The prepared membrane pellet was used fresh for studies of glutamate and diazepam binding and after freezing and thawing for studies of kainate binding.

Binding of ligands

The specific binding of [^{3}H]glutamate, [^{3}H]kainate and [^{3}H]GABA to brain synaptic membranes was determined by the methods of Larder and McLennan (27), London and Coyle (28) and Varga et al. (41), respectively. The incubations were terminated by centrifugation at 6500 g for 15 min. The specific [^{3}H]diazepam binding was determined by the filtration method of Tallman et al. (39). The possible minor modifications of the methods are indicated in the legends to Figures and Tables. The substances to be tested were first preincubated with synaptic membranes for 10 min at 0°C and then [^{3}H]glutamate, [^{3}H]kainate or [^{3}H]GABA were added to a final concentration of 20 nM and [^{3}H]diazepam to 1 nM. Specific binding was defined as the difference in binding of the radioligand in the presence and absence of 1 mM nonradioactive glutamate or GABA, 100 μM kainate and 10 μM diazepam. IC_{50} values with their 95 % confidence intervals were calculated by inverse prediction using logarithmic displacement plots. It was confirmed with an amino acid analyzer that the γ-glutamyl peptides tested were not hydrolyzed in the incubation medium.

Amino acid uptake

Amino acid uptake by synaptosomal preparations was measured essentially as earlier described (18, 20). The synaptosomal suspension (0.8 mg protein/sample) was first preincubated for 10 min in Krebs-Ringer-Hepes-glucose medium and then labelled GABA, taurine or glutamate (all 10 μM, 30 MBq/l) were added. The peptides or other effectors tested were also present only during the final incubations, in the case of GABA for 5 min, and of glutamate and taurine for 10 min. The incubations were terminated by filtration through Whatman GF/B filters (18) and the amounts of intrasynaptosomally penetrated taurine, GABA or glutamate calculated as in Kontro and Oja (19).

Amino acid efflux

The cerebral cortex slices were first preloaded for 30 min with 10 μM labelled taurine, GABA or glutamate and then superfused for 50 min as previously described (24), from 30 min onward with medium containing 50 mM K^+ ions and supplemented with the peptides. In GABA efflux experiments all the media contained 10 μM aminooxyacetic acid to prevent the metabolic breakdown of labelled GABA. The superfused slices and the collected 2-min superfusate fractions (0.5 ml) were measured for radioactivity. The efflux rate constants for the time intervals of 20-30 min (k_1), of 34-50 min (k_2) for taurine and 32-40 min (k_2) for GABA and glutamate were computed from linear regression of the logarithmic contents of labelled amino acids in slices versus superfusion time (23). The constants k_2 are given as percentages of the baseline efflux constant k_1 of each individual slice.

Table 1. Inhibition of [^{3}H]glutamate binding by peptides

Displacer	Percentage inhibition		
	0°C	37°C	37°C + $CaCl_2$
γ-Glu-tau	54 ± 4	29 ± 4	58 ± 3
γ-D-Glu-tau	22 ± 3	22 ± 3	24 ± 4
γ-Glu-AMS	61 ± 6	32 ± 3	75 ± 6
γ-Glu-homotau	67 ± 4	48 ± 5	65 ± 3
γ-Glu-N-methyltau	60 ± 3	48 ± 12	71 ± 4
γ-Glu-gly	82 ± 7	65 ± 4	85 ± 7
γ-Glu-asp	91 ± 2	96 ± 3	95 ± 3
α,γ-Glu-ditau	20 ± 4	24 ± 3	28 ± 2
α-Glu-tau	72 ± 7	35 ± 3	77 ± 11
α-Glu-asp	47 ± 6	45 ± 4	42 ± 4
β-Asp-gly	65 ± 8	66 ± 3	92 ± 5
β-Asp-tau	24 ± 3	22 ± 2	35 ± 4
N-Acetylasp-glu	55 ± 9	38 ± 5	57 ± 8

The binding was determined in 50 mM Tris-citrate buffer, pH 7.4, in the presence or absence of 2.5 mM $CaCl_2$. The displacers were tested at 1 mM concentration. Samples were incubated in ice water at 0°C for 60 min or in shaking incubator at 37°C for 30 min, with 20 nM of [^{3}H]glutamate. The binding in control experiments was 413 ± 11, 351 ± 14 and 1472 ± 57 nmol/kg protein at 0°C, 37°C and 37°C with $CaCl_2$, respectively. Results are means (±SEM) of at least three determinations in triplicate. γ-Glu-AMS = γ-glutamylaminomethylsulfonate.

RESULTS

Binding experiments

Sodium-independent [^{3}H]glutamate binding to synaptic membranes displaceable by 1 mM unlabelled glutamate at 37°C was increased 4.2 times by $CaCl_2$. The binding was inhibited by the acidic peptides (1 mM) at both 0°C and 37°C (Table 1). Among the most potent displacers were γ-glutamyl-aspartate, γ-glutamyl-glycine, β-aspartyl-glycine and α-glutamyl-taurine, which was more effective than the γ-glutamyl-taurine. γ-D-Glutamyl-taurine was considerably less effective. $CaCl_2$ enhanced the inhibition of most of the peptides tested at 37°C, approximately doubling it with both α- and γ-glutamyl-taurine but not with γ-D-glutamyl-taurine. In general, the percentage inhibitions at 0°C without Ca^{2+} and at 37°C with Ca^{2+} were very much alike.

The displacement curves of glutamate binding at varying concentrations of the most effective dipeptides and glutamate are shown in Fig. 1, and the estimated IC_{50} values for glutamate, γ-glutamyl-aspartate and α-glytamyl-taurine in Table 2. For the other compounds tested the values were almost in the mM range. All glutamate-containing peptides were effective displacers of kainate binding except γ-D-glutamyl-taurine and α,γ-glutamyl-ditaurine (Table 3). γ-Glutamyl-aspartate was again the most effective, as it was in the case of glutamate binding. β-Aspartyl peptides did not inhibit kainate binding but N-acetylaspartyl-glutamate was a moderate inhibitor. Glutamate was, as expected, a strong displacer.

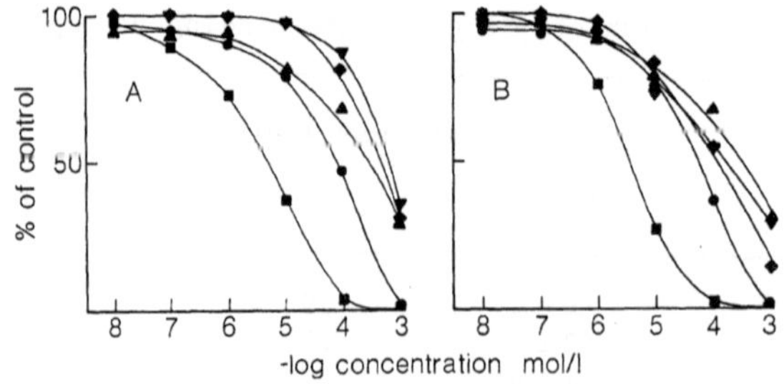

Fig. 1. Displacement of 20 nM [^{3}H]glutamate binding at 37°C by acidic dipeptides at varying concentrations (A) without and (B) with 2.5 mM $CaCl_2$. (■) Glutamate, (•) γ-glutamyl-aspartate, (▲) α-glutamyl-taurine, (▼) γ-glutamyl-glycine, (◆) β-aspartyl-glycine.

Table 2. IC_{50} values of glutamate binding

Displacer	No $CaCl_2$	$CaCl_2$ added
Glutamate	4.19 (3.09 - 5.69)	4.12 (3.59 - 4.73)
γ-Glu-asp	38.6 (30.9 - 48.2)	45.9 (38.7 - 54.3)
α-Glu-tau	216 (171 - 274)	242 (182 - 320)

The values with their 95 % confidence intervals calculated from data shown in Fig. 1 are given in μM.

In GABA binding, homotaurine and γ-glutamyl-homotaurine were the strongest inhibitors, followed by γ-glutamyl-GABA, GABA-glycine and GABA-taurine (Table 4). γ-Glutamyl-aspartate, α-glutamyl-aspartate and γ-glutamyl-glycine were slightly inhibitory. Other peptides listed in Table 1, but not shown here, had only minor effects (less than 10%) on GABA binding or none at all. The most potent displacers of GABA binding also stimulated [^{3}H]diazepam binding (Fig. 2), GABA itself exhibiting the strongest effect, followed by homotaurine. An addition of a γ-glutamyl group to either homotaurine or GABA (data on γ-glutamyl-GABA not shown) rendered these molecules strikingly less stimulatory.

Amino acid uptake

Most of the peptides slightly inhibited the uptake of taurine, GABA and glutamate when tested at the relatively high concentration of 1 mM (Table 5). In taurine uptake the inhibition caused by γ-glutamylaminomethylsulfonate, γ-glutamyl-homotaurine, α,γ-glutamyl-ditaurine, and α-glutamyl-aspartate was nearly 25 %, the other peptides being less effective; β-aspartyl-taurine and β-aspartyl-N-methyltaurine being without significant effect. The inhibition of GABA uptake was generally smaller, γ-glutamyl-taurine, γ-glutamyl-N-methyltaurine and β-aspartyl-taurine being ineffective. On the other hand, GABA-taurine caused an inhibition of 22 %.

Table 3. Inhibition of [^{3}H]kainate binding by peptides

Displacer	Percentage inhibition
γ-Glu-tau	62 ± 3
γ-D-glu-tau	10 ± 2
γ-Glu-N-methyltau	61 ± 3
γ-Glu-AMS	59 ± 6
γ-Glu-homotau	62 ± 3
γ-Glu-gly	72 ± 1
γ-Glu-asp	94 ± 3
α,γ-Glu-ditau	5 ± 3
α-Glu-tau	86 ± 2
α-Glu-asp	58 ± 2
β-Asp-tau	5 ± 2
β-Asp-gly	0
N-Acetylasp-glu	59 ± 3
Glutamine	93 ± 2
Glutamate	99 ± 2

The binding was assessed in 0.05 mM Tris-citrate buffer (pH 7.1, 0°C, 60 min) with 20 nM [^{3}H]kainate. The test compounds were added to a final concentration of 1 mM 10 min before commencing the binding assays. The binding in control experiments was 456 ± 4 nmol/kg protein. Results are means (±SEM) of four determinations in triplicate. γ-Glu-AMS = γ-glutamyl-aminomethylsulfonate.

Table 4. Inhibition of [^{3}H]GABA binding by peptides

Displacer	Percentage inhibition
γ-Glu-homotau	100
γ-Glu-gly	15 ± 4
γ-Glu-asp	29 ± 2
γ-Glu-GABA	93 ± 1
α-Glu-asp	16 ± 1
GABA-tau	78 ± 1
GABA-gly	85 ± 2
Homotaurine	98 ± 1

The binding was assessed at the [^{3}H]GABA concentration of 20 nM. The displacers (1.0 mM) were added 10 min before commencing the assays of 30 min. Binding in control experiments was 199 ± 8 nmol/kg protein. Results (±SEM) are means of three experiments in triplicate.

In glutamate uptake, β-aspartyl-taurine was the most effective inhibitor, followed by γ-glutamyl-N-methyltaurine, and β-aspartyl-glycine. Of the amino acids, taurine and GABA inhibited taurine uptake, glutamate uptake was strongly affected by L-glutamate, L- and D-aspartate, but less so by D-glutamate, whereas GABA uptake was significantly inhibited only by GABA itself (Table 5).

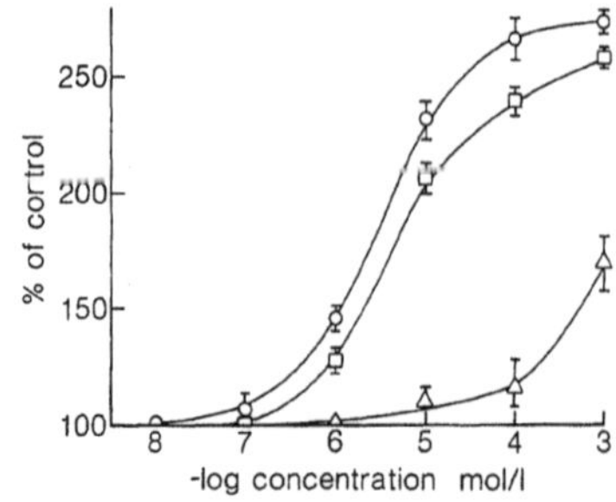

Fig. 2. Stimulation of 1 nM [^{3}H]diazepam binding by (o) GABA, (□) homotaurine and (Δ) γ-glutamyl-homotaurine at varying concentrations. Binding in control experiments (100 %) was 35.5 ± 3.7 nmol/kg protein (n=6).

Table 5. Effect of small acidic peptides on the uptake of taurine, GABA and glutamate in mouse brain synaptosomal preparations

Effector (1 mM)	Uptake % of control Taurine	 GABA	 Glutamate
γ-Glu-tau	89 ± 2** (8)	105 ± 3 (4)	91 ± 2** (8)
γ-D-Glu-tau	88 ± 5** (4)	86 ± 1** (4)	94 ± 1 (4)
γ-Glu-AMS	77 ± 1** (4)	83 ± 3** (4)	93 ± 2* (4)
γ-Glu-homotau	77 ± 1** (4)	81 ± 1** (4)	94 ± 1 (4)
γ-Glu-N-methyltau	90 ± 1** (4)	97 ± 1 (4)	81 ± 0** (4)
γ-Glu-asp	82 ± 2** (8)	85 ± 2** (8)	92 ± 1 (8)
α,γ-Glu-ditau	79 ± 3** (4)	81 ± 2** (4)	88 ± 1** (4)
α-Glu-tau	96 ± 3 (4)	85 ± 3** (4)	83 ± 2** (4)
α-Glu-asp	76 ± 2** (4)	93 ± 1* (4)	94 ± 2 (4)
β-Asp-tau	96 ± 7 (4)	105 ± 3 (4)	78 ± 2** (4)
β-Asp-gly	89 ± 4* (4)	84 ± 2** (4)	82 ± 0** (4)
β-Asp-N-methyltau	99 ± 3 (4)	85 ± 1** (4)	99 ± 2 (4)
N-Acetylasp-glu	89 ± 1** (4)	91 ± 4 (4)	97 ± 4 (4)
GABA-tau	99 ± 1 (4)	78 ± 1** (8)	87 ± 4** (4)
Taurine	25 ± 0** (8)	97 ± 1 (4)	92 ± 1 (9)
GABA	39 ± 4** (8)	32 ± 5** (4)	83 ± 3** (5)
Glutamate	79 ± 1** (4)	92 ± 3 (8)	11 ± 0** (8)
D-Glutamate	84 ± 4** (4)	93 ± 2 (8)	61 ± 1** (8)
Aspartate	84 ± 2** (4)	91 ± 2 (8)	5 ± 0** (8)
D-Aspartate	86 ± 2** (4)	88 ± 1** (4)	5 ± 0** (8)

Uptake in the control incubations without effectors (μmol/s per kg wet weight): taurine 0.0163 ± 0.0004 (n=32), GABA 0.338 ± 0.005 (n=30), glutamate 0.262 ± 0.010 (n=27). The number of independent experiments is given in parentheses. Significance of differences from the corresponding controls: *$p<0.05$, **$p< 0.01$. Means (±SEM) are shown. γ-Glu-AMS = γ-glutamylamino-methylsulfonate.

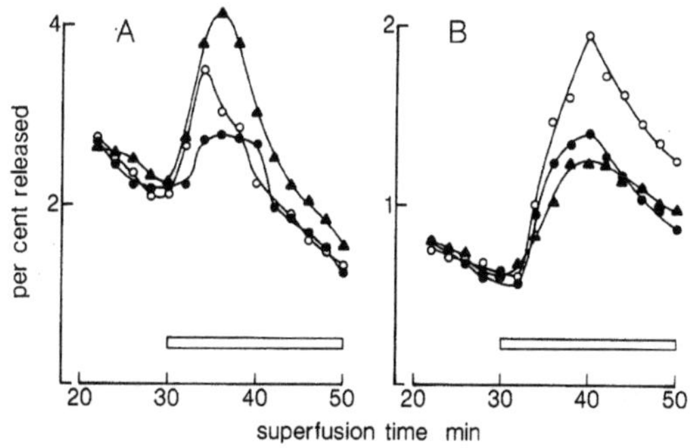

Fig. 3. K^+-stimulated release (o) of glutamate (A) and taurine (B) in per cent/min from mouse cerebral cortex slices and the effects of 0.01 mM (•) and 0.1 mM (▲) γ-glutamyl-taurine added to superfusion medium together with potassium. The period of stimulation by 50 mM K^+ is indicated by the horizontal bars. Single representative experiments are shown.

Amino acid efflux

Potassium stimulation of taurine release from cerebral cortex slices exhibited a strikingly prolonged time course, whereas glutamate release reacted to potassium ions much more rapidly (Fig. 3). GABA release resembled in this respect the release of glutamate (data not shown). γ-Glutamyl-taurine affected taurine and glutamate release but not GABA release. Stimulated glutamate release was enhanced by 0.1 mM γ-glutamyl-taurine, the 0.01 mM concentration delayed the maximal response (Fig. 3A), whereas the 0.001 mM concentration was ineffective (Table 6). The inhibition of taurine release was also dependent on the dipeptide concentration in medium, but in a peculiar manner: the inhibition was greater at 0.1 mM than at 0.01 mM (Fig. 3B), but the 0.001 mM concentration was the most effective (Table 6). The other peptides did not interfere with glutamate release at the 0.01 mM concentration. α-Glutamyl-taurine, α-glutamyl-aspartate, γ-glutamyl-aspartate and N-acetylaspartyl-glutamate slightly inhibited stimulated taurine release. γ-Glutamyl-homotaurine caused a marked stimulation of GABA release and GABA-taurine a moderate increase (Table 6). Also 1.0 mM taurine affected potassium-stimulated GABA release, causing an inhibition.

DISCUSSION

Herein we studied the possible interferences of a series of peptides containing putative amino acid transmitters with the high affinity uptake, depolarization-induced release and receptor binding of glutamate, GABA and taurine. A number of the tested peptides, α- and γ-glutamyl-taurine, β-aspartyl-taurine, β-aspartyl-glycine, γ-glutamyl-aspartate, γ-glutamyl-glycine, and N-acetylaspartyl-glutamate, have been recognized as endogenous constituents of brain tissue (5, 15, 31, 36, 42). They could thus participate in the regulation of synaptic functions in vivo. To date, intrahippocampal injections of N-acetylaspartyl-glutamate have caused prolonged seizure activity (44) and amygdaloid kindling correspondingly increased the

Table 6. Effect of small acidic peptides on the potassium-stimulated release of taurine, GABA and glutamate from mouse cerebral cortex slices

Concentration (mM) in medium from 30 to 50 min		Efflux rate constants (% of k_1) Taurine k_2 (34–50 min)	GABA k_2 (32–40 min)	Glutamate k_2 (32–40 min)
Control (50 mM K^+)		264 ± 9 (23)	1605 ± 71 (19)	170 ± 6 (7)
γ-Glu-tau	0.10	191 ±12** (5)	–	209 ± 5** (5)
γ-Glu-tau	0.01	210 ±15* (5)	1622 ±108 (4)	163 ± 6 (5)
γ-Glu-tau	0.001	155 ±12** (6)	–	173 ± 6 (5)
γ-Glu-homotau	0.01	251 ±19 (5)	2199 ±125** (4)	–
γ-Glu-asp	0.01	218 ± 7* (4)	–	168 ± 8 (5)
γ-Glu-GABA	0.01	–	1747 ±158 (5)	–
α-Glu-tau	0.01	227 ±11* (6)	–	–
α-Glu-asp	0.01	175 ± 9** (5)	–	174 ± 7 (5)
N-Acetylasp-glu	0.01	190 ± 8** (5)	–	167 ± 5 (5)
GABA-tau	0.01	278 ±23 (5)	1891 ±121* (5)	–
Glutamate	1.0	258 ±28 (5)	–	–
Glutamate	0.01	–	–	164 ± 6 (6)
Taurine	1.0	288 ±24 (4)	1028 ± 88** (5)	162 ±14 (3)
Taurine	0.01	–	–	158 ± 7 (5)
Taurine	0.001	206 ± 9* (5)	1703 ± 67 (4)	–
GABA	0.001	218 ±11* (5)	1696 ±106 (5)	–

The slices were preloaded for 30 min with labelled taurine, GABA or glutamate (all 10 μM) in Krebs-Ringer-Hepes-glucose medium, pH 7.4, and then superfused for 50 min, from 30 min onwards with 50 mM K^+ (control) and with the above effectors. The results are given as percentages of the rate constants (k_1, 20–30 min) for basal efflux of each slice. They were (means ±SEM) (1.92 ± 0.06) x $10^{-3}min^{-1}$ (taurine, n=50), (0.37 ± 0.01) x $10^{-3}min^{-1}$ (GABA, n=42) and (10.33 ± 0.20) x $10^{-3}min^{-1}$ (glutamate, n=48). The number of independent experiments is given in parentheses. Significance of differences from the corresponding controls: *$p<0.05$, **$p<0.01$. Means (±SEM) are shown.

levels of this dipeptide in the entorhinal cortex (34). Its binding to a subpopulation of excitatory receptor sites has been observed (17). On the other hand, some peptides in the D-form not found in nervous tissue have been used as neurochemical tools; for instance, γ-D-glutamyl-taurine, γ-D-glutamylaminomethylsulfonate (14) and γ-D-glutamyl-glycine (7) in the characterization of different subtypes of excitatory receptor sites.

In general, effects on the binding of glutamate and GABA or their agonists to synaptic membranes, as well as on the stimulus-induced release of taurine and GABA, were more pronounced than on the uptake of the above amino acids. The inhibition of uptake was invariably relatively small considering the high concentration of peptides in incubation medium. Dipeptides may be too large to fit to transport sites of single amino acid molecules. For instance, it has been shown that the transport sites of taurine accept only molecules with ionized electronegative and -positive groups ideally separated by a carbon chain with a length of two atoms (20, 26). GABA and glutamate transport sites also exhibit comparable selectivity (35). Despite the flexibility of peptide structures, such stereochemical requirements are met only with difficulty.

At least three subtypes of excitatory amino acid receptors have been differentiated in the brain (8). They are designated N-methyl-D-aspartate, quisqualate and kainate receptors according to the most potent and selective agonists known. L-glutamate serves as a ligand for all of them. In our assays all tested acidic peptides displaced glutamate binding, though at relatively high concentrations. γ-Glutamyl-aspartate was the most potent displacer in all experimental conditions. The sodium-independent specific binding of glutamate is modified by calcium and chloride ions (11). $CaCl_2$ increases the number of available binding sites (2) but in general the sites inhibited by the peptides increased even more. Uncertainties in classification of calcium- and chloride-dependent binding sites to the above three receptor types (11) preclude safe inferences from the differences in the behavior of peptides in our experiments.

It has been reported that N-methyl-D-aspartate, kainate and quisqualate-activated receptors are antagonized by γ-D-glutamyl-glycine in that order of decreasing potency (7). The receptors specific for kainate are apparently only a fraction of all glutamate binding sites (13). The binding of kainate was strongly inhibited by several γ-glutamyl peptides. Although γ-D-glutamyl-taurine and γ-D-glutamylaminomethylsulfonate are considered relatively selective antagonists of kainate/quisqualate receptors (14), several other glutamyl peptides were here considerably more effective in kainate binding than γ-D-glutamyl-taurine. The efficacy of γ-glutamyl-taurine and that of γ-glutamyl-homotaurine or γ-glutamylaminomethylsulfonate were essentially the same, but two taurine residues in α,γ-glutamyl-ditaurine rendered the tripeptide considerably less efficient. β-Aspartyl peptides had no effect on kainate binding, though β-aspartyl-glycine was a potent displacer in glutamate binding assays. The glutamyl residue seems to be an essential constituent of peptides active in kainate binding. The present inhibition patterns do not exclude the possibility that a glutamyl oligopeptide could act as the so far unknown natural ligand for the kainate receptors.

In GABA binding assays only those peptides were inhibitory which contain GABA or the sulfonic acid analog of GABA, homotaurine. An enhancement of benzodiazepine binding by GABA agonists has been reported by several authors (e.g., 16, 39). The relative potency of GABA agonists in enhancing benzodiazepine binding correlates positively with their inhibition in the binding of GABA itself (3, 25). The same phenomenon was discernible in the present study. For instance, γ-glutamyl-homotaurine was the most potent in displacing GABA from the receptor sites and causing a significant enhancement in diazepam binding. Taurine has been shown to antagonize the enhancement of flunitrazepam binding caused by GABA and to interfere with GABA binding (29). However, the glutamyl-taurine peptides were entirely ineffective on GABA and diazepam binding.

The neurotransmitters could inhibit their own release by activating specific autoreceptors localized at presynaptic membranes, as has been described for GABA and the glutamate analog D-aspartate (1, 33). Of the peptides tested, only γ-glutamyl-homotaurine and GABA-taurine enhanced potassium-stimulated GABA release in the same way as GABA analogs and agonists (22). These peptides did not influence potassium-stimulated taurine release, whereas the others were already inhibitory at relatively low concentrations from 0.001 to 0.1 mM. Stimulated taurine release is also inhibited by a number of taurine analogs and derivatives (21, 22). The effects have often been more pronounced at low concentrations (e.g., 0.001 mM) of the effectors. This was the case also here in taurine release with γ-glutamyl-taurine, which in contrast increased glutamate release. The concept of a negative feed-back regulation of transmitter release by autoreceptors (4) may not be the only explanation of the peptide effects, since the proposed antagonists of taurine, GABA and glycine (TAG, picrotoxin and strychnine, respectively) have also been found to inhibit stimulus-evoked taurine release (21).

In conclusion, the present results combined with previous biochemical and electrophysiological data call for further testing of the hypothesis that small acidic peptides are involved in neurotransmission in the central nervous system.

ACKNOWLEDGEMENTS

The skilful technical assistance of Miss Riitta Mero, Miss Paula Kosonen, Miss Jaana Parmonen and Mr. Ari-Pekka Viirtelä and the grants from the European Science Foundation and the Research Council for Medicine of the Academy of Finland to Dr. Vince Varga are gratefully acknowledged.

REFERENCES

1. Arbilla, S., Kamal, L., and Langer, S.Z., 1979, Presynaptic GABA autoreceptors on GABAergic nerve endings of the rat substantia nigra, Eur. J. Pharmac., 57: 211-217.
2. Baudry, M., Siman, R., Smith, E.W., and Lynch, G., 1983, Regulation by calcium ions of glutamate receptor binding in hippocampal slices, Eur. J. Pharmac., 90: 161-166.
3. Braestrup, C., Nielsen, M., Krogsgaard-Larsen, P., and Falch, E., 1979, Partial agonists for brain GABA/benzodiazepine receptor complex, Nature, 280: 331-333.
4. Chesselet, M.-F., 1984, Presynaptic regulation of neurotransmitter release in the brain: facts and hypothesis, Neuroscience, 12: 347-375.
5. Cheung, S.T., and Lim, R., 1979, Isolation of γ-glutamylaspartic acid and α-aspartylalanine from pig brain, Biochim. Biophys. Acta, 586: 418-424.
6. Curatolo, A., d'Arcangelo, P., Lino, A., and Brancati, A., 1965, Distribution of N-acetyl-aspartic and N-acetyl-aspartyl-glutamic acids in nervous tissue, J. Neurochem., 12: 339-342.
7. Davis, J., and Watkins, J.C., 1981, Differentiation of kainate and quisqualate receptors in the cat spinal cord by selective antagonism with γ-D-(and L-)glutamylglycine, Brain Res., 206: 172-177.
8. Fagg, G.E., 1985, L-Glutamate, excitatory amino acid receptors and brain function, Trends Neurosci., 8: 207-210.

9. Feuer, L., 1977, Theoretical background of the recognition of a new bioactive substance, Litoralon, isolated from the parathyroid. Further theoretical considerations, Biologia (Budapest), 25: 3-33.
10. Feuer, L., 1981, Biological effects of gamma-L-glutamyl taurine (glutaurine): A new parathyroid hormone, in: "The Effects of Taurine on Excitable Tissues", S.W. Schaffer, S.I. Baskin and J.J. Kocsis, eds., Spectrum Publications, New York, pp. 31-39.
11. Foster, A.A., and Fagg, G.E., 1984, Acidic amino acid binding sites in mammalian neuronal membranes: their characteristics and relationship to synaptic receptors, Brain Res. Rev., 7: 103-163.
12. Hajós, F., 1975, An improved method for the preparation of synaptosomal fraction in high purity, Brain Res., 93: 485-489.
13. Hall, J.G., Hicks, T.P., and McLennan, H., 1978, Kainic acid and the glutamate receptor, Neurosci. Lett., 8: 171-175.
14. Jones, A.W., Smith, D.A.S., and Watkins, J.C., 1984, Structure-activity relations of dipeptide antagonists of excitatory amino acids, Neuroscience, 13: 573-581.
15. Kanazawa, A., Kakimoto, Y., Nakajima, T., Sano, I., Shimizu, H., and Takesada, M., 1965, Isolation and identification of γ-L-glutamylglycine from bovine brain, Biochim. Biophys. Acta, 97: 460-464.
16. Karobath, M., and Sperk, G., 1979, Stimulation of benzodiazepine receptor binding by γ-aminobutyric acid, Proc. Natl Acad. Sci. USA, 76: 1004-1006.
17. Koller, K.J., and Coyle, J.T., 1985, The characterization of the specific binding of [^{3}H]-N-acetylaspartylglutamate to rat brain membranes, J. Neurosci., 5: 2882-2888.
18. Kontro, P., 1984, Comparison of taurine, hypotaurine and β-alanine uptake in brain synaptosomal preparations from developing and adult mouse, Int. J. Dev. Neurosci., 2: 465-470.
19. Kontro, P., and Oja, S.S., 1978, Taurine uptake by rat brain synaptosomes, J. Neurochem., 30: 1297-1304.
20. Kontro, P., and Oja, S.S., 1984, Binding and uptake of taurine and GABA in developing and adult mouse brain, Acta Univ. Tamper. [B], 21: 60-67.
21. Kontro, P., and Oja, S.S., 1987, Effects of the anticonvulsive taurine derivative, taltrimide, on membrane transport and binding of GABA and taurine in the mouse brain, Neuropharmacology, 26: 19-23.
22. Kontro, P., and Oja, S.S., 1987, Taurine and GABA release from mouse cerebral cortex slices: effects of structural analogues and drugs, Neurochem. Res., 12: 475-482.
23. Korpi, E.R., and Oja, S.S., 1979, Efflux of phenylalanine from rat cerebral cortex slices as influenced by extra- and intracellular amino acids, J. Neurochem., 32: 789-796.
24. Korpi, E.R., and Oja, S.S., 1984, Comparison of two superfusion systems for study of neurotransmitter release from rat cerebral cortex slices, J. Neurochem., 43: 236-242.
25. Krogsgaard-Larsen, P., Hjeds, H., Curtis, D.R., Lodge, D., and Johnston, G.A.R., 1979, Dihydromuscimol, thiomuscimol and related heterocyclic compounds as GABA analogues, J. Neurochem., 32: 1717-1724.
26. Lähdesmäki, P., and Oja, S.S., 1973, On the mechanism of taurine transport at brain cell membranes, J. Neurochem., 20: 1411-1417.
27. Larder, A.P., and McLennan, H., 1983, Binding sites for L-glutamate in the central nervous system of the rat, Neurochem. Res., 9: 393-403.
28. London, E., and Coyle, J.T., 1978, Specific binding of [^{3}H]kainic acid to receptor sites in rat brain, Mol. Pharmac., 15: 492-505.
29. Malminen, O., and Kontro, P., 1986, Modulation of the GABA-benzodiazepine receptor complex by taurine in rat brain membranes, Neurochem. Res., 11: 85-94.
30. Marnela, K.-M., Timonen, M., and Lähdesmäki, P., 1984, Mass spectrometric analyses of brain synaptic peptides containing taurine, J. Neurochem., 43: 1650-1653.

31. Marnela, K.-M., Morris, H.R., Panico, M., Timonen, M., and Lähdesmäki, P., 1985, Glutamyl-taurine is the predominant synaptic taurine peptide, J. Neurochem., 44: 752-754.
32. Marnela, K.-M., Varga, V., Dibó, G., and Lähdesmäki, P., 1987, Position of the peptide linkage in glutamyl-taurine from calf brain synaptic vesicles, J. Neurochem., 48, in press.
33. McBean, G.J., and Roberts, P.J., 1981, Glutamate-preferring receptors regulate the release of D-[^{3}H]aspartate from rat hippocampal slices, Nature, 291: 593-594.
34. Meyerhoff, J.L., Koller, K.J., Walczak, D.D., and Coyle, J.T., 1985, Regional brain levels of N-acetyl-aspartyl-glutamate: the effect of kindled seizures, Brain Res., 346: 392-396.
35. Oja, S.S., and Korpi, E.R., 1983, Amino acid transport, in: "Handbook of Neurochemistry, 2nd edn, Vol 5", A. Lajtha, ed., Plenum Press, New York, pp. 311-337.
36. Reichelt, K.L., and Fonnum, F., 1969, Subcellular localization of N-acetyl-aspartyl-glutamate, N-acetylglutamate and glutathione in brain, J. Neurochem., 16: 1409-1416.
37. Reichelt, K.L., and Kvamme, E., 1973, Histamine-dependent formation of N-acetyl-aspartyl peptides in mouse brain, J. Neurochem., 21: 849-859.
38. Sano, I., 1970, Simple peptides in brain, Int. Rev. Neurobiol., 12: 235-263.
39. Tallman, J.F., Thomas, J.W., and Gallager, D.W., 1978, GABAergic modulation of benzodiazepine binding site sensitivity, Nature, 274: 383-385.
40. Török, K., Varga, V., Somogyi, J., Feuer, L., and Gulyás, J., 1981, Formation of γ-glutamyl-taurine in the rat brain, Neurosci. Lett., 27: 145-149.
41. Varga, V., DeFeudis, F.V., Ossola, L., Geffard, M., and Mandel, P., 1980, Binding of muscimol and GABA in subfractions of crude membrane fraction of rat brain, Biochem. Pharmac., 29: 1077-1079.
42. Varga, V., Török, K., Feuer, L., Gulyás, J., and Somogyi, J., 1985, γ-Glutamyltransferase in the brain and its role in formation of γ-L-glutamyl-taurine, in: "Progress in Clinical and Biological Research, Vol. 179, Taurine: Biological Actions and Clinical Perspectives", S.S. Oja, L. Ahtee, P. Kontro and M. K. Paasonen, eds., Alan R. Liss, New York, pp. 115-125.
43. Whittaker, V.P., and Barker, L.A., 1972, The subcellular fractionation of brain tissue with special reference to the preparation of synaptosomes and their component organelles, in: "Methods of Neurochemistry, Vol. 2", R. Fried, ed., Marcel Dekker, New York, pp. 1-52.
44. Zaczek, R., Koller, K., Cotter, R., Heller, D., and Coyle, J.T., 1983, N-Acetylaspartylglutamate: an endogenous peptide with high affinity for a brain "glutamate" receptor, Proc. Natl Acad. Sci. USA, 80: 1116-1119.
45. Zukin, S.R., Young, A.B., and Snyder, S.H., 1974, Gamma-aminobutyric acid binding to receptor sites in the rat central nervous system, Proc. Natl Acad. Sci. USA, 71: 4802-4807.

SECTION V: CALCIUM-TAURINE MODULATORY MECHANISM

FROM HEART TO HYPOTHESIS: A MECHANISM FOR THE CALCIUM MODULATORY ACTIONS OF TAURINE

Ryan J. Huxtable

Department of Pharmacology, College of Medicine
University of Arizona, Tucson, Arizona 85724

Taurine is ubiquitous in animals. In mammals, numerous taurine-dependent phenomena have been described (Table 1). The increasing intensity of interest in this simple but recalcitrant compound is indicated by the various symposia on taurine and the monographs published on it within the last 15 years (5,6,17,18,19,20,24,32,38,44).

Two physiological functions have been established for taurine: lipid digestion, and osmoregulation. The first is mediated via one of the few metabolic pathways available to taurine: hepatic conjugation with cholic acid to form detergent bile salts such as taurocholate and taurochenodeoxycholate. This is phylogenetically a late appearing phenomenon, occurring first in fishes such as sturgeon, carp and goldfish, but being absent from primitive fishes such as hagfish, sharks, and lungfish (reviewed in (19)). Osmoregulation, on the other hand, is a function that spans the phylogenetic ladder from the single-celled protozoa through mammalian heart and brain (4,49,50). This is clearly a major biological function for taurine. In addition, taurine has numerous effects that appear to be mediated via calcium (Table 1). Direct chemical interaction of calcium and taurine has been excluded (28). No general mechanism has been proposed for these calcium modulatory actions.

There is an obvious reductionist attraction in developing a unified hypothesis to account for the bewilderingly diverse effects of taurine on calcium-dependent phenomena. Any general hypothesis must explain the wide species to species variation in the cardiac content of taurine, ranging from 4 M per gram wet weight in the cow to 35 M per gram wet weight in the rat. Other substances of general significance to cellular function, such as adenine nucleotide and inorganic cations, do not show this variation. There is also need to explain the tissue to tissue variation within species, and the marked developmental decrease in taurine levels in mammalian brain . Furthermore, although taurine is abundant in the animal kingdom, it is low in the bacterial kingdom, and absent from the plant kingdom (none of the reports on its possible presence in plants is convincing (9,34,36) - even HPLC analyses can be erroneous due to large amount of phosphoglycerol derivatives eluting with similar retention times to taurine). A generalized hypothesis must also account for the membrane stabilizing actions of taurine, including its inhibitory effects on neurotransmitter and hormone release (10,42,45). Finally, a mechanism of action must be ubiquitous, in keeping with the ubiquitous distribution of taurine in animals.

TABLE 1. Some of the Biological Actions of Taurine

Cardiovascular System	Antiarrhythmic Positive inotropy at low calcium Negative inotropy at high calcium Potentiation of digitalis inotropy Antagonism of calcium paradox Hypotensive (central and peripheral action) Retardation of lesion development in calcium overload cardiomyopathy
Brain	Anticonvulsant Modulator of neuronal excitability Maintenance of cerebellar function
Eye	Maintenance of structure and function of photoreceptors, outer segments, and tapetum lucidum
Liver	Bile salt synthesis
Reproductive System	Sperm motility factor
Muscle	Muscle membrane stabilizer
General	Modulation of neurotransmitter and hormone release Osmoregulation Stimulation of glycolysis and glycogenesis Attenuation of hypercholesterolemia Cell proliferation

This list is not exclusive. For details, see the various monographs on taurine (5,6,17,18,19,20,24,32,38,44).

A paradoxical phenomenon is the opposing distributions of taurine and calcium. In general, intracellular levels of taurine are high, and extracellular low, the concentration ratios ranging from 100:1 to 50,000:1. Calcium is largely excluded from the cell cytoplasm, extracellular concentrations being higher than intracellular. In excitable cells, extracellular to intracellular ratios range from 10^3:1 to 10^6:1, depending on the contractile or secretory state of the cell. The mM concentrations at which taurine is typically found in cells imply that physiologically significant intracellular phenomena for taurine involve low affinity interactions, whereas physiologically significant extracellular phenomena involve high affinity interactions. The opposite statements hold for calcium. The effects of taurine on calcium are indirect, taurine being a poor chelator for the cation. It has been calculated by ^{13}C NMR spectroscopy that less then 2% of the calcium in the heart cell cytoplasm is bound to taurine (28).

I wish to advance the outlines of a hypothesis to account for the calcium modulatory actions of taurine; a hypothesis rooted in the physicochemical properties of taurine. These properties permit taurine to interact uniquely with membrane phospholipids, thereby altering membrane cation binding characteristics. This hypothesis was elucidated from work in my laboratory on the cardiac actions of taurine, which has resulted in a

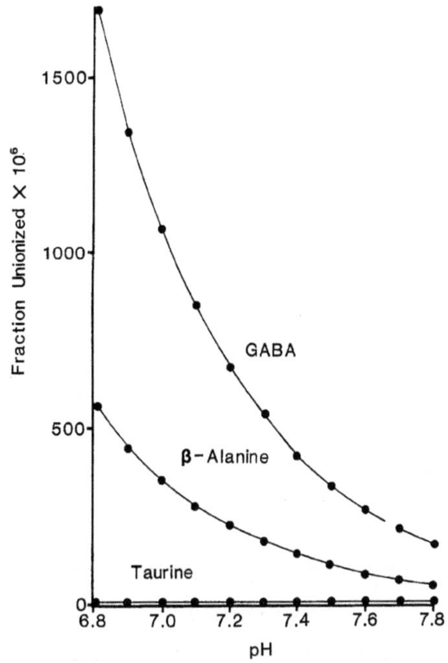

Figure 1. Effect of pH on degree of ionization of the acid function of taurine (pKa 1.5; pI 5.29), beta-alanine (pKa 3-6; pI 10.36) and GABA (pKa 4.03; pI 7.30). The number of unionized functions per 10^6 molecule is shown.

generalized insight into the interactions of taurine and biological membranes. This article discusses the following points:

1. The basis of the cellular ability to maintain uniquely high concentration gradients for taurine.
2. The evolution of an osmoregulatory function for taurine.
3. The physicochemical basis for interactions with phospholipid components of cell membranes.
4. Experimental findings on the binding of taurine to natural and artificial membranes.
5. Experimental findings on the influence of taurine on the binding of calcium to natural and artificial membranes.

MAINTENANCE OF GRADIENTS FOR TAURINE

In general, transport systems for taurine are specific for beta amino acids, and are both energy- and sodium-dependent. Taurine is unique amoung the biological alpha omega amino acids in having a highly acidic sulfonate group. The pKa is 1.5, compared with 4.03 for GABA and 3.6 for beta-alanine. Over the range of physiological pH, there is a significantly higher proportion of unionized acid functions in GABA and beta-alanine than there are in taurine (Fig. 1). This high degree of zwitterionic character in taurine renders the molecule exceedingly lipophobic, giving it a low diffusion coefficient through undamaged liproprotein membranes. Thus, whereas beta-alanine transport in the heart has a clear nonsaturable component, indicating diffusion, taurine transport is largely saturable, indicating carrier transport (Fig. 2). This suggests an explanation for the

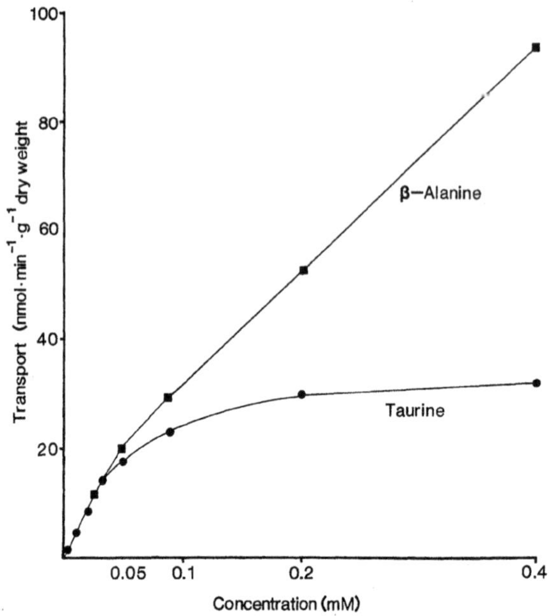

Figure 2. Transport of beta-alanine and taurine into the Langendorff-perfused rat heart. There is a nonsaturable component to the transport of beta-alanine, indicating considerable membrane diffusion. Data from Refs. (22,24).

unusually high concentration gradiants maintained for taurine. For brain cells, this may be as high as 300-500:1. for Ehrlich ascites cells on culture the gradient ranges up to 2000:1 (8), and for HeLa cells up to 7000:1 (41).

The nonsaturable diffusion exhibited by beta-alanine would make maintenance of mM concentrations within the cell excessively expensive in terms of the energy needed to pump substrate to balance the loss by diffusion. It is noteworthy that the free amino acid in greatest abundance in the brain, glutamate, also has a high ionic character (isoelectric point 3.24).

OSMOREGULATION

Osmoregulation is a function that has been unequivocally established for taurine: In structures ranging from unicellular organisms to mammalian heart (50) and brain (49), taurine is the osmotically active substance present in highest concentration. The largest shift in osmolar equivalents within a cell in response to osmotic stress is contributed by taurine. This is true of mammalian organs and of marine organisms responding to a change in salinity. The ability to adapt rapidly to osmotic stresses in such a manner is a *sine qua non* for survival for cells lacking a rigid cell wall. Taurine is well suited for such a purpose because of its poor liposolubility (it cannot "leak" from a cell) and metabolic inertness. High affinity transport systems for taurine respond sensitively to a variety of signals, including changes in osmolarity (4), and changes in cyclic AMP concentration (22,23).

I

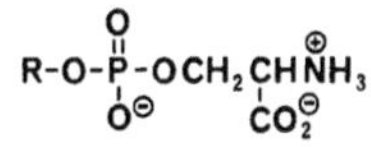

II

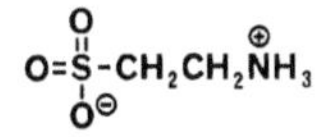

III

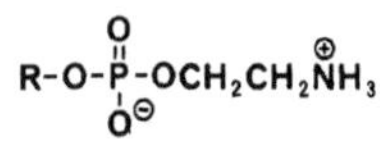

Figure 3. Relationship between the structure of taurine (II) and the charged head groups of the acidic phospholipid, phosphatidylserine (I), and the neutral phospholipid, phosphatidylethanolamine (III). There are considerable electronic and steric similarities between II and III.

Taurine appears to be essential for growth and survival of mammalian cells. Thus, lymphoblastoid cells do not grow in the absence of taurine (14), myocytes have poor viability in the absence of taurine (15), and taurine depletion causes degenerative changes in retinal cells (48). Extracellular taurine also appears to be the hormonal agent signalling aggregation in slime moulds (43).

Why should taurine be essential for mammalian cells when it is absent from plants and low in bacteria? The most obvious difference between the cells of these kingdoms and animal cells is the presence of a cell wall in the former. This relieves the plasma membrane, when one is present, of the responsibility of preserving the cell from osmotic stresses. Bacterial and plant cell membranes also differ from animal plasma membranes in lipid and phospholipid composition. The neutral phospholipid, phosphatidylcholine, for example, is absent from most bacteria (3).

There are intriguing ionic and structural similarities between taurine and the zwitterionic head groups of the neutral phospholipids, phosphatidylcholine and phosphatidylethanolamine (Fig. 3). A low energy interaction of taurine with these neutral phospholipids in membranes was postulated some years ago (16). Calcium and other cations bind to the acidic membrane phospholipids, phosphatidylserine and phosphatidylinositol. We have suggested, and have presented evidence supporting the suggestion, that ion pair or dipole-dipole interactions of taurine with the neutral phospholipids increases the affinity of calcium to high affinity binding sites on the acidic phospholipids (27). It is postulated that the changes in membrane confrontation induced by taurine are responsible for a number of the biological actions of this compound, in addition to modulation of cation binding and transport (27). For example, Lombardini (35) has found that taurine shifts the membrane transition temperature for Ca^{2+} transport in retinal membranes from $17.9^{\circ}C$ to $25.4^{\circ}C$. There is, in addition, a significant lowering of activation energy. Several groups have reported that taurine reduces lipid peroxidation (2,37). Taurine antagonizes the deleterious effects on Ca^{2+} binding and transporting ability of cleaving phospholipid head groups from muscle sacroplasmic reticulum (21).

THE BINDING OF TAURINE TO SARCOLEMMA AND PHOSPHOLIPID VESICLES

Evidence for low energy interactions of taurine with membranes has been obtained from binding studies. Taurine-binding proteins have been isolated

TABLE 2. Low affinity binding of taurine to sarcolemma and phospholipid vesicles

Preparation	Bmax ($nmol.mg^{-1}$ phospholipid)	Kd (mM)	Hill Coefficient	Molar ratio (taurine:lipid)
Rat heart sarcolemma[a]	2777±174	19.2±4.0	1.9±0.3	2:1
Mixed phospholipid vesicles[b]	418±26	63.7±12.4	1.5±0.1	
Mixed phospholipid vesicles + 50% cholesterol	139±18	28.8±10.0	1.5±0.1	
Phosphatidylcholine vesicles	1215±410	59.9±24.0	1.9±0.2	1:1
Phosphatidylethanol-amine vesicles	736±13	70.1±12.2	1.5±0.3	0.5:1
Phosphatidylserine vesicles	1362±457	107±17.0	3.8±1.4	1:1
Phosphatidylinositol vesicles		Insignificant binding observed		

[a]Binding was 661±43 nmol mg^{-1} protein; phospholipid content of sarcolemma was 313 nmol mg^{-1} protein. [b]Containing phosphatidylcholine (45%); phosphatidylethanolamine (47.5%); phosphatidylinositol (5%) and phosphatidylserine (2.5%). Data are reported as means ± SD for 4-6 preparations. Adapted from Ref. (27).

from the heart, but their binding affinities, structure-activity requirements, and sodium dependency suggest that they represent transport proteins (7,30,31).

We have described specific low affinity binding of taurine to the sarcolemma, established by equilibrium dialysis (46). In Tris buffer, the sarcolemma exhibited a Bmax of 661 nmol $taurine.mg^{-1}$ protein (2677 $nmol.mg^{-1}$ phospholipid), and a Kd of 19.2 mM. The low affinity is consonant with the concentrations of taurine present in rat heart. Binding exhibited a Hill coefficient of 1.90, indicating positive cooperativity. The binding of taurine was unaffected by beta-alanine or guanidinoethane sulfonate, establishing that - as indicated by the widely differing affinities - the binding site was not related to the taurine transporting system. Furthermore, transport is absolutely dependent on sodium and is unaffected by calcium. Binding, however, is inhibited by the cations sodium, calcium and potassium.

Essentially the same binding characteristics are seen for taurine binding to mixed phospholipid vesicles, containing the individual phospholipids in the same proportion as present in sarcolemma (Table 2). Thus, saturable, low affinity binding is seen with a Hill coefficient of 1.5. The inclusion of 50% cholesterol in the vesicles significantly increased affinity, indicating that fluidity and other membrane properties modify taurine binding. When vesicles of individual phospholipids were examined, the neutral phospholipids showed the highest binding affinities. Phosphatidylinositol showed no significant binding of taurine. This, again, is supportive of the view that primary membrane interactions of taurine involve the neutral phospholipids (Fig. 3).

Table 3. Inhibition of low affinity taurine binding by structural analogs of taurine and cations

Compound	Percent of Control
Analogs	
hypotaurine	37±9*
Beta-alanine	97±8
guanidinoethane sulfonate	93±8
Cations	
Ca^{2+}	19±13*
Na^{+}	64±9**
K^{+}	68±7**

Results are expressed as means ± SEM of 4-6 separate sarcolemmal preparations. Taurine and inhibitor concentration was measured in the same preparations. Statistics were evaluated by analysis of variance "F" test, followed by Student's t-test for paired samples. Adapted for Ref. 2810. from ref. (46). * Significantly different from taurine alone (P<0.01). ** Significantly different from taurine alone (P<0.05).

The positive cooperativity observed, indicating that the binding of one molecule of taurine modifies binding to an adjacent site, and the general antagonism by cations of taurine binding (Table 3) suggest that taurine is binding on, or close to, a cation binding site. The interaction would appear to be mutual. The binding of a cation to an acidic phospholipid modifies the binding of taurine to an adjacent neutral phospholipid, and vice versa.

THE EFFECT OF TAURINE ON CALCIUM BINDING TO SARCOLEMMA AND PHOSPHOLIPID VESICLES

Calcium binding to cardiac sarcolemma shows two components. In Krebs-Henseleit buffer, the low affinity component has a Kd of 3.94 mM and a Bmax of 92 $nmol.mg^{-1}$ protein. The high affinity component has a Kd of 0.03 mM and a Bmax of 9.4 $nmol.mg^{-1}$ protein (47). Taurine is without effect on the low affinity component. However, in a modified Krebs-Henseleit buffer containing low sodium and high potassium concentrations to simulate intracellular conditions, taurine significantly enhances the high affinity binding of calcium. The degree of stimulation deceases over the calcium concentration range 10^{-6}M to 10^{-4}M, suggesting an alteration in affinity rather than binding capacity. High sodium concentration, however, antagonizes this effect. Indeed, in standard Krebs-Henseleit buffer containing 145 mM sodium, taurine (10 mM) significantly decreases calcium binding at 10^{-6}M calcium from 0.13 $nmol.mg^{-1}$ protein to 0.09 $nmol.mg^{-1}$ protein. ATP (1 mM) antagonized the action of sodium. Thus, the addition of taurine to a Krebs-Henseleit buffer containing 1 mM ATP resulted in a significant stimulation of calcium binding at 10^{-6}M calcium.

The inhibitory action of sodium on the taurine-induced stimulation of calcium binding may have functional significance in the heart. The depolarization which initiates the action potential in ventricular myocytes is caused by an inward sodium current. The increased sodium concentration on the inner aspect of the cell membrane would act to raise cytoplasmic calcium concentrations both via direct displacement of calcium from its binding sites, and by antagonism of the stimulated binding induced by taurine. This

Table 4. Effect of taurine on calcium binding to sarcolemma and Phospholipid Vesicles

Preparation	Calcium Binding (pmol.mg^{-1} phospholipid) Control	20 mM Taurine	% Change[a]
Rat Heart Sarcolemma	514±162	708±98	+38
Phosphatidylserine vesicles	596±150	823±204	+38
Phosphatidylinositol vesicles	292±78	272±76	
Phosphatidylcholine vesicles	16±10	20±20	
Phosphatidylethanolamine vesicles	11±16	6±10	
Mixed vesicles[b]	20.7±5.6	38.3±6.8	+85
(Mixed vesicles-calculated)[c]	17.2	18.9	(+10)

Conditions: Tris, pH 7.4 (20 mM); NaCl (5 mM); KCl (110 mM); $MgCl_2$ (1 mM); $CaCl_2$ (1 µM). Data are means ± SD for 4-6 paired preparations. [a] Only shown when statisically significant, except for calculated percent change for mixed vesicles. [b] Containing phosphatidylcholine (18.45%); phosphatidylethanolamine (19.475%); phosphatidylinositol (2.05%); phosphatidylserine (1.025%); and cholesterol (59%). [c] Calculated binding from values for individual phospholipids. Adapted from Ref. (27).

phenomenon could be involved in the increase in cytoplasmic calcium concentration needed to initiate contraction. The antagonism by calcium (i.e. the decrease in taurine-induced stimulation with increasing calcium concentration) may also have functional significance. It has been well established that the most marked effects of taurine in the heart are seen at physiologically low levels of calcium (reviewed in (26)). The actions of taurine are minimal at physiological calcium concentrations. Furthermore, calcium antagonism may be a component of calcium-induced calcium release in cardiac cells.

The stimulatory effects of taurine on the high affinity binding of calcium to the sarcolemma can be mimicked in artificial phospholipid vesicles (Table 4). In a Tris buffer of the indicated ionic composition, taurine (20 mM) at 1 µM calcium stimulates calcium binding by 38%. Calcium binding can also be observed to artificial vesicles of the acidic phospholipids phosphatidylserine and phosphatidylinositol. The neutral phospholipids, phosphatidylcholine and phosphatidylethanolamine, however, exhibit little calcium binding capacity (Table 4). Taurine stimulates the binding of calcium to vesicles of phosphatidylserine, but is without effect on vesicles of phosphatidylinositol. Mixed phospholipid vesicles were also examined, which contained the individual phospholipids in the same proportion present in rat heart sarcolemma (51). The calcium binding capacity at 1 μM calcium is in good agreement with that calculated from the binding characteristics of the individual phospholipids. However, the effect of taurine was much greater than predicted on the basis of experiments with phosphatidylserine vesicles. This indicates that the action of taurine on the calcium binding abilities of individual phospholipids is not additive. The interaction of taurine with the neutral, low calcium binding phospholipids, phosphatidylcholine and phosphatidylethanolamine, leads to a modification of calcium binding to the acidic phospholipids.

NMR SPECTROSCOPY OF PHOSPHOLIPID VESICLES

^{31}P NMR spectroscopy offers the promise of examining the electronic environment of the phosphorus atom in a nondestructive manner. The NMR spectra of calcium complexes can not be studied directly, as calcium is a diamagnetic ion which does not produce marked shifts in phosphorus signals. We chose instead to study the binding of the paramagnetic shift reagent, europium. Europium, a transition metal, is a calcium analog which produces upfield shifts and line broadening of phosphorus NMR signals. A further constraint was the need to use mM quantities of europium. For the high affinity calcium binding data of Table 4, less than one in one thousand binding sites on phosphatidylserine is occupied, an undetectable proportion by NMR.

^{31}P NMR studies clearly show that taurine modifies the interaction of phospholipids with europium. Fig. 4 shows the spectrum obtained in Tris buffer from vesicles prepared from 1:1 phosphatidylcholine and phosphatidylserine. In the presence of europium ion, signals from phospholipids on the outside of the vesicles are shifted upfield and broadened, whereas signals from the europium-free intravesicular phospholipids are unaltered (Fig. 4B). Outside, the signal from the cation-binding, acidic phospholipid, phosphatidylserine is shifted further upfield than the signal for phosphatidylcholine, and is broadened to the point of near undetectability. When 20 mM taurine is added to the sample (Fig. 4C), the phosphatidylserine signal is narrowed, and moved downfield.

Further evidence for a modification in cation-phospholipid interactions derives from studies with the shift reagents, cobalt. This paramagnetic ion broadens ^{31}P phospholipid signals to the point of undetectability. Mixed vesicles of phosphatidylcholine and phosphatidylserine are prepared in the presence of cobalt, which thus becomes trapped in intravesicular space. The chelating agent, EDTA, is then added to bind extravesicular cobalt. This allows only the signals from phospholipids on the outer aspect of the vesicle to be seen (Fig. 5). Addition of europium ion to such a preparation results in both signals being shifted, the phosphatidylserine more so than the phosphatidylcholine (Figure 5B). Again, further addition of taurine produces a downfield shift (shielding) of the phosphatidylserine signal (Figure 5C).

These studies provide physical evidence for the ability of taurine to modify cation binding in mixed vesicles. Other factors, such as membrane fluidity, presumedly also modify the interaction, and require appropriate study.

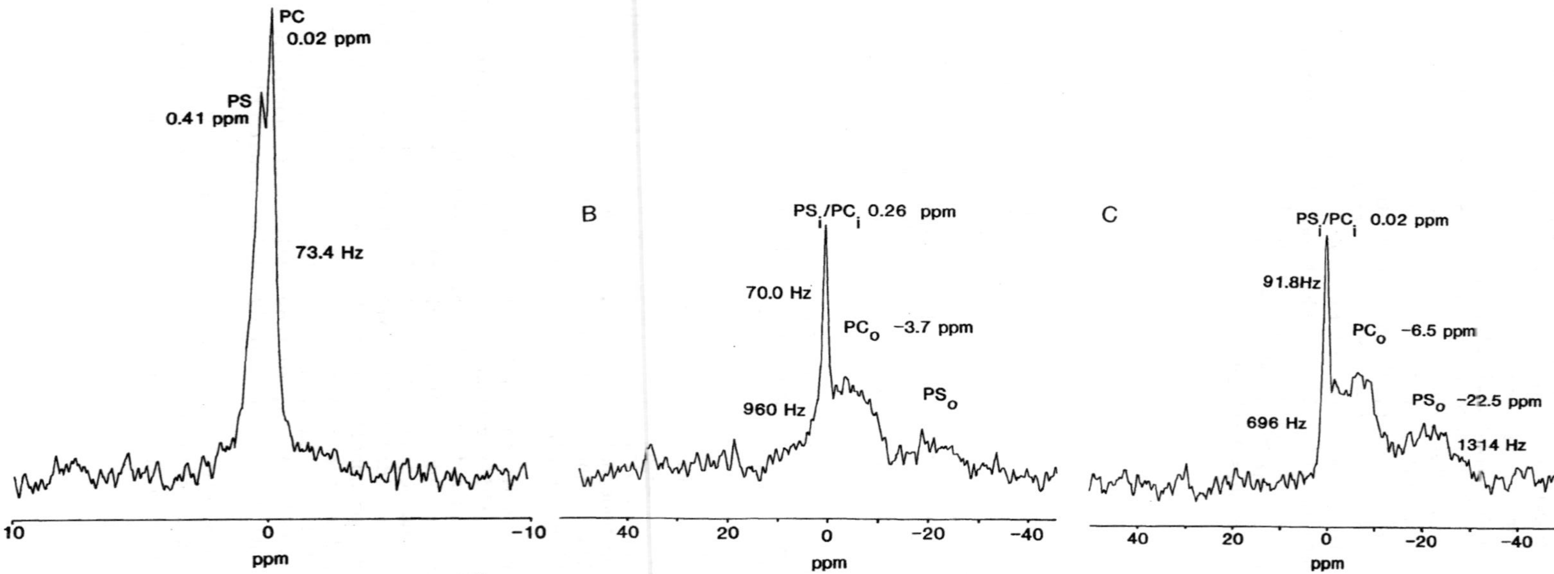

Figure 4. ${}^{31}P$ NMR spectra of vesicles prepared from a 1:1 mixture of phosphatidylcholine (PC) and phosphatidylserine (PS). A: In 10 mM Tris HCl. pH 7.4. B: With the addition of 2 mM europium chloride. Signals originating from phospholipids on the outer leaflet of the vesicles are broadened and shifted upfield. The signals from the acidic phospholipid, phosphatidylserine (PS_o) and affected more than those from the neutral phospholipid, phosphatidylcholine (PC_o). C: With the addition of 2 mM $EuCl_3$ and 20 mM taurine. Signals from the exterior phospholipids are narrowed, and the PS_o signal is shifted downfield.

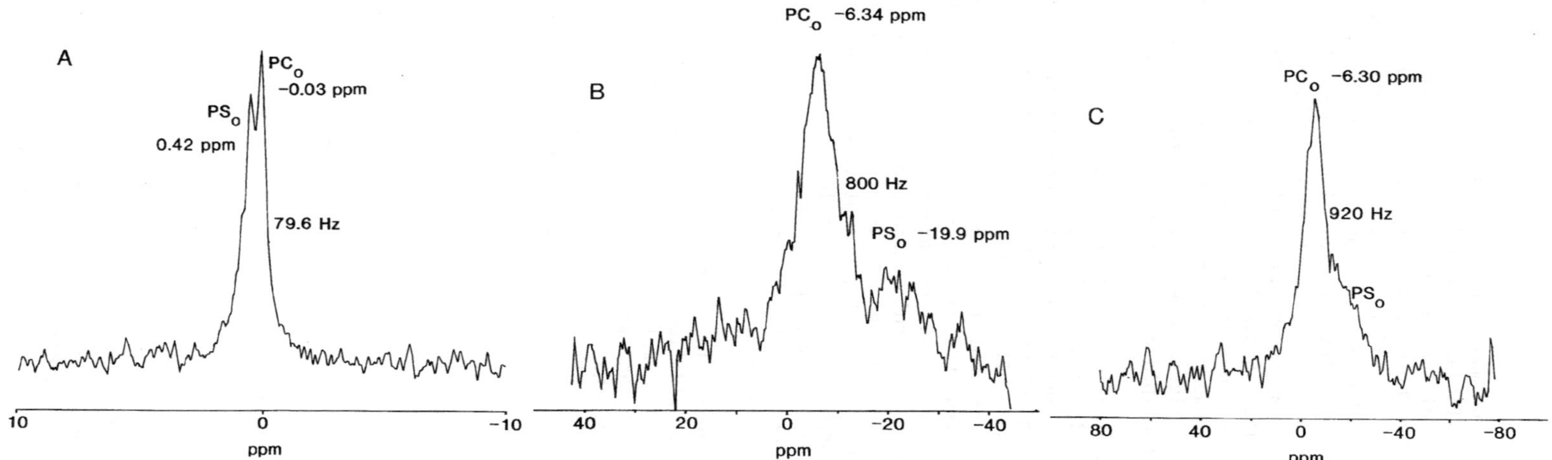

Figure 5. ^{31}P NMR spectra of vesicles containing $CoCl_2$ (3 mM) prepared from a 1:1 mixture of phosphatidylcholine (PC) and phosphatidylserine (PS). A: The shift reagent cobalt broadens signals, effectively erasing them from the measured spectrum. EDTA (5 mM) in the incubation medium chelates cobalt on the exterior of the vesicles, allowing only the signals from phospholipids in the outer leaflet to be detected. B: The addition of $EuCl_3$ (5 mM) shifts and broadens signals due to PC_o and PS_o signal, a particular, is shifted downfield.

DISCUSSION

All the actions of taurine on modifying calcium binding to biological membranes can be mimicked in artificial phospholipid vesicles. We conclude, therefore, that phospholipids in biological membranes are a major site of action for taurine. As seen from Table 2, the affinity with which taurine interacts with phospholipid vesicles and membranes is a function of the phospholipid composition. Thus, the inclusion of 50% cholesterol in mixed phospholipid vesicles increases the affinity from 64 mM to 29 mM. Presumably, other factors, such as the degree of unsaturation of fatty acids in phospholipids, and presence of sphingomyelins, also affect the affinity of interaction. This allows a putative basis for the variation of taurine content in brain development, and the species-to-species variation in cardiac content. There are marked developmental changes in phospholipid composition of the brain (11)(Huxtable, unpublished observations), and marked species variations in sarcolemmal phospholipid composition in the heart.

Phospholipids are important calcium binding sites in the cell. They serve to regulate calcium entry in to the cell, and also provide a reservoir of bound calcium in the cell that may be released under various conditions. Taurine modulates this cation-binding ability. Taurine is uniquely suited for this role. Other low MW substances with structures related to the charged portion of neutral phospholipids are shown on Fig. 6. All these structures, including the phosphate derivatives obtained by hydrolysis of phospholipids, are dibasic: that is to say, they contain two exchangable acidic protons. Thus, they are acidic amino acids with ionic character and pIs radically different from taurine and the charged head groups of neutral phospholipids. Beta-alanine and aminoethane phosphonate, for example, have a remarkably geometric correspondence to taurine (Fig. 7 and Table 5). Their ionic characters, however, are quite different. Ionization constants could not be located for aminoethane phosphonate. However, in view of its dibasicity, its isoelectric point is presumable around 3.

The sulfonate sulfur in taurine is at an oxidation state of +4. Further oxidation to sulfate, at oxidation state +6, is energetically highly favorable, releasing a further 200 kJ mol^{-1}. However, the transition barrior for this oxidation is high, and mammals lack metabolic pathways for performing the reaction. Taurine, therefore, is inert both chemically and metabolically. Its ionic character allows it to be uniquely accumulated. It is transported into the via an active transport system which is modulated in

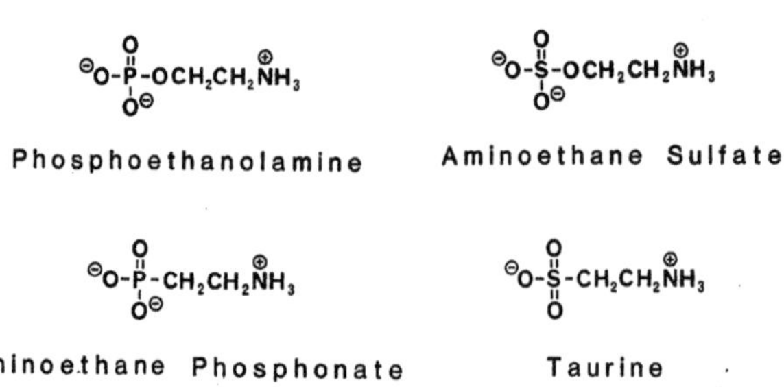

Figure 6. Structures of some low MW isosteric analogs of taurine. The ionic properties of these analogs, however, differ considerably from those of taurine. Taurine is a monobasic acid. The analogs shown are dibasic acids.

Table 5. Structural and ionic properties of taurine, beta-alanine and amino ethanephosphonate

Average bond angles (degrees)

	HNH	CNH	Around C_1	Around C_2	OXO	CXO
Aminoethane phosphonate	111.0	108.0	112.0	111.9	111.7	107.1
Taurine	110.3	108.7	109.3[a]	109.3[b]	112.5	106.2
Beta-Alanine		108.4	112.4[a]	108.4[b]	127.0	116.5

Bond lengths (nm)

	C-X	X-O[c]	C-C	C-N	N-H[c]	C_1-H[c]	C_2-H[c]
Aminoethyl phosphonate	0.180	0.153	0.151	0.148	0.095	0.102	0.102
Taurine	0.178	0.146	0.152	0.148	0.085	0.096	0.095
Beta-Alanine	0.155	0.129	0.155	0.148			

Ionization Constants

	pK_a	pK_b	Isoelectric Point
Aminoethane phosphonate	?	?	
Taurine	1.5	9.08	5.29
Beta-Alanine	3.6	10.36	6.99
Aminomethane sulfonate	?	5.75	

[a] $X-C_1-C_2$; [b] $N-C_2-C_1$; [c] Averages. For numbering see Figure 7. Bond lengths and bond angles: Aminoethane phosphonate (40); taurine (39); and beta-alanine (29). Ionization constants from (1).

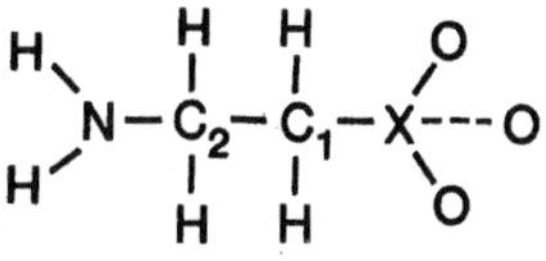

Figure 7. Type structure for the beta-aminoethane oxyacids taurine (X=S), beta-alanine (X=C) and aminoethane phosphonate (X=P). See Table 5.

a number of ways, including adrenergically (22,23). How its intracellular actions are regulated will presumably be the subject of much further investigation. Already, however, the phospholipid theory allows an understanding of the calcium dependency of taurine. The most marked effects are seen at physiologically low calcium concentrations. Regulation is achieved, among other means, by cation (calcium, sodium) antagonism of taurine modulation. The use of taurine for membrane modulation integrates economically with its use in osmoregulation.

A primary intracellular site of action of taurine is in keeping with its distribution, and is also in agreement with the side of action of the alpha adrenergic effect (13) and the opiate actions (33).

It should be stressed that chemicals are not single-minded. The postulation of a function for a biological molecule does not prohibit the molecule from simultaneously expressing other functions. Taurine is both an osmoregulator and a calcium modulator. It may have additional functions. For example, a small fraction of taurine in the brain may serve as a neurotransmitter. The demonstration of cysteine sulfinic acid decarboxylase activity in some neurons (Wu, this volume) may be an indication of a small pool of taurine involved in neural function.

The direct effects of taurine on phospholipid components in membranes may modify other lipid-dependent phenomena, such as the operation of ion channels, the regulation of membrane bound enzymes, and protein phosphorylation processes. Examples are available for all of these. Taurine is abundant in mammalian cells. In keeping with this, taurine has intracellular actions expressed at low affinities. Like water itself, another abundant biological molecule, taurine is an enabler. It helps to maintain the appropriate conditions which enable other phenomena and functions to be expressed.

In general, taurine is absent from organisms containing cell walls, such as plants and bacteria. One can speculate that the lack of need for a cytosoluble osmoregulator in these organisms precluded the devlopment of other functions for taurine.

There are many experiments clearly suggested by this theory, and many details of mechanism and regulation to explore. Doubtless, many will find this theory provocative, and will design experiments to test it. Despite this, I believe this theory to be fundamentally sound because it is based on fundamental physicochemical principles. One question that will be raised is how does this concept account for the specificity of effect seem with taurine? Again, I think an analogy can be drawn with water. This abundant substance is intimately involved in the mechanism of innumerable specific phenomena. Highly specific effects can derive from local application of a general property. The actions of taurine can be channeled and focussed by other factors, including cyclic AMP and noradrenergic systems in general, and bicarbonate, calcium, sodium, ATP and other substances. Phospholipids are present in all biological membranes, intracellular and cellular. However, the phospholipid composition varies from membrane to membrane, permitting specificity of site of action to be expressed. Furthermore, a phenomenon may be expressed at two or more sites but may have physiological significance only at the one site. Thus, modulation of high affinity calcium binding to the plasma membrane is of physiological significance in the brain and the heart, but may not have significance in the liver or fast-twitch muscle, even though it occurs. The retinal degeneration seen in taurine-deprived mammals may also be an example of local significance of a generalized phenomenon. Rod outer segments in the retina undergo a volume change in response to light (12). Deprivation of an osmoregulator (i.e. taurine) may therefore, have rapid pathological consequences.

ACKNOWLEDGEMENTS

Work from the author's laboratory was supported by the Flinn Foundation (Arizona Heart Association). Experimental findings from my laboratory are based on the work of a number of people, especially that of James Chubb (calcium transport), Leslie Sebring (calcium binding), and Shoichiro Shindo (NMR). I am grateful to these fellow adventurers for the continual intellectual stimulation they have provided.

REFERENCES

1. Alberts, A., and Serjeant, E.P., 1962, "Ionization Constants of Acids and Bases: A Laboratory Manual", Metheun and Co., London.
2. Alvarez, J.G., and Storey, B.T., 1983, Taurine, Hypotaurine, epinephrine and albumin inhibit lipid peroxidation in rabbit spermatozoa and protect against loss of motility, Biol. Reprod. 29:548-555.
3. Ambron, R.T., and Pieringer, R.A., 1973, Phospholipids in micro-organisms, "Form and Function of Phospholipids", Ansell, G.B., Hawthorne, J.N., and Dawson, R.M.C., eds., Elsevier, pp. 289-331.
4. Atlas, M., Bahl, J.J., Roeske, W., and Bressler, R., 1984, In vitro osmoregulation of taurine in fetal mouse hearts, J. Mol. Cell. Cardiol. 16:311-320.
5. Barbeau, A., and Huxtable, R., eds., 1978, "Taurine and Neurological Disorders", Raven Press, New York, pp. 468.
6. Cavallini, D., Gaull, G.E., and Zappia, V., 1980, "Natural Sulfur Compounds: Novel Biochemical and Structural Aspects", Plenum Press, New York City.
7. Chovan, J.P., Kulakowski, E.C., Sheakowski, S., and Schaffer, S.W., 1980, Calcium regulation by the low-affinity taurine binding sites of cardiac sarcolemma, Mol. Pharmacol. 17:295-300.
8. Christensen, H.N., Hess, B., and Riggs, T.R., 1954, Concentration of taurine, β-alanine, and triiodothyronine by ascites carcinoma cells, Cancer Res. 14:124-127.
9. Close, R., 1960, Free amino acids of some fungi, Nature 185:609.
10. Collu, R., Charpenet, G., and Clermont, M.J., 1978, Antagonism by taurine of morphine induced growth hormone secretion, Can. J. Neurol. Sci. 5:139-142.
11. Cuzner, M.L., and Davison, A.N., 1968, The lipid composition of rat brain myelin and subcellular fractions during development, Biochem. J. 106:29-34.
12. Enoch, J.M., Scandrett, J., and Toby, F.L., 1973, A study of the outer segments, Vision Res. 13:171-183.
13. Franconi, F., Stendardi, I., Failli, P., Antonini, G., Bennardini, F., Matucci, R., Manzini, S., and Giotti, A., 1983, Taurine antagonizes the alpha-adrenergic positive inotropic effect of phenylephrine, "In Sulfur Amino Acids: Biochemical and Clinical Aspects", Kuriyama, K., Huxtable, R.J., and Iwata, H., eds., Alan R. Liss, Inc., New York, pp. 51-60.
14. Gaull, G.E., Wright, C.E., and Tallan, H.H., 1983, Taurine in human lymphoblastoid cells: uptake and role in proliferation, "In Sulfur Amino Acids: Biochemical and Clinical Aspects", Kuriyama, K., Huxtable, R.J., and Iwata, H., eds., Alan R. Liss, Inc., New York, pp. 297-304.
15. Hunter, E.G., 1986, Adult ventricular myocytes isolated from CHF 146 and CHF 147 cardiomyopathic hamsters, Can. J. Physiol. Pharm. 64:1503-1506.
16. Huxtable, R.J., 1976, Metabolism and function of taurine in the heart, "In Taurine", Huxtable, R.J., and Barbeau, A., eds., Raven Press, New York, pp. 99-119.

17. Huxtable, R.J., 1980, Does taurine have a function? Fed. Proceed. 39:2678-2679.
18. Huxtable, R.J., 1981, Insights on function: metabolism and pharmacology of taurine in the brain, "In The Role of Peptides and Amino Acids as Neurotransmitters", Lombardini, J.B., and Kenny, A., eds., A.R. Liss Inc., New York, pp. 53-97.
19. Huxtable, R.J., 1986, Biochemistry of Sulfur, Plenum Press, New York, pp. 445.
20. Huxtable, R., and Barbeau, A., eds., 1976, "In Taurine", Raven Press, New York, pp. 398.
21. Huxtable, R.J., and Bressler, R., 1973, Effect of taurine on a muscle intracellular membrane, Biochim. Biophys. Acta 323:573-583.
22. Huxtable, R., and Chubb, J., 1977, Adrenergic stimulation of taurine transport by the heart, Science 198:409-411.
23. Huxtable, R.J., Chubb, J., and Azari, J., 1980, Physiological and experimental regulation of taurine content in the heart, Fed. Proceed. 39:2685-2690.
24. Huxtable, R.J., Laird, H.E., and Lippincott, S.E., 1979, The transport of taurine in the heart and the rapid depletion of tissue taurine content by guanidinoethyl sulfonate, J. Pharmacol. Exptl. Therap. 211:465-471.
25. Huxtable, R.J., and Pasantes-Morales, H., eds., 1982, "Taurine in Nutrition and Neurology", Plenum Press, New York, pp. 551.
26. Huxtable, R.J., and Sebring, L., 1982, Cardiovascular actions of taurine, Sulfur Amino Acids 5:29-51.
27. Huxtable, R.J., and Sebring, L.A., 1986, Towards a unifying theory for the action of taurine, TIPS, 7:481-485.
28. Irving, C.S., Hammer, B.E., Danyluk, S.S., and Klein, P.D., 1982, Coordination and Binding of Taurine as Determined by Nuclear Magnetic Resonance Measurements on ^{13}C-labeled Taurine, In: "Taurine in Nutrition and Neurology", Huxtable, R.J., and Pasantes-Morales, H., eds., Plenum Press, New York, pp. 5-17.
29. Jose, P., and Pant, L.M., 1965, The crystal and molecular structure of β-alanine, Acta Cryst. 18:806-810.
30. Kulakowski, E.C., Maturo, J., and Schaffer, S.W., 1981, Solubilization and characterization of cardiac sarcolemmal taurine-binding proteins, Arch. Biochem. Biophys. 210:204-209.
31. Kulakowski, E.C., and Schaffer, W.W., 1983, Partial purification of a high affinity taurine binding protein by affinity chromatography, Biochem. Pharmacol. 32:753-756.
32. Kuriyama, K., Huxtable, R.J., and Iwata, H., eds., A.R. Liss, Inc., 1983, "Sulfur Amino Acids: Biochemical and Clinical Aspects", New York, pp. 485.
33. Kuriyama, K., and Yoneda, Y., 1978, Morphine induced alterations of γ-aminobutyric acid and taurine contents and L-glutamate decarboxylase activity in rat spinal cord and thalamus: possible correlates with analgesic action of morphine, Brain Res. 148:163-179.
34. Lahdesmaki, P., 1986, Determination of taurine and other acidic amino acids in plants, Phytochemistry 25:2409-2411.
35. Lombardini, J.B., 1985, Taurine effects on the transition temperature in Arrhenius plots of ATP-dependent calcium ion uptake in rat retinal membrane preparation, Biochem. Pharmacol. 34:3741-3745.
36. Marquardt, P., and Vogg, G., 1952, Pharmakologische und chemische Untersuchungen uber Wirkstoffe in Bienenpollen, Arzneimittel. Forsch. 2:267-271.
37. Nakashima, T., Taniko, T., and Kuriyama, K., 1982, Therapeutic effect of taurine administration on carbon tetrachloride-induced hepatic injury, Jap. J. Pharmacology 32:583-590.
38. Oja, S.S., Ahtee, L., Kontro, P., and Paasonen, M.K., eds., 1985, "Taurine: Biological Actions and Clinical Perspective", A.R. Liss, Inc., New York, pp. 483.

39. Okaya, Y., 1966, Refinement of the crystal structure of taurine, 2-aminoethylsulfonic acid. An example of computer-controlled experimentation, Acta Crys. 21:726-735
40. Okaya, Y., 1966, Crystal structure of the stable modification of 2-aminoethyl phosphonic acid, ciliatine, Acta Cryst. 20:712-715.
41. Piez, K.A., and Eagle, H., 1958, The free amino acid pool of cultured human cells, J. Biol. Chem. 231-533-545.
42. Price, M.T., Olney, J.W., Mitchell, M.V., Fuller, T., and Cicero, T.J., 1978, Luteinizing hormone releasing action of N-methyl aspartate is blocked by GABA or taurine but not by dopamine antagonists, Brain Res. 158:461-465.
43. Raizuda, M.K., and Murti, C.R.K., 1973, Binding of taurine to Hartmannella culbertsoni membranes and the synthesis of cyclic AMP, Curr. Sci. 42:202-204.
44. Schaffer, S.W., Baskin, S.I., and Kocsis, J.J., eds., 1981, "The Effects of Taurine on Excitable Tissues", Spectrum Press, pp. 448.
45. Scheibel, J., Elsasser, T., and Ondo, J.G., 1981, A neuromodulatory role for taurine in controlling prolactin secretion in female rats, Psychoneuroendocrinology 6:139-144.
46. Sebring, L., and Huxtable, R.J., 1985, Taurine modulation of calcium binding to cardiac sarcolemma, J. Pharmacol. Exptl. Therap. 232:445-451.
47. Sebring, L., and Huxtable, R.J., 1986, Low affinity binding of taurine to phospholipids and cardiac sarcolemma, Biochim. Biophys. Acta 884:559-556.
48. Sturman, J.A., Wen, G.Y., Wisniewski, H.M., and Neuringer, M.D., 1984, Retinal degeneration in primates raised on a synthetic human infant formula, Int. J. Devl. Neuroscience 2:121-129.
49. Thurston, J.H., Hauhart, R.E., and Dirgo, J.A., 1980, Taurine: a role in osmotic regulation of mammalian brain and possible clinical significance, Life Sciences 26:1561-1568.
50. Thurston, J.H., Hauhart, R.E., and Naccarato, E.F., 1981, Taurine: possible role in osmotic regulation of mammalian heart, Science 214:1373-1374.
51. Tibbits, G.F., Sasaki, M., Ikeda, M., Shimada, K., Tsuruhara, T., and Nagatomo, T., 1981, Characterization of rat myocardial sarcolemma, J. Mol. Dell. Cardiol. 13:1051-1061.

PARTICIPANTS

H. Aihara
Research Center
Taisho Pharmaceutical Co., Ltd.
Yoshiono-cho 1-403
Ohmiya, Saitama 330
Japan

S. Allo
University of South Alabama
School of Medicine
Department of Pharmacology
Mobile, AL

N. Awata
The Third Department of Internal Medicine
Osaka University Medical School
1-1-50 Fukushima, Osaka 533
Japan

J. Azuma
The Third Department of Internal Medicine
Osaka University Medical School
1-1-50 Fukushima, Osaka 533
Japan

A. Baba
Department of Pharmacology
Faculty of Pharmaceutical Sciences
Osaka University
1-6 Yamada-oka
Suita, Osaka, 0565
Japan

A. Bartolini
Dipartimento di Farmacologia Preclinica e Clinica
"Mario Aiazzi Mancini"
Universita degli Studi di Firenze
Firenze, Italy

F. Bennardini
Dipartimento di Farmacologia Preclinica e Clinica
"Mario Aiazzi Mancini"
Universita di Firenze
Firenze, Italy

H. Bohles
Department of Pediatrics
University of Erlangen
F.R. Germany

J.P. Bolam
Department of Pharmacology
Oxford University
South Parks Road
Oxford, UK

S.H. Bryant
Department of Pharmacology and
Cell Biophysics
College of Medicine,
University of Cincinnati
Cincinnati, OH

W.E.S. Carr
Cornelius Vanderbilt Whitney Laboratory for Experimental Marine
Biology and Medicine of the
University of Florida
Route 1, Box 121
St. Augustine, FL

A. Castellucci
"Malesci S.p.A. - Instituto Farmacobiologico"
Pharmacology Department
Via Porpora, 22 Firenze
Italy

R.W. Chesney
University of California-Davis
School of Medicine
Davis, CA

D.J. Clarke
Department of Pharmacology
Oxford University
South Parks Road
Oxford, UK

D. Conte-Camerino
Cattedra di Farmacologia
e Farmacognosia
Facolta di Farmacia
Universita di Bari, Italy

S. Dabbagh
University of California-Davis
School of Medicine
Davis, CA

K.H. Deeg
Kinderklinik und Poliklinik
der Universitat Erlangen-Nurnberg
Universitats-Kinderklinik
Loschgestrasse 15, 0-8520 Erlangen
West Germany

R.M. del Rio
Depto. Investigacion
Hosp. "Ramon y Cajal"
Crta. Colmenar Km 9
28034-Madrid, Spain

L. Della Corte
Dipartimento Farmacologia
Preclinica e Clinica
Universita degli Studi di Firenze
Viale G.B. Morgagni
65, 50134 Firenze

A.L. Eicher
University of Oregon
Health Sciences Center
Portland, OR

P. Failli
Dipartimento di Farmacologia
Preclinica e Clinica
"Mario Aiazza-Mancini"
Universita di Firenze
Firenze, Italy

R.G. Fariello
Department of Neurology
Thomas Jefferson University Hospital
Philadelphia, PA

A. Fazzini
Dipartimento di Farmacologie
Preclinica e Clinica
"Mario Aiazza-Mancini"
Universita di Firenze
Firenze, Italy

J.H. Fellman
University of Oregon
Health Sciences Center
Portland, OR

F. Franconi
Cattedra di Saggi e Dosaggi
Farmacologici
Facolta di Farmacia
Universita di Sassari
Via Muroni, 23
Sassari, Italy

J. French
Department of Pathological
Neurobiology
Institute for Basic Research
in Developmental Disabilities
Staten Island, NY

A.L. Friedman
University of California-Davis
School of Medicine
Davis, CA

M. Frosini
Instituto di Scienze Farmacologiche
Universita di Siena
Via E.S. Piccolomini
170 - Villa Betlem, 53100
Siena, Italy

T. Fukuda
Department of Pharmacology
Faculty of Medicine
Kogoshima University
Kogoshima 890, Japan

Y. Fukui
Faculty of Pharmaceutical Sciences
Department of Pharmacology
Osaka University
1-6 Yamada-oka,
Suita Osaka, 0565
Japan

G.E. Gaull
Department of Nutritional Sciences
The NutraSweet Company
Skokie, IL

B. Gillam
New York State Institute for
Basic Research in Developmental
Disabilities
Staten Island, NY

A. Giotti
Department of Preclinical
and Clinical Pharmacology
"Mario Aiazzi Manzini"
Universita di Firenze
Firenze, Italy

R.A. Gleeson
Cornelius Vanderbilt Whitney
Laboratory for Experimental Marine
Biology and Medicine of the
University of Florida
Route 1, Box 121
St. Augustine, FL

T.R. Green
University of Oregon
Health Sciences Center
Portland, OR

J. Gulyas
Institute of Organic Chemistry
Eotvos Lorand University
Muzeum krt. 6-8, H-1088
Budapest, Hungary

N. Gusowski
University of California-Davis
School of Medicine
Davis, CA

T. Hamaguchi
The Third Department of Internal
Medicine
Osaka University Medical School
1-1-50 Fukushima, Osaka 533
Japan

A. Hamberger
Institute of Neurobiology
University of Goteborg
P.O.B. 33031
S-400 33 Goteborg
Sweden

A.T. Hanretta
Department of Pharmacology
Texas Tech University
Health Sciences Center
Lubbock, TX

H. Harada
The Third Department of Internal
Medicine
Osaka University Medical School
1-1-50 Fukushima, Osaka 533
Japan

A.S. Herranz
Depto. Investigacion
Hosp. "Ramon y Cajal"
Crta. Colmenar Km 9
28034-Madrid, Spain

O. Herrevas
Depto. Investigacion
Hosp. "Ramon y Cajal"
Crta. Colmenar Km 9
28034-Madrid, Spain

K. Hongoh
Research Center
Taisho Pharmaceutical Co., Ltd.
Yoshiono-cho 1-403
Ohmiya, Saitama 330
Japan

Y. Hosokawa
Division of Maternal and Child
Nutrition
National Institute of Nutrition
1-23-1, Toyama, Shinjuku-ku
Tokyo, Japan

R.J. Huxtable
Department of Pharmacology
College of Medicine
University of Arizona
Tucson, AZ

H. Imaki
Department of Developmental
Biochemistry
Institute for Basic Research
in Developmental Disabilities
Staten Island, NY

H. Iwata
Faculty of Pharmaceutical Sciences
Department of Pharmacology
Osaka University
1-6 Yamada-oka,
Suita Osaka, 0565
Japan

K. Izumi
Department of Pharmacology
Faculty of Medicine
Kagoshima University
Kagoshima 890
Japan

F.F. Johansen
Institute of Neuropathology
University of Copenhagen
Denmark

B. Ki Kim
Faculty of Pharmaceutical Sciences
Department of Pharmacology
Osaka University
1-6 Yamada-oka,
Suita Osaka, 0565
Japan

M. Kimori
Department of Pharmacology
Kyoto Prefectural University
of Medicine
Kawaramachi-Hirokoji
Kamikyo-Ku, Kyoto 602
Japan

M. Kishi
Department of Pharmacology
Kyoto Prefectural University
of Medicine
Kawaramachi-Hirokoji
Kamikyo-Ku, Kyoto 602
Japan

S. Kishimoto
The Third Department of Internal
Medicine
Osaka University Medical School
1-1-50 Fukushima, Osaka 533
Japan

T. Koja
Department of Pharmacology
Faculty of Medicine
Kogoshima University
Kogoshima 890, Japan

P. Kontro
Department of Biomedical Sciences
University of Tampere
Box 607
SF-33101, Tampere 10
Finland

Y. Koyama
Department of Pharmacology
Faculty of Pharmaceutical
Sciences
1-6 Yamada-oka
Suita, Osaka, 0565
Japan

E.C. Kulakowski
Office of Program Planning and
Evaluation
National Heart Lung and
Blood Institute
Bldg. 31, Room 5A06
Bethesda, MD

M. Kurachi
Research Center
Taisho Pharmaceutical Co., Ltd.
Yoshiono-cho 1-403
Ohmiya, Saitama 330
Japan

K. Kuriyama
Department of Pharmacology
Kyoto Prefectural University
of Medicine
Kawaramachi-Hirokoji
Kamikyo-ku, Kyoto 602
Japan

P. Lahdesmaki
Department of Biochemistry
University of Oulu
SF-90570 Oulu, Finland

A. Lehmann
Institute of Neurobiology
University of Goteborg
P.O.B. 33031
S-400 33 Goteborg
Sweden

J. Lerma
Depto. Investigacion
Hosp. "Ramon y Cajal"
Crta. Colmenar Km 9
28034-Madrid, Spain

C.-T. Lin
Department of Physiology
The Pennsylvania State University
Hershey, PA

S. Lippincott
Department of Pediatrics
University of Wisconsin
Madison, WI

J.B. Lombardini
Department of Pharmacology
Texas Tech University
Health Sciences Center
Lubbock, TX

R. Lopez-Escalera
Instituto de Fisiologia Cellular
Universidad Nacional Autonoma de
Mexico
Apartado Postal 70-600
04510 Mexico, D.F. Mexico

J.-W. Lui
Department of Physiology
The Pennsylvania State University
Hershey, PA

S. Madsen
Anatomical Institute
University of Oslo
Karl Johansgate 47
N-0162 Oslo 1
Norway

M. Mambrini
Cattedra di Farmacologia
e Farmacognosia
Facolta di Farmacia
Universita di Bari, Italy

S. Manzini
"Malesci S.p.A. - Instituto
Farmacobiologico"
Pharmacology Department
Via Porpora, 22 Firenze
Italy

K.-M. Marnela
Department of Biomedical Sciences
University of Tampere
Box 607, SF-33101 Tampere
Finland

R. Matucci
Dipartimento di Farmacologia
Preclinica e Clinica
"Mario Aiazzi Mancini"
Universita degli Studi di Firenze
Firenze, Italy

J. Maturo, III
Biology Department
C.W. Post College
P.O. Greenvale, NY

D. Michalk
Kinderklinik und Poliklinik
der Universitat Erlangen-Nurnberg
Universitats-Kinderklinik
Loschgestrasse 15, 0-8520 Erlangen
West Germany

D. Mitolo-Chieppa
Instituto di Farmacologia
Facolta di Medicina
Universita di Bari, Italy

D. Morandini
Dipartimento di Farmacologia
Preclinica e Clinica
"Mario Aiazzi Mancini"
Universita degli Studi di Firenze
Firenze, Italy

R.C. Moretz
Department of Pathological
Neurobiology
Institute for Basic Research
in Developmental Disabilities
Staten Island, NY

H. Morimoto
Department of Pharmacology
Faculty of Pharmaceutical
Sciences
1-6 Yamada-oka
Suita, Osaka, 0565
Japan

M. Mozaffari
University of South Alabama
School of Medicine
Department of Pharmacology
Mobile, AL

E. Munekata
Institute of Applied Biochemistry
University of Tsukaba
Ibaraki 305, Japan

T. Nakanishi
Department of Neurology
Institute of Clinical Medicine
University of Tsukuba
Ibaraki 305, Japan

M. Neuringer
Oregon Regional Primate Research
Center
Beaverton, OR

Y. Nomura
Department of Pharmacology
Institute of Pharmaceutical
Sciences
Hiroshima University
School of Medicine
Kasumi 1-2-3
Hiroshima 734, Japan

S. Ohkuma
Department of Pharmacology
Kyoto Prefectural University
of Medicine
Kawaramachi-Hirokoji
Kamikyo-Ku, Kyoto 602
Japan

H. Ohta
The Third Department of Internal
Medicine
Osaka University Medical School
1-1-50 Fukushima, Osaka 533
Japan

S.S. Oja
Department of Biomedical Sciences
University of Tampere
Box 607
SF-33101 Tampere 10
Finland

O.P. Ottersen
Anatomical Institute
University of Oslo
Karl Johansgate 47
N-0162 Oslo 1
Norway

T. Palackal
Department of Developmental
Biochemistry
Institute for Basic Research
in Developmental Disabilities
Staten Island, NY

M. Palmi
Instituto di Scienze Farmacologiche
Universita di Siena
Via E.S. Piccolomini
170 - Villa Betlem
53100 Siena
Italy

H. Pasantes-Morales
Instituto de Fisiologia Celular
Universidad Nacional Autonoma de
Mexico
Apartado Postal 70-600
04510 Mexico, D.F. Mexico

A. Picones
Instituto de Fisiologia Celular
Universidad Nacional Autonoma de
Mexico
Apartado Postal 70-600
04510 Mexico, D.F. Mexico

L. Polenzani
Dipartimento di Farmacologia
Preclinica e Clinica
"Mario Aiazzi Mancini"
Universita degli Studi di Firenze
Firenze, Italy

O. Quesada
Instituto de Fisiologia Celular
Universidad Nacional Autonoma de
Mexico
Apartado Postal 70-600
04510 Mexico, D.F. Mexico

R. Ringeisen
Kinderklinik und Poliklinik
der Universitat Erlangen-Nurnberg
Universitats-Kinderklinik
Loschgestrasse 15, 0-8520 Erlangen
West Germany

A. Sawamura
The Third Department of Internal
Medicine
Osaka University Medical School
1-1-50 Fukushima, Osaka 533
Japan

S.W. Schaffer
University of South Alabama
School of Medicine
Department of Pharmacology
Mobile, AL

T. Segawa
Department of Pharmacology
Institute of Pharmaceutical
Sciences
Hiroshima University
School of Medicine
Kasumi 1-2-3
Hiroshima 734, Japan

G.P. Sgaragli
Instituto di Scienze Farmacologiche
Universita di Siena
Via E.S. Piccolomini
170 - Villa Betlem
51300 Siena, Italy

I. Shimazaki
Department of Pharmacology
Institute of Pharmaceutical
Sciences
Hiroshima University
School of Medicine
Kasumi 1-2-3
Hiroshima 734, Japan

A.D. Smith
Department of Pharmacology
Oxford University
South Parks Road
Oxford, UK

J.M. Solis
Depto. Investigacion
Hosp. "Ramon y Cajal"
Crta. Colmenar Km 9
28034-Madrid, Spain

I. Stendardi
Dipartimento di Farmacologia
Preclinica e Clinica
"Mario Aizzi-Mancini"
Universita di Firenze
Firenze, Italy

E. Stjeskal-Lorenz
University of California-Davis
School of Medicine
Davis, CA

J. Storm-Mathisen
Anatomical Institute
University of Oslo
Karl Johansgate 47
N-0162 Oslo 1
Norway

J.A. Sturman
Department of Developmental
Biochemistry
Institute for Basic Research
in Developmental Disabilities
Staten Island, NY

K. Takihara
The Third Department of Internal
Medicine
Osaka University Medical School
1-1-50 Fukushima, Osaka 533
Japan

H.H. Tallan
New York State Institute for
Basic Research in Developmental
Disabilities
Staten Island, NY

Y. Tanaka
The Third Department of Internal
Medicine
Osaka University Medical School
1-1-50 Fukushima, Osaka 533
Japan

F. Tittor
Kinderklinik und Poliklinik
der universitat Erlangen-Nurnberg
Universitats-Kinderklinik
Loschgestrasse 15, 0-8520 Erlangen
West Germany

H.G. Trapido-Rosenthal
Cornelius Vanderbilt Whitney
Laboratory for Experimental Marine
Biology and Medicine of the
University of Florida
Route 1, Box 121
St. Augustine, FL

Z. Vadasz
Institute of Biochemistry
Department I
Semmelweis University
Medical School, Box 260, H-1444
Budapest, Hungary

V. Varga
Department of Biomedical Sciences
University of Tampere, Box 607
SF-33101 Tampere, Finland

E. Wagner-Thiessen
Department of Pathology
University of Erlangen
F.R. Germany

A. Watanabe
Research Center
Taisho Pharmaceutical Co., Ltd.
Yoshiono-cho 1-403
Ohmiya, Saitama 330
Japan

H.M. Wisniewski
Department of Pathological
Neurobiology
Institute for Basic Research
in Developmental Disabilities
Staten Island, NY

C.E. Wright
New York State Institute for
Basic Research in Developmental
Disabilities
Staten Island, NY

T.R. Wright
New York State Institute for
Basic Research in Developmental
Disabilities
Staten Island, NY

J.-Y. Wu
Department of Physiology
The Pennsylvania State University
Hershey, PA

K. Yamaguchi
Division of Maternal and Child
Nutrition
National Institute of Nutrition
1-23-1, Toyama, Shinjuku-ku
Tokyo, Japan

M. Yoshida
Department of Neurology
Institute of Clinical Medicine
University of Tsukuba
Ibaraki 305, Japan

I. Zelikovic
University of California-Davis
School of Medicine
Davis, CA

ABBREVIATIONS

ACh	-	Acetylcholine
ADR	-	Adriamycin
AP	-	Action potential
BSA	-	Bovine serum albumin
BUN	-	Blood urea nitrogen
CA	-	Cysteic acid
CaM	-	Calmodulin
CD	-	Cysteine dioxygenase
CNS	-	Central nervous system
CPK	-	Creatine phosphokinase
CSA	-	Cysteine sulfinic acid
CSAD	-	Cysteine sulfinic acid decarboxylase
CSAD-IR	-	Cysteine sulfinic acid decarboxylase immunoreactivity
CSF	-	Cerebrospinal fluid
DAME	-	(D-Ala^2, Met^5) enkephalinamide
DEAE	-	Diethylaminoethyl cellulose
DFP	-	Diisopropylfluorophosphate
DIDS	-	4,4'-Diisothiocyanatostilbene 2,2'-disulfonic acid
ECG	-	Electrocardiogram
EDL	-	Extension digitorum longus muscle
EEG	-	Electroencephalogram
ERG	-	Electroretinogram
GABA	-	γ-Aminobutyric acid
GABA-LI	-	GABA-like immunoreactivity
GAD	-	Glutamic acid decarboxylase
GES	-	Guanidinoethane sulfonate
GFAP	-	Anti-glial fibrillary acid protein
GSH	-	Glutathione
GSH-PX	-	Glutathione peroxidase
GUT	-	Glutamate oxaloacetate transaminase
h	-	Hour
HRP	-	Horseradish peroxidase
icv	-	Intracerebroventricular
ip	-	Intraperitoneal
ISO	-	Isoproterenol
iv	-	Intravenous
LDH	-	Lactate dehydrogenase
LTD4	-	Lukotriene D4
min	-	Minute
MW	-	Molecular weight
OPA	-	O-Phthaldehyde
OS	-	Action potential overshoot
PAF	-	Platelet activating factor
PAP	-	Peroxidase anti-peroxidase
PBS	-	Dulbecco's phosphate buffered saline
PEA	-	Phosphoethanolamine

PPP	-	Platelet-poor plasma
PRP	-	Platelet-rich plasma
PTZ	-	Pentylene tetrazole
s	-	Second
SN	-	Substantia nigra
ST	-	Neostriatum
TAG	-	("Taurine antagonist") 6-Aminomethyl-3-methyl-4H-1,2,4-Benzothiadiazine 1,1-dioxide
TAU-LI	-	Taurine-like immunoreactivity
W-7	-	N-(6-Aminohexyl)-5-chloro-1-naphthalene sulfonamide hydrochloride
WBC	-	White blood cells
WDS	-	Wet dog shakes
WGA	-	Wheat germ agglutinin

INDEX

9 780306 426650